Handbook of Research on Machine Learning–Enabled IoT for Smart Applications Across Industries

Neha Goel
Raj Kumar Goel Institute of Technology, India

Ravindra Kumar Yadav
Raj Kumar Goel Institute of Technology, India

A volume in the Advances in Systems Analysis, Software Engineering, and High Performance Computing (ASASEHPC) Book Series

Published in the United States of America by
IGI Global
Engineering Science Reference (an imprint of IGI Global)
701 E. Chocolate Avenue
Hershey PA, USA 17033
Tel: 717-533-8845
Fax: 717-533-8661
E-mail: cust@igi-global.com
Web site: http://www.igi-global.com

Library of Congress Cataloging-in-Publication Data

Names: Goel, Neha, 1983- editor. | Yadav, Ravindra Kumar, 1973- editor.
Title: Handbook of research on machine learning-enabled IoT for smart applications
    across industries / edited by Neha Goel, Ravindra Kumar Yadav.
Description: Hershey, PA : Engineering Science Reference, [2023] | Includes
    bibliographical references and index.
Identifiers: LCCN 2023005805 (print) | LCCN 2023005806 (ebook) | ISBN
    9781668487853 (h/c) | ISBN 9781668487877 (eISBN)
Subjects: LCSH: Internet of things. | Machine learning. | Industrial
    management--Data processing.
Classification: LCC TK5105.8857 .M33 2023 (print) | LCC TK5105.8857
    (ebook) | DDC 006.3--dc23/eng/20230222
LC record available at https://lccn.loc.gov/2023005805
LC ebook record available at https://lccn.loc.gov/2023005806

This book is published in the IGI Global book series Advances in Systems Analysis, Software Engineering, and High Performance Computing (ASASEHPC) (ISSN: 2327-3453; eISSN: 2327-3461)

British Cataloguing in Publication Data
A Cataloguing in Publication record for this book is available from the British Library.

All work contributed to this book is new, previously-unpublished material. The views expressed in this book are those of the authors, but not necessarily of the publisher.

For electronic access to this publication, please contact: eresources@igi-global.com.

Advances in Systems Analysis, Software Engineering, and High Performance Computing (ASASEHPC) Book Series

Vijayan Sugumaran
Oakland University, USA

ISSN:2327-3453
EISSN:2327-3461

MISSION

The theory and practice of computing applications and distributed systems has emerged as one of the key areas of research driving innovations in business, engineering, and science. The fields of software engineering, systems analysis, and high performance computing offer a wide range of applications and solutions in solving computational problems for any modern organization.

The **Advances in Systems Analysis, Software Engineering, and High Performance Computing (ASASEHPC) Book Series** brings together research in the areas of distributed computing, systems and software engineering, high performance computing, and service science. This collection of publications is useful for academics, researchers, and practitioners seeking the latest practices and knowledge in this field.

COVERAGE

- Software Engineering
- Performance Modelling
- Virtual Data Systems
- Parallel Architectures
- Engineering Environments
- Storage Systems
- Computer System Analysis
- Distributed Cloud Computing
- Metadata and Semantic Web
- Enterprise Information Systems

IGI Global is currently accepting manuscripts for publication within this series. To submit a proposal for a volume in this series, please contact our Acquisition Editors at Acquisitions@igi-global.com or visit: http://www.igi-global.com/publish/.

Titles in this Series

For a list of additional titles in this series, please visit: http://www.igi-global.com/book-series/advances-systems-analysis-software-engineering/73689

Neuromorphic Computing Systems for Industry 4.0
S. Dhanasekar (Department of ECE, Sree Eshwar Engineering College, India) K. Martin Sagayam (Karunya Institute of Technology and Sciences, India) Surbhi Vijh (KIET Group of Institutions, India) Vipin Tyagi (Jaypee University of Engineering and Technology, India) and Alex Norta (Tallinn University of Technology, Estonia)
Engineering Science Reference • © 2023 • 300pp • H/C (ISBN: 9781668465967) • US $270.00

Business Models and Strategies for Open Source Projects
Francisco José Monaco (Universidade de São Paulo, Brazil)
Business Science Reference • © 2023 • 300pp • H/C (ISBN: 9781668447857) • US $270.00

Principles, Policies, and Applications of Kotlin Programming
Duy Thanh Tran (University of Economics and Law, Ho Chi Minh City, Vietnam & Vietnam National University, Ho Chi Minh City, Vietnam) and Jun-Ho Huh (Korea Maritime and Ocean University, South Korea)
Engineering Science Reference • © 2023 • 457pp • H/C (ISBN: 9781668466872) • US $215.00

Concepts and Techniques of Graph Neural Networks
Vinod Kumar (Koneru Lakshmaiah Education Foundation (Deemed), India) and Dharmendra Singh Rajput (VIT University, India)
Engineering Science Reference • © 2023 • 247pp • H/C (ISBN: 9781668469033) • US $270.00

Cyber-Physical System Solutions for Smart Cities
Vanamoorthy Muthumanikandan (Vellore Institute of Technology, Chennai, India) Anbalagan Bhuvaneswari (Vellore Institute of Technology, Chennai, India) Balamurugan Easwaran (University of Africa, Toru-Orua, Nigeria) and T. Sudarson Rama Perumal (Rohini College of Engineering and Technology, India)
Engineering Science Reference • © 2023 • 300pp • H/C (ISBN: 9781668477564) • US $270.00

Adaptive Security and Cyber Assurance for Risk-Based Decision Making
Tyson T. Brooks (Syracuse University, USA)
Engineering Science Reference • © 2023 • 243pp • H/C (ISBN: 9781668477663) • US $225.00

Novel Research and Development Approaches in Heterogeneous Systems and Algorithms
Santanu Koley (Haldia Institute of Technology, India) Subhabrata Barman (Haldia Institute of Technology, India) and Subhankar Joardar (Haldia Institute of Technology, India)
Engineering Science Reference • © 2023 • 323pp • H/C (ISBN: 9781668475249) • US $270.00

701 East Chocolate Avenue, Hershey, PA 17033, USA
Tel: 717-533-8845 x100 • Fax: 717-533-8661
E-Mail: cust@igi-global.com • www.igi-global.com

List of Contributors

Table of Contents

Detailed Table of Contents

The internet of things (IoT) has revolutionized various aspects of human life, such as smart cities, home automation, and energy efficiency. However, fully realizing the potential of IoT necessitates overcoming significant challenges. This article provides a comprehensive evaluation of IoT from both technological and societal perspectives, delving into its architecture, applications, and critical difficulties. It highlights the contributions of different IoT components and emphasizes the value of IoT and data analytics. The findings contribute to a better understanding of IoT applications and challenges, benefiting both consumers and researchers in the field.

Chapter 2

Today, it is feasible to observe how quickly electronic devices are becoming connected to the internet. Electronic devices that are connected to the internet can be managed or observed from any location in the world. Many of the challenges have been made easier by internet connectivity for technological gadgets. A good judgement may be made by spotting certain trends using IoT and machine learning (ML) technologies. Its application areas can be expanded much farther than they are now by combining IoT and ML algorithms. ML uses a variety of algorithms, and while analysing them to choose the best one for a certain electronic device, runtime complexity, memory needs, and accuracy are taken into consideration. In comparison to other ML algorithms, support vector machine, random forest, and k-nearest neighbour have higher runtime complexity, a smaller memory requirement, and higher accuracy. In this chapter, the aforementioned topics have all been covered. The different ML algorithms and IoT pattern recognition application areas are covered in this chapter.

Chapter 3

The use of artificial intelligence and machine learning in the oil and gas industry has steadily increased over the last decade. The focus of this chapter is understanding what type of external factors influence oil and gas firms to implement AI. A systematic review of the literature found that external pressures to achieve business goals and public pressures to reduce environmental impact continue to rise. This chapter leverages institutional theory to look at external isomorphic pressures in the form of government regulations and competitive pressures that influence the adoption of AI. The chapter concludes by reviewing opportunities for future research in the areas of AI within the oil and gas industry.

Generative adversarial networks (GANs) have transformed machine learning and created new research and application areas. GANs are now used for data augmentation, picture, audio, text-to-image, and 3D object production thanks to IoT. These applications could make IoT devices more personalized, efficient, and productive by collecting and using data. GANs are also employed in healthcare and finance for IoT applications, enabling new research. GANs have many benefits, however stability issues during training, interpretability issues, ethical concerns, and more must be addressed. GANs could transform the IoT market, and hardware and infrastructure improvements are projected to increase their influence. GANs will open up many new research and development avenues for IoT devices.

Chapter 5

Internet of things solutions with machine learning capabilities is a hot research area in industries, including agriculture. They can be used for data analysis and further forecasting the big data and intelligent applications in farming. In traditional farming, the main obstacles are disease prediction, automatic irrigation, energy harvesting, and constant monitoring. Today, farmers' cultivation of their crops has changed by introducing automated harvesters, drones, autonomous tractors, sowing, and weeding. Smart farming with ML-enabled IoT systems can improve crop harvesting decisions. The main topic of this chapter is to provide an ML-enabled IoT solution for smart agriculture. The MIoT solution in agriculture allows farmers to use predictive analytics to help them make better harvesting decisions. Designing a MIoT system for smart agriculture can assist farmers in improving yields, planning more effective irrigation, and making harvest forecasts by monitoring essential data like humidity, air temperature, and soil quality via remote sensors.

N. A. Natraj, Symbiosis International University, India
Sriya Mitra, Symbiosis International University, India
Giri Gundu Hallur, Symbiosis International University, India

The emergence of IoT in the healthcare sector can bring about a surge of innovation and development. This qualitative research paper is aimed to explore the growth, adoption and threats that the emergence of IoT in the healthcare sector holds. IoT has various capabilities that pose an opportunity to provide solutions for a digitised healthcare system that would also aid in battling pandemics such as COVID. IoT has applications in multiple places, such as improving hospital treatment systems, tracking different medical parameters, identifying irregularities. The correct usage of IoT may aid in dealing with a variety of medical issues such as speed, cost, and complexity. It makes the surgeon's job easier by lowering risks and improving overall performance. The growing use of connected devices in healthcare is a major driver for the IoT-Health market.This research identifies the threats and the prospects of IoT in the future of healthcare.

Chapter 7

Jyoti, Department of Engineering and Technology, Guru Nanak Dev University, Regional Campus, Jalandhar, India
Sheetal Kalra, Department of Engineering and Technology, Guru Nanak Dev University, Regional Campus, Jalandhar, India
Amit Chhabra, Department of Computer Engineering and Technology, Guru Nanak Dev University, Amritsar, India

The increasing popularity of internet of things (IoT), dissimilar networks, distributed devices, and applications has turned out to be a major call for the identification of novel security threats and tracing malicious network behaviours. An intrusion detection system (IDS) is a self-defense tool for preventing several types of cyberattacks. Latest machine learning (ML) methods are becoming the backbone for constructing intelligent IDS that are highly data driven. This chapter proposes decision tree based IDS for the dataset NSL-KDD. A novel approach has been developed for the ranking of security features. The proposed system has been validated against performance evaluation metrics consisting of recall, precision, accuracy, and F score. The results produced by the proposed system are compared with well-known ML methods including logistic regression, support vector machines and K-nearest neighbor in order to analyse the efficiency.

Kunal Dhibar, Bengal College of Engineering and Technology, India
Prasenjit Maji, Dept of CSD, Dr. B.C. Roy Engineering College, Durgapur, India

Throughout many real-world investigations, outliers are prevalent. Even a few aberrant data points can cause modeling misspecification, biased parameter estimate, and poor forecasting. Outliers in a time series are typically created at unknown moments in time by dynamic intervention models. As a result, recognizing outliers is the starting point for every statistical investigation. Outlier detection has attracted significant attention in a variety of domains, most notably machine learning and artificial intelligence. Anomalies are classified as strong outliers into point, contextual, and collective outliers. The most significant difficulties in outlier detection include the narrow boundary between remote sites and natural areas, the propensity of fresh data and noise to resemble genuine data, unlabeled datasets, and varying interpretations of outliers in different applications.

Kavita Srivastava, Institute of Information Technology and Management, GGSIP University,
India

This chapter discusses the emerging paradigm of edge computing and its potential to optimize performance and enhance user experience in modern computing systems. The chapter begins by introducing the concept of edge computing, its definition, and its various applications. It then highlights the key benefits of edge computing. The chapter also delves into the various architectures and models of edge computing. It examines the challenges associated with edge computing, such as resource constraints, security, and privacy issues. The authors then provide an overview of the technologies and tools used in edge computing. They discuss how these technologies can be used to optimize performance and enhance user experience in edge computing systems. The chapter also presents several use cases and real-world applications of edge computing, including smart homes, autonomous vehicles, and healthcare systems. The authors examine the benefits and challenges of using edge computing in these domains and provide insights into how to optimize performance and enhance user experience.

Shahad P., Indian Institute of Information Technology, Kottayam, India
Ebin Deni Raj, Indian Institute of Information Technology, Kottayam, India

This chapter covers various technical features available for building a smart healthcare system. Its different components are included as separate sections along with introduction and conclusion parts. There are many other sections incorporated here, namely infrastructure development, streaming data management, cloud computing and machine learning techniques for data analysis. The advancements in sensor technology and internet of things could enable real-time monitoring of several body parameters. The values from these sensors connected to the body can be Integrated for analysis. Which can enhance our health monitoring system. The way of harvesting information from these fast-moving data is also explored here in this chapter. Further, the role of machine learning techniques to build intelligence to the system is discussed in detail. This chapter is organized as a systematic study to form a smart healthcare system which are accepted by the people because of its analytical accuracy.

Chapter 11

Mohan Raj C. S., Hindusthan College of Arts and Science, India
A. V. Senthil Kumar, Hindusthan College of Arts and Science, India
Meenakshi Sharma, University of Petroleum and Energy Studies, India
Ibrahiem M. M. El Emary, King Abdulaziz University, Saudi Arabia
Rohaya Latip, Universiti Putra Malaysia, Malaysia
Saifullah Khalid, Civil Aviation Research Organisation, India
Chandrashekar D. V., T.J.P.S. College, India

Industry 4.0 refers to the phase of transition that is taking place, enabling automation and data interchange in industrial technologies and processes. Fog computing architecture can provide real-time processing, nearby storage, extremely low latency, dependability, large data rates, and other requirements for industrial Internet of Things (IIoT) applications. In the context of IoT applications, fog infrastructure and protocols are the main areas of interest. The phrase "fog computing," sometimes known "edge cloud," is a new paradigm. Between edge devices and Cloud Core, it adds another layer. Along with providing computing, storage, and networking capabilities, it also fills a need left by the cloud. The main features of fog computing are covered in this chapter, along with current research on the subject and a focus on the difficulties encountered when creating its architectural design.

Chapter 12

Claudio Urrea, University of Santiago of Chile, Chile

Construction and manufacturing are performed in environments where automation is highly appreciated as it improves security and efficiency. This article proposes a new cloud computing-based strategy for coordinating multi-robot systems aimed to transport and manipulate heavy objects like a beam. The proposed solution is applied to a team of dynamically modeled selectively compliance assembly robot arms (SCARAs) with four degrees-of-freedom (4-DoF), simulated in virtual reality modeling language (VRML) and coordinated by a cloud server. The SCARAs are organized on a leader/follower configuration, so that the leader performs a computed trajectory, constantly reporting its position to the remote server and this, in turn, resends this information to the followers. The SCARAs are modeled using MatLab and they communicate through a transmission control protocol (TCP) with a Python process running in the cloud server to demonstrate the feasibility of incorporating cloud resources in a multi-robot system, testing their performance through simulation.

Agriculture is mostly practiced in rural areas where there is less population and no proper scouting. Unmanned aerial vehicle (UAVs) can reduce human involvement in agriculture and solve many issues such as monitoring water levels, detecting crop disease, controlling the consumption of water and many more. UAVs application has contributed to many areas of agriculture such as insecticide as well as fertilizer prospecting and spraying, seed planting, weed recognition, soil mapping using aerial imaging, crop forecasting and so forth. Through these methods, crops can be cultivated without making excess use of water and chemicals which keep them safe and strong. Further, UAVs are replacing the man-made aircrafts because of their peculiar feature of capturing high resolution imagery below cloud level and its flexibility to work on different geographical locations. The multifunctioning UAVs reduce time and increase productivity. Therefore, this chapter provides a review on a smarter agricultural system using UAVs in order to enhance food productivity.

Chapter 14

Artificial intelligence (AI) techniques play an important role in predicting, diagnosing, and monitoring the emission of hazardous or toxic gases emitted from industries or various places. Also, it helps in continuous monitoring of environmental pollutants based on time series. Leakage of gas or toxic gases leads to health-related issues to human beings. Deep learning is a part of AI, and several deep learning algorithms are used to detect the emission of toxic gases in industries, indoors, refineries, etc. This chapter gives a detailed survey on different sensors and different deep learning algorithms used for diagnosing emission of hazardous gases.

The internet of things (IoT) is the network of sensors, devices, processors, and software, enabling connection, communication, and data transfer between devices. IoT is able to collect and analyze large amounts of data which can then be used to automate daily tasks in various fields. IoT holds the potential to revolutionise and create many opportunities in multiple industries like smart cities, smart transport, etc. Autonomous vehicles are smart vehicles that are able to navigate and move around on their own on a well-planned road.

    Kavitha Rajamohan, CHRIST University (Deemed), India
    Sangeetha Rangasamy, CHRIST University (Deemed), India
    Alvis Abreo, CHRIST University (Deemed), India
    Rohit Upadhyay, CHRIST University (Deemed), India
    Raison Sabu, CHRIST University (Deemed), India

A smart city is designed using acceptable internet of things (IoT) technologies that solve urban life problems and provide quality of life to the residents. IoT refers to a network of physical devices that are capable of gathering and sharing data and expediting numerous functions without human assistance. IoT supports smart home builders and managers by providing an efficient ecosystem in terms of less operating cost and improvising residence services. In recent days, the initiative of smart homes/buildings/cities is increasing gradually around the globe. The inclined population in an urban area also expects well-managed automated services in their everyday life.

    Asha Rajiv, School of Sciences, Jain University (Deemed), India
    Abhilash Kumar Saxena, Teerthanker Mahaveer University, India
    Digvijay Singh, Dev Bhoomi Uttarakhand University, India
    Aishwary Awasthi, Sanskriti University, India
    Dharmesh Dhabliya, Symbiosis Law School, Symbiosis International University, Pune, India
    R. K. Yadav, Raj Kumar Goel Institute of Technology, India
    Ankur Gupta, Vaish College of Engineering, India

This study emphasises the need for energy efficiency in buildings, focusing primarily on the heating, ventilation, and air conditioning (HVAC) systems, which consume 50% of building energy. A predictive system based on artificial neural networks (ANNs) was created to generate short-term forecasts of indoor temperature using data from a monitoring system in order to reduce this energy use. The technology seeks to estimate inside temperature in order to determine when to start the heating, ventilation, and air conditioning system, potentially reducing energy use dramatically. The chapter describes the system's code implementation, which includes data pre-processing, model training and evaluation, and result visualisation. In terms of evaluation metrics, the model performed well and revealed the potential for large energy savings in buildings.

Using time series data obtained from accelerometer and gyroscope sensors on an iPhone 6s, the authors address the topic of human activity and attribute detection. The collection contains time series data from 24 subjects who completed six activities in 15 trials. The aim is to appropriately identify the six activities using machine learning techniques. The usage of a convolutional neural network (CNN) for the categorization of human activity and attribute identification data obtained from accelerometer and gyroscope sensors on an iPhone 6s is proposed in this research. The collection contains time series data from 24 subjects who completed six activities in 15 trials. The study begins by pre-processing the data by transforming the folders into class labels and plotting the time series data. The time-series data is made up of multivariate data from both the accelerometer and gyroscope sensors, totaling 12 characteristics. The accelerometer sums up two acceleration vectors, gravity, and user acceleration, which may be distinguished using core motion tracking technology.

Chapter 19

Due to the limited resources available in the connected IoT devices, it is difficult to assess the reliability of the system. IoT reliability desires favourable issues for measuring analysis with proposing, solving, qualifying considering, besides giving IoT responsiveness arrangements is, accordingly, essential for the enormous preparation of IoT innovation over all areas of society. In this unique situation, the most important reason is to propose and investigate survey responsiveness instruments for IoT outcomes, upholding the use of repetitive courses then gadgets that benefit as much as possible from cell phones, as key parts of IoT. Reliability for IoT has networking and various formula-based evaluations that has magnified node to node connection at various levels. The authors go through reliability measurements and models in this chapter. A detailed investigation for the sake to quantify reliability in the IoT based devices has been discussed alongside issues associated with it.

Kavitha Rajamohan, CHRIST University (Deemed), India
Sangeetha Rangasamy, CHRIST University (Deemed), India
M. Ajai Kumar, CHRIST University (Deemed), India
Bidura Sarkar, CHRIST University (Deemed), India
Jubaraj Mukherjee, CHRIST University (Deemed), India

The network of physical items/things/objects, that are implanted with sensors, software, and other networking technologies to communicate and exchange data with other devices and systems through the internet, is referred to as the Internet of Things (IoT). Environmental control in the context of IoT systems refers to the use of connected devices and sensors to manage and regulate various aspects of the environment, such as temperature, lighting, air quality, water quality, and more. The goal is to create an intelligent environment that is more efficient, comfortable, and sustainable.

Sangeetha Rangasamy, CHRIST University (Deemed), India
Kavitha Rajamohan, CHRIST University (Deemed), India
V. S. Lavan, CHRIST University (Deemed), India
C. Mayur, CHRIST University (Deemed), India
Mary F. Lalitha, CHRIST University (Deemed), India

Technological development has led us to various paths. Though each technological change has evolved in its own space and time, it resulted in amenities and a pragmatic approach to problem-solving. One such technology is the "Internet of Things" - IoT which was coined in the year 1999 by a computer scientist called Kevin Ashton. Despite its start in the late 90s, it has come a long way to achieve the status of 10 Billion active IoT devices by the end of 2021, with the IoT solutions costing an estimated economic value of 4-11 trillion by 2025.

Chapter 22

Sandeep Bhatia, Galgotias University, India
Zainul Abdin Jaffery, Jamia Millia Islamia, India
Shabana Mehfuz, Jamia Millia Islamia, India
Neha Goel, Raj Kumar Goel Institute of Technology, India

IoT refers to a network of interconnected devices which are able to share data with each other. Each device in a network has a unique address. The placement of devices called nodes in any network has always been a matter of concern associated with its routing, shortest path, power dissipation by nodes and topology etc. The integration of a wireless sensor network with the internet of things is an area of research among the researchers and provides a valid approach for networking in any IoT enabled system. WSN- IoT integration finds applications in smart agriculture. In this chapter the authors will discuss the methods of WSN-IoT integration and its benefits among smart agriculture. The authors will discuss the application using Lora WAN to enhance better crop yield. LoRa WAN is a low cost and long-range device which is suitable for smart agriculture application for sending collected data to a remote location which we cannot communicate through Zigbee, wi-fi, Bluetooth and other short-range devices. The researchers will explain the routing methods and protocols associated with smart agriculture.

Chapter 23

    Sandeep Bhatia, Galgotias University, India
    Neha Goel, Raj Kumar Goel Institute of Technology, India
    Soniya Verma, KIET Group of Institutions, India

Broadband cellular networking technologies 5G/6G/7G for establishing connections between users and sending data over the internet are sometimes referred to as the "internet of things" or embedded items. IoT brings about a huge amount of traffic as a result of the strong interaction among the millions of connected things available at the time of deployment to particular applications. The increase in the popularity of the internet of things has been likened to an increase in the number of barriers. In this study, the current use of 5G edge network infrastructure with IoT-enabled and hybrid and multi-cloud deployments need 6G and 7G technologies, and the authors examine the current state of IoT as well as potential conditions and challenges that may influence IoT acquisition. In this chapter there is a comparison of different communication technology in the context of IoT. At the end of the chapter, the impact of 5G, 6G, and beyond 7G technologies have been discussed in the preface of IoT.

Preface

Machine learning (ML) and internet of things (IoT) are the top technologies used by businesses to increase efficiency, productivity, and increase their competitiveness in this fast-paced digital era transformation.

Machine learning (ML) is the key tool for fast processing and decision-making applied to smart city applications and next-generation IOT devices, which require ML to satisfy their working objective.

IOT technology has proven as an efficient in solving many real world problems and machine learning algorithms combine with IOT means the fusion of product and intelligence to achieve better automation, efficiency, productivity, and connectivity. Emerging IoT trends, including Internet of Behavior (IoB), pandemic management, edge and fog computing, and connected autonomous vehicles with a primary focus on machine learning to develop futuristic and sustainable solutions. Despite IOT's ability to transform our present-day societies into smarter and more sustainable ones, it has to overcome a set of challenges, e.g., technological, individual, business, and those related to our societies.

Machine learning gives a brain to IoT-enabled systems to grasp the insight from data produced by millions of IOT objects. IoT-based ML consists of algorithms that will learn from the colossal amount of data. ML tasks can be seen from IoT perspectives to predict from a treasure trove of data (1) data quality and (2) pattern recognition. Machine learning Algorithms can also be used to enhance data quality. Machine learning Algorithms are used to identify outliers and impute data before training Machine learning Algorithms for prediction.

The importance of Machine learning for IoT's success and diverse Machine learning powered IOT applications. Machine learning developments in IoT is classified from three perspectives: data, application, and industry.

The actual objective of ML in IoT is to bring complete automation by enhancing learning that facilitates intelligence through smarter objects . ML gives IoT-enabled systems the potential to mimic human-like decisions after training from the data and further improve their understanding of our surroundings. Some important research areas in domain of smart homes, agriculture, warning systems, smart shopping, smart gadgets, smart cities, intelligent roads, health care, fire systems, threat-identification systems, tracking, and surveillance with machine learning-enabled IoT.

This book gives insight into the current state-of-art research and developments in IoT with a specific focus on ML-related developments. Therefore, this book will focus on research and researchers working in the fields of image processing, biomedical sciences, medical instrumentation, etc.

This book identifies emerging IoT trends that will use the machine at its core to develop futuristic and sustainable solutions, helps the readers to identify future opportunities in IOT-based ML research, hence this book is useful for Designers and engineers working in the fields of image processing, students and educators of higher education, librarians, researchers, and academicians.

Overall, the upcoming pages will present a diverse range of topics that have been explored by prominent scholars and professionals in 23 chapters. These topics address the problem and challenges in energy, industry, and healthcare and solutions proposed for Machine Learning Enabled IOT and new algorithms in machine learning and also addresses their accuracy for existing real-time problems. Furthermore, the following pages will also include detailed discussions and instances related machine learning applications.

Chapter 1, "Challenges in Various Applications Using IoT," draws an awareness to the unique set of challenges that must be addressed to fully leverage the potential of IoT. IoT offers tremendous potential for various applications; there are significant challenges that must be addressed to realize its full potential. Organizations must ensure that they are addressing security and privacy concerns, ensuring interoperability, and building scalable and reliable systems to fully leverage the benefits of IoT. One of the primary challenges in various applications using IoT is ensuring the security and privacy of the data collected and transmitted. Another significant challenge is ensuring the interoperability of different IoT devices and systems. Finally, there are concerns about the scalability and reliability of IoT systems. As the number of connected devices continues to grow, it becomes increasingly challenging to manage and maintain them effectively. This can lead to issues with system downtime, increased maintenance costs, and a lack of scalability to meet future demands.

Chapter 2, "Pattern Recognition by IoT Systems of Machine Learning," explores multiple IoT-based ML algorithms for pattern identification. With IoT and machine learning (ML) technology, some activities may be detected and a good judgment may be derived. By merging IoT with ML algorithms, its application areas may be extended considerably farther than they are currently. Runtime complexity, memory requirements, and accuracy are taken into account while comparing several ML algorithms to select the optimum one for a particular electronic device. Support Vector Machine(SVM), Random Forest, and K-Nearest Neighbour (K-NN) offer higher runtime complexity, less memory demand, and more accuracy when compared to other ML algorithms. All of the aforementioned challenges have been addressed in this chapter

Chapter 3, "Institutional Pressures on the Oil and Gas Industry: The Role of Machine Learning," focuses on type of external factors influence oil and gas firms to implement AI. This chapter leverages institutional theory to look at external isomorphic pressures in the form of government regulations and competitive pressures that influence the adoption of AI. The chapter concludes by reviewing opportunities for future research in the areas of AI within the oil and gas industry.

Chapter 4, "Generative Adversarial Networks: A Game Changer," explores new fields of study and application of GANs in a variety of contexts, including data augmentation, image, audio, text-to-image, and 3D object creation. The use of GANs in the healthcare and financial sectors for IoT applications has opened the door to new areas of study. While GANs have a lot going for them, there are still several challenges that need to be worked out before they can be widely adopted, such as training stability, interpretability, ethical considerations, and more. Increases in computing power and network capacity are expected to expand GANs' potential to reshape the Internet of Things (IoT) industry. Several new lines of inquiry and development for Internet of Things (IoT) devices will be made possible by GANs.

Chapter 5, "Machine Learning-Enabled Internet of Things Solution for Smart Agriculture Operations," introduces smart farming to ensure greater yields and efficiency. Traditional farming practices are challenging due to climatic fluctuations and extremities, growing population, urbanization, loss of crops to various diseases, and difficulty detecting weeds. These challenges are motivated to apply a technology-based solution to increase the quality and quantity of the crops. Advanced technologies can transform inadequate farming methods into smart farming to ensure greater yields and efficiency.

Cyber-physical systems (CPS), robotics/drones, sensors/IOT devices, agricultural data management, AI, and ML approaches significantly transform process chains in agriculture and forestry. ML and IoT applications contain sensor data quantities, and intelligent processing could bridge the holes between the cyber and physical systems.

Chapter 6, "An Investigative Study on Internet of Things in Healthcare," explores the growth, adoption and threats that the emergence of IoT in the healthcare sector holds. IoT has various capabilities that pose an opportunity to provide solutions for a digitized healthcare system that would also aid in battling pandemics such as COVID-19. IoT has applications in multiple places, such as improving hospital treatment systems, tracking different medical parameters, identifying irregularities. The correct usage of IoT may aid in dealing with a variety of medical issues such as speed, cost, and complexity. It makes the surgeon's job easier by lowering risks and improving overall performance. This chapter also looks at other aspects of IoT such as fog computing which reduces latency while communicating to remote servers leading to an acceleration of medical services. By enabling patients to spend more time comfortably at home and only communicate with care facilities and medical experts when necessary, IoT also helps patients participate with their care and feel more satisfied. The growing use of connected devices in healthcare is a major driver for the IOT-Health market. In recent years, wearable sensors have grown significantly. The development of numerous technologies is required for end-to-end health data networking, which should allow for dependable communication between a patient and a healthcare professional. Finally, this research identifies the threats and the prospects of IoT in the future of healthcare.

Chapter 7, "Machine Learning-Based Threat Identification Systems," explained the basics of IoT, ML and threat identification system, i.e., IDS. ML based decision tree approach is used while designing IDS utilizing KDD dataset. Distributed Intelligent Systems are becoming popular with the advancements in Artificial Intelligence (AI). A huge amount of data is travelling over millions of distributed heterogeneous networks. IoT and cloud services lead the users to deal with enormous data on daily basis. As the horizon of devices is increasing so is the concern towards security breaches. In today's world cyber security services are essential because of massive collection of crucial data over computers and other devices which is used in government sector, business, healthcare, military and financial organizations.

Chapter 8, "Future Outlier Detection Algorithm for Smarter Industry Application Using ML and AI," investigates outlier identification that must be addressed in a variation of usage, which include fake prevention and detection (e.g., potentially malicious utilization credit and debit cards or even different kinds of monetary transactions), healthcare information analyzation (e.g., capable of recognizing dynamics are changing to therapeutic interventions among patient populations), fault detection in production processes, and detection of network intrusions, among others. Additionally, the presence of outliers affects several data processing tasks, necessitating the limitation or elimination of outlier observations. Recognizing outliers in multidimensional data is a tough task that gets increasingly difficult when dealing with high-dimensional datasets.

Chapter 9, "Edge Computing," is a research study on edge computing that involves processing and analyzing data closer to where the data is actually originated. The term "edge" refers to the outer layer of an organization's network where data is generated, and the processing and analysis of that data occurs. Edge computing aims to reduce data processing time and bandwidth requirements by moving data processing closer to where the data is generated. Edge computing has emerged as a result of the increasing demand for real-time, high-performance, and low-latency processing of data generated by Internet of Things (IoT) devices, autonomous systems, and other emerging technologies. By processing data closer to the source of data generation, edge computing offers several advantages over traditional cloud computing models, including reduced latency, improved data security, and enhanced reliability. Edge computing represents a significant shift in the way organizations process and analyze data, and it has the potential to transform industries ranging from healthcare to manufacturing to transportation.

Chapter 10, "The Role of Wireless Body Area Networks in Smart Healthcare Systems in the Context of Big Data and AI," covers various technical features available for building a smart healthcare system. The various of smart healthcare system are added as different sections here. The introduction session, infrastructure development session, streaming data management session and cloud computing session are added as first four sessions. The role of machine learning for data analysis process is discussed in last session before conclusion. The developments in sensor technology and internet of things could enable real-time monitoring of several body parameter. Integrating values from sensors of connected smart devices can improve health monitoring system. Harvesting information from these fast-moving data is explored here for advancing smart healthcare system. The machine learning algorithms helps to enhances intelligence of the system. This chapter is organized as a systematic study to transform healthcare industry to smart healthcare industry. Analytical accuracy through Artificial intelligence decides acceptance of smart healthcare industry among people.

Chapter 11, "Significance of Fog Computing to Machine Learning-Enabled IoT for Smart Applications Across Industries," discusses fog computing architecture that can provide real-time processing, nearby storage, extremely low latency, dependability, large data rates, and other requirements for industrial Internet of Things (IIoT) applications. The main features of fog computing are covered in this chapter, along with current research on the subject and a focus on the difficulties encountered when creating its architectural design.

Chapter 12, "New Cloud Computing-Based Strategy for Coordinating Multi-Robot Systems," deals with the development of a strategy for coordinating Selective Compliance Articulated Robot Arms (SCARAs) supported by the Google Cloud platform, which aim to complete the task of piling up blocks. The derivation of the coordination task to a cloud server provides a powerful computing resource without the need of having it embedded in the robot or in a local data center, hence preventing the hardware from exposure to the hazardous environment of the construction field. The objective of this application is to demonstrate the feasibility of modeling a communication and coordination service for SCARA with construction purposes, in a real cloud platform and with a leader/follower mode. To present the results, a 3D simulation is provided in Virtual Reality Modeling Language (VRML).The simulation is applied to the dynamic model of SCARA robots with one redundant prismatic joint on its base.

Chapter 13, "Impact of UAVs in Agriculture," provides the utilization of multiple UAVs systems for agriculture which increase the overall work efficiency and productivity. UAVs and their control systems together constitute the Unmanned Aerial System (UAS) which primarily relies on the utilization of Machine Learning (ML) techniques through which high resolution image data is obtained at more frequent periodic intervals when compared to images taken by the satellite. There are different aspects where UAVs especially in precision agriculture as it comprises of various processes such as soil mapping, water management, weather monitoring, etc.

Chapter 14, "A Survey on Diagnosis of Hazardous Gas Emission Using AI Techniques," gives a detailed survey on different sensors and different deep learning algorithms used for diagnosing emission of hazardous gases. Artificial Intelligence (AI) techniques plays an important role in predicting, diagnosing and monitoring the emission of Hazardous or toxic gases emitted from industries or various places. Also, it helps in continuous monitoring of environmental pollutants based on time series. As leakage of gas or toxic gases leads to health-related issues to human beings. Deep learning is a part of AI and several deep learning algorithms are used to detect the emission of toxic gases in industries, indoors, refineries, etc.

Chapter 15, "IoVST: Internet of Vehicles and Smart Traffic – Architecture, Applications, and Challenges," explores the Internet of Vehicles and Smart Traffic. The Internet of Things (IOT) is the network of sensors, devices, processors, and software, enabling connection, communication, and data transfer between devices. IOT is able to collect and analyze large amounts of data which can then be used to automate daily tasks in various fields. IOT holds the potential to revolutionize and create many opportunities in multiple industries like smart cities, smart transport, etc. Autonomous vehicles are smart vehicles that are able to navigate and move around on their own on a well-planned road.

Chapter 16, "Smart Cities: Redefining Urban Life Through IoT," discusses how IoT supports smart home builders and managers by providing an efficient ecosystem in terms of less operating cost and improvising residence services. A smart city is designed using acceptable Internet of Things (IoT) technologies that solve urban life problems and provide quality of life to the residents. IoT refers to a network of physical devices that are capable of gathering and sharing data and expediting numerous functions without human assistance.

Chapter 17, "IoT and Machine Learning on Smart Home-Based Data and a Perspective on Fog Computing Implementation," emphasizes the need for energy efficiency in buildings, focusing primarily on the Heating, Ventilation, and Air Conditioning (HVAC) systems, which consume 50% of building energy. A predictive system based on Artificial Neural Networks (ANNs) was created to generate short-term forecasts of indoor temperature using data from a monitoring system in order to reduce this energy use. The technology seeks to estimate inside temperature in order to determine when to start the heating, ventilation, and air conditioning system, potentially reducing energy use dramatically. The paper describes the system's code implementation, which includes data pre-processing, model training and evaluation, and result visualization. In terms of evaluation metrics, the model performed well and revealed the potential for large energy savings in buildings.

Chapter 18, "Activity Recognition and IoT-Based Analysis Using Time Series and CNN," address the topic of human activity and attribute detection. The collection contains time series data from 24 subjects who completed six activities in 15 trials. The aim is to appropriately identify the six activities using machine learning techniques. The usage of a convolutional neural network (CNN) for the categorization of human activity and attribute identification data obtained from accelerometer and gyroscope sensors on an iPhone 6s is proposed in this research. The collection contains time series data from 24 subjects who completed six activities in 15 trials. The study begins by pre-processing the data by transforming the folders into class labels and plotting the time series data. The time-series data is made up of multivariate data from both the accelerometer and gyroscope sensors, totaling 12 characteristics. The accelerometer sums two acceleration vectors, gravity and user acceleration, which may be distinguished using Core Motion tracking technology.

Chapter 19, "A Comprehensive Review of IoT Reliability and Its Measures: Perspective Analysis," is a detailed investigation to quantify reliability in the IoT-based devices along with challenges and issues associated with it. Due to the limited resources available in the connected IOT devices, it is difficult to assess the reliability of the system. IoT reliability desires favourable issues for measuring analysis with Proposing, solving, qualifying considering, besides giving IoT responsiveness arrangements is, accordingly, essential for the enormous preparation of IoT innovation over all areas of society. In this unique situation, the most important reason is to propose, investigate, in addition that has with survey responsiveness instruments for IoT outcomes, upholding the use of repetitive courses then gadgets that benefit as much as possible from cell phones, as key parts of IoT. Reliability for IoT has networking and various formula-based evaluations that has magnified node to node connection at various levels. We go through reliability measurements and models in this chapter.

Chapter 20, "Sustainable IoT for Smart Environmental Control," creates an intelligent environment that is more efficient, comfortable, and sustainable. The network of physical items/things/objects, that are implanted with sensors, software, and other networking technologies to communicate and exchange data with other devices and systems through the internet, is referred to as the Internet of Things (IoT). Environmental control in the context of IoT systems refers to the use of connected devices and sensors to manage and regulate various aspects of the environment, such as temperature, lighting, air quality, water quality, and more

Chapter 21, "Evolutionized Industry With the Internet of Things," is a research study on the internet of things. Technological development has led us to various paths. Though each technological change has evolved in its own space and time, it resulted in amenities and a pragmatic approach to problem-solving. One such technology is the "Internet of Things" - IoT which was coined in the year 1999 by a computer scientist called Kevin Ashton. Despite its start in the late 90s, it has come a long way to achieve the status of 10 billion active IoT devices by the end of 2021, with the IoT solutions costing an estimated economic value of 4-11 trillion by 2025.

Chapter 22, "Integration of WSN and IoT: Wireless Network Architecture and Protocols – A Way to Smart Agriculture," explains the routing methods and protocols associated with smart agriculture. IOT refers to a network of interconnected devices which are able to share data with each other. Each device in a network has a unique address. The placement of devices called nodes in any network has always been a matter of concern associated with its routing, shortest path, power dissipation by nodes and topology etc. The integration of a wireless sensor network with the internet of things is an area of research among the researchers and provides a valid approach for networking in any IoT enabled system. WSN- IoT integration finds applications in smart agriculture. In this chapter, the methods of WSN-IoT integration and its benefits among smart agriculture are discussed. The application using Lora WAN to enhance better crop yield also discussed. LoRa WAN is a low cost and long-range device which is suitable for smart agriculture application for sending collected data to a remote location which cannot communicate through Zigbee, wi-fi, Bluetooth and other short-range devices.

Chapter 23, "The Current Generation 5G and Evolution of 6G to 7G Technologies: The Future IoT," examines a comparison of different communication technology in the context of IoT. Broad band cellular networking technologies 5G/6G/7G for establishing connections between users and sending data over the internet are sometimes referred to as the "internet of things" or embedded items. IoT brings about huge amount of traffic as a result of the strong interaction among the millions of connected things available at the time of deployment to particular application. The increase in the popularity of the Internet of Things has been likened to an increase in the number of barriers. In this study, the current use of 5G edge network infrastructure with IOT enabled, and hybrid and multi-cloud deployments need 6G and 7G technologies, and the author examine the current state of IoT, as well as potential conditions and challenges that may influence IoT acquisition. At the end of the chapter the impact of 5G, 6G, and beyond 7G technologies have been discussed in the preface of IoT.

Finally, the book can contribute to the subject matter in several ways:

This book is useful to classify IoT-related research and development work into three major perspectives (classes): data, application, and industry.

This book gives insight into the current state-of-art research and developments in IoT with a specific focus on ML-related developments.

This book identifies emerging IoT trends that will use the machine at its core to develop futuristic and sustainable solutions.

This book helps the readers to identify future opportunities in IoT-based ML research.

This book includes applications of machine learning-enabled IoT like smart grid, smart cities, smart traffic, smart home, smart health care system, smart environmental control and smart social applications.

Include key design element some classification of future outlier detection algorithm that adopt machine learning for IoT application.

We believe these chapters will generate ideas for future research efforts as well as for the development of practical policies that will influence both global and national prosperity. All suggestions to improve this work for future publication efforts would be welcomed.

Neha Goel
Raj Kumar Goel Institute of Technology, India

Ravindra Kumar Yadav
Raj Kumar Goel Institute of Technology, India

Chapter 1
Challenges in Various Applications Using IoT

Rajeshwari Sissodia
Hemvati Nandan Bahuguna Garhwal University, Srinagar, India

ManMohan Singh Rauthan
Hemvati Nandan Bahuguna Garhwal University, Srinagar, India

Varun Barthwal
Hemvati Nandan Bahuguna Garhwal University, Srinagar, India

ABSTRACT

The internet of things (IoT) has revolutionized various aspects of human life, such as smart cities, home automation, and energy efficiency. However, fully realizing the potential of IoT necessitates overcoming significant challenges. This article provides a comprehensive evaluation of IoT from both technological and societal perspectives, delving into its architecture, applications, and critical difficulties. It highlights the contributions of different IoT components and emphasizes the value of IoT and data analytics. The findings contribute to a better understanding of IoT applications and challenges, benefiting both consumers and researchers in the field.

1. INTRODUCTION

The Internet of Things (IoT) has revolutionized our interaction with the world, providing us with a vast network of connected devices and sensors. This interconnected web enables the collection and analysis of massive amounts of real-time data, empowering us to make informed decisions and optimize our operations like never before. The applications of IoT span across diverse domains, including healthcare, transportation, energy, agriculture, and many others, promising groundbreaking advancements in each field. However, alongside these opportunities, a distinct set of challenges emerges that must be effectively addressed to fully harness the potential of IoT.

DOI: 10.4018/978-1-6684-8785-3.ch001

One of the foremost challenges encountered in various IoT applications pertains to the security and privacy of the data that is collected and transmitted. With an ever-growing number of connected devices, the potential attack surface for cybercriminals expands, putting sensitive data at risk. In parallel, the increasing prevalence of data breaches has brought privacy concerns to the forefront, necessitating robust measures to ensure compliance with data protection regulations and maintain the trust of individuals and organizations involved.

Another significant challenge lies in achieving seamless interoperability among diverse IoT devices and systems. As the technology continues to evolve rapidly, it is essential to establish efficient communication protocols and standards that enable devices to seamlessly collaborate. Failure to address this challenge may result in data silos, inefficient processes, and reduced productivity, hindering the realization of IoT's full potential.

Moreover, the scalability and reliability of IoT systems pose additional hurdles. With the exponential growth of connected devices, effectively managing and maintaining these networks becomes increasingly complex. Ensuring uninterrupted system operation, minimizing downtime, containing maintenance costs, and establishing scalability to meet future demands are all crucial considerations for organizations embracing IoT.

In conclusion, while the advent of IoT presents unparalleled opportunities across various applications, it also brings forth significant challenges that require careful attention. Organizations must proactively address security and privacy concerns, establish interoperability standards, and construct scalable and reliable systems to fully capitalize on the transformative potential of IoT. By overcoming these obstacles, we can embark on a future where IoT technology empowers us to optimize processes, enhance decision-making, and unlock innovative solutions across industries and sectors.

The emergence of the IoT has opened up a realm of unparalleled opportunities across diverse applications. However, it is crucial to acknowledge and address the significant challenges that accompany this technological revolution. To fully harness the transformative potential of IoT, organizations must adopt a proactive approach to tackle the pressing issues of security and privacy. Robust measures and protocols should be implemented to safeguard the data collected and transmitted, mitigating the risks posed by cyber threats. Furthermore, establishing interoperability standards is paramount to ensure seamless communication and collaboration among a myriad of IoT devices and systems. By overcoming these hurdles, organizations can construct scalable and reliable IoT systems that are capable of meeting evolving demands. In doing so, we can unlock a future where IoT technology empowers us to optimize processes, enhance decision-making capabilities, and pave the way for innovative solutions across industries and sectors.

2. LITERATURE SURVEY

Al-Fuqaha et al. (2015) examines the challenges of IoT-enabled smart city applications. The authors identify challenges such as the heterogeneity and complexity of the IoT system, privacy and security concerns, and the need for appropriate data analytics tools. The authors also highlight the potential benefits of IoT for smart cities, including improved resource utilization, reduced pollution, and enhanced public services.

Goyal et al. (2018) discusses the challenges of IoT-based supply chain management. The authors identify challenges such as data security, interoperability, integration of various IoT devices, and the need for real-time monitoring and decision-making. The authors also highlight the potential benefits of using IoT for supply chain management, including improved efficiency, visibility, and agility.

Al-Turjman et al. (2018) discusses the challenges and opportunities of using IoT in healthcare systems. The author identifies challenges such as data privacy and security concerns, the need for real-time monitoring and decision-making, and the need for appropriate regulations and standards. The author also highlights the potential benefits of IoT in healthcare, including improved patient care, reduced costs, and enhanced patient outcomes.

Kaarlela et al. (2019) examines the challenges of IoT for service innovation in healthcare. The authors identify several key challenges, including the complexity of the IoT system, interoperability issues, data privacy and security concerns, and the need for appropriate regulations and standards.

Kaddoum et al. (2019) analyzes the challenges and opportunities of IoT-based agriculture. The authors identify challenges such as the need for reliable and robust sensor networks, data privacy and security concerns, and the need for appropriate data analytics tools. The authors also highlight the potential benefits of using IoT for agriculture, such as improved crop yields, reduced water usage, and enhanced crop quality.

Li et al. (2019) discusses the challenges and opportunities of using IoT in transportation systems. The authors identify challenges such as the need for real-time data analytics, interoperability issues, and the need for appropriate regulations and standards. The authors also highlight the potential benefits of IoT in transportation, including improved safety, reduced congestion, and enhanced efficiency.

Yang et al. (2019) discusses the challenges and opportunities of using IoT in manufacturing systems. The authors identify challenges such as data privacy and security concerns, the need for real-time data analytics, and the need for appropriate regulations and standards. The authors also highlight the potential benefits of IoT in manufacturing, including improved efficiency, reduced costs, and enhanced quality control.

Sharma, A.et al. (2019) discusses the challenges and opportunities of using IoT in agriculture. The authors identify challenges such as the need for appropriate regulations and standards, data privacy and security concerns, and the need for appropriate data analytics tools. The authors also highlight the potential benefits of IoT in agriculture, including improved crop yields, reduced resource waste, and enhanced efficiency.

Bhattacharyya et al. (2019) discusses the challenges and opportunities of using IoT in energy management. The authors identify challenges such as data privacy and security concerns, the need for interoperability, and the need for appropriate regulations and standards. The authors also highlight the potential benefits of IoT in energy management, including improved energy efficiency, reduced costs, and enhanced grid stability.

Yang et al. (2019) discusses the challenges and opportunities of using IoT in retail. The authors identify challenges such as data privacy and security concerns, the need for interoperability, and the need for appropriate regulations and standards. The authors also highlight the potential benefits of IoT in retail, including improved customer experience, enhanced inventory management, and reduced costs.

Raza et al. (2020) discusses done systematic review identifies and analyzes the challenges and opportunities of IoT for smart grids. The authors find that the main challenges include scalability, interoperability, reliability, security, and data management. They also identify several opportunities, such as the potential for real-time monitoring, control, and optimization of the grid.

Kabassi et al. (2020) discusses the challenges and opportunities of using IoT in education. The authors identify challenges such as the need for appropriate regulations and standards, data privacy and security concerns, and the need for appropriate data analytics tools. The authors also highlight the potential benefits of IoT in education, including improved student engagement, enhanced teacher performance, and enhanced learning outcomes.

Singh et al. (2020) discusses the challenges and opportunities of using IoT in supply chain management. The authors identify challenges such as data privacy and security concerns, the need for interoperability, and the need for appropriate regulations and standards. The authors also highlight the potential benefits of IoT in supply chain management, including improved inventory management, reduced costs, and enhanced supply chain visibility.

Hussain et al. (2021) provides a comprehensive review of the challenges and opportunities of using IoT in healthcare systems. The authors identify challenges such as data privacy and security concerns, the need for appropriate data analytics tools, and the need for appropriate regulations and standards. The authors also highlight the potential benefits of IoT in healthcare systems, including improved patient care, reduced costs, and enhanced medical research.

Marouf et al. (2021) discusses the challenges and opportunities of using IoT in intelligent transportation systems. The authors identify challenges such as data privacy and security concerns, the need for interoperability, and the need for appropriate regulations and standards. The authors also highlight the potential benefits of IoT in transportation systems, including improved traffic management, reduced congestion, and enhanced public safety.

Al-Rubaie et al. (2021) provides a comprehensive review of the challenges and opportunities of using IoT in smart agriculture. The authors identify challenges such as the need for appropriate regulations and standards, data privacy and security concerns, and the need for appropriate data analytics tools. The authors also highlight the potential benefits of IoT in agriculture, including improved crop yields, reduced resource waste, and enhanced efficiency.

Kumar et al. (2022) provides a comprehensive review of the challenges and opportunities of using IoT in smart grids. The authors identify challenges such as data privacy and security concerns, the need for interoperability, and the need for appropriate regulations and standards. The authors also highlight the potential benefits of IoT in energy management, including improved energy efficiency, reduced costs, and enhanced grid stability.

Jindal et al. (2022) provides a comprehensive review of the challenges and opportunities of using IoT in smart manufacturing. The authors identify challenges such as the need for appropriate regulations and standards, data privacy and security concerns, and the need for appropriate data analytics tools. The authors also highlight the potential benefits of IoT in manufacturing, including improved efficiency, reduced costs, and enhanced quality control.

Arvind et al. (2023) reviews the security challenges in IoT-based healthcare applications, including confidentiality, integrity, authentication, and availability. The authors propose a security framework to address these challenges.

Zhang et al. (2023) explores the interoperability challenges in IoT-based smart homes, including device heterogeneity, data heterogeneity, and communication heterogeneity. The authors propose a framework to enable seamless interoperability among different IoT devices.

Overall, these recent studies provide valuable insights into the challenges and opportunities of using IoT in various applications. The challenges identified in these studies include data privacy and security concerns, interoperability issues, and the need for appropriate regulations and standards. The potential benefits of using IoT in these applications include improved efficiency, reduced costs, enhanced quality control, and enhanced customer experience.

3. IOT ARCHITECTURE AND TECHNOLOGIES

The IoT systems are structured based on a comprehensive architecture consisting of five core levels: perception, network, middleware, applications, and business. Each layer plays a crucial role in the system's functionality and contributes to the overall operation and management of IoT devices and data.

At the foundation of the IoT architecture lies the perception layer, which encompasses physical items such as sensors, RFID chips, and barcodes. These components act as the network's building blocks by gathering data from the environment and transmitting it to the network layer.

The network layer facilitates the transfer of data collected at the perception layer to the information processing system. This layer employs various communication mediums, including cables and wireless technologies like 3G/4G, Wi-Fi, and Bluetooth, to transmit the data efficiently.

Above the network layer is the middleware layer, which plays a critical role in processing the data received from the network layer. Its primary responsibility is to analyze and interpret the data, enabling the system to make informed decisions based on the outcomes of ubiquitous computing.

The application layer leverages the processed data from the middleware layer for system-wide device administration. This layer utilizes the insights derived from the data to facilitate various applications and services within the IoT system.

Sitting atop the architecture is the business layer, which encompasses the applications and services of the IoT system. The business layer provides a graphical representation of the data and statistics collected by the application layer. These representations are then utilized for strategic planning and goal-oriented decision-making processes.

It's worth noting that the IoT architecture can be tailored to specific use cases, allowing for customization and optimization based on specific requirements and objectives.

In addition to the layered architecture, an IoT system consists of various functional blocks that enable a wide range of operations. These functional blocks include sensing mechanisms, authentication and identity management, control and management functionalities, and more. Each of these blocks plays a crucial role in handling I/O activities, addressing connectivity challenges, managing data processing, facilitating audio/video monitoring, and ensuring efficient storage management within the IoT system.

By combining the layered architecture and functional blocks, IoT systems are designed to provide comprehensive and efficient solutions for various applications and industries, enabling the realization of the full potential of the Internet of Things.

Figure 1. Module of IoT system

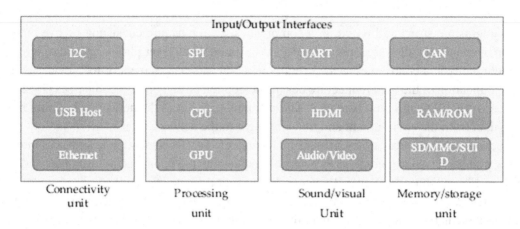

All of these building blocks work together to make an effective IoT system. Despite the fact that a number of reference architectures together with technical requirements have been presented, these remain far from the ideal architecture for the IoT on a global scale. That's why it's crucial to continue working on the development of an appropriate architecture that might meet the requirements of the IoT on a global scale. The IoT system architecture is depicted in Fig. 2. Figure 3 illustrates how the IoT is sensitive to the settings of a certain application.

Figure 2. Structure of IoT

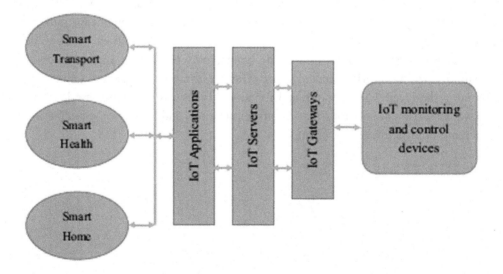

The IoT gateway plays a crucial role in IoT communication by establishing a connection between IoT servers and IoT devices used in a variety of use cases. Key design concerns for an effective IoT architecture in a diverse environment are scalability, modularity, interoperability, and openness. The goal of the IoT architecture's design should be to meet the needs of big data analytics and storage, user-friendly applications, cross-domain interactions, and the integration of several systems with the possibility of easy and scalable management. Also, the design needs to be extensible so that more IoT devices can join in the system's intelligence and automation. In addition, a new difficulty has emerged due to the huge amount of data generated by the connectivity between IoT sensors and devices. As a result, the IoT system needs an efficient architecture to handle the large amounts of streaming data. Cloud and fog/edge computing are two common IoT system topologies that help in managing, monitoring, and analyzing massive amounts of data. Therefore, as can be seen in Figure 3, a contemporary IoT architecture consists of four distinct phases.

Figure 3. Four stage IoT architecture to deal with massive data

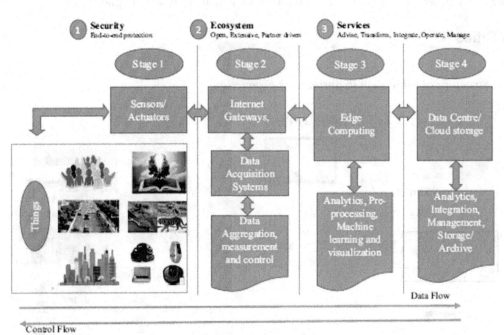

In the first phase of the design, sensors and actuators play a crucial role. The environment, living beings, animals, technology, vehicles, structures, etc. all make up what we call "the real world." Sensors pick up on the signals and data flow from these physical objects, which are then converted into information that can be analyzed. Actuators can also alter the world around them by, for example, adjusting the temperature, slowing the pace of a vehicle, or turning off the lights and music. So, the first stage helps gather potentially important data from the real world for the second. In Stage 2, you will work with sensors, actuators, gateways, and data loggers. In this phase, the huge amounts of data created in phase 1 are consolidated and optimized in a format appropriate for processing. The third and last level, edge computing, is where the huge amounts of data are sent once they have been compiled and organized. Edge computing is the use of IoT technology and large amounts of processing power from dispersed locations around the world, and it can be defined as a distributed open architecture. It's ideal for IoT devices because it's a very effective method for processing data in real time. Third-stage edge computing solutions process large volumes of data and offer a range of features, including the ability to visualize that data, combine it with data from other sources, analyze it with machine learning techniques, etc. There are a number of crucial steps in the final stage, including in-depth processing and analysis and delivering feedback to increase the system's accuracy and precision. At this point, everything is being done on a remote server in the cloud or at a data center. For this massive stream of data, a big data framework like Hadoop or Spark can be employed, and with the aid of machine learning techniques, more accurate and trustworthy IoT systems can be built to match current demands.

4. IOT APPLICATIONS

There are many different ways to use the IoT, some of which are briefly explained below.

Smart Cities

Figure 4. Structure of IoT system in agriculture production

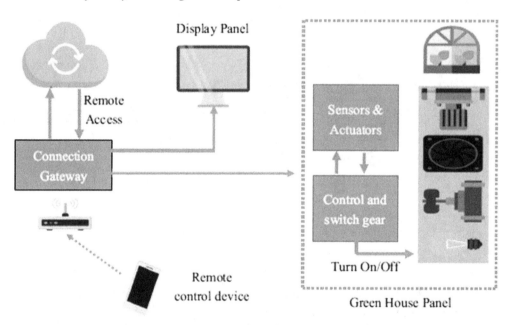

Through the IoT, smart cities can better manage their resources, become more resilient to transitory faults and calamities, and incentivize productive behavior as shown in figure 4. Today's urban problems can be alleviated in a number of ways with the help of the IoT. These include smart and weather-adaptive lighting, gas leakage monitoring, smart parking with variable pricing and automated car parking advice, and so on. By spotting potential threats and providing vital data on crowd behavior and citizen requirements, ubiquitous vision can provide a previously unimaginable level of safety and protection. Vision in the IoT enables a more dynamic match between the demand for services and the provision of those services, in addition to providing ubiquitous and augmented surveillance. For the same reasons, it can be put to use in the construction of real-time noise urban maps with the goal of reducing noise pollution at crucial periods and pinpointing noise occurrences for increased security. The smart watering of parks and other outdoor areas is another potential application of the IoT. The IoT will also improve the safety (e.g., structural monitoring of bridges) and security of smart infrastructures (e.g., automated identification of unattended bags and suspicious behavior). The IoT architecture for smart metering and monitoring will facilitate precise automated meter reading and the distribution of bills for utilities such as electricity, gas, and water to consumers. LED streetlights with built-in motion detectors can be used to dynamically switch on when there is vehicle or pedestrian traffic in the area and off when there is not. The city can save electricity (and money), and residents can feel safe by avoiding the creation of dark zones around them. In addition to the static information that is already available, "smart tourism" promises to provide visitors with a quick understanding of the city's accessibility and the crowdiness of different areas and to provide them with dynamic suggestions on tours that are tailored to their mood. Finding potentially harmful and inappropriate garbage that would require a different technique for disposal is one way to improve the efficiency of waste management.

Healthcare

In many ways, the healthcare industry is a prime candidate for the implementation of IoT technologies. They can be used to improve on existing forms of assisted living, for example. Patients will be fitted with sensors that will monitor their vital signs in real time. These readings will include temperature, heart rate, glucose, and oxygen supply. Patient activity in their homes will be monitored using data gathered by other devices. Data will be collected locally and sent to faraway medical facilities that can do sophisticated remote monitoring and take swift action as needed. The data from such disparate devices working together could paint a more complete picture of a patient's health, allowing doctors to intervene proactively when they see warning signs of an impending health decline. Furthermore, the availability of big data from a huge number of patients presents a once-in-a-generation chance to investigate correlations, develop tools and models for predicting diagnosis and prompt therapy, and streamline and improve drug discovery. The same logic applies to the care of the elderly and disabled; continuous, unobtrusive monitoring allows for superior, highly responsive, and highly predictive care, all while protecting individuals' autonomy and relieving hospital resources. Care costs can be reduced because of remote supervision's capacity to increase the number of people one expert can look after at once.

Transportation System

By integrating cutting-edge sensor, data, and network technologies, an intelligent transportation system will optimize the administration and control of public transportation. The benefits of intelligent transportation are numerous and varied, and they include electronic highway tolls that are always on, mobile emergency command and scheduling, enforcement of transportation laws, monitoring of violations of vehicle rules, lower emissions, theft protection, fewer accidents, shorter delays, and fewer environmental impacts. The increasing prevalence of sensors and/or actuators in all urban transportation modes (automobiles, railways, buses, and bicycles) is creating a network of mobile sensors. Transportation infrastructure (including roads, trains, and the commodities they carry), as well as the goods themselves, are increasingly fitted with labels and sensors that relay data to central command centers. In addition to keeping tabs on shipments in transit, this also paves the way for novel solutions like rerouting vehicles in the transportation sector to reduce congestion or providing tourists with accurate transit details. Vehicles can also be fitted with additional sensors to measure environmental factors including pollution, humidity, and temperature. As a result, the concept of "smart vehicles" emerges, wherein such data, if collected and transmitted correctly, can help to make road transit greener, smarter, and safer. For instance, public transit drivers may benefit from eco-driving tips like those that seek to lessen their dependence on fossil fuels and cut down on emissions. Mobile apps like Google Traffic collect data on traffic conditions from users and use that information to improve the app's performance. By using smart traffic light systems, cities can make driving and cycling easier on their residents. For instance, an intelligent traffic light arrangement may be possible by merging data from cyclists' smartphones with traffic data received from various sensors installed in a city's traffic signal infrastructure.

Retail and Logistics

Monitoring storage conditions along the supply chain, tracking products for traceability, and processing payments based on geographic location or activity duration for public transportation, theme parks, etc. are just a few examples of the many advantages that can be obtained by integrating the IoT in retail and supply chain administration. Several retail-related uses for the IoT have been identified, including shopper navigation based on a specified list, biometrics-based self-checkout, the identification of allergens in products, and automated stock replenishment. The IoT is used in logistics to keep tabs on everything from the moment a shipment leaves a supplier to the moment it arrives at its destination. This includes checking in on the condition of the shipment, tracking the fleet, checking in on any insurance claims, searching for specific items in a large area like a warehouse, and even emitting warnings if any containers containing flammable goods are too close to containers storing explosives.

Smart Energy

The smart grid is a relatively new form of intelligent power system that has the potential to enhance energy efficiency, lessen the grid's environmental effect, boost the supply's security and dependability, and cut down on transmission losses. Integrating IoT into smart grids enables fault detection and monitoring, consumption tracking, and other functions. To achieve this goal of energy savings, other associated solution groups anticipate heat and energy management in buildings and homes. Manufacturing businesses can boost their energy effectiveness and competitiveness at the energy production level by collecting data on energy usage with the use of IoT technology. In order to monitor, report, and notify operational personnel in real time, facilities energy management uses a combination of data systems, operational technology, and advanced metering. These management systems are quite capable of providing real-time insight into the operation of buildings and other facilities. They also allow data flows from a wide range of building equipment and other subsystems, and they can provide a dashboard view of energy consumption levels with varying degrees of granularity. Home energy management (HEM) is the process of adjusting a home's heating, cooling, and lighting systems based on factors such as the number of occupants, the time of day, the weather, and the cost of the corresponding utility. The system enhances the efficiency of home energy generation and utilization. With the help of a home area network and energy management sensors, the HEM system can adapt to fluctuations in the mains power supply and maximize efficiency. Together, these measures can make a significant dent in residential energy consumption and carbon dioxide emissions. The IoT also has the potential to be used for remote home appliance control and the prevention of burglaries.

Smart Agriculture

Through the use of communication infrastructure, such as mobile phone text messages, a network of various sensors may collect data, process that data, and then alert the farmer to the specific areas of land that require special attention. Seed, fertilizer, and pest control methods that can adapt to local conditions and signal when to be applied could fall under this category. Agronomists will be able to better understand plant development models and implement efficient agricultural strategies if they are aware of soil conditions and climate changes. As unfavorable farming conditions are avoided, agricultural output will skyrocket. The IoT plays an important role in water management by allowing for the study of river and sea water's suitability for agricultural and drinkable uses, the detection of liquid existence outside tanks and pressure variants along pipes, and the monitoring of water influences throughout river systems, dams, and reservoirs.

Environmental Monitoring

The IoT can improve environmental programmers like recycling and e-waste disposal. Monitoring air pollution is a crucial use of IoT: To reduce the amount of carbon dioxide (CO_2) released by industrial processes, vehicular pollution, and agricultural hazardous gas emissions, Monitoring of flammable gases and preventative fire conditions to establish warning zones: It's critical to keep an eye on the weather so you can be prepared for anything. Shock waves, early warning systems, maintaining safe drinking water standards, and preventing pollution from waste and chemicals dumped into rivers and the ocean are two benefits of monitoring water quality. Monitoring river levels, dams, and reservoirs on wet days is part of conservation efforts made possible by GPS and GSM module-equipped collars that allow wild animals to be tracked and their locations relayed via short message service (SMS).

Security & Emergencies

Perimeter access control, liquid presence, radiation levels, explosive and hazardous gases, etc. are only a few examples of how IoT technologies have expanded the field of security and emergency preparedness. Perimeter access control is used for a variety of purposes, including, but not limited to, detecting and preventing unauthorized entry into restricted areas, conducting surveillance in those areas, keeping tabs on individuals and property, performing asset tracking, alerting authorities to potential problems, and so on. In order to avoid costly breakdowns and corrosion, liquid detection is employed in server rooms, warehouses, and other high-security areas of buildings. The third IoT application monitors radiation levels around nuclear power plants to create leakage alerts, while the fourth is used to monitor gas levels and leakages in industrial settings, chemical factory settings, and mines. Predicting the onset of tsunamis and earthquakes by identifying vibrations or other natural calamities and taking suitable actions in advance is made possible through the combination of sensors as well as their independent coordination and modelling.

5. ADVANTAGES AND CHALLENGES

There are several potential advantages and challenges associated with the Internet of Things (IoT). Here are some important ones:

Advantages

i.   **Efficiency**: The IoT can increase efficiency by automating tasks, reducing waste, and optimizing resource utilization.
ii.   **Convenience:** The IoT can make our lives more convenient by enabling remote control of devices and providing real-time data on our surroundings.
iii.   **Cost Savings**: The IoT can lead to cost savings by reducing energy usage, minimizing equipment downtime, and improving supply chain management.
iv.   **Personalization**: The IoT can provide personalized experiences by tailoring services and products to individual users' preferences and needs.
v.   **Innovation**: The IoT can drive innovation by enabling the creation of new products and services that were previously impossible.

Challenges

Privacy

Privacy is a crucial topic in the context of the Internet of Things (IoT), as it involves the collection and processing of massive amounts of personal data. Some important points to consider when discussing privacy in the context of IoT include:

A.   What types of personal data are collected by IoT devices?
B.   Who has access to this data, and how is it used?
C.   What measures are in place to ensure the security of this data?
D.   What are the implications of the use of this data for individual privacy and autonomy?

Security

Security is also a critical concern when it comes to IoT, as the vast number of connected devices creates a large attack surface for potential hackers. Some important questions to consider when discussing security in the context of IoT include:

A.   What are the most significant security risks associated with IoT devices?
B.   How can these risks be mitigated?
C.   What security standards and best practices should be followed in the development and deployment of IoT devices?
D.   What are the implications of a security breach in the context of IoT?

Interoperability

Interoperability refers to the ability of different IoT devices to communicate and work together seamlessly. Some important questions to consider when discussing interoperability in the context of IoT include:

A. What are the benefits of interoperability in the context of IoT?
B. What are the challenges associated with achieving interoperability among different IoT devices?
C. What standards and protocols should be followed to ensure interoperability among different IoT devices?
D. How can interoperability be ensured while still maintaining privacy and security?

Standardization

Standardization involves the development and implementation of common technical standards for IoT devices to ensure compatibility and interoperability. Some important questions to consider when discussing standardization in the context of IoT include:

A. Why is standardization important for IoT?
B. What are the challenges associated with developing and implementing standards for IoT devices?
C. What organizations are responsible for developing IoT standards, and what are their roles?
D. What is the impact of non-standardization on the growth and adoption of IoT?

Law and Regulation

Law and regulation play a critical role in ensuring the responsible development and deployment of IoT devices. Some important questions to consider when discussing law and regulation in the context of IoT include:

A. What are the current laws and regulations that apply to IoT devices?
B. Are these laws and regulations sufficient, or do they need to be updated to reflect the unique characteristics of IoT?
C. Who is responsible for enforcing these laws and regulations?
D. What are the potential implications of inadequate regulation of IoT devices?

Liberties

Liberties refer to the individual rights and freedoms that should be protected in the context of IoT. Some important questions to consider when discussing liberties in the context of IoT include:

A. What are the potential threats to individual liberties posed by the use of IoT devices?
B. How can individual liberties be protected while still promoting the development and deployment of IoT devices?
C. What measures should be taken to ensure that the use of IoT devices does not infringe on individual liberties?

D.   How can individuals be educated and empowered to protect their liberties in the context of IoT?

6. MEASURSE TO OVERCOME IOT CHALLENGES

Developers working with IoT face several challenges related to data management, security, connectivity, interoperability, and device management. Here are some measures that developers can take to overcome these challenges in the future:

a)   **FOCUS ON SECURITY:** Security is one of the most significant challenges associated with IoT. Developers should focus on building robust security mechanisms into their IoT solutions, such as using secure protocols, implementing strong authentication, and encryption, and keeping the software and firmware up to date with security patches.

b)   **EMBRACE INTEROPERABILITY:** IoT devices from different manufacturers use different protocols and communication technologies. Developers should focus on developing solutions that can communicate with different devices, platforms, and technologies.

c)   **IMPLEMENT EFFECTIVE DEVICE MANAGEMENT:** Managing a large number of IoT devices can be challenging, especially when it comes to software updates, device configuration, and troubleshooting. Developers should focus on building effective device management tools that can streamline these processes and ensure that devices are functioning correctly.

d)   **OPTIMIZE DATA MANAGEMENT:** The amount of data generated by IoT devices can be overwhelming. Developers should focus on optimizing data management by designing systems that only collect the data needed for specific use cases, implementing data compression techniques, and using cloud-based data storage solutions.

e)   **ENHANCE CONNECTIVITY:** IoT devices need to be connected to the internet to function correctly. Developers should focus on building reliable and secure connectivity solutions, such as mesh networks or cellular connectivity that can ensure that devices are always connected.

f)   **ADOPT A COLLABORATIVE APPROACH:** Developers working with IoT devices should adopt a collaborative approach to solving problems. By collaborating with other developers, manufacturers, and industry groups, developers can share knowledge and expertise and develop solutions that are more efficient, secure, and interoperable.

So, developers working with IoT devices should focus on building secure, interoperable, and well-managed solutions that optimize data management and connectivity. By adopting a collaborative approach to problem-solving, developers can help overcome the challenges associated with IoT and unlock the full potential of this transformative technology.

7. BIG DATA IN IOT

The IoT is a network of interconnected electronic gadgets and sensors. IoT networks are fast growing and expanding, leading to a surge in the deployment of new sensors and devices. There is a tremendous amount of data flow between these gadgets and the internet. The volume and velocity of this data make it a prime candidate for the label "big data." As IoT-based networks continue to proliferate, new challenges emerge, including those related to data management, collection, storage, processing, and analytics. The IoT big data framework for smart buildings is helpful for handling a variety of smart building concerns, including oxygen level management, smoke and hazardous gas measurement, and light level monitoring. The framework can easily gather information from the many in-building sensors and analyse the data for use in making decisions. Furthermore, an IoT-based cyber physical system equipped with information analysis and knowledge acquisition methodologies might boost industrial production. Smart cities face the challenge of dealing with traffic congestion. An IoT-based traffic management system can collect data in real time from sensors and devices put in traffic lights. Many bits of data regarding patients' health conditions are generated every second by the IoT sensors used in healthcare analysis. If you want to make quick, well-informed decisions based on a vast amount of data, you need to be able to analyse that data in real time, and big data technology is the ideal option for that. The IoT and big data analytics can facilitate the transition from conventional to cutting-edge methods in manufacturing. When combined with big data analysis methods, the data produced by the sensors could prove useful in a number of decision-making scenarios. In addition, the use of cloud computing and analytics can improve the development and conservation of energy at a lower cost and to the delight of customers. The massive amounts of streaming data produced by IoT devices necessitate efficient storage and subsequent analysis to enable timely decisions. If you have a large dataset, deep learning is a powerful tool that can help you make sense of it and get reliable findings. This means that the convergence of IoT, big data analytics, and deep learning is crucial to the progress of a technologically advanced civilization.

8. CONCLUSION

The Internet of Things (IoT) offers immense potential to connect devices, people and their environment in new ways, enabling analysis of the world at new levels of detail. However, this comes with significant risks to society and infrastructure, particularly in relation to security, privacy, vulnerability management, and interoperability. To address these challenges, there needs to be a greater focus on the standardization of IoT communication protocols and technology enablers, as well as the development of new techniques and enablers in communication/middleware systems, high-performance embedded and computing technologies, and wireless sensor networks. Additionally, the issues of security, privacy, and interoperability should be prioritized in any IoT design, build, and implementation, and device users must be made accountable for any problems that occur. Finally, there is a need for reconsideration of legislation and regulation in a networked society, particularly around data generated by IoT devices and the power it gives to those that possess it, with a focus on transparency and industry-driven standards.

REFERENCES

Al-Fuqaha, A., Guizani, M., Mohammadi, M., Aledhari, M., & Ayyash, M. (2015). Internet of things: A survey on enabling technologies, protocols, and applications. *IEEE Communications Surveys and Tutorials*, *17*(4), 2347–2376. doi:10.1109/COMST.2015.2444095

Al-Rubaie, A. J., Aziz, A. A., Yousif, A. A., Hadi, S. S., & Bader, A. H. (2021). Challenges and Opportunities in IoT-Based Smart Agriculture: A Comprehensive Review. *IEEE Access : Practical Innovations, Open Solutions*, *9*, 127071–127106. doi:10.1109/ACCESS.2021.3105695

Al-Turjman, F. (2018). Challenges and Opportunities in IoT-Based Healthcare Systems: A Review. *IEEE Access : Practical Innovations, Open Solutions*, *6*, 40027–40043. doi:10.1109/ACCESS.2018.2858027

Arvind, R., Shukla, A., & Yadav, S. (2023). Security Challenges in IoT-based Healthcare Applications: A Review. *Journal of Medical Systems*, *47*(2), 27. doi:10.100710916-022-01803-5

Bhattacharyya, S., Bose, S., Majumder, S., Sengupta, S., & Bhattacharyya, S. P. (2019). Challenges and opportunities in IoT-based energy management: A review. *Journal of Ambient Intelligence and Humanized Computing*, *10*(2), 717–734. doi:10.100712652-018-1006-9

Goyal, S., & Singh, A. (2018). Challenges of IoT-Based Supply Chain Management: A Review. *International Journal of Supply Chain Management*, *7*(6), 223–231.

Hussain, M., Al-Fuqaha, A., Guizani, M., & Mohammadi, M. (2021). Challenges and Opportunities of IoT-Based Healthcare Systems: A Comprehensive Review. *IEEE Access : Practical Innovations, Open Solutions*, *9*, 44622–44654. doi:10.1109/ACCESS.2021.3078443

Jindal, A., Goyal, S., Singh, A., & Singh, P. (2022). Challenges and Opportunities in IoT-Based Smart Manufacturing: A Comprehensive Review. *IEEE Access : Practical Innovations, Open Solutions*, *10*, 27035–27050. doi:10.1109/ACCESS.2022.3071264

Kaarlela, A., & Halonen, R. (2019). Challenges of IoT for Service Innovation in Healthcare. *IEEE Access : Practical Innovations, Open Solutions*, *7*, 161426–161439. doi:10.1109/ACCESS.2019.2953786

Kabassi, K., & Virvou, M. (2020). Challenges and opportunities in IoT-based education: A review. *Computers & Education*, *144*, 103698. doi:10.1016/j.compedu.2019.103698

Kaddoum, G., Diab, H., & Gani, A. (2019). Challenges and Opportunities in IoT-Based Agriculture: A Review. *IEEE Access : Practical Innovations, Open Solutions*, *7*, 174373–174390. doi:10.1109/ACCESS.2019.2954717

Kumar, S., Singh, R., Kaur, R., & Kumar, S. (2022). Challenges and Opportunities in IoT-Based Smart Grids: A Comprehensive Review. *IEEE Transactions on Industrial Informatics*, *18*(2), 1176–1190. doi:10.1109/TII.2021.3072124

Li, M., Li, Y., & Li, X. (2019). Challenges and opportunities in IoT-based transportation systems: A review. *IEEE Internet of Things Journal*, *6*(5), 8249–8261. doi:10.1109/JIOT.2019.2937666

Marouf, M., Salam, Z. A., & Al-Fuqaha, A. (2021). Challenges and Opportunities in IoT-Based Intelligent Transportation Systems: A Review. *IEEE Internet of Things Journal*, *8*(4), 2564–2585. doi:10.1109/JIOT.2020.3042138

Raza, M., Hussain, M., & Ali, A. (2020). Challenges and Opportunities of IoT for Smart Grid: A Systematic Review. *IEEE Access : Practical Innovations, Open Solutions*, *8*, 108704–108723. doi:10.1109/ACCESS.2020.3002175

Sharma, A., & Singh, S. (2019). Challenges and Opportunities in IoT-Based Agriculture: A Review. In P. Kumar, P. Singh, H. Le Duc, & S. Pandey (Eds.), *Internet of Things: Challenges and Opportunities* (pp. 85–94). Springer. doi:10.1007/978-981-13-3014-7_9

Singh, A. K., & Srivastava, S. (2020). Challenges and Opportunities in IoT-Based Supply Chain Management: A Review. In Internet of Things and Big Data Analytics Toward Next-Generation Intelligence (pp. 223-247). Springer. doi:10.1007/978-981-15-4702-6_11

Yang, B., Luo, Y., & Wu, L. (2019). Challenges and Opportunities in IoT-Based Manufacturing Systems: A Review. *IEEE Access : Practical Innovations, Open Solutions*, 7, 35709–35723. doi:10.1109/ACCESS.2019.2903235

Yang, J., Yu, W., Yang, H., Lv, H., Zhang, X., & Xiao, L. (2019). Challenges and opportunities in IoT-based retail: A review. *IEEE Internet of Things Journal*, 6(5), 8165–8180. doi:10.1109/JIOT.2019.2910298

Zhang, Y., Zhang, Y., & Chen, Y. (2023). Interoperability Challenges in IoT-based Smart Homes: A Survey. *Journal of Ambient Intelligence and Humanized Computing*, 14(2), 931–944. doi:10.100712652-022-03571-x

Chapter 2
Pattern Recognition by IoT Systems of Machine Learning

Shyam Sihare

https://orcid.org/0000-0003-2096-8273

Dr. A.P.J. Abdul Kalam Government College, Silvassa, India

ABSTRACT

Today, it is feasible to observe how quickly electronic devices are becoming connected to the internet. Electronic devices that are connected to the internet can be managed or observed from any location in the world. Many of the challenges have been made easier by internet connectivity for technological gadgets. A good judgement may be made by spotting certain trends using IoT and machine learning (ML) technologies. Its application areas can be expanded much farther than they are now by combining IoT and ML algorithms. ML uses a variety of algorithms, and while analysing them to choose the best one for a certain electronic device, runtime complexity, memory needs, and accuracy are taken into consideration. In comparison to other ML algorithms, support vector machine, random forest, and k-nearest neighbour have higher runtime complexity, a smaller memory requirement, and higher accuracy. In this chapter, the aforementioned topics have all been covered. The different ML algorithms and IoT pattern recognition application areas are covered in this chapter.

INTRODUCTION

The term "internet of things" refers to a network of things with embedded software, sensors, transmitters, receivers, and other technologies. All IoT devices are capable of carrying out a variety of complex calculations with the help of ML algorithms. Selecting the appropriate ML algorithms is essential for pattern recognition in IoT devices (Gupta & Quamara, 2020). Faster processing, less space and power consumption would be possible with a more efficient ML algorithm. This can be supplemented by ongoing research into biometric security and human motion detection (Kim et al., 2016; Sihare, 2017a; Chen et al., 2022).

Choosing the right ML algorithm for IoT devices is a very challenging process. An IoT device is a hardware machine that follows a computer-like architecture that includes memory, processing units,

DOI: 10.4018/978-1-6684-8785-3.ch002

input and output units, and a data transfer bus, whereas ML involves building algorithms according to the architecture and organization of IoT devices (Burd et al., 2018). Combining pattern recognition and human activity with IoT devices has become very important today for security and safety purposes (Merenda et al., 2020). IoT devices are such cutting-edge technology that every piece of hardware we own must be controlled remotely by connecting it to the Internet. If IoT devices are combined with ML, then hardware machines will work intelligently; there is no doubt about it. Today's era is one of technologies, and the size of each technology is also getting smaller (Ferozkhan & Anandharaj, 2021). Hardware, ML, and algorithm development are going through a new phase of change. The integration of ML algorithms for pattern recognition with IoT devices has grown in popularity as the demand for these algorithms has increased (Batra et al., 2019). Due to the compactness of different types of sensors and their size, monitoring human activity can be made more robust and efficient. Wearable sensors can monitor human activity and body changes (Mukhopadhyay, 2014).

ML algorithms built on the IoT are widely used to track and manage human activity. To select the optimal algorithm from a collection of ML algorithms, the processing efficiency, memory needs, and energy consumption criteria are given significant weight (Mahdavinejad et al., 2018; Aloraini et al., 2022). Wearable sensors are installed in the IoT device to detect various human activities using predetermined datasets, and human activity is then tracked in accordance with the entered dataset (Larranaga et al., 1996). A pre-existing dataset is fed into the IoT machine, and the computing performance of each is assessed in order to choose the optimal ML algorithm (Singh et al., 2013).

The dataset used for all ML algorithms is identical, and the algorithm with the highest computing speed and accuracy is chosen. Among the algorithms, gradient boosting, random forest, SVM, and K-NN have the maximum processing capacity and efficiency (Cho et al., 2019; Chkirbene et al., 2020). Accuracy and zero percent error rates are crucial in the field of security. IoT and ML algorithms are therefore employed to identify various patterns. Security systems can be strengthened and made more potent by integrating ML algorithms with IoT devices to recognize various human actions, like a biometric gadget that can identify a person by their finger print or face recognition (Boutaba et al., 2018). IoT systems are employed in smart homes so that the owner can control and monitor security from anywhere. In a smart home, the security of the property can be effectively maintained by connecting every electronic device to the Internet (Abdulla et al., 2020).

A smart home is outfitted with wireless technology that allows users to monitor it and electronically connect to IoT-based devices from a distance without regard for geography. IoT uses the Internet, which greatly expands the potential for cyberattacks (Zaidan A. A. & Zaidan B. B., 2020; Shafiq et al., 2020). By identifying and fixing any flaws in personal data before sending it over the Internet, ML is a technique that effectively reduces cyberattacks. By distinguishing between proper and improper users of the home, the current ML algorithm increases the security of the data generated by the IoT (Tahsien et al., 2020; Din et al., 2019). There is currently work being done to connect existing ML algorithms to the IoT in order to apply them in various contexts. Additionally, efforts are being made to further improve the security of IoT devices and ML algorithms (Outchakoucht et al., 2017; Ehatisham-ul-Haq et al., 2018).

As a crucial component of any smart home setup, IoT devices are required. Numerous threats, including data mining, denial-of-service attacks, information breaches, and other cyberattacks, are present (Tyagi & Goyal, 2020; Malathi et al., 2021). ML algorithms can be used to analyze the data being sent over the network and detect anomalies if they are present (Prabhakar et al., 2022; Sihare, 2018; Nweke et al., 2019).

Both smartphones and smartwatches include a capability called "activity-aware human context detection." According to one study, a variety of human behavioral states can be used to identify and forecast typical actions (Ehatisham-ul-Haq & Azam, 2020). Extrasensory's data showed that Random Forest was the best method for applying behavioural inference to find patterns in human behaviour (Asim et al., 2020; Niebla-Montero et al., 2022). Additionally, it was shown that dynamic activities like walking had a somewhat lower likelihood of producing reliable predictions (Kitani et al., 2012; Sihare, 2017b). Components for remote monitoring can detect human movement by using remote viewing and alerting (Alam et al., 2012; Menter et al., 2021; Sihare, 2022).

Section 2 explores in detail on the literature review. The benefits, drawbacks, and application areas of ML and the IoT have been covered in section 3. The use of IoT systems and associated difficulties are covered in section 4. In section 5, the relationship between machines and IoT is then demonstrated. The K-Nearest Neighbour algorithm, the SVM algorithm, and the random forest are the three principal machine learning algorithms for pattern identification that are covered in section 6. The relationship between automated pattern recognition and machine learning is discussed in Section 7. In section 8, pattern recognition and IoT have been thoroughly explored, including several pattern recognition applications. The final component, section 8, wraps up the chapter.

LITERATURE REVIEW

A cloud-centric vision for the global adoption of IoT is presented by Gubbi et al. (2013), along with important supporting technologies and application fields. The necessity for the convergence of the WSN (wireless sensor network), the Internet, and distributed computing is emphasized by Gubbi et al., who also propose a cloud implementation utilizing Aneka (Gubbi et al., 2013). The IoT can give disabled individuals the help and support they need to live well and take part in social and economic activities (Domingo, 2012). IoT was explored by Miorandi et al. (2012), who came up with creative technical solutions that might transform IoT from a research idea into a reality (Miorandi et al., 2012). Like every field, ML includes a wealth of "folk knowledge" that is essential for success but can be challenging to find. Domingos concentrated on how ML is applied in a variety of applications, including Online search, spam filters, recommender systems, ad placement, credit scoring, fraud detection, stock trading, and medicine creation (Domingos, 2012).

Both the rapidly developing fields of ML and the IoT are undergoing constant evolution. Future industrial market changes might be totally brought about by one of these techniques. Modern ML methods in the IoT and WSNs were highlighted by de Souza et al. (2020). He discussed into great length on the issues of using ML in WSNs and the IoT. The discussion of a few ML algorithms, including supervised, unsupervised, and reinforcement learning, is then expanded to include the IoT and impending difficulties (de Souza et al., 2020). Moreover, Lakshmanna et al. (2022) looked at data created by the IoT and ML, including data heterogeneity, velocity, volume, and other factors, and discussed the forthcoming challenges related to these issues (Lakshmanna et al., 2022). In their study paper, Mahdavinejad et al. (2018) discussed different ML and IoT application domains, including healthcare, smart homes, transportation, etc. Also, he addressed into considerable length on the security and related challenges (Mahdavinejad et al., 2018). Shah et al. (2022) addressed current energy conservation and smart building concerns by tying ML and the IoT together. He thoroughly examined the use of several ML methods, including neural networks, decision trees, and SVMs, in current energy conservation and smart building (Shah et al., 2022).

Zhang et al. (2021) attempted to relate pattern recognition with ML methods, and he somewhat succeeded in doing so. To apply ML algorithms for pattern recognition and establish the size of a dataset for pattern recognition, it was necessary to employ classification, clustering, and feature selection parameters. Also, he added variability to the training dataset in an effort to find various patterns (Zhang et al., 2021). By his book, Bishop (2006) attempted to illustrate pattern recognition and ML. He used ML to develop a useful mathematical formula for pattern recognition. The mathematical analysis of ML and pattern recognition is further described in depth. Deep-learning approaches for image-based cancer detection and prevention were highlighted by Hu et al. in their study from 2018. Recurrent neural networks (RNNs), autoencoders, and neural networks are employed for this (Hu et al., 2018). Using decision trees, SVMs, and neural networks, Boughaci and Alkhawaldeh (2020) conducted a comparison study between credit scores and ML and described the problems and potential benefits (Boughaci & Alkhawaldeh, 2020). Gaurav et al. (2022) examined data analysis, clustering, classification, and anomaly detection to thoroughly investigate the IoT and pattern recognition. By relating privacy and associated variability to ML and the IoT, further discussion was generated (Gaurav et al., 2022). For pattern recognition, Mohammadi et al. (2018) employed contemporary deep learning techniques including convolutional neural networks (CNNs), RNNs, and autoencoders (Mohammadi et al., 2018). In order to identify various patterns of human activity, Ramanujam et al. (2021) combined an IoT sensor network with a ML system (Ramanujam et al., 2021).

ADVANTAGES, DISADVANTAGES, AND APPLICATIONS OF ML AND IoT

IoT and ML are two technological areas that are quickly growing and have the potential to transform a variety of businesses. Many benefits, drawbacks, and uses of ML with IoT are listed below:

1) Advantages of ML

1. ML can be used to automate a variety of processes that would often be done manually. For instance, data analysis, pattern recognition, and decision-making might indeed benefit from the use of ML algorithms.
2. ML algorithms can learn from data and improve in accuracy over time. This suggests that when more data is analysed by an algorithm, the predictions it makes get more accurate.
3. ML can forecast future events using previous data. This could be advantageous in a variety of fields, including marketing, healthcare, and finance.

2) Disadvantages of ML

1. If the training data for ML algorithms contains bias and contradictions, the data may be turned biased by modifying it appropriately. Because of this, the information will not be processed in a way that allows for accurate judgments and predictions. Transparency of training data becomes crucial in ML algorithms since manipulation of data will cause information to be processed in accordance with user expectations.

2. The training data is used to construct ML algorithms, which may need high-performance computer equipment to operate. It's also feasible that no corporation will be able to finance such expensive computer systems. Several firms lack enough computing resources.

3. Nowadays, certain ML algorithms lack transparency, which makes it difficult to make appropriate conclusions after processing the training data. The aforementioned questions must be answered, particularly in the context of medical machine learning algorithms. Because even a minor error might result in a life being lost, the ML algorithm employed in the healthcare industry is obligated to make decisions that are 100% correct.

3) Applications of ML

1) ML algorithms may be used in the banking and insurance industries to detect fraud by recognizing suspicious transactions and trends.

2) By examining consumer data and behaviour, ML algorithms may be utilized to deliver tailored customer help.

3) ML algorithms may be used, among other things, to assess medical images and aid in disease detection.

The IoT is a networked system of connected things with unique IDs that can share data without needing to directly interact with other people or machines. A few IoT benefits, drawbacks, and applications are listed below:

1) Advantages of IoT

1) By the automation of operations and the elimination of waste, IoT has the potential to boost productivity in a variety of businesses.

2) IoT devices could make it feasible to track a range of activities in real-time, assisting organizations in identifying issues and taking prompt action.

3) IoT data can offer insights that allow businesses to make better decisions.

4) By analysing data and behaviour, IoT can provide customers a more customized and seamless experience.

5) By automating activities and reducing waste, IoT might aid in cost reduction.

2) Disadvantages of IoT

1) Cyberattacks on IoT devices might endanger the security of both individuals and businesses.

2) Because IoT devices generate a lot of data, privacy concerns might develop, especially if the information is misused.

3) IoT system creation and maintenance can be challenging and need specialist skills.

4) The interoperability of IoT devices may be constrained by incompatibility with other gadgets.

3) Applications of IoT

1) IoT devices may be used to automate household duties such as turning on lights and regulating temperatures.
2) The IoT may be utilized to improve city traffic control, trash management, and energy efficiency.
3) The IoT may be utilized to automate numerous activities in businesses such as manufacturing and logistics.
4) IoT may be used to remotely monitor patients, track medication compliance, and give physicians and nurses real-time notifications.
5) Among other things, IoT may be used to monitor soil conditions, weather patterns, and crop health in order to optimize agricultural yields and decrease waste.
6) The IoT can be used to monitor vehicle performance, optimize routing and scheduling, and improve road safety.

IoT SYSTEMS APPLICATION AND ITS CHALLENGES

The phrase "IoT" describes actual physical objects that may connect to other systems and devices through the Internet to exchange data. These objects may incorporate sensors, processing power, software, and other technologies. The gadgets do not have to be linked to the open Internet; they just need to be addressable and individually addressable (Gokhale et al., 2018).

IoT devices are a subset of the significant concept of home automation, which may include lighting, heating, and air conditioning in addition to media and security systems and video systems. Long-term benefits might include reducing energy use by ensuring that lights and appliances turn off automatically or by increasing resident consumption awareness. The Internet of Medical Things (IoMT) is one of the IoT applications used in medicine and healthcare. The device can enable systems for remote health monitoring and emergency alerts (Schwartz et al., 2015).

In the IoT, a person's pulse rate could be monitored, an animal could have a biochip transponder implanted, or a car may have sensors that warn the driver when the tyre pressure is low. The majority of tasks are carried out by gadgets without the necessity for human interaction, despite the fact that people can interact with them. Companies can monitor and interact with their operations in real time because of the IoT, a network of linked gadgets. Some firms are already benefiting from this technology, and it will spread as more companies realise its value. ZigBee is a wireless network that mostly functions in industrial settings, consumes little power, and has slow data speeds (Lee et al., 2007).

Figure 1. IoT applications with various ML techniques
(Courtesy: https://www.stl.tech/blog/what-are-the-applications-of-iot/)

IoT has a variety of real-world applications, including manufacturing and industrial IoT as well as consumer and corporate IoT. With billions of linked devices, the IoT requires access to billions of data points, all of which must be secured. IoT poses a threat to vital infrastructure, such as transportation, power, and financial services (Kimani et al., 2019).

Networked sensors are used for measurement, automated control, plant optimization, health and safety management, and other purposes. IoT and smart grid integration can improve energy efficiency and lower carbon emissions. One of the primary areas where IoT may play a significant role is in monitoring and regulating the operation of urban and rural infrastructure, such as bridges and wind turbines (Kaygusuz, 2009).

The success of smart city programmes depends on the government, since new legislation will make it easier for municipalities to implement IoT. A close working relationship between technology developers and the authorities in charge of maintaining local assets is required for resources to be publicly accessible. The Internet of Military Things (IoMT) may be significantly impacted by urban warfare in the future. This includes the use of advanced technology suited for the battlefield, such as sensors, weaponry, vehicles, robots, and wearable biometrics for people (Puliafito et al., 2019).

The Social Internet of Things (SIoT), a novel subset of the IoT, emphasises the importance of links between IoT devices and interpersonal contact. In order to accomplish a number of goals that are useful to both machine and human users in daily life, SIoT devices connect with one another. System security is a significant barrier to both SIoT and the protocols used to establish mutual confidence amongst IoT devices. IoT items require the next-generation Internet Protocol (IPv6) in order to expand to the exceedingly large address spaces required by the industry (Sfar et al., 2018).

IoT presents privacy challenges that need to be addressed by research and policymakers. Technology already influences moral decision-making, which influences human agency, privacy, and autonomy. The rapid growth of the IoT has allowed for the potential network connectivity of billions of objects.

The problems are caused by the configuration of the devices holding the database warehouse, not by the devices themselves (Figure 1) (Kakandwar et al., 2023).

These issues have been widely acknowledged in manufacturing and in businesses that have started their digital transition. Attacks that target IoT devices are on the rise and involve fault injection. Numerous gadgets lack a comprehensive security patching procedure due to their low cost and consumer-centric design. Many IoT devices are unable to employ firewalls or other fundamental security mechanisms like cryptosystems directly (Tran et al., 2023).

ML AND IoT

The three primary types of ML techniques are regression, classification, and clustering. Regression is appropriate given the continuous nature of the data. Through categorization, an unknown collection of objects may be divided into a predetermined number of classes. A cluster is defined as a group of data points or objects that are unique from one another yet similar to other cluster members. The goal of "deep learning," a subfield of ML, is to model the human brain. The process of making the IoT intelligent and analysing the massive amounts of data created by billions of such devices is used by several industries, including self-driving vehicles, wearable technology, industrial automation, healthcare, and retail shopping (Figure 2).

The IoT generates vast volumes of data in real time. ML could be able to help in spotting behavioural tendencies that could provide particular business insights. The proposed ML4IoT framework may be used to build a variety of ML models, each with a unique methodology (Pramod et al., 2021).

Record analytics may be used to compare and improve goals, enable improved decision-making, and then offer a perspective to track development over time. ML might be useful when you know what you want but are unaware of the informational factors that can affect your decision. IoT ML may need a major degree to make a name for itself, depending on its nature. By gathering information from specialised sensors or on machines, machine-mastery calculations are able to "feel" what is customary for a machine and identify the moment something odd begins to happen (Cabanac, 1971; Yang & Shami, 2022).

Edge ML and Tiny ML are two examples of the next generation of ML for industrial IoT. Edge ML is the process of directly running ML algorithms on edge devices. TinyML improves ML models for usage on small, inexpensive, and low-power microcontrollers, advancing the edge of ML. Currently, exploratory ML is the term used to describe it (Warden & Situnayake, 2019).

Figure 2. Correlations between ML and IoT, and related applications
(Courtesy: https://data-flair.training/blogs/iot-and-machine-learning/)

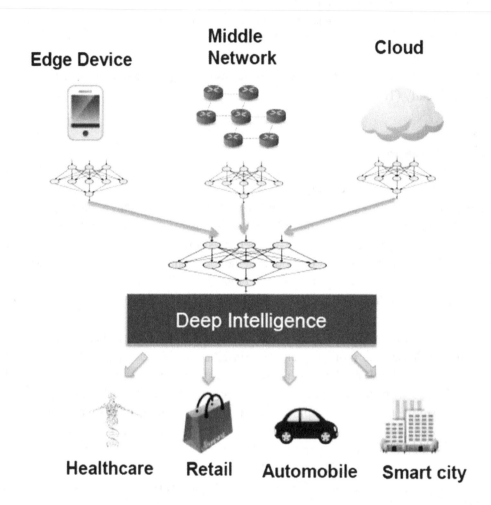

ML ALGORITHMS FOR PATTERN RECOGNITION

Linear regression, logistic regression, decision trees, SVM algorithms, Naive Bayes algorithms, KNN algorithms, K-means algorithms, and random forest algorithms are only a few of the numerous algorithms used in ML. Image PR can be done using any of the aforementioned algorithms. Since accuracy, memory usage, and runtime complexity are given more weight when choosing ML algorithms. Because of this, we have covered the same ML algorithms in this section, which are well-liked for their runtime complexity, memory needs, and accuracy in pattern identification.

4) Pattern Recognition by K-Nearest Neighbour Algorithm

A non-parametric supervised learning classifier is the K-Nearest Neighbour approach. It employs proximity to classify or predict groups of certain data points. Although it may be used for both classification and regression problems, classification problems are the ones it is most frequently utilised for. The K-Nearest

Neighbour method is one of the earliest in data science to be investigated (KNN). It is often employed in data mining, PR, fundamental recommender systems, and other applications (Popescul et al., 2013).

KNN retains all of its training data in memory, making it an example-based or memory-based learning strategy. The k parameter of a K-NN algorithm indicates the number of neighbours that will be looked at in order to classify a certain query point. Due to the possibility of overfitting or underfitting brought on by different values, defining K may need careful balance. You may choose the optimal k value for your dataset with the help of a cross-validation strategy (Ala'raj et al., 2020).

Figure 3. Prior to and following dataset classification using KNN
(Courtesy: https://www.javatpoint.com/k-nearest-neighbor-algorithm-for-machine-learning)

The most fundamental ML and image classification method now in use is called the K-Nearest Neighbour Classifier. The distinction between the feature vectors or images is used to categorise the data.

Its goal is to use an image dataset's raw pixel intensities to train a K-NN classifier, which will subsequently be used to classify unidentified images. This behaviour stands in stark contrast to the majority of ML techniques, which compel us to trade off approximation accuracy for either space or temporal complexity. The k-NN approach just employs distances in an n-dimensional space; it does not "learn" anything. In high-dimensional feature spaces, distances are usually unexpected. By comparing each unknown data point to each data point in the training set, the K-NN approach classifies unknown data points (Figure 3) (Ahmed et al., 2020).

The categorization is made overall based on whatever category receives the most votes. By using parameterized learning, we really learn from our incoming data and identify underlying patterns.

5) Pattern Recognition by SVM Algorithm

SVM, or supervised ML, is a technique that may be used to solve classification and regression issues. Finding an n-dimensional hyperplane that accurately classifies the input points is the aim of the SVM technique. It gets difficult to visualise facilities with more than three. The SVM kernel is a function that converts a low-dimensional input space into a high-dimensional space. It works best in situations when there is non-linear separation (Song et al., 2011).

The method provides features to minimise outliers and select the ideal hyperplane to increase margin. Many strategies and approaches have been proposed and developed by researchers for using SVM to PR issues. A typical foundation for classification is the aim of a procedure. Object detection is a technique for locating specific sorts of items in a photo or video. These things include, among others, animals, plants, and automobiles (Figure 4).

Figure 4. Pattern recognition by SVM algorithm
(Courtesy: https://www.javatpoint.com/machine-learning-support-vector-machine-algorithm)

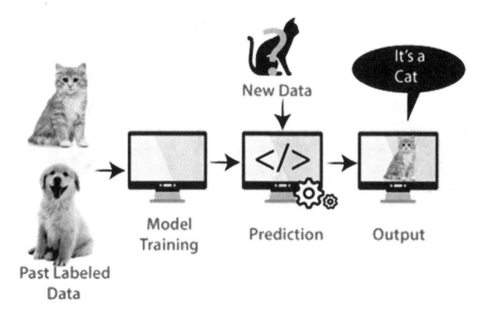

Many extremely effective face recognition algorithms have been developed using SVM. With a 3% error rate, one SVM classifier beat the opposition on the ORL face database. SVM was used to compare two global face recognition techniques. SVMs are often employed in discriminating classification. SVMs do not use traditional thresholding criteria; instead, they decide whether a model is accepted or rejected.

Using audio-visual data studies and the objective of properly recognising the first four English numerals, performance was assessed. In terms of classification accuracy, testing time efficiency, and training time efficiency, SVM surpasses conventional risk reduction algorithms. SVM is commonly employed and has shown to be successful in a range of classification applications.

Consider a group of images that we have in our control. It is necessary to determine the precise pattern from the group of images, as well as the real identities of the images and their locations in the outside world. Let's examine the dataset below, which has a range of colours, edges, and vertices.

Table 1. A dataset containing groupings of images in a certain limited set is used to determine a specific shape

Image Name	No. of Vertices	No. of Edges	Colours used
Image1	4	5	5
Image2	5	8	4
Image3	7	9	8
Image4	8	20	6
Image5	15	36	9

We require the following decision imperative to distinguish the negative and positive vertices to the query image from the collection of finite set images:

$$\vec{X} \cdot \vec{w} - c \geq 0 \tag{1}$$

Negative effects on c must be converted to positive effects on b for Eq. (1) to be rewritten as,

$$\vec{X} \cdot \vec{w} + b \geq 0 \tag{2}$$

So,

$$y = \begin{cases} +1 \, if \vec{X} \cdot \vec{w} + b \geq 0 \\ -1 \, if \vec{X} \cdot \vec{w} + b < 0 \end{cases} \tag{3}$$

w stands for vector normal, while b is its offset.

Moreover, we must determine distance (d) to ensure that positive or negative points do not cross the margined line because,

for all the red points $\vec{X} \cdot \vec{w} + b \leq -1$

for all the green points $\vec{X} \cdot \vec{w} + b \geq 1$ \hfill (4)

The Eq. (4) value should always be true for points to be correctly classified, as demonstrated by,

$$y_i \left(\vec{X} \cdot \vec{w} + b \right) \geq 1 \tag{5}$$

We must use the dot product of the two vectors to find the projection of a vector onto another vector,

$$\left(x_2 - x_1\right) \cdot \frac{\vec{w}}{w} \tag{6}$$

$$x_2\vec{w} - x_1\vec{w} \cdot \frac{\vec{w}}{w} \tag{7}$$

As the support vectors in this instance, x_2 and x_1, are placed on the hyperplane and adhere to the drawn margin lines, it is simple to identify the needed pattern point from the collections of the finite image set. Hence, Eqs. (6) and (7) have been further investigated for $y=1$.

$$1 \cdot \left(\vec{w} \cdot x_1 + b\right) = 1 \tag{8}$$

$$\vec{w} \cdot x_1 = 1 - b \tag{9}$$

Similarly, the Eqs. (6) and (7) are investigated at the negative point of $y = -1$ as,

$$-1 \cdot \left(\vec{w} \cdot x_2 + b\right) = 1 \tag{10}$$

$$\vec{w} \cdot x_2 = -b - 1 \tag{11}$$

Eq. (7) is kept in Eqs. (9) and (11) as,

$$\frac{1 - b + b + 1}{w} = \frac{2}{w} = d \tag{12}$$

The equation we must thus maximize is,

$$argmax\left(w^*, b^*\right)\frac{2}{w}, such\ that\ y_i\left(\vec{X} \cdot \vec{w} + b\right) \geq 1 \tag{13}$$

By using Eq. (13) we may maximize the point needed for pattern recognition from a set of images by optimizing the function. We can calculate the necessary point by collecting the vertices, edges, and colours from the information shown in Table 1 and then creating a margin line to divide the red and green points. In this case, green points are needed to extract a pattern from a collection of various images, whereas red points are those that are not necessary for pattern recognition.

6) Pattern Recognition by Random Forest

Random Forest is a well-liked ML method for supervised learning. The concept of ensemble learning, a method for resolving complex problems by combining several classifiers, serves as its basis. By increasing the number of trees in the forest, overfitting is prevented while accuracy is improved. Random Forest is capable of handling both classification and regression problems. It has the ability to handle large datasets with several dimensions. Thus, overfitting is lessened while model precision is increased (Almomani et al., 2021).

Figure 5. Pattern recognition by random forest algorithm
(Courtesy: https://www.javatpoint.com/machine-learning-random-forest-algorithm)

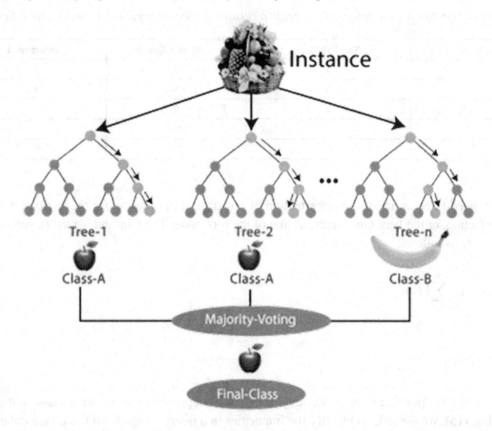

Classification and regression issues may be tackled with random forests. It can overcome the shortcomings of single decision trees while keeping their benefits. The construction of each of the forest's decision trees has an impact on how well random forests perform. A randomly selected subset of predictor variables was used to divide each node, resulting in a decreased error rate and a reduction in the link between trees. It is easy to redesign the model without the variables that add nothing to the research by recognising these characteristics (Figure 5).

To create each tree, the values of one predictor variable from the out-of-bag data are randomly rearranged. Outliers are data that deviates from other data in the same class. By locating classes that include substantial outliers, the training data may be re-evaluated.

Random forests are still not often used for picture classification and regression, despite their increasing popularity. Applications of random forest classification often include cloud and shadow detection, mapping of land cover and land cover change, and cloud and shadow analysis. It has been compared to the efficacy of other ML techniques including boosting and SVMs (Adhya et al., 2022).

Imagine that we have a collection of images in a certain set. The specific pattern from the collection of images must be identified, and each image's genuine name and whereabouts in the real world must also be known. Let's look at the dataset below, which has a collection of vertices, edges and colours.

Table 2. Dataset containing groupings of images in a certain limited set used to determine a specific shape

Image Name	No. of Vertices	No. of Edges	Colours used
Image1	4	5	5
Image2	5	8	4
Image3	7	9	8
Image4	8	20	6
Image5	15	36	9

We only need to be aware of the imperfection of our dataset in order to use the attribute with the least imperfection, or the least Gini index, as that of the root node. The Gini index may be represented mathematically as follows:

$$Gini\ Index = 1 - \sum_{i=1}^{n} (P_i)^2 \tag{14}$$

$$= 1 - [(P_+)^2 + (P_-)^2] \tag{15}$$

P_+ represents the positive class probability, whereas P_- represents the negative class probability. By using Eq. (15), we are able to identify the impurities in a query image from the image collections included in Table 2's image collections.

AUTOMATIC PATTERN RECOGNITION AND ML

A ML algorithm is a tool for discovering patterns in data. PR is done by these algorithms in two stages. Making predictions for unseen or unobservable items follows the development or creation of the model in the first step. Using the platforms and technologies indicated below, you may create ML models by analysing a lot of data from many sources, including chatbots. The two basic types of image processing are digital image processing and analogue image processing.

Digital image processing makes use of smart ML algorithms to enhance the quality of photos taken from distant sources, including satellites. Utilizing PR, even the tiniest concealed or undetected data may be located and anticipated. Our daily lives are greatly influenced by patterns. It aids in identifying the trend in the supplied data so that an accurate analysis can be performed. ML algorithms are used, among other things, to anticipate the presence of barriers and alert users in zero-miss scenarios (Villegas-Ch et al., 2020).

ML and PR are inextricably linked. Through PR, an attempt is made to imitate the neural networks of the human brain. The foundation of "machine learning," a subtype of artificial intelligence (AI), is the notion that computers are capable of learning from data, spotting patterns, and making decisions with little to no human input. Finding and understanding patterns is a crucial part of mathematical thinking and problem resolution. Understanding patterns is essential because they help us organise our work and make it more accessible (Wild & Pfannkuch, 1999).

Figure 6. Recognition of numerical patterns from a set of datasets, machine learning processing on the training dataset, data filtering, and production of useful results

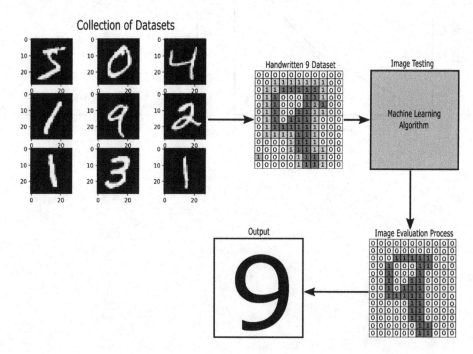

The next steps in the recognition process include data gathering, preparation, and noise reduction. Systems for surveillance and traffic monitoring employ PR to identify specific cars, trucks, or buses. Geologists can use it to locate and identify certain types of rocks and minerals. Analysts can apply ML to recognise and interpret temporal trends in seismic array data.

Biometric, colour, picture, and face recognition are just a few of the input types that may be employed with this kind of recognition. Seismic analysis, image analysis, and computer vision are just a few of the businesses that have employed it. An algorithm is used to create a statistical model that can predict the pattern's probabilistic nature. When patterns are hard to spot, these techniques are helpful.

When working with multidimensional things like seismic data and pictures, they are crucial. The following sections provide descriptions of several of the algorithms used in PR. The primary method of PR enhances both data regularity and PR. It has been employed by computer vision to carry out operations like object recognition and medical imaging. PR technology has the ability to advance and assist a wide range of digital technologies.

Figure 7. (a) Analysis of the differences between image accuracy and image cycle. It is predicted that image accuracy will rise as the number of cycles of images in the machine learning algorithm increases. The best possible filtering of picture noise is achieved by raising the image cycle to the required quality (b) Analyzing the differences between the image cycle and entropy loss. Entropy loss will rise for image filtration as you increase the number of cycles (Figure 6 dataset analysis).

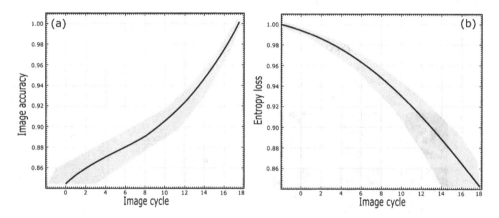

Figure 8. Recognition of image and colure patterns from a set of datasets, machine learning processing on the training dataset, data filtering, and production of useful results

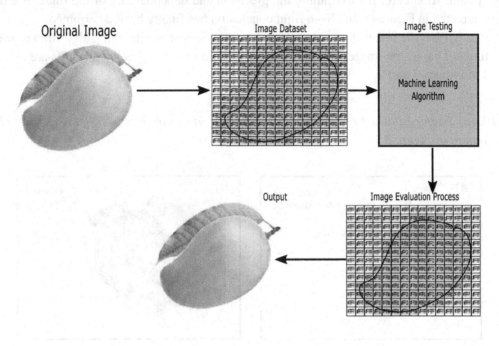

Figure 9. (a) Analysis of the differences between image accuracy and image cycle. It is predicted that image accuracy will sharply rise as the number of cycles of images in the machine learning algorithm increases. The best possible filtering of picture noise is achieved by raising the image cycle to the required quality (b) Analyzing the differences between the image cycle and entropy loss. Entropy loss will very sharply rise for image filtration as you increase the number of cycles (Figure 8 dataset analysis).

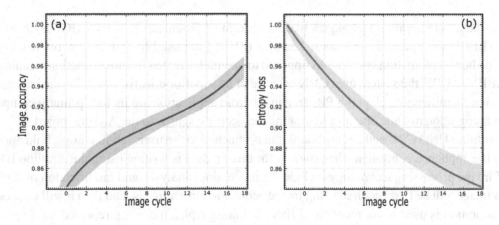

The required binary numeric dataset is derived from the image pattern with the help of data acquisition. Then, with the use of feature extraction, the associated features—which are written as handwritten numbers—are retrieved. The picture dataset was then pre-processed, the final image would be normalized, and the shape would be filtered. The matching pattern has been chosen or categorised using a machine learning system. After carefully examining the precision and dependability of the findings attained—which are depicted in Figures 6 and 8—a right conclusion has finally been determined.

To correctly filter the image dataset (training dataset), the machine learning algorithm repeats itself multiple times. That is, the image dataset is cycled through until the predicted results are obtained, as illustrated in Figures 7 and 9.

Figure 10. (a) Training dataset classifier expected output of Figure 6; (b) Training dataset classifier expected output of Figure 8

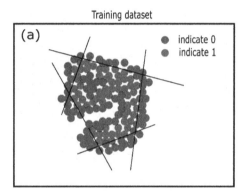 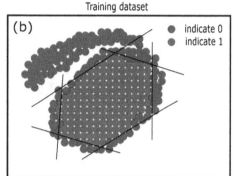

PATTERN RECOGNITION AND IoT

This section will look at various IoT devices, ML algorithms, and related PR applications. This section has addressed a variety of issues to help you understand the broad range of applications for PR.

1) Pattern Recognition for Home Automation

At the moment, hand-crafted techniques like Histogram of Oriented Gradients (HOG), Local Binary Patterns (LBP), and Speeded Up Robust Features (SURF) are used to extract features from gesture-recognition home automation systems. As opposed to manually created feature extraction techniques like HOG, LBP, or SURF, the system uses CNN because of its superior effectiveness. The device controller, hand motion identification, clapping PR, and an Android application are its four primary components. The clap recognition module operates by linking the acoustic sensor to the Arduino board.

The Arduino UNO CPU and a relay board that is attached to the user's chosen household appliances make up the appliance controller. The program for this module is written using the Arduino IDE. The fields of image processing technology include the processing, analysis, and modification of digital images. An image will be first acquired using image acquisition instruments, and then it will be processed. Image restoration is used to fix recognised flaws in photographs, including repeated noise, geometric distortion, and out-of-focus pictures, among others (Iyer & Sharma, 2019).

CNN is widely used to create innovative results. They are used to create feature maps by using kernels to convolutionize the signals from the image. These networks have little weight and require very little setup time because kernels from the same map are shared. The Arduino UNO open source circuit board contains a microprocessor. Programming it is easy because of the Arduino Integrated Development Environment (ADE).

Additionally, the system contains a capability that recognises clap patterns and allows users to automate instrument control using clap patterns. Future development will be built on the IoT, which will allow us to automate our equipment from anywhere on the globe.

Figure 11. IoT-enabled smart home automation, ML algorithms, and related mechanisms have been utilised to identify patterns for security and safety reasons
(Courtesy: https://voltoraindustries.com.au/how-to-transform-your-home-into-a-smart-home/)

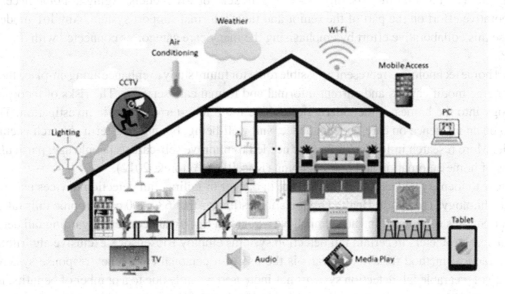

Speech training and voice testing are the two fundamental procedures in home automation. Voice training starts with speech acquisition. During setup, the microphone is connected to the speech acquisition device using the sound card input port on the laptop. The device is speaker-dependent, so the user must record his voice before using it (Figure 11).

These samples are contrasted with the training signals to validate the necessary speech. This research uses the vector quantization methodology, one of the several feature matching methods available in Matlab. The three elements of PR are optimal selection, pattern learning, and PR. Pattern training is the process of creating patterns that can be used by a pattern-matching algorithm. The second type of reference model involves a statistical evaluation of the behaviour of a collection of tokens.

Pattern matching is a technique for calculating the likelihood of speech using statistical knowledge from probabilistic models. Pattern matching is the process of comparing the similarity of two speech patterns. A computer can now recognise and comprehend human speech using voice recognition technology. The basic concept is to regulate power by adjusting the duty cycle. The received control characters are compared to a set of preset characters by the microcontroller.

The speech recognition module must first be trained with voice instructions before employing the recognition feature. There are various potential uses for the speech-operated system. The system's two main components are speech recognition and an electrical control system for smart home equipment. In the Matlab environment, the Command Line Language (MFCC) language is used to implement speech recognition.

2) Pattern Recognition in Aging and Disability

Because of intelligent technology and the IoT, older individuals may age gracefully. The intricacy of the carer networks involved in ageing in place is simplified by actor network theory (ANT). Despite the fact that ANT-based technologies are commonly employed in health care, they typically disregard the impact of the built environment. It is vital to address the real physical environment of housing while building smart technology models, digital service models, or IoT models. Aging at home necessitates a collaborative effort on the part of the senior and their informal support system. Any IoT model must recognise this collaborative effort by emphasising the many care categories connected with developing technology.

Smart home technologies represent a possible route for future study to enhance aging-in-place therapies such as home modifications and current informal and formal care services. The risks of incorporating technology into the homes of the elderly necessitate more long-term scientific investigation. There is likely to be an influence on care, independence, and well-being, but no long-term research is currently available. More research in IoT technology is needed to improve self-care as people age, particularly in the areas of home automation and housekeeping (Kocejko & Wtorek, 2012).

In order to identify and assist elders who are in danger of falling, fall detection devices use warning system technology. The World Health Organization estimates that 684,000 catastrophic falls take place each year. Studies in industry and academia have revealed a significant concern regarding fall detection. Tri-axial accelerometers in certain fall detection systems employ Biosensor's exclusive algorithms.

One practical method for avoiding falls is the use of a personal emergency response system. The popularity of wearable fall detection systems has increased recently due to a number of benefits, including their portability, low cost, ability to save energy, and lack of intrusiveness.

Deep reinforcement learning, which is based on psychological and neurological ideas about how humans respond to changing conditions, is another field of fall detection research. The Smart IoT Gateway sends fall data to cloud services, which then store it in a document-oriented database. The model is rebuilt and trained using an application programming interface and a cloud-hosted ML platform with representational state transfer. Edge computing and AI collaborate extremely effectively to detect human falls.

Crosswalks give pedestrians a way to manipulate the timing of traffic signals and ask for additional time to cross the street. The Seeing AI programme uses a smartphone's camera to identify activity around a user. In order to understand someone's emotions, it can also read facial expressions. For the less capable, smart speakers, refrigerators, stoves, and other domestic appliances are becoming increasingly useful. Some people's lives may be made simpler by wearables made possible by IoT technology. In order to prevent misuse, it is crucial that we ensure that smart gadgets are sufficiently safe as they become more popular.

3) Pattern Recognition in Healthcare and Medical Field

The IoT has changed every aspect of our lives by making it possible to fully monitor and analyse health-related concerns in real-time. IoT integration in the healthcare industry has led to the development of smart applications like mobile healthcare and sophisticated healthcare monitoring systems. Mobile health, which is leading this change, is the main engine for progress. This structure is novel for distinguishing the capabilities of ambulatory biomedical equipment, wearable sensors, and mobile electronics like smartphones and tablets.

One of the most important areas of m-healthcare is the research of human behaviour, which is done using the recommended HAR (human activity recognition) system. A reliable and precise HAR model is built for IoT-enabled HAR systems, utilising data mining and ML techniques (Figure 12).

By the end of 2020, it's predicted that there will be 162 million wearable medical gadgets. Interest in architectures and approaches that leverage edge and fog computing to make up for the drawbacks of the cloud in smart health has recently increased. IoMT systems may meet the strictest standards, as shown by the intelligent mapping of computational and resource management operations across nodes. "Edge/Fog Health" systems use the proper computing paradigms to distribute the processing and storing of health sensor data among a number of nodes positioned at various levels of proximity to customers. In order to provide doctors with information, scientists have been working on technologies for more than ten years that allow for remote patient monitoring. A variety of smarter solutions that make use of both network and software platform domains may now be created thanks to the IoT technology's recent development (Deng et al., 2023).

Several contributions creating IoT frameworks for smart health monitoring have been proposed throughout the previous decade. The study addresses numerous applications of smart health systems and employs a variety of technologies and sensors. The bulk of contributions in this area are from fog nodes that act as local servers, gathering and storing health data. method to respond quickly to service needs.

The hospital created a sophisticated CT scanner that made use of computer vision. It uses deep learning and neural networks to send data directly to the cloud or a network of connected clouds. Most people now think that AI is better than doctors at diagnosing patients.

Continuous data inputs may be transformed into continuous outputs using ANNs. An easy-to-use but powerful ML classification and PR algorithm is the K-nearest-neighbour method. A discrete hyperplane is produced by the classification method SVM in order to divide the shown data points. Although SVMs are accurate at simulating non-linear decision boundaries, their accuracy requires a lengthy training period. In comparison to other classification algorithms, SVM is less prone to overfitting.

Figure 12. IoT-compatible application, ML algorithms and associated processes have been used in the healthcare and medical fields to find patterns and diagnose dangerous diseases
(Courtesy: https://dida.do/blog/pattern-recognition-in-medical-imaging)

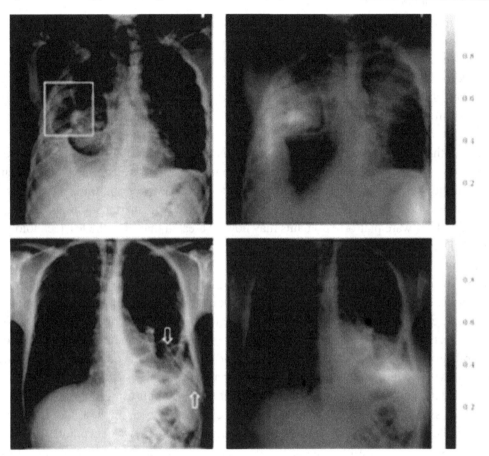

Using a least-cost complexity sorting method, the algorithms CART and C4.5 provide classification and prediction models. An ensemble-based ML method called the random forest algorithm. The values of the tree's sample random vector are used to divide up each tree's nodes. Internal estimates keep track of the inaccuracy, power, correlation, and contribution of each tree to the model.

4) Pattern Recognition in Transportation

Numerous ML techniques on the IoT have been used to tackle the route optimization challenge. The data is processed by the MCC (master control centre), which may alter each route's speed restriction based on factors including weather, pollution levels, and traffic flow. By doing this, traffic congestion is reduced, and accidents are reduced. Google introduced a brand-new service in 2009 that would integrate traffic information into Google Maps. The GPS, accelerometer, and gyroscope sensors are all found in mobile devices. Only the Maps application may be used by end users' mobile devices to submit anonymized data about their speed and position.

Parking apps have been created to track parking availability and give customers reservation alternatives. Several IoT sensors work together to identify cars in parking lots and send that information to a central server. Parking lot sensors employ ultrasonic sensors to track the availability of parking spaces, and a WiFi module gathers and transfers the data to a cloud server (Saha et al., 2019).

The management system and SSL (smart street lights) communicate over the NB-IoT network. To enhance public safety in an emergency, cameras and environmental detection devices will also be embedded inside the lampposts.

Because many components of transportation are directly impacted by road conditions, it is necessary to identify road anomalies for smart transportation to function. Traffic congestion, vehicle damage, and accidents can all result from poor road conditions.

To speed up communication, the protocol supports the MAC, MAC, physical, and network levels. It is based on social network theory for use in smart transportation applications. In Matlab simulations, the proposed method outperforms the existing strategy. The physical layer of WSNs is made up of nodes, while the network layer allows for routing from outer rings to a fixed access point (Figure 13).

Figure 13. ML algorithms, IoT-compatible applications, and related procedures have been applied in the transportation sector for various pattern recognition tasks
(*https://viso.ai/deep-learning/pattern-recognition/*)

5) Pattern Recognition in Industrial

Time series information from industrial processes may help to improve throughput, operator safety, and product quality while decreasing downtime. Through the identification of patterns, ML has emerged as a well-known method for maximising the value of operational data. Digital analytics will be the primary driver of the last 20% increase in operational productivity. Operational data can be subjected to pattern learning for real-time condition monitoring and predictive analysis. These predictive systems offer actionable insight into the present and upcoming health of production systems by recognising antecedent patterns of undesirable occurrences or operating circumstances (Chen, 2015).

It could be required to use enough signals to adequately define the system's activity. Because patterns are complicated, finding them takes work. Without the assistance of data scientists or outside consultants, ML chooses historical data, discovers patterns, and constructs and evaluates models without their assistance.

6) Pattern Recognition in Manufacturing

One approach that makes use of AI technology to comprehend designs is PR. Public relations cover all interconnected business sectors, including those in healthcare, banking, commerce, production, aviation, and many more. It may assist in attracting clients by doing thorough research on what the market wants and providing the machine instructions. In Industry 4.0, data communication is crucial for public relations. Predictive analytics techniques may be used to determine how customers feel about the items. Public relations techniques can be used to generate and compile feedback and ideas. The future of public relations techniques in the industries that produce clouds is bright.

Modern sensing technologies are being employed to increase information visibility and system controllability as manufacturing processes become more complicated. Several sensors can be used in industrial IoT to continuously track machine conditions and send data to the cloud. To date, no new tools for system diagnostics, prognostics, or optimization have been developed using sensing data from large-scale IoT machine networks. The Internet has advanced into a new era of smart and connected networks of industrial things, replacing hardwired computer networks with wireless, human-linked networks. This development ushers in a new paradigm for smart manufacturing when coupled with developments in cloud computing, virtual reality, and big data analytics (Abiodun et al., 2019).

For digital performance management, energy management, cost analysis, quality control, and supply chain optimization, the MES (manufacturing execution system) offers a foundational framework. The architecture of conventional MES systems has lately undergone substantial modification because of IoT technology. Manufacturing informatics and control based on sensors may be divided into four areas. There is real-time data interchange between the database and the sensing equipment. The transformation of raw data into several domains, including the frequency domain, wavelet domain, and state-space domain, is facilitated by data visualisation and representation. Production systems often employ a number of sensors, each with a variable sensitivity to certain operational aspects, to gather homogeneous or heterogeneous data. The information-transfer flow in real-time sensor signals as well as the rise of non-linear dynamics in the underlying processes must be considered in multi-sensor data fusion solutions.

Two conventional methods for identifying the underlying cause are failure mode and effects analysis and statistical models driven by engineering (such as stream of variation analysis and probabilistic graph models). Data-driven models recreate the degrading behaviour of the underlying process using real-time sensor information. The terms "Industrial IoT" and "Cyberphysical Systems" have received a lot of attention during the past six years. Making complicated industrial choices can be tough, but cloud computing and analytics can assist. In the IoT ecosystem, cloud databases and Internet-based data flow effectively encourage interactions between people and machines.

Predictive manufacturing will be significantly impacted by the production of cyberphysical systems. An IoT-based system was developed to track and examine how much energy is used throughout the selective laser sintering procedure. An energy-saving control system was created to optimise each construction project. It is feasible to execute the best energy management strategies for ongoing operations using the offered procedures.

Building IoT systems for worker safety and ergonomics as well as environmental and physical qualities is the subject of a lot of study in the construction industry. An IoT concept called the "Physical Internet" was first discussed in the logistics and industrial supply chain fields.

7) Pattern Recognition in Agriculture

Automation and the IoT are real-world solutions to human problems. Remote crop and soil health monitoring may be done with any tool in any field. Through the Android and iOS app platforms, farmers, students, and supporters may collaborate. Data on temperature, precipitation, humidity, wind speed, insect infestation, and soil composition are collected using agricultural IoT apps. This information might be used to automate farming processes, increase crop quality and quantity, lower risk and waste, and remove some of the labour required for crop management (Raj et al., 2015).

8) Pattern Recognition in Energy Conservation

IoT-based energy management solutions may benefit all parties involved in the power supply chain, including power companies, dealers, and users. IoT energy efficiency has a wide range of applications that are readily accessible to customers. Smart lighting, learning thermostats, and sensor-based HVAC systems are all available to automatically maintain zones in top condition. These systems can conserve energy by dynamically changing regimes in response to changing circumstances.

Energy companies employ digital energy management systems, which come with sensors, metres, controllers, analytical tools, and other IoT applications. The industry's growing demand exemplifies how even the most basic IoT energy control devices may yield savings with little effort on the part of the user (Panda, Sengupta & Roy, 2016).

IoT is increasingly being used for green energy use and energy management. A residential boiler or a large piece of industrial machinery with sensor capabilities can anticipate overheating, damage risk, and line blockages and provide load information. The IoT may be used to control and enhance the functioning of the power grid. Users of solar and other renewable energy sources may more efficiently manage the clean energy they create, minimise surplus, and ensure peak performance with the help of smart storage systems.

9) Ecosystem Protection Using Pattern Recognition

Smart agriculture monitoring (SAM) systems frequently span a large geographic area and are connected to the Internet. They are also mobile. Various studies on smart water pollution monitoring (SWPM) methodology and systems, including wireless sensors, IoT, and ML, have been carried out. Devices for detecting air and water pollution are under development.

ML was used to analyse and assess the data that the smart sensor nodes collected. The PM2.5, O3, and SO2 forecasting models in particular were utilised to rate the quality of the urban air (Arzoumanian et al., 2005).

10) Additional Pattern Recognition Application Fields

Pattern recognition is used to differentiate and properly identify individuals and objects in images and videos. It is utilized in security, surveillance, and self-driving automobiles.

1) Voice to text conversion uses pattern recognition technology. It is present in virtual assistants, dictation software, and voice-activated gadgets.
2) With the use of pattern recognition, people are identified by their distinctive biological traits, such as their features, irises, and fingerprints. It is utilized in security, border control, and authentication systems.
3) Pattern recognition is used in medical diagnostics to examine patterns in medical data such as electrocardiograms, magnetic resonance imaging, and blood tests. It is applied to diagnose diseases and predict how they will progress.
4) Financial analysis makes use of pattern recognition to spot trends in stock prices, interest rates, and market movements. It is employed in financial forecasting as well as investment analysis.
5) To analyse text input and provide meaningful data, such as sentiment analysis, pattern classification, and text categorization identification, natural language processing uses pattern recognition. Search engines, chatbots, and translators all make use of it.
6) Pattern recognition is used in video games to build intelligent agents that can detect and adapt to player behaviour. Both AI and gaming bots utilize it.

These are just a handful of the numerous applications for pattern recognition that are employed in several industries. Many potential new applications for pattern recognition are expected as technology advances.

CONCLUSION

This chapter discussed the application areas of IoT, ML algorithms, and patterns generated from them. After reviewing different literature, we came to the conclusion that the above-mentioned technology can solve the current problems of human beings to some extent. With the help of pattern recognition (PR) technology, the security of different electronic gadgets can be increased. PR has been proven to be effective with ML algorithms. The efficiency of PR is multiplied by choosing efficient ML algorithms and IoT devices. After reviewing the literature related to the above areas, it is known that SVM, K-Nearest Neighbour, and Random Forest are some of the ML algorithms that give effective results with IoT.

REFERENCES

Abdulla, A. I., Abdulraheem, A. S., Salih, A. A., Sadeeq, M. A., Ahmed, A. J., Ferzor, B. M., & Mohammed, S. I. (2020). Internet of things and smart home security. *Technol. Rep. Kansai Univ*, *62*(5), 2465–2476.

Abiodun, O. I., Jantan, A., Omolara, A. E., Dada, K. V., Umar, A. M., Linus, O. U., & Kiru, M. U. (2019). Comprehensive review of artificial neural network applications to pattern recognition. *IEEE Access : Practical Innovations, Open Solutions*, *7*, 158820–158846. doi:10.1109/ACCESS.2019.2945545

Adhya, D., Chatterjee, S., & Chakraborty, A. K. (2022). Performance assessment of selective machine learning techniques for improved PV array fault diagnosis. *Sustainable Energy. Grids and Networks*, *29*, 100582.

Ahmed, N., Rafiq, J. I., & Islam, M. R. (2020). Enhanced human activity recognition based on smartphone sensor data using hybrid feature selection model. *Sensors (Basel)*, *20*(1), 317. doi:10.339020010317 PMID:31935943

Ala'raj, M., Majdalawieh, M., & Abbod, M. F. (2020). Improving binary classification using filtering based on k-NN proximity graphs. *Journal of Big Data*, *7*(1), 1–18. doi:10.118640537-020-00297-7

Alam, M. R., Reaz, M. B. I., & Ali, M. A. M. (2012). A review of smart homes—Past, present, and future. *IEEE transactions on systems, man, and cybernetics, part C (applications and reviews)*, *42*(6), 1190-1203.

Almomani, O., Almaiah, M. A., Alsaaidah, A., Smadi, S., Mohammad, A. H., & Althunibat, A. (2021, July). Machine learning classifiers for network intrusion detection system: comparative study. In *2021 International Conference on Information Technology (ICIT)* (pp. 440-445). IEEE. 10.1109/ICIT52682.2021.9491770

Aloraini, F., Javed, A., Rana, O., & Burnap, P. (2022). Adversarial machine learning in IoT from an insider point of view. *Journal of Information Security and Applications*, *70*, 103341. doi:10.1016/j.jisa.2022.103341

Arzoumanian, Z., Holmberg, J., & Norman, B. (2005). An astronomical pattern-matching algorithm for computer-aided identification of whale sharks Rhincodon typus. *Journal of Applied Ecology*, *42*(6), 999–1011. doi:10.1111/j.1365-2664.2005.01117.x

Asim, Y., Azam, M. A., Ehatisham-ul-Haq, M., Naeem, U., & Khalid, A. (2020). Context-aware human activity recognition (CAHAR) in-the-Wild using smartphone accelerometer. *IEEE Sensors Journal*, *20*(8), 4361–4371. doi:10.1109/JSEN.2020.2964278

Batra, G., Jacobson, Z., Madhav, S., Queirolo, A., & Santhanam, N. (2019). *Artificial-intelligence hardware: New opportunities for semiconductor companies*. McKinsey and Company.

Bishop, C. M., & Nasrabadi, N. M. (2006). *Pattern recognition and machine learning*. Springer.

Boughaci, D., & Alkhawaldeh, A. A. (2020). Appropriate machine learning techniques for credit scoring and bankruptcy prediction in banking and finance: A comparative study. *Risk and Decision Analysis*, *8*(1-2), 15–24. doi:10.3233/RDA-180051

Boutaba, R., Salahuddin, M. A., Limam, N., Ayoubi, S., Shahriar, N., Estrada-Solano, F., & Caicedo, O. M. (2018). A comprehensive survey on machine learning for networking: Evolution, applications and research opportunities. *Journal of Internet Services and Applications*, *9*(1), 1–99. doi:10.118613174-018-0087-2

Burd, B., Barker, L., Divitini, M., Perez, F. A. F., Russell, I., Siever, B., & Tudor, L. (2018, January). Courses, content, and tools for internet of things in computer science education. In *Proceedings of the 2017 ITiCSE conference on working group reports* (pp. 125-139). 10.1145/3174781.3174788

Cabanac, M. (1971). Physiological Role of Pleasure: A stimulus can feel pleasant or unpleasant depending upon its usefulness as determined by internal signals. *Science*, *173*(4002), 1103–1107. doi:10.1126cience.173.4002.1103 PMID:5098954

Chen, C. H. (Ed.). (2015). *Handbook of pattern recognition and computer vision*. World Scientific.

Chen, Z., Feng, X., & Zhang, S. (2022). Emotion detection and face recognition of drivers in autonomous vehicles in IoT platform. *Image and Vision Computing*, *128*, 104569. doi:10.1016/j.imavis.2022.104569

Chkirbene, Z., Eltanbouly, S., Bashendy, M., AlNaimi, N., & Erbad, A. (2020, February). Hybrid machine learning for network anomaly intrusion detection. In *2020 IEEE International Conference on Informatics, IoT, and Enabling Technologies (ICIoT)* (pp. 163-170). IEEE. 10.1109/ICIoT48696.2020.9089575

Cho, G., Yim, J., Choi, Y., Ko, J., & Lee, S. H. (2019). Review of machine learning algorithms for diagnosing mental illness. *Psychiatry Investigation*, *16*(4), 262–269. doi:10.30773/pi.2018.12.21.2 PMID:30947496

de Souza, P. S. S., Rubin, F. P., Hohemberger, R., Ferreto, T. C., Lorenzon, A. F., Luizelli, M. C., & Rossi, F. D. (2020). Detecting abnormal sensors via machine learning: An IoT farming WSN-based architecture case study. *Measurement*, *164*, 108042. doi:10.1016/j.measurement.2020.108042

Deng, L., Cheng, F., Gao, X., Yu, W., Shi, J., Zhou, L., Zhang, L., Li, M., Wang, Z., Zhang, Y.-D., & Lv, Y. (2023). Hospital crowdedness evaluation and in-hospital resource allocation based on image recognition technology. *Scientific Reports*, *13*(1), 299. doi:10.103841598-022-24221-6 PMID:36609446

Din, I. U., Guizani, M., Rodrigues, J. J., Hassan, S., & Korotaev, V. V. (2019). Machine learning in the Internet of Things: Designed techniques for smart cities. *Future Generation Computer Systems*, *100*, 826–843. doi:10.1016/j.future.2019.04.017

Domingo, M. C. (2012). An overview of the Internet of Things for people with disabilities. *journal of Network and Computer Applications, 35*(2), 584-596.

Domingos, P. (2012). A few useful things to know about machine learning. *Communications of the ACM*, *55*(10), 78–87. doi:10.1145/2347736.2347755

Ehatisham-ul-Haq, M., & Azam, M. A. (2020). Opportunistic sensing for inferring in-the-wild human contexts based on activity pattern recognition using smart computing. *Future Generation Computer Systems*, *106*, 374–392. doi:10.1016/j.future.2020.01.003

Ehatisham-ul-Haq, M., Azam, M. A., Naeem, U., Amin, Y., & Loo, J. (2018). Continuous authentication of smartphone users based on activity pattern recognition using passive mobile sensing. *Journal of Network and Computer Applications*, *109*, 24–35. doi:10.1016/j.jnca.2018.02.020

Ferozkhan, A. B., & Anandharaj, G. (2021). The Embedded Framework for Securing the Internet of Things. *Journal of Engineering Research*, *9*(2).

Gaurav, A., Psannis, K., & Peraković, D. (2022). Security of cloud-based medical internet of things (miots): A survey. [IJSSCI]. *International Journal of Software Science and Computational Intelligence*, *14*(1), 1–16. doi:10.4018/IJSSCI.285593

Gokhale, P., Bhat, O., & Bhat, S. (2018). Introduction to IOT. *International Advanced Research Journal in Science. Engineering and Technology*, *5*(1), 41–44.

Gubbi, J., Buyya, R., Marusic, S., & Palaniswami, M. (2013). Internet of Things (IoT): A vision, architectural elements, and future directions. *Future Generation Computer Systems*, *29*(7), 1645–1660. doi:10.1016/j.future.2013.01.010

Gupta, B. B., & Quamara, M. (2020). An overview of Internet of Things (IoT): Architectural aspects, challenges, and protocols. *Concurrency and Computation*, *32*(21), e4946. doi:10.1002/cpe.4946

Hu, Z., Tang, J., Wang, Z., Zhang, K., Zhang, L., & Sun, Q. (2018). Deep learning for image-based cancer detection and diagnosis– A survey. *Pattern Recognition*, *83*, 134–149. doi:10.1016/j.patcog.2018.05.014

Iyer, R., & Sharma, A. (2019). IoT based home automation system with pattern recognition. *International Journal of Recent Technology and Engineering*, *8*(2), 3925–3929. doi:10.35940/ijrte.B2060.078219

Kakandwar, S., Bhushan, B., & Kumar, A. (2023). Integrated machine learning techniques for preserving privacy in Internet of Things (IoT) systems. In *Blockchain Technology Solutions for the Security of Iot-Based Healthcare Systems* (pp. 45–75). Academic Press. doi:10.1016/B978-0-323-99199-5.00012-4

Kaygusuz, K. (2009). Energy and environmental issues relating to greenhouse gas emissions for sustainable development in Turkey. *Renewable & Sustainable Energy Reviews*, *13*(1), 253–270. doi:10.1016/j.rser.2007.07.009

Kim, Y. K., Wang, H., & Mahmud, M. S. (2016). Wearable body sensor network for health care applications. In *Smart textiles and their applications* (pp. 161–184). Woodhead Publishing. doi:10.1016/B978-0-08-100574-3.00009-6

Kimani, K., Oduol, V., & Langat, K. (2019). Cyber security challenges for IoT-based smart grid networks. *International Journal of Critical Infrastructure Protection*, *25*, 36–49. doi:10.1016/j.ijcip.2019.01.001

Kitani, K. M., Ziebart, B. D., Bagnell, J. A., & Hebert, M. (2012, October). Activity forecasting. In *European conference on computer vision* (pp. 201-214). Springer, Berlin, Heidelberg.

Kocejko, T., & Wtorek, J. (2012). Gaze pattern lock for elders and disabled. In *Information Technologies in Biomedicine* (pp. 589–602). Springer. doi:10.1007/978-3-642-31196-3_59

Lakshmanna, K., Kaluri, R., Gundluru, N., Alzamil, Z. S., Rajput, D. S., Khan, A. A., Haq, M. A., & Alhussen, A. (2022). A review on deep learning techniques for IoT data. *Electronics (Basel)*, *11*(10), 1604. doi:10.3390/electronics11101604

Larranaga, P., Poza, M., Yurramendi, Y., Murga, R. H., & Kuijpers, C. M. H. (1996). Structure learning of Bayesian networks by genetic algorithms: A performance analysis of control parameters. *IEEE Transactions on Pattern Analysis and Machine Intelligence*, *18*(9), 912–926. doi:10.1109/34.537345

Lee, J. S., Su, Y. W., & Shen, C. C. (2007, November). A comparative study of wireless protocols: Bluetooth, UWB, ZigBee, and Wi-Fi. In *IECON 2007-33rd Annual Conference of the IEEE Industrial Electronics Society* (pp. 46-51). IEEE.

Mahdavinejad, M. S., Rezvan, M., Barekatain, M., Adibi, P., Barnaghi, P., & Sheth, A. P. (2018). Machine learning for Internet of Things data analysis: A survey. *Digital Communications and Networks*, *4*(3), 161–175. doi:10.1016/j.dcan.2017.10.002

Malathi, C., & Padmaja, I. N. (2021). Identification of cyber attacks using machine learning in smart IoT networks. *Materials Today: Proceedings*.

Menter, Z., Tee, W. Z., & Dave, R. (2021). Application of Machine Learning-Based Pattern Recognition in IoT Devices. In *Proceedings of International Conference on Communication and Computational Technologies* (pp. 669-689). Springer, Singapore. 10.1007/978-981-16-3246-4_52

Merenda, M., Porcaro, C., & Iero, D. (2020). Edge machine learning for ai-enabled iot devices: A review. *Sensors (Basel)*, *20*(9), 2533. doi:10.339020092533 PMID:32365645

Miorandi, D., Sicari, S., De Pellegrini, F., & Chlamtac, I. (2012). Internet of things: Vision, applications and research challenges. *Ad Hoc Networks*, *10*(7), 1497–1516. doi:10.1016/j.adhoc.2012.02.016

Mohammadi, M., Al-Fuqaha, A., Sorour, S., & Guizani, M. (2018). Deep learning for IoT big data and streaming analytics: A survey. *IEEE Communications Surveys and Tutorials*, *20*(4), 2923–2960. doi:10.1109/COMST.2018.2844341

Mukhopadhyay, S. C. (2014). Wearable sensors for human activity monitoring: A review. *IEEE Sensors Journal*, *15*(3), 1321–1330. doi:10.1109/JSEN.2014.2370945

Niebla-Montero, Á., Froiz-Míguez, I., Fraga-Lamas, P., & Fernández-Caramés, T. M. (2022). Practical Latency Analysis of a Bluetooth 5 Decentralized IoT Opportunistic Edge Computing System for Low-Cost SBCs. *Sensors (Basel)*, *22*(21), 8360. doi:10.339022218360 PMID:36366060

Nweke, H. F., Teh, Y. W., Mujtaba, G., & Al-Garadi, M. A. (2019). Data fusion and multiple classifier systems for human activity detection and health monitoring: Review and open research directions. *Information Fusion*, *46*, 147–170. doi:10.1016/j.inffus.2018.06.002

Outchakoucht, A., Hamza, E. S., & Leroy, J. P. (2017). Dynamic access control policy based on blockchain and machine learning for the internet of things. *International Journal of Advanced Computer Science and Applications*, *8*(7). Advance online publication. doi:10.14569/IJACSA.2017.080757

Panda, P., Sengupta, A., & Roy, K. (2016, March). Conditional deep learning for energy-efficient and enhanced pattern recognition. In 2016 Design, Automation & Test in Europe Conference & Exhibition (DATE) (pp. 475-480). IEEE. doi:10.3850/9783981537079_0819

Popescul, A., Ungar, L. H., Pennock, D. M., & Lawrence, S. (2013). Probabilistic models for unified collaborative and content-based recommendation in sparse-data environments. *arXiv preprint arXiv:1301.2303*.

Prabhakar, P., Arora, S., Khosla, A., Beniwal, R. K., Arthur, M. N., Arias-Gonzáles, J. L., & Areche, F. O. (2022). Cyber Security of Smart Metering Infrastructure Using Median Absolute Deviation Methodology. *Security and Communication Networks*, *2022*, 2022. doi:10.1155/2022/6200121

Pramod, A., Naicker, H. S., & Tyagi, A. K. (2021). Machine learning and deep learning: Open issues and future research directions for the next 10 years. *Computational analysis and deep learning for medical care: Principles, methods, and applications*, 463-490.

Puliafito, C., Mingozzi, E., Longo, F., Puliafito, A., & Rana, O. (2019). Fog computing for the internet of things: A survey. *ACM Transactions on Internet Technology*, *19*(2), 1–41. doi:10.1145/3301443

Raj, M. P., Swaminarayan, P. R., Saini, J. R., & Parmar, D. K. (2015). Applications of pattern recognition algorithms in agriculture: A review. *International Journal of Advanced Networking and Applications*, *6*(5), 2495.

Ramanujam, E., Perumal, T., & Padmavathi, S. (2021). Human activity recognition with smartphone and wearable sensors using deep learning techniques: A review. *IEEE Sensors Journal*, *21*(12), 13029–13040. doi:10.1109/JSEN.2021.3069927

Saha, R., Tariq, M. T., Hadi, M., & Xiao, Y. (2019). Pattern recognition using clustering analysis to support transportation system management, operations, and modeling. *Journal of Advanced Transportation*, *2019*, 2019. doi:10.1155/2019/1628417

Schwartz, T., Stevens, G., Jakobi, T., Denef, S., Ramirez, L., Wulf, V., & Randall, D. (2015). What people do with consumption feedback: A long-term living lab study of a home energy management system. *Interacting with Computers*, *27*(6), 551–576. doi:10.1093/iwc/iwu009

Sfar, A. R., Natalizio, E., Challal, Y., & Chtourou, Z. (2018). A roadmap for security challenges in the Internet of Things. *Digital Communications and Networks*, *4*(2), 118–137. doi:10.1016/j.dcan.2017.04.003

Shafiq, M., Tian, Z., Sun, Y., Du, X., & Guizani, M. (2020). Selection of effective machine learning algorithm and Bot-IoT attacks traffic identification for internet of things in smart city. *Future Generation Computer Systems*, *107*, 433–442. doi:10.1016/j.future.2020.02.017

Shah, S. F. A., Iqbal, M., Aziz, Z., Rana, T. A., Khalid, A., Cheah, Y. N., & Arif, M. (2022). The role of machine learning and the internet of things in smart buildings for energy efficiency. *Applied Sciences (Basel, Switzerland)*, *12*(15), 7882. doi:10.3390/app12157882

Sihare, S. (2022). Future Digital Marketing Revolutionizing E-Commerce. *Towards Excellence, 14*(1).

Sihare, S. R. (2017a). Image-based digital marketing. *International Journal of Information Engineering and Electronic Business*, *9*(5), 10. doi:10.5815/ijieeb.2017.05.02

Sihare, S. R. (2017b). Role of m-Banking for Indian Rural Consumers, its Adaptation Strategies, and Challenges: Consumer Behavior Analysis. *International Journal of Information Engineering & Electronic Business*, *9*(6). doi:10.5815/ijieeb.2017.06.05

Sihare, S. R. (2018). Roles of E-content for E-business: Analysis. *International Journal of Information Engineering & Electronic Business*, *10*(1). Advance online publication. doi:10.5815/ijieeb.2018.01.04

Singh, V. K., Piryani, R., Uddin, A., & Waila, P. (2013, January). Sentiment analysis of textual reviews; Evaluating machine learning, unsupervised and SentiWordNet approaches. In *2013 5th international conference on knowledge and smart technology (KST)* (pp. 122-127). IEEE.

Song, S., Zhan, Z., Long, Z., Zhang, J., & Yao, L. (2011). Comparative study of SVM methods combined with voxel selection for object category classification on fMRI data. *PLoS One*, *6*(2), e17191. doi:10.1371/journal.pone.0017191 PMID:21359184

Tahsien, S. M., Karimipour, H., & Spachos, P. (2020). Machine learning based solutions for security of Internet of Things (IoT): A survey. *Journal of Network and Computer Applications*, *161*, 102630. doi:10.1016/j.jnca.2020.102630

Tran, M. Q., Amer, M., Abdelaziz, A. Y., Dai, H. J., Liu, M. K., & Elsisi, M. (2023). Robust fault recognition and correction scheme for induction motors using an effective IoT with deep learning approach. *Measurement*, *207*, 112398. doi:10.1016/j.measurement.2022.112398

Tyagi, A. K., & Goyal, D. (2020, June). A survey of privacy leakage and security vulnerabilities in the internet of things. In *2020 5th International conference on communication and electronics systems (ICCES)* (pp. 386-394). IEEE. 10.1109/ICCES48766.2020.9137886

Villegas-Ch, W., Román-Cañizares, M., & Palacios-Pacheco, X. (2020). Improvement of an online education model with the integration of machine learning and data analysis in an LMS. *Applied Sciences (Basel, Switzerland)*, *10*(15), 5371. doi:10.3390/app10155371

Warden, P., & Situnayake, D. (2019). *Tinyml: Machine learning with tensorflow lite on arduino and ultra-low-power microcontrollers*. O'Reilly Media.

Wild, C. J., & Pfannkuch, M. (1999). Statistical thinking in empirical enquiry. *International Statistical Review*, *67*(3), 223–248. doi:10.1111/j.1751-5823.1999.tb00442.x

Yang, L., & Shami, A. (2022). IoT data analytics in dynamic environments: From an automated machine learning perspective. *Engineering Applications of Artificial Intelligence*, *116*, 105366. doi:10.1016/j.engappai.2022.105366

Zaidan, A. A., & Zaidan, B. B. (2020). A review on intelligent process for smart home applications based on IoT: Coherent taxonomy, motivation, open challenges, and recommendations. *Artificial Intelligence Review*, *53*(1), 141–165. doi:10.100710462-018-9648-9

Zhang, X., Wang, L., & Su, Y. (2021). Visual place recognition: A survey from deep learning perspective. *Pattern Recognition*, *113*, 107760. doi:10.1016/j.patcog.2020.107760

ADDITIONAL READING

Goyal, L. M., Saba, T., Rehman, A., & Larabi-Marie-Sainte, S. (Eds.). (2021). *Artificial Intelligence and Internet of Things: Applications in Smart Healthcare*. CRC Press. doi:10.1201/9781003097204

Hareesha, K. S. (2021). *IoT in healthcare and ambient assisted living* (G. Marques & A. K. Bhoi, Eds.). Springer.

Hassaballah, M., & Awad, A. I. (Eds.). (2020). *Deep learning in computer vision: principles and applications*. CRC Press. doi:10.1201/9781351003827

Kurniawan, A. (2018). *Learning AWS IoT: Effectively manage connected devices on the AWS cloud using services such as AWS Greengrass, AWS button, predictive analytics and machine learning*. Packt Publishing Ltd.

Madhumathy, P., Kumar, M. V., & Umamaheswari, R. (Eds.). (2021). *Machine Learning and IoT for Intelligent Systems and Smart Applications*. CRC Press. doi:10.1201/9781003194415

Nakamatsu, K., Kountchev, R., Aharari, A., El-Bendary, N., & Hu, B. (Eds.). (2021). *New Developments of IT, IoT and ICT Applied to Agriculture: Proceedings of ICAIT 2019*. Springer Singapore.

Ng, J., & Shah, S. (2020). *Hands-On Artificial Intelligence for Banking: A practical guide to building intelligent financial applications using machine learning techniques*. Packt Publishing Ltd.

Shrivastava, G., Peng, S. L., Bansal, H., Sharma, K., & Sharma, M. (Eds.). (2020). *New age analytics: Transforming the internet through machine learning, IoT, and trust modeling*. CRC Press. doi:10.1201/9781003007210

KEY TERMS AND DEFINITIONS

ANN (Artificial Neural Network): It is a hardware and/or software system that replicates how neurons in the human brain function.

Artificial Intelligence: The approach through which computers may be programmed to think like humans.

Internet of Things: Computers that have been incorporated into everyday objects and are connected to one another over the internet to share data.

Machine learning: The application of statistical models and algorithms to analyse data patterns and draw inferences, allowing computer systems to adapt and learn without explicit instructions.

Pattern recognition: Pattern recognition is a sort of data analysis that involves finding patterns and regularities in data automatically. Data that may be used includes text, images, music, and other fixed attributes. Systems for pattern recognition are quick and precise at spotting well-known patterns.

Smart home: A home with technology that can be controlled remotely via a computer or smartphone, including heating, lighting, and other appliances.

Smart watch: A touchscreen-equipped mobile device that is intended to be worn on the wrist.

Wearable devices: Wearable products are those that may be incorporated into clothing or worn as accessories on the body. They are items that use software and electronic components to control them.

Chapter 3
Institutional Pressures on the Oil and Gas Industry:
The Role of Machine Learning

LaTrelle Annette Bolding
https://orcid.org/0009-0000-4838-7476
Gupta College of Business, University of Dallas, Irving, USA

ABSTRACT

The use of artificial intelligence and machine learning in the oil and gas industry has steadily increased over the last decade. The focus of this chapter is understanding what type of external factors influence oil and gas firms to implement AI. A systematic review of the literature found that external pressures to achieve business goals and public pressures to reduce environmental impact continue to rise. This chapter leverages institutional theory to look at external isomorphic pressures in the form of government regulations and competitive pressures that influence the adoption of AI. The chapter concludes by reviewing opportunities for future research in the areas of AI within the oil and gas industry.

INTRODUCTION

Artificial Intelligence (AI) permeates our everyday lives. Most AI goes unnoticed and is used without question - such as facial recognition to identify friends on social media or voice recognition in home devices to play music or to shop online. AI surrounds us in industries such as financial services, healthcare, energy, consumer services, and telecom to name a few. Within these industries, AI is used for several diverse business functions including customer service, manufacturing, marketing, research & development, and supply chain management (Ransbotham, Kiron, Gerbert, & Reeves, 2017). The most notable industries using AI include tourism and hospitality, manufacturing, healthcare and the oil and gas sector.

The tourism and hospitality industry currently uses a form of AI called chatbots that respond with different answers based on responses from customers. Successfully implemented, this type of decision-tree logic used by chatbots is indistinguishable from human interaction (Ukpabi, Aslam, & Karjaluoto, 2019). The airline giant, Airbus, uses AI to quickly address disruption issues in their manufacturing facilities.

DOI: 10.4018/978-1-6684-8785-3.ch003

Their AI system is able to accurately match these disruptions to previously used solutions in almost real time (Ransbotham et al., 2017). This cuts down on unprofitable downtime and allows manufacturing to resume quickly. Similarly, healthcare utilizes AI as a complement to physicians' knowledge in order to more quickly analyze and diagnose patients. This allows the physician to focus on patient treatment by using contextual understanding and eliminate any machine biases (Krakowski, Luger, & Raisch, 2022).

Recently, the oil and gas industry (OGI) has seen a considerable upshift in AI activity (WhatsNext, 2022). For example, there is growing use of AI technology to more accurately predict oil well performance through simulations before wells are brought online. This utilization of AI to perform simulations cuts down on computation time that normally takes weeks or months, to as little as a few hours (Ibrahim, et al., 2022). OGI has substantial power on economies around the globe, and implementation of AI can help optimize production, reduce maintenance cost, and mitigate environmental impacts (Gupta & Shah, 2022). To achieve successful AI implementation, firms must have large amounts of usable, historical data to program algorithms for machine learning to take place. Big Data (BD) is a critical component needed for the advancement of the analytic processes in the application of machine learning (ML) (Nguyen, Gosine & Warrian, 2020). As AI increases in commercial applications, the attraction of AI increases to company shareholders (Su, 2019). Implementation of these systems creates new opportunities for real-time information (Nguyen et al., 2020) that can be automatically adjusted by an AI system for the most productive optimization and create the best profit margin available (Gupta & Shah, 2022).

Firms across industries are actively exploring the AI space to improve efficiencies and reduce costs (Gupta & Shah, 2022). Yet, the OGI industry is unique in that the pressures placed on these firms to be more environmentally sustainable are considerably greater than what firms experience in other industries (Lellis, 2022). The OGI faces immense environmental, social and governance (ESG) pressures from stakeholders (Brown, 2022). ESG performance is intimately related to AI and this relationship not only impacts valuation, but also risk assessments (Sætra, 2022). For companies that are aiming for NetZero targets and strategies, AI and data related emissions must be a part of these strategies (Sætra, 2022). Therefore, it is suggested to look beyond efficiency explanations and more closely examine how extra-organizational factors prompt firms in the OGI space to invest resources in AI. This chapter focuses on understanding the factors prompting oil and gas companies to adopt artificial intelligence and machine learning. It is implied that pressures to adopt AI come in the form of external pressures from government regulations, and from increasing AI technology developed by competitors. It is suggested that both of these extra-organizational factors create isomorphic pressures for firms in the OGI space to actively develop AI technologies.

This research is intended to help scholars understand how each of these external forces can impact a firm's decision to implement AI into their organization. Understanding the factors leading to implementation could help firms better plan for the future of sustainability and lessen the environmental impact currently being made. Looking through the lens of institutional theory, it is argued that the pressures exerted by the external environment to implement AI will greatly determine the success of firms in the OGI. Leveraging new technologies such as AI will increase the standard required by government entities and also change what shareholders come to expect.

THEORETICAL BACKGROUND

In order to understand the current literature regarding artificial intelligence and machine learning in OGI, a systematic literature review was conducted. The aim of the review is two-fold. First, it is an effort to identify how artificial intelligence and machine learning is currently being used within the OGI, and second is to identify the reasoning behind implementation apart from increased revenue and efficiency. A search was conducted using the EBSCO database, using a combination of the key words: *oil and gas, energy, upstream, midstream, downstream, artificial intelligence,* and *machine learning* to first identify how the information is currently being used. An additional search was conducted using a combination of the keywords *oil and gas, energy, regulatory pressures, environmental pressures, competitive pressures, strategy,* and *isomorphism* to understand what pressures firms face and how artificial intelligence and machine learning play a role, if at all. Results from these searches were carefully reviewed and included peer reviewed journal articles, books and book chapters, trade publications and periodicals written by companies currently using this technology. Since the goal of the chapter is not a technical one in nature, but an understanding behind current use and implementation, only articles meeting that criteria are included in the theoretical background; each of which is talked about in detail next. First, the background of artificial intelligence will be discussed, followed by the rise of machine learning in the oil and gas industry, and finishing with the environmental pressures that OGI firms face.

Artificial Intelligence (AI)

Artificial intelligence as an organized field of research started in 1956 with the Dartmouth Summer Research Project on Artificial Intelligence (Chalmers, MacKenzie & Carter, 2021). Artificial intelligence is defined as "a system's ability to interpret external data correctly, to learn from such data, and to use those learnings to achieve specific goals and tasks through flexible adaptation" (Kaplan & Haenlein, 2019: 15). AI, at the most advanced stage, can learn to solve complex problems and perform tasks previously only done by humans (Taddy, 2018). AI is used as an umbrella term to describe all tasks carried out by computers; primarily search algorithms, machine learning, and integrating statistical analysis (Kumar & Bhatt, 2013). Terms like AI, machine intelligence, machine learning, deep learning, and cognitive computing, are used interchangeably in the literature to express nearly the same ideas by different experts (Caner & Bhatti, 2020). They are not, however, the same. It is recognized that there are many different aspects to AI that are made up of super complex functions. At the most basic concept, all AI systems are made up of three broad pillars: domain structure, data generation and machine learning (Chalmers et al., 2021). These pillars act in conjunction with one another and AI cannot successfully function without all three dimensions building on one another. The first pillar is domain structure. This is where complex problems are broken down in smaller, simpler tasks - it is where the 'rules' are written (Taddy, 2018). The second pillar, data generation, is where all available information is gathered and allows for a steady stream of new and useful information to flow into the algorithms (Taddy, 2018). The last pillar of AI is machine learning. Machine learning (ML) looks for pattern recognition in which to predict future outcomes based on historical data. (Taddy, 2018). The ML learning process is the focus of this chapter.

ML takes place through processing big data using programmed algorithms to perform tasks. Big Data refers to ample amounts of information and gets the distinction of big data once the data set reaches a Petabyte (=1024 Terabytes) or Exabyte (=1024 Petabyte) (Nguyen et al., 2020). Vast amounts of data are created and stored each day. It is this historical data that companies use to teach machines how to process information and 'learn'. ML makes and uses calculations based on this historical data to make predictions about future performance (Gupta & Shah, 2022). However, no amount of calculations or predictive algorithms can provide viable solutions without sufficient data to draw and learn from. Even worse, bad data is paralyzing (Ransbotham et al., 2017).

ML has three main components: unsupervised learning, supervised learning, and reinforced learning. Unsupervised learning is simply the process of looking for exceptions in patterns within data clusters (Caner & Bhatti, 2020). Machines can learn to identify one blue dot out of 100 red ones. Supervised learning is the most common form of machine learning and looks for relationships between inputs and outputs (LeCun, Bengio & Hinton, 2015). With supervised learning, you can teach a system to recognize the difference between a cat and a dog through training specific parameters and adjusting as needed (LeCun et al., 2015). Reinforced learning takes place only through deep learning and requires 100's to 1000's times more inputs to make viable decisions. Reinforced learning is a goal-oriented model that allows algorithms to learn through trial and error (Chui, Manyika, & Miremadi, 2018). This trial-and-error process is the building foundation for machine learning. Through this process there are several potential benefits for the oil and gas industry including reduction in cost and downtime, improvement in profitability, and sustainability with environmental concerns (Gupta & Shah, 2022; Habson, 2021).

Oil and Gas Industry: The Growth in Machine Learning

Firms can be classified as either be upstream, midstream, downstream, or integrated – which means they hold space in more than one sector of the OGI. According to the Library of Congress, firms in the OGI space will be classified into three major categories based on the space they hold in the sector and their market capitalization. They are classified as supermajor, major or independents. Supermajor integrated companies (often known as 'the majors' or 'big oil') hold space in all three segments of the OGI market and have a market capitalization of $100 billion or more. These firms include household names like ExxonMobil, Shell, and BP and are most likely to be international operators. Major integrated companies can have space in one, two or all three segments of the industry but only have market capitalization between $10 - $100 billion. Lastly, independent companies focus on only one segment of the industry. They include firms that refine less than 75,000 barrels per day (on average) and have less than $5 million in retail sales of oil and gas per year. As of 2010, there were over 9,000 independent producers in the OGI operating in the United States (IPAA, 2022). According to Independent Petroleum Association of America (IPAA), these independent producers account for 91% of the oil wells in the US and produce 83% of America's oil, and 90% of America's natural gas (IPAA, 2022).

The OGI is made up of three large sectors: upstream, midstream, and downstream. Each of these sectors in turn, is made up of subsections. The upstream sector includes a) exploration and discovery of new fields containing petroleum products that can be exploited for profit (exploration), b) development of these fields with drilling activities to extract oil and gas to bring to the surface (development), and c) removal of hydrocarbons through a set of facilities to be sent to the midstream market (production) (Cifarelli, Wagner & Balossino, 2019). The upstream sector is "specifically compelling as it is the most capital-concerted and significant of the three sections in the OGI business" (Gupta & Shah, 2022: 50986).

ML in OGI has primarily been at the supervised level of learning. However, implementation of the reinforced learning aspect of ML can accelerate lessons learned from past failures and improve predictions of future performance (Gupta & Shah, 2022). Studies have shown that implementation of AI and ML can increase productivity (Choubey & Karmakar, 2020), diminish hazards (Gupta & Shah, 2022), and change the source of competitive advantage (Krakowski et al., 2022). For simplicity, all references in the remainder of this chapter to AI or ML are referencing specifically the reinforced learning aspect of machine learning.

Upstream

There are several activities within the OGI where ML could prove to be beneficial. For example, when exploring new territory to drill, the ability for a company to predict a wells' performance is a critical element in the decision to create a new well, therefore most companies use simulation to best predict positive results (Ibrahim et al., 2022). This simulation process is currently a tedious, time-consuming effort that takes hours to days to be compiled into usable information. Machine learning can speed up this process by a factor of 1000+ (Gupta & Shah, 2022). Ibrahim and colleagues (2022) were able to utilize historical information from mature wells to train an AI system to forecast viability based on geological features.

In addition to exploration, ML could also prove to be beneficial with development and production of these oil wells. Once the well is drilled, the development process includes flowing these wells to the surface by adjusting pressure and choke size. Depending on the temperature, oil to gas ratio and water saturation, the variation of flow can vary substantially. A recent study using a form of ML in a field of stable wells was able to estimate production with a first-year accuracy of over 90% and production values up to 5 years with an approximate 75% accuracy (Ibrahim, et al, 2022). This information can be used to help increase the longevity of the oil well and create the best possible extraction rate so that the resources are utilized to the fullest extent possible (Habson, 2021). This is a cost-effective way for ML to predict how long an oil well will be sustainable and when it is time to abandon it.

In production wells, artificial lift technologies are sometimes used when producing wells are in harsh environments and need supplemental lift in order to bring fluid to the surface and sustain production flowrate (Pham et al., 2021). The most common types of artificial lift are the electrical submersible pump (ESP) (Abdalla et al., 2022; Pham et al., 2021) and gas lift (Rojas Soares et al., 2022). ESP pumps use electric motors to build kinetic energy around the fluid leading to pressure that eventually lifts fluid through the pump. ESP ML optimization relies on data from multiple sources to build algorithms for increased production (Brown, 2022). On the other hand, gas lift technology uses high pressure natural gas to help production by injecting gas to displace the fluid. In both cases, restrictions can be put on the algorithms used to improve production so that results in change cannot exceed specified parameters. This allows for acceptance or rejection with human intervention to gradually build confidence in the programming as firms move towards a more autonomous solution (Brown, 2022). This can eventually enhance the rate of production and minimize the uplifting cost (Brelsford, 2018).

Along with production benefits, ML implementation can help reduce the environmental impact through gas hydrate reduction. Gas hydrates form around low weight gas such as ethane, or carbon dioxide and are formed when water combines with these gasses under low temperatures and high pressure, resulting in ice-like structures (Fossen, 2022). Decomposition of these gas hydrates release large amounts of methane. Methane is considered a greenhouse gas. Traditional methods for handling gas hydrates are through heat and chemical injection, further compounding the environmental impact. ML can be used to identify the hydrate-active components, develop correlations between hydrate properties and oil composition, and drastically reduce the need for chemicals to break down these hydrates (Fossen, 2022).

Currently, a mid-size company operating in the Permian Basin portion of the United States has used ML to increase their number of automated processes by 113% (Brown, 2022). They use smart cameras and other IOT devices to detect events related to pressure relief valve emissions. These devices are paired with surveillance equipment and are able to detect emission events up to 20 minutes before they happen. With early detection such as this, product will stay in the pipelines and prevent adverse events (Brown, 2022).

Midstream

AI could also prove beneficial to firms in the midstream sector. Midstream companies act as the intermediary between upstream producers and downstream refineries. The main aspect of midstream is gathering and transportation of oil and gas through pipelines. In addition to transportation, midstream companies also include gas plants that take gas obtained from upstream producers and separates it into natural gas liquids (NGLs), liquified petroleum gas (LPG) and residue gas. Transportation of these liquids and gasses can be corrosive to pipelines due to the nature of the composition of the gas. To mitigate the chance of failure in these pipelines, AI can be incorporated to establish an optimal flow pattern to help increase transportation efficiency and therefore decrease wear and tear on the pipeline (Choubey, & Karmakar, 2021). By reducing downtime and cost of transportation, firms could potentially save $100s of millions in value annually (Gupta & Shah, 2022).

Another portion of the midstream industry are cryogenic gas plants. These plants are used to separate other hydrocarbons within the natural gas such as methane, ethane, propane, butane, pentane and natural gasoline (Zhu et al., 2020). Extraction of these different components utilize external refrigeration, turbo-expansion, and absorption (Zhu et al., 2020) brought into the plant through compressors. These compressors work my putting the gas under immense pressure and compressing it down so that it can be sent through the above-mentioned extraction vessels. ML can be implemented to detect early rod drop failure within these compressors saving downtime and lost revenue to these plants (Hidalgo-Mompeán et al., 2021). Additionally, ML can be used to within the vessels to enhance configuration, recovery rate, and energy consumption as well as enabling extraction of heavier portions of the raw gas not currently utilized (Zhu et al., 2020).

Downstream

The final sector of OGI is downstream. The downstream segment includes petroleum refining, distribution, marketing, and retail of petroleum products. This is the final segment of OGI before public consumption and disposal. This sector consumes 4% of total global energy (Patel, Prajapati, Mahida, & Shah, 2020) and almost 30% of total global emissions (Lyons, et al., 2021). AI could potentially help companies predict plant emissions in advance and help achieve their sustainability goals (WhatsNext, 2022). Refineries are able to utilize AI in such a way to reduce fossil fuel consumption and electricity by operating more efficiently (Patel et al., 2020). Additionally, AI allows refineries to forecast and plan for maintenance and repair, avoiding unexpected failures and downtime (Olaizola et al., 2022). AI helps firms collect and fully understand their data while optimizing and fine-tuning algorithms to meet specific business strategies (Brown, 2022).

Oil and Gas Industry: Environmental Pressures

As the preceding section has illustrated, OGI firms across the spectrum of upstream, midstream, and downstream segments stand to benefit from the adoption of AI and ML. However, aside from the operational benefits, there is little known of the factors that prompt OGI firms to ultimately get into the AI and ML space. Currently, the insinuation that OGI companies are confronting immense pressure from regulating agencies to mitigate environmental effects from their operations, and competitive pressures to remain a relevant player in the industry (Gupta & Shah, 2022) are the subsequent tipping point that push OGI firms to ultimately act upon the pressures to take claim in the AI and ML space.

OGI is currently the most regulated industry in the United States and currently has over 37,000 federal restrictions between extraction and manufacturing (Lellis, 2022). In addition to state governing agencies, there are seven federal agencies that solely regulate the OGI, and numerous other agencies that share regulatory responsibilities across other firms (Library of Congress). OGI firms, directly or indirectly, account for over 40% of global greenhouse gas emissions (Lyons, et al., 2021) and face enormous pressures from governing bodies to reduce their carbon footprint on the environment (Habson, 2021). Increasing sustainability through emission reduction practices, will create a more sustainable strategy for firms as the tolerance for environmental impact becomes increasingly low (Lyons, et al., 2021).

Environmental management systems (EMSs) are growing in popularity among firms, along with the strategies needed to improve environmental performance (Obeidat et al., 2020). Failure of any part of the OGI can result in negative impacts on air and water, and on fragile ecosystems including soil, plant, and animal communities (Nejati et al., 2017). Additionally, OGI has been under pressure to address global issues such as climate change (Wright & Nyberg, 2017) and urged to implement extraction and processing procedures using green strategies (Fayyazi et al., 2015; Nejati et al., 2017). AI can help find solutions to environmental problems. Specifically, predictive analytics can be both a major revenue driver and contribute to environmental, social and governance (ESG) measures (Brown, 2022). These predictive analytics can be used to prevent adverse events and optimize production. One upstream firm currently uses these analytics to fine tune pumps, making them more efficient, and therefore reducing their environmental impact (Brown, 2022). As firms become more strategically aligned with outside suppliers and manufacturers concerning their carbon footprint, those who do not make that adjustment may have a hard time remaining relevant in the industry (Adner & Kapoor, 2010). In the following sections, these pressures are discussed in detail, along with how these pressures promote the adoption of AI / ML in the OGI industry.

PRESSURES FOR IMPLEMENTATION

Institutional theory is an approach that studies the social, economic, and political dynamics of firms and how they can play a role in a host of organizational behaviors, including the adoption of an innovation (Iwuanyanwu, 2021). Institutional isomorphism is a useful tool for understanding the pressures that infiltrate organizations (DiMaggio & Powell, 1983). There are three channels in which isomorphic change can occur. They include 1) coercive isomorphism that come from political influence and governmental issues, 2) mimetic isomorphism resulting from responses to uncertainty and 3) normative isomorphism stemming from professionalization (DiMaggio & Powell, 1983). The focus is exclusively on governmental regulation and competitive pressures in the OGI because the authors feel that these two external factors are the largest influential forces prompting firms to enter into the AI / ML space.

Figure 1. Conceptual framework

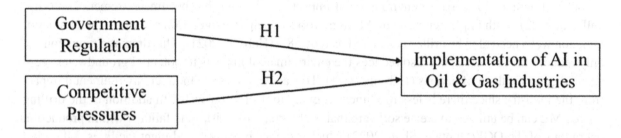

Government Regulation

In the OGI, there is ever increasing pressure from federal and state regulatory bodies requiring OGI firms to implement safer practices. With the OGI being responsible (either directly or indirectly) for close to half of all greenhouse gas emissions on a global scale (Lyons, et al., 2021) it is no wonder why there are ever increasing regulations regarding the safe practices of these firms. There is a surplus of regulatory agencies that implement more rules and regulations with each passing year. Looking at these agencies through an institutional theory lens, they create coercive isomorphic pressures on firms. Coercive iso-morphism includes external pressure by governments or regulatory agencies to conform to the 'standard' way of operations (Iwuanyanwu, 2021). To help with regulatory requirements, AI implementation can determine the most effective maintenance schedule for production facilities, pipelines and plants (Gupta & Shah, 2022) that will reduce the probability of an environmental violation. It is proposed that there will be a positive relationship between governmental pressures and the implementation of AI.

The federal government has several agencies that monitor and regulate standards required for the OGI to adhere to. Such agencies include the Environmental Protection Agency (EPA), Occupational Safety and Health Administration (OSHA), Bureau of Land Management (BLM), Department of Transportation (DOT), Bureau of Ocean Energy Management (BOEM) and Pipeline and Hazardous Materials Safety Administration (PHMSA) to name a few. In addition to federal agencies, each state has regulating bodies that have their own set of rules and regulations to follow. For example, Texas has the Railroad Commission, Louisiana has the Department of Natural Resources, Oklahoma has the Corporation Commission, New Mexico has the Oil Conservation Division, each with their own set of requirements. If a firm operates in more than one state, or crosses jurisdictions with pipelines, they must follow not only the regulations from the federal government, but also both sets of state regulating bodies that the pipeline crosses.

Recently, there have been newsworthy incidents about oil & gas companies having plant explosions, ruptured pipelines, and offshore leaks on oil rigs. In July 2022, a natural gas plant in Oklahoma exploded, resulting in evacuations and loss of gas available for residential customers. November 2019, the Keystone pipeline spilled over 9,000 barrels of oil after a previous leak in 2017, and another in 2011. And who can forget the catastrophe of the Deepwater Horizon explosion in the Gulf of Mexico? This incident resulted in eleven fatalities, 4 million barrels of oil being spilled, and almost $14 billion in penalties and restoration to the EPA alone. This incident took 87 days to contain. Incidents like these shed a bad light on the OGI and cost stakeholders millions of dollars in lost revenue and incurred fees. With the implementation of AI, anomalies in pipeline strength can be detected through sensor-based tools and can predict when failures may happen (Choubey & Karmakar 2021). Prevention of critical issue failures can not only save investors' money but prevent environmental catastrophes before they happen.

Public tolerance for negative environmental impacts by the OGI has become extremely low (GlobalData, 2022). With implementation of AI frameworks to help in exploration activities, the number of injection wells needed to be drilled can be reduced by 58% (Habson, 2021). This dramatic reduction not only reduces drilling costs, but also reduces the environmental impacts to potential groundwater seepage and geological stratifications (Habson, 2021). This also reduces the indirect environmental impact from the industry since there is less machinery running to drill fewer wells. In addition to the drilling aspect, ML can be utilized to create software that is able to predict risks and failures within production operations of the OGI (Gupta & Shah, 2022) which includes pipelines, and plant facilities. ML could potentially save the OGI upwards of $50 billion in the coming decade (Gupta & Shah, 2022) not only from decreasing non profitable downtime, but also through avoidance of penalties and fees from environmental effects that AI can help mitigate.

Competitive Pressures

Similar to governmental regulations, competitive pressures can also be a driver for companies to make changes to their overall strategy. Competitive advantages are not permanent, and firms must remain flexible in order to stay relevant (Phuoc, 2022). These isomorphic pressures are explained within institutional theory. Isomorphism is a process that forces one participant within an environment to implement changes seen by others (Hong, Kim & Kwon, 2022). Viewing these competitive pressures as mimetic isomorphic factors show how a competitor's activities can influence others in the same market to adopt similar processes (Iwuanyanwu, 2022). Within the OGI, all firms are connected to some degree. Each upstream producer sends to one or more midstream firm, and midstream firms sell to each other as well as the downstream firms. Since not all midstream firms operate within the same exact scope, they have the ability to sell gas and gas byproducts to other midstream firms for alternative uses. One firm cannot operate without the support and influence of others within the industry. The connection of these firms in this way supports the isomorphic factors that hold influence on others within the industry. Looking at these isomorphic pressures, it is proposed that there will be a positive relationship between competitive pressures and implementation of AI.

"AI changes the sources of competitive advantage" (Krakowski et al., 2022: 2). Those that successfully implement use of AI technologies in the OGI sectors could shift the sources of competitive advantage since these technologies can process large amounts of information much quicker than humans. Machines never get tired or emotional, and their calculations are always correct (Krakowski et al., 2022). With the implementation of AI, resources can be distributed throughout a firm without additional costs (Krakowski et al., 2022). Firms are using their advanced data analytics and data science as strategic tools to advance their digital transformation and generate more profit and shareholder value (Brown, 2022). These competitive advantages can create an shift in the industry to set new standards and influence with competitors.

Implementation of AI leads to complementation of machine and human functions (Krakowski, et al., 2022). When machines are functioning with high levels of machine learning, this enables humans to focus on tasks that cannot be done by machines. For example, in the upstream sector of OGI, when there is an issue, AI can tell the operator exactly what the issue is and how to rectify it. Without AI, an operator only sees that there is an issue and starts working backwards through a significant checklist to try and identify and correct the problem. This complementation of abilities allows the partitioning of tasks between human and machines to create a new source of competitive advantage (Krakowski et al., 2022).

AI implementation alone may not create a competitive advantage, but alongside other innovators in the firm's environment can embed them within an ecosystem of interdependent innovations (Adner & Kapoor, 2010). Understanding how one section or subsection of the OGI impacts another is critical to understanding the importance of firms creating a network that meet each other's needs and allows all members of the network to operate at their highest potential and create the most value for stakeholders. In the interest of being environmentally friendly, OGI firms that are optimizing reduction in emissions, will likely turn to suppliers who are acting with the same environmental interests (Lyons, et al., 2021). AI innovation can change the structure of the OGI industry and allow firms that successfully implement this technology to transform the competitive environment (Phuoc, 2022).

DISCUSSION

This chapter focuses on understanding the factors prompting oil and gas companies to adopt artificial intelligence and machine learning, apart from the revenue and efficiency drivers. A review of the literature only identifies how AI is currently being used, and how it can be used in the future. There has been no direct link made between the known efficiencies of the current technology and external forces driving firms to be more efficient. There is an understanding that firms always want to make more money, but implementation of these systems is extremely costly. So, the question remains: How do external forces play a role that ultimately leads firms into adopting artificial intelligence and machine learning? More information is needed to understand the relationship between these external forces and implementation of artificial intelligence.

Pressures to adopt AI come in the form of extra-organizational forces in the form of government regulations, and competition from other firms in OGI. Through the lens of institution theory, these are both isomorphic pressures facing firms within the industry. With increasing regulations from federal and state agencies, OGI firms need to adopt and actively develop AI technologies to help meet those standards and the higher ESG initiatives being set by shareholders. Meeting those standards will increase firms' sustainability and competitive advantage. It is important to note that all firms face external pressures. OGI firms ultimately act on these pressures because of the enormous regulations and competitiveness in the industry. There will always be enormous pressure on OGI firms to become more 'green' and environmentally friendly since incidents within the industry can be catastrophic to the environment. Implementation of AI / ML can help identify failures before they happen - saving time, money, and restoration efforts. Additionally, AI can help meet ESG standards by optimizing performance and allowing firms to evaluate the impacts of their decisions, both monetarily and through social impact.

Practical Implications

Increased productivity, reduced downtime, and mitigation of environmental impacts are all benefits of a successfully implemented AI strategy in the OGI (Gupta & Shah, 2022). In the upstream sector, ML can speed up simulation processes when looking for appropriate places to drill successful wells (Ibrahim, et al., 2022). Based on historical data, ML can predict well performance and longevity of wells once drilled. Predicting extraction rate will help fully utilize the resources available from these wells (Habson, 2021). The midstream sector can utilize ML to optimize flow through pipelines reducing wear and tear and risk of failure (Choubey & Karmakar, 2021) as well as find new ways to process the gas more efficiently (Zhu et al., 2020).

Along with production and profit benefits, there are many in the OGI looking to reduce their carbon footprint to net zero over the next several decades (Adballat, 2021). Gas hydrate reduction can come from ML analyzing hydrate-active components and reducing the need for additional chemicals, thus reducing their emissions and carbon footprint (Fossen, 2022). Downstream applications of ML can help predict plant emissions in advance to help achieve their sustainability goals. ML will also allow more efficient operations that can reduce fossil fuel consumption used in the daily operations of plants and refineries.

CONCLUSION

As AI increases across industries, OGI can greatly benefit from its implementation. Reduction in downtime, costs, and environmental impacts are some of the major benefits that could create a sustainable advantage for the industry. Understanding why extra-organizational factors prompt organizations to get into the AI space is important so that the right technology is used and implemented so that firms are able to better plan for the future of sustainability and lessen the environmental impact currently being made.

REFERENCES

Abdalla, R., Samara, H., Perozo, N., Carvajal, C. P., & Jaeger, P. (2022). Machine Learning Approach for Predictive Maintenance of the Electrical Submersible Pumps (ESPs). *ACS Omega*, 7(21), 17641–17651. doi:10.1021/acsomega.1c05881 PMID:35664599

Adballat, A. J. (2021, November 2). *How AI is helping oil and gas companies achieve net zero*. Energy Connects. https://www.energyconnects.com/opinion/thought-leadership/2021/november/how-ai-is-helping-oil-and-gas-companies-achieve-net-zero/

Adner, R., & Kapoor, R. (2010). Value creation in innovation ecosystems: How the structure of technological interdependence affects firm performance in new technology generations. *Strategic Management Journal*, 31(3), 306–333. doi:10.1002mj.821

Brelsford, R. (2018, June 6). Repsol launches Big Data, AI project at Tarragona Refinery. *Oil & Gas Journal*. https://www.ogj.com/refining-processing/refining/operations/article/17296578/repsol-launches-big-data-ai-project-at-tarragona-refinery

Brown, B. (2022, November 21). *High IQ it meets ESG at Laredo Petroleum: Cover story: Magazine*. The American Oil & Gas Reporter Magazine. https://www.aogr.com/magazine/cover-story/high-iq-it-meets-esg-at-laredo-petroleum

Caner, S., & Bhatti, F. (2020). A conceptual framework on defining businesses strategy for artificial intelligence. *Contemporary Management Research, 16*(3), 175–206. doi:10.7903/cmr.19970

Chalmers, D., MacKenzie, N. G., & Carter, S. (2021). Artificial intelligence and entrepreneurship: Implications for venture creation in the fourth industrial revolution. *Entrepreneurship Theory and Practice, 45*(5), 1028–1053. doi:10.1177/1042258720934581

Choubey, S., & Karmakar, G. (2021). Artificial intelligence techniques and their application in oil and gas industry. *Artificial Intelligence Review, 54*(5), 3665–3683. doi:10.100710462-020-09935-1

Chui, M., Manyika, J., & Miremadi, M. (2018). What AI can and can't do (yet) for your business. [What-AI-can-and-cant-do-yet-for-your-business.pdf] [mckinsey.com]. *The McKinsey Quarterly, 1,* 97–108.

Cifarelli, L., Wagner, F., & Balossino, P. (2020). The oil & gas upstream cycle: Development and production. *EPJ Web of Conferences, 246,* 1–12. doi:10.1051/epjconf/202024600009 doi:10.1051/epjconf/202024600001

DiMaggio, P. J., & Powell, W. W. (1983). The iron cage revisited: Institutional isomorphism and collective rationality in organizational fields. *American Sociological Review, 48*(2), 147–160. doi:10.2307/2095101

Fayyazi, M., Shahbazmoradi, S., Afshar, Z., & Shahbazmoradi, M. R. (2015). Investigating the barriers of the green human resource management implementation in oil industry. *Management Science Letters, 5*(1), 101–108. doi:10.5267/j.msl.2014.12.002

Fossen, M. (2022, May 11). *Machine learning can reduce the environmental impact of oil production.* #SINTEFblog. https://blog.sintef.com/sintefenergy/gas-technology/machine-learning-can-reduce-the-environmental-impact-of-oil-producti on/

GlobalData Thematic Research. (2022, February 14). *Ai is key for O&G companies targeting sustainability.* Offshore Technology. https://www.offshore-technology.com/comment/ai-for-og-sustai nability/

Gupta, D., & Shah, M. (2022). A comprehensive study on artificial intelligence in oil and gas sector. *Environmental Science and Pollution Research International, 29*(34), 50984–50997. doi:10.100711356-021-15379-z PMID:34378133

Habson, K. (2021, October 31). *Cognitive AI to boost sustainability for the oil and gas sector.* Gulf Business. https://gulfbusiness.com/cognitive-ai-to-boost-sustainabilit y-for-the-oil-and-gas-sector/

Hidalgo-Mompeán, F., Fernández, J. F. G., Cerruela-García, G., & Márquez, A. C. (2021). Dimensionality analysis in machine learning failure detection models. A case study with LNG compressors. *Computers in Industry, 128,* 103434. doi:10.1016/j.compind.2021.103434

Hong, S., Kim, S. H., & Kwon, M. (2022). Determinants of digital innovation in the public sector. *Government Information Quarterly, 39*(4), N.PAG. doi:10.1016/j.giq.2022.101723

Ibrahim, N. M., Alharbi, A. A., Alzahrani, T. A., Abdullah, M. A., Ibrahim, A. A., Hameed, A. M., Albabtain, A. S., Alqahtani, D. A., Alsawwaf, M. K., & Almuqhim, A. A. (2022). Well performance classification and prediction: Deep learning and machine learning long term regression experiments on oil, gas, and water production. *Sensors (Basel)*, *22*(14), 1–22. doi:10.339022145326 PMID:35891005

Iwuanyanwu, C. C. (2021). Determinants and impact of artificial intelligence on organizational competitiveness: A study of listed American companies. *Journal of Service Science and Management*, *14*(5), 502–529. doi:10.4236/jssm.2021.145032

Kaplan, A., & Haenlein, M. (2019). Siri, Siri, in my hand: Who's the fairest in the land? On the interpretations, illustrations, and implications of artificial intelligence. *Business Horizons*, *62*(1), 15–25. doi:10.1016/j.bushor.2018.08.004

Krakowski, S., Luger, J., & Raisch, S. (2022). Artificial intelligence and the changing sources of competitive advantage. *Strategic Management Journal*, *1*, 1–28. doi:10.1002mj.3387

Kumar, S., & Bhatt, A. (2013). Foundations of artificial intelligence. *International Journal of Technical Research and Applications*, *1*(4), 52–56.

LeCun, Y., Bengio, Y., & Hinton, G. (2015). Deep learning. *Nature*, *521*(7553), 436–444. doi:10.1038/nature14539 PMID:26017442

Lellis, C. (2022, February 8). *10 most regulated industries in the U.S.* Perillon. https://www.perillon.com/blog/10-most-regulated-industries-in-the-us

Library of Congress. (n.d.). *Oil and Gas Industry: A Research Guide: U.S. Regulatory Agencies*. Library of Congress. https://guides.loc.gov/oil-and-gas-industry/laws/agencies

Lyons, M., Appathurai, S., Vasquez, M., Bolikowski, L., Alcalá, P., Carducci, F., & Tarabelloni, N. (2022, October 14). *The AI angle in solving the oil and gas emissions challenge*. BCG Global. https://www.bcg.com/publications/2021/ai-in-oil-and-gas-emissions-challenge

Moyano-Fuentes, J., & Martínez-Jurado, P.-J. (2016). The influence of competitive pressure on manufacturer internal information integration. *International Journal of Production Research*, *54*(22), 6683–6692. doi:10.1080/00207543.2015.1131866

Nejati, M., Rabiei, S., & Jabbour, C. J. C. (2017). Envisioning the invisible: Understanding the synergy between green human resource management and green supply chain management in manufacturing firms in Iran in light of the moderating effect of employees' resistance to change. *Journal of Cleaner Production*, *168*, 163–172. doi:10.1016/j.jclepro.2017.08.213

Nguyen, T., Gosine, R. G., & Warrian, P. (2020). A systematic review of big data analytics for oil and gas industry 4.0. *IEEE. IEEE Access : Practical Innovations, Open Solutions*, *8*(1), 61183–61201. doi:10.1109/ACCESS.2020.2979678

Obeidat, S. M., Al Bakri, A. A., & Elbanna, S. (2020). Leveraging 'green' human resource practices to enable environmental and organizational performance: Evidence from the Qatari oil and gas industry. *Journal of Business Ethics*, *164*(2), 371–388. doi:10.100710551-018-4075-z

Olaizola, I. G., Quartulli, M., Unzueta, E., Goicolea, J. I., & Flórez, J. (2022). Refinery 4.0, a review of the main challenges of the industry 4.0 paradigm in oil & gas downstream. *Sensors (Basel)*, *22*(23), 9164. doi:10.339022239164 PMID:36501863

Patel, H., Prajapati, D., Mahida, D., & Shah, M. (2020). Transforming petroleum downstream sector through big data: A holistic review. *Journal of Petroleum Exploration and Production Technology*, *10*(6), 2601–2611. doi:10.100713202-020-00889-2

Pham, S. T., Vo, P. S., & Nguyen, D. N. (2021). Effective electrical submersible pump management using machine learning. *Open Journal of Civil Engineering*, *11*(1), 70–80. doi:10.4236/ojce.2021.111005

Phuoc, N. V. (2022). The critical factors impacting artificial intelligence applications adoption in Vietnam: A structural equation modeling analysis. *Economies*, *10*(129), 1–16. doi:10.3390/economies10060129

Ransbotham, S., Kiron, D., Gerbert, P., & Reeves, M. (2017). Reshaping business with artificial intelligence. *MIT Sloan Management Review*, *59*(1), 1–17.

Rojas Soares, F. D., Secchi, A. R., & Bezerra de Souza, M. Jr. (2022). Development of a nonlinear model predictive control for stabilization of a gas-lift oil well. *Industrial & Engineering Chemistry Research*, *61*(24), 8411–8421. doi:10.1021/acs.iecr.1c04728

Sætra, H. S. (2022). The AI ESG protocol: Evaluating and disclosing the environment, social, and governance implications of artificial intelligence capabilities, assets, and activities. *Sustainable Development*, *1*. Advance online publication. doi:10.1002d.2438

Su, J. (2019, February 12). Venture capital funding for artificial intelligence startups hit record high in 2018. *Forbes*.

Taddy, M. (2019). The technological elements of artificial intelligence. In A. Agrawal, J. Gans, & A. Goldfarb (Eds.), *The Economics of Artificial Intelligence: An Agenda* (61–87). National Bureau of Economic Research Conference Report series. Chicago: University of Chicago Press. 10.7208/chicago/9780226613475.003.0002

Ukpabi, D. C., Aslam, B., & Karjaluoto, H. (2019). Chatbot adoption in tourism services: A conceptual exploration. In *Robots, artificial intelligence, and service automation in travel, tourism and hospitality*. Emerald Publishing Limited. doi:10.1108/978-1-78756-687-320191006

WhatsNext. (2022, September 30). *Artificial Intelligence and Oil & Gas Industry – Partnership towards Sustainable Future*. Exponential Transformation Ecosystems. https://whatnextglobal.com/insights/artificial-intelligence-and-oil-gas-industry-partnership-towards-sustainable-future/

Who are America's independent producers? (n.d.). Independent Petroleum Association of America. https://www.ipaa.org/independent-producers/#:~:text=There%20are%20about%209%2C000%20independent%20oil%20and%20natural,oil%20and%2090%20percent%20of%20America%E2%80%99s%20natural%20gas

Wright, C., & Nyberg, D. (2017). An inconvenient truth: How organizations translate climate change into business as usual. *Academy of Management Journal*, *60*(5), 1633–1661. doi:10.5465/amj.2015.0718

Zhu, W., Chebeir, J., & Romagnoli, J. A. (2020). Operation optimization of a cryogenic NGL recovery unit using deep learning based surrogate modeling. *Computers & Chemical Engineering, 137*, 106815. doi:10.1016/j.compchemeng.2020.106815

Chapter 4
Generative Adversarial Networks:
A Game Changer – GAN for Machine Learning and IoT Applications

A. Manikandan

Department of ECE, Amrita School of Engineering, Amrita Vishwa Vidyapeetham, Kollam, India

T. Sanjay

(iD) https://orcid.org/0000-0001-8591-6314

Software Engineer Analyst, JPMorgan Chase & Co., Bengaluru, India

ABSTRACT

Generative adversarial networks (GANs) have transformed machine learning and created new research and application areas. GANs are now used for data augmentation, picture, audio, text-to-image, and 3D object production thanks to IoT. These applications could make IoT devices more personalized, efficient, and productive by collecting and using data. GANs are also employed in healthcare and finance for IoT applications, enabling new research. GANs have many benefits, however stability issues during training, interpretability issues, ethical concerns, and more must be addressed. GANs could transform the IoT market, and hardware and infrastructure improvements are projected to increase their influence. GANs will open up many new research and development avenues for IoT devices.

GENERATIVE ADVERSARIAL NETWORKS – A GAME CHANGER

Issues with data creation, processing, and interpretation have arisen as a result of the IoT's rapid development. Traditional algorithms for data analysis and manipulation have reached their performance limits in terms of efficiency, accuracy, and scalability, necessitating the exploration of new approaches. In this setting, GANs have emerged as a potent method for both producing and interpreting data for Internet of Things (IoT) applications. To create high-quality synthetic data, GANs employ a game-theoretic ap-

DOI: 10.4018/978-1-6684-8785-3.ch004

proach in contrast to traditional algorithms by combining the efforts of two neural networks (a generator and a discriminator). This novel architecture outperforms the state-of-the-art in terms of accuracy, scalability, and generality.

Since GANs are capable of processing massive amounts of data, they are increasingly being used in the development of cutting-edge IoT systems. As the Internet of Things industry grows, the need for trustworthy and effective data gathering and processing technologies becomes increasingly pressing. This chapter will introduce GANs and discuss their effect on the Internet of Things industry as well as its benefits, disadvantages, and future possibilities. The goal is to provide a stable foundation for researchers and professionals interested in exploring GANs' potential for IoT. Two neural networks, called a "generator" and a "discriminator," make up GANs, a type of machine learning model (Goodfellow et al., 2020). New data samples that are supposed to resemble real-world data are generated by the generator network.

The discriminator network, on the other hand, evaluates the models and attempts to distinguish them from the real data. The discriminator's goal is to improve its ability to detect fake data, whereas the generator's goal is to provide more realistic examples. These two networks are "competed" for improvement by being trained simultaneously. Eventually, the generator can use this competitive process to synthesis new samples that are like real-world data.

In 2014, Ian Goodfellow and his coworkers first introduced the idea of GANs. In game theory, GANs are based on the idea of two opponents competing to fulfil their goals, as in the two-player games that inspired them. Two players are involved in GANs, the generator and the discriminator. Data samples are created by the generator, and their authenticity is determined by the discriminator. Computer vision was one of the earliest uses of GANs; the generator was taught to produce realistic images. The early versions of GANs had some problems, such as mode collapse (where the generator only created a small number of data variations) and instability when being trained. Improvements in the stability and robustness of GANs in the years after its introduction have led to their growing use in a wide range of areas, including computer vision, natural language processing, and audio synthesis. To solve particular problems and expanding GAN capabilities, new variants of GANs have been developed, such as Conditional GANs, Wasserstein GANs, and Style GANs. Several intriguing new uses have been discovered for GANs recently, and they have become a popular and prominent technique in deep learning. Many applications have been found for GANs, including but not limited to image-to-image translation, text-to-image synthesis, video production, and audio synthesis. Their impact on machine learning and AI is likely to continue to rise in the years to come.

The term "Internet of Things" is used to describe the global infrastructure that allows sensors, cameras, and other "smart" devices to collect, transmit, and process data in real time (IoT). The goal of machine learning enabled IoT is to use the capabilities of machine learning algorithms to assess and make sense of the massive volumes of data generated by these devices in order to improve decision-making and offer new, cutting-edge services. Since GANs can generate new data samples that may be used to train and test ML models, they are well-suited for Machine Learning enabled IoT environments. In the field of computer vision for the IoT, for instance, GANs could be utilized to supply artificial data for item identification and recognition algorithms. In cases when real-world data is scarce or inaccessible, GANs may be used to train and evaluate machine learning models by generating new samples of data. GANs can be utilized in IoT applications not just to generate new data but also to supplement existing data. Audio synthesis for the Internet of Things, for instance, can benefit from the employment of GANs by creating fresh speech samples for the purpose of training speech recognition algorithms. As a result, these algorithms may be more accurate and resistant to data variances. In the context of Machine Learning powered IoT, GANs also play an important role in image-to-image translation. As an illustration, GANs can be used to enhance the decision-making information provided by low-resolution photographs acquired by Internet of Things (IoT) sensors by transforming them into high-resolution images.

Architecture

Generator Network

Several fully connected or convolutional layers, with an activation function such as ReLU or leaky ReLU applied after each layer, make up the standard architecture of a GAN's generator network (Razavi-Far et al., 2022). For input, the generator network maps a random noise vector, also called the input noise, to a data sample.

The fully connected layer in a generator network can be mathematically represented as follows:

$$z_{out} = f(W.z_{in} + b)$$

where z_{in} is the input noise vector, W is the weight matrix, b is the bias vector, f is the activation function and z_{out} is the output of the layer.

Convolutional layers in general, can be represented as:

$$z_{out} = f(conv(z_{in}) + b)$$

where conv is the convolution operation, W is the weight tensor, b is the bias tensor, f is the activation function, and z_{out} is the layer's output. The generator network is trained using backpropagation to update its weights and biases to produce more realistic data samples over time.

A GAN's generator network architecture consists of several convolutional or fully connected layers, with an activation function applied after each layer. The generator network uses the mathematical formula for fully connected or convolutional layers to translate a random noise vector to a data sample.

Discriminator Network

In a GAN, the discriminator network decides whether or not a sample is genuine. Binary classifiers like this one accept an input sample and return a probability that the sample is genuine. Typically, a discriminator network will be built with several layers of artificial neurons, each of which will abstract the data further. Typically, the input to the discriminator is a feature vector derived from an actual or synthetic sample. The discriminator network ends with a single synthetic neuron, which returns a probability between 0 and 1 based on whether or not the input sample is genuine.

The output can be mathematically represented as:

$$p = f(x)$$

where p is the output probability, f is the activation function, and x is the input to the discriminator network.

The activation function used in the final layer is often a sigmoid function, which maps the input to a value between 0 and 1.

The discriminator network weights and biases are updated during the training process to improve the network's identification of fake samples. This is achieved by minimizing the binary cross-entropy loss between the true labels and the predicted probabilities.

The architecture of a discriminator network can be summarized as follows:

- Input: a feature vector extracted from a real or generated sample
- Multiple layers of artificial neurons transform the input data into increasingly abstract representations
- Final layer: a single artificial neuron outputting a value between 0 and 1, representing the probability that the input sample is real
- Activation function: sigmoid function ($1/(1+e^{-x})$)
- Loss function: binary cross-entropy loss ($-(y*\log(p)+(1-y)*\log(1-p))$) where y is the true label (0 for fake samples and 1 for real samples)

In summary, the discriminator network in a GAN takes an input sample and outputs a probability that the sample is real. Its architecture consists of multiple layers of artificial neurons and a final layer with a sigmoid activation function. During training, the network weights are updated to minimize the binary cross-entropy loss between the actual labels and the predicted probabilities.

Objective Function

In a Generative Adversarial Network (GAN), the objective function is made up of the generator loss and the discriminator loss. GAN works by training a generator and a discriminator to the point where the generated data cannot be distinguished from the actual thing. On the other hand, the discriminator improves its ability to spot sham data.

The generator loss measures how well the generator can and biases fool the discriminator. It is defined as the negative of the expected log probability that the discriminator assigns to the real label for a sample generated by the generator. Mathematically, the generator loss is given by:

$$L_G -E[\log(D(G(z)))]$$

where L_G is the generator loss, D(G(z)) is the probability that the discriminator assigns to the real label for a sample generated by the generator, and G(z) is the generator's output for a given input noise vector z, and E is the expectation over the generated samples.

The discriminator's ability to tell the difference between authentic and spoofed samples is quantified by the discriminator loss. It is the difference between the discriminator's predicted negative log probability for the genuine label for a real sample and the expected positive log probability for the false label for a fake sample. Mathematically, the discriminator loss is given by:

$$L_D = -E[\log(D(x))] - E(\log(1 - D(G(z))))]$$

where L_D is the discriminator loss, D(x) is the probability that the discriminator assigns to the real label for a real sample x, D(G(z)) is the probability that the discriminator assigns to the real label for a sample generated by the generator, and E is the expectation over the real and generated samples.

Finding the Nash Equilibrium between the generator and discriminator, where the generator creates samples that are indistinguishable from actual data and the discriminator cannot discriminate between genuine and fake samples, is the goal of a GAN. Namely, the Nash Equilibrium is attained when the generator loss is minimized and the discriminator loss is maximized. In practice, training has involved alternating between updating the generator to reduce its loss and the discriminator to increase its gain. This process is repeated until convergence is reached, at which point the generator's samples are indistinguishable from real data to the discriminator.

Training Process

A Generative Adversarial Network (GAN) learns by repeating a training process that involves modifying the network's parameters at each stage. Finding a Nash Equilibrium between the generator and discriminator is the target, where the former generates samples that can't be distinguished from genuine data and the latter can't tell the difference between actual and fake data.

The training process is usually carried out in two stages:

- Generator Training: In this stage, the generator is updated to minimize the generator loss, which is a measure of how well the generator can fool the discriminator. The generator loss is defined as the negative of the expected log probability that the discriminator assigns to the real label for a sample generated by the generator. To update the generator parameters, the gradients of the generator loss concerning the parameters are computed using backpropagation and used to update the parameters using an optimization algorithm, such as Adam, SGD, or RProp.
- Discriminator Training: In this stage, the discriminator is updated to maximize the discriminator loss, which is a measure of how well the discriminator can distinguish between real and fake samples. The discriminator loss is defined as the expected negative log probability that the discriminator assigns to the real label for a real sample and the expected positive log probability that the discriminator assigns to the fake label for a fake sample. To update the discriminator parameters, the gradients of the discriminator loss concerning the parameters are computed using backpropagation and used to update the parameters using an optimization algorithm, such as Adam, SGD, or RProp.

Once convergence is reached, the stages of training the generator and discriminator are repeated. After this point, the samples produced by the generator look very close to genuine data, making it impossible for the discriminator to tell them apart. While training, the generator and discriminator are not updated at the same time to prevent instability. Instead, the generator is updated after a predetermined number of discriminator updates and the discriminator is updated after a predetermined number of generator updates. It is also crucial to keep an eye on things during training to prevent the generator from collapsing if it becomes too powerful relative to the discriminator.

Applications in the Internet of Things (IoT)

Image Synthesis for Object Detection in IoT Cameras

Generative adversarial networks (GANs) are useful for image synthesis in a number of Internet of Things (IoT) applications, including object recognition in security and surveillance cameras and in industrial quality assurance inspections. These programmes' cameras are used to quickly find objects of interest (Yang, J et al., 2020). When there isn't enough labelled data, however, training object identification models can be difficult. However, GANs can be used to circumvent this issue by generating fake images for use in training object identification models. After learning the underlying data distribution, the GAN can generate new images that are like real-world photographs. These fabricated images can be annotated to improve the accuracy of object recognition software in real-world settings.

Furthermore, GANs' ability to generate new synthetic images could bolster existing data sets. Object recognition models can benefit from increased generalization performance if they are made more robust to data variation. To compensate for the dearth of annotated data and improve object detection performance, the generated synthetic images can be added to the training dataset.

Using GANs for image synthesis and data augmentation can improve object identification performance, which in turn can lead to more accurate and trustworthy inspection, security, and surveillance systems. Consequently, these systems may improve in efficacy and utility, allowing them to be implemented in more areas.

Image-to-Image Translation for Sensor Data Enhancement

Here, GANs are used to generate improved copies of sensor data, including images captured by cameras, to augment the data for further processing and analysis (Beaulieu, M. et al., 2018). Noise, blur, and low resolution are just a few of the factors that can lower the quality of sensor data. This may make it difficult to extract useful information from the sensor data for later processing and analysis. GANs can be used to improve the quality of sensor data by making improved copies of the data.

In the context of sensor data enhancement, the generator network is trained to generate improved versions of the input sensor data free from noise, blur, and other degradation factors. The discriminator network is trained to differentiate between the original sensor data and the improved synthetic data generated by the generator network. Through this process, the generator network learns to create improved versions of the sensor data that are similar to the original data in terms of structure and content but are free from degradation factors. The enhanced sensor data generated by the GAN can then be used for further processing and analysis, leading to improved performance and results.

Overall, the use of GANs for image-to-image translation in sensor data enhancement offers a flexible and effective solution for improving the quality of sensor data in IoT applications. By generating enhanced versions of the sensor data, GANs can help to overcome the challenges posed by degradation factors and provide high-quality data for further analysis and processing.

Video Generation for IoT Security Surveillance

GANs have the potential to significantly alter the status quo of video production for IoT surveillance systems. To that end, GANs are employed here to produce fabricated footage for the purposes of testing and educating computer vision algorithms employed in security surveillance. The use of GAN-generated synthetic videos for training and testing computer vision algorithms has several advantages over the use of real-world videos (Chen, D. et al., 2021). To begin with, it is possible to generate synthetic videos with a wide range of aesthetic options, such as a selection of alternative lighting, camera positions, and settings. In this way, computer vision algorithms can be trained and tested in a wide variety of settings, which can ultimately lead to better results and greater stability.

Using synthetic videos generated by GANs has several advantages over using real videos for training and testing computer vision algorithms. They are:

- Synthetic videos can be generated with various variations, including different lighting conditions, camera angles, and background scenes. This allows for the training and testing of computer vision algorithms under various conditions, which can lead to improved performance and robustness.
- Synthetic videos allow for the generation of large datasets that can be used for training deep-learning models. Deep learning models typically require large datasets for training, and the age of synthetic videos can provide a cost-effective and scalable solution for building such datasets.
- It can help overcome the privacy and security concerns associated with using real videos in security surveillance. Synthetic videos can be generated in a controlled environment, and the content and characteristics of the videos can be easily controlled and modified. This makes it possible to use synthetic videos for testing and evaluating computer vision algorithms without risking the privacy and security of individuals.

Text-to-Image Synthesis for Creating Virtual Environments for IoT Simulations

GANs are used to generate images based on textual descriptions, such as those describing the layout and objects in a virtual environment. The use of text-to-image synthesis with GANs has several advantages over other methods of creating virtual environments for IoT simulations. It provides a flexible and scalable solution for generating a large number of virtual environments with different configurations and objects. This allows for the simulation of various scenarios and environments, which can help test and evaluate IoT systems and applications (Li, R. et al., 2020).

Text-to-image synthesis with GANs can help overcome traditional methods' limitations for creating virtual environments, such as manual modeling or computer graphics techniques. These methods can be time-consuming, require specialized skills, and may not be able to easily generate complex environments with a large number of objects and variations.

GANs can help to improve the realism of virtual environments for IoT simulations. GANs can generate images that are visually similar to real-world environments, which can help create more accurate and realistic simulations. This can be particularly important for testing and evaluating IoT systems and applications that require real-world scenarios and environments.

Overall, the impact of GANs in text-to-image synthesis for creating virtual environments for IoT simulations has been significant. By providing a flexible, scalable, and realistic solution for creating virtual environments, GANs have enabled the development of advanced and effective IoT simulations that can be used to test and evaluate IoT systems and applications.

Audio Synthesis for Voice-Controlled IoT Devices

GANs are used to generate synthetic audio signals, such as speech and music, based on input data, such as text or musical scores. The use of audio synthesis with GANs has several advantages over other methods of creating synthetic audio signals for voice-controlled IoT devices (Cai, Z. et al., 2021). It provides a flexible and scalable solution for generating many synthetic audio signals with different features and properties. This can be useful for testing and evaluating voice-controlled IoT devices and their performance under various conditions.

Audio synthesis with GANs can help overcome traditional methods' limitations for creating synthetic audio signals, such as manual recording or computer-generated signals. These methods can be time-consuming, require specialized skills, and may not be able to easily generate complex audio signals with a large number of variations. GANs can help to improve the realism of synthetic audio signals for voice-controlled IoT devices. GANs can generate audio signals that are acoustically similar to real-world signals, which can help create more accurate and realistic simulations. This can be particularly important for testing and evaluating voice-controlled IoT devices that require real-world speech or music signals.

By providing a flexible, scalable, and realistic solution for creating synthetic audio signals, GANs have enabled the development of advanced and practical tests and evaluations for voice-controlled IoT devices. This has helped improve the performance and reliability of these devices and paved the way for developing new and innovative voice-controlled IoT applications.

Variations of GAN for IoT Applications

Conditional GANs for Personalization in IoT Devices

Conditional Generative Adversarial Networks (CGANs) have emerged as a promising solution for personalization in IoT devices. Unlike regular GANs, CGANs allow control over the generated output by incorporating additional information, or conditions, into the generation process. This can help to achieve more personalized and relevant results for each user (Cao, X. et al., 2022).

The basic architecture of a CGAN is similar to a regular GAN, with two main components: a generator and a discriminator. However, in a CGAN, the generator and discriminator receive additional input information or conditions during the training process.

The mathematical formula for the objective function in a CGAN can be represented as follows:

$$L(D,G) = \min(\max(D(x,y),\ 1 - G(z|y)))$$

where L is the objective function, D is the discriminator, G is the generator, x is the real data sample, y is the condition information, and z is the random noise input.

The generator in a CGAN is trained to generate output data samples similar to the real data samples and are conditioned on the input information y. The discriminator in a CGAN is trained to distinguish between the real data samples and the generated data samples, while also taking into account the condition information y. In the context of IoT devices, CGANs can be used to achieve personalization in various applications, such as speech and music synthesis, image generation, and data augmentation. For example, a CGAN can generate personalized speech signals for a voice-controlled IoT device by conditioning the generation process on the user's age, gender, or accent.

Overall, CGANs offer a powerful solution for personalization in IoT devices by allowing control over the generated output and incorporating additional information into the generation process. With the increasing demand for personalization in IoT applications, CGAN has the potential to play a significant role in the development of advanced and effective IoT devices.

Wasserstein GANs for Anomaly Detection in IoT Sensor Data

Wasserstein Generative Adversarial Networks (WGANs) is a type of GANs that aim to address some of the limitations of traditional GANs, such as mode collapse and instability during training. In the context of IoT sensor data, WGANs can be used for anomaly detection by training the network on a large dataset of regular sensor readings and then using it to detect abnormal readings (Lu, H., et al., 2021).

The objective function for a WGAN can be represented as follows:

$$\min(G) = -E[D(x)]$$

$$\max(D) = E[D(x) - E[D(G(z))]$$

where G represents the generator, D represents the discriminator, x represents the real data sample, and z represents the random noise input. E represents the expectation operator, and D(x) and D(G(z)) represent the discriminator's output for the real data sample and the generated data sample, respectively.

The generator G is trained to generate data samples that have high values of D(G(z)), while the discriminator D is trained to have high values of D(x) and low values of D(G(z)). The objective function balances these objectives and encourages the generator to produce data samples that are as similar as possible to real data samples while encouraging the discriminator to distinguish between real and generated data samples.

Once the network has been trained, the discriminator can detect anomalies in new sensor readings. A new reading with a low value of D(x) can be classified as an anomaly. The use of WGANs for anomaly detection in IoT sensor data can improve the accuracy and efficiency of IoT-based surveillance and security systems, as well as other industrial applications that rely on sensor data.

StyleGANs for the Generation of Synthetic Data for IoT Device Testing

Style Generative Adversarial Networks (StyleGANs) are a type of GANs specifically designed for image generation tasks. In the context of IoT device testing, StyleGANs can be used to generate synthetic data to test IoT devices' functionality and performance (Song, S. et al., 2022).

The objective functions for a StyleGAN can be represented as follows:

$$\min(G) = -E[D(x, w_i)]$$

$$\max(D) = E[D(x, w_i)] - E[D(G(z, w_i), w_i)]$$

where G represents the generator, D represents the discriminator, x represents the real data sample, z represents the random noise input, and w_i represents the learned style representation. E represents the expectation operator, and D (x, w_i) and D (G (z, w_i), w_i) represent the discriminator's output for the real data sample and the generated data sample, respectively.

The generator G is trained to generate data samples that have high values of D (G (z, w_i), w_i), while the discriminator D is trained to have high values of D (x, w_i) and low values of D (G (z, w_i), w_i). The objective function balances these objectives and encourages the generator to produce data samples that are as similar as possible to real data samples while encouraging the discriminator to distinguish between real and generated data samples.

Once the network has been trained, the generator can be used to produce synthetic data that can be used to test IoT devices. The use of StyleGANs for synthetic data generation in IoT testing can potentially improve the efficiency and accuracy of IoT device testing by providing a large and diverse set of synthetic data samples that can be used to test different scenarios and conditions.

BigGANs for 3d Object Generation for IoT AR/VR Applications

BigGANs (Big Generative Adversarial Networks) is a type of GAN designed for high-resolution image generation, making them particularly well-suited for applications in IoT-powered Augmented Reality (AR) and Virtual Reality (VR). In these applications, BigGANs can be used to generate synthetic 3D objects that can be used to test and train object detection and recognition algorithms in IoT devices (Jaiswal, A. et al., 2021).

The objective function of a BigGAN can be defined as:

Minimize:

$$L_G - E[\log(D(G(z)))]$$

Maximize:

$$L_D = -E[\log(D(x))] - E(\log(1 - D(G(z))))]$$

where x is an actual image and z is a random noise vector. The generator network tries to minimize L_G, while the discriminator network tries to maximize L_D. The two networks play a minimax game, where the generator tries to produce images that are indistinguishable from real images while the discriminator tries to distinguish between real and generated images correctly.

Training BigGANs is a computationally intensive process that requires large amounts of data and high-powered GPUs. Despite these challenges, BigGANs have shown promising results in generating high-resolution images of 3D objects, making them a powerful tool for IoT AR/VR applications.

Limitations and Challenges in IoT

Mode Collapse in IoT Edge Devices

GAN mode collapse is a common problem in training Generative Adversarial Networks (GANs). It occurs when the generator network of a GAN generates limited variations of the same image, instead of producing a diverse range of ideas. This can lead to a lack of diversity in the generated data and can impact the performance of GANs in various applications, including IoT edge devices.

In IoT edge devices, mode collapse can be particularly problematic due to limited computational resources and the need for GANs to produce diverse data to support various IoT applications (Liu, Y. et al., 2021). For example, in video surveillance or security applications, mode collapse can result in the generation of only limited variations of images, making it difficult for GANs to detect and classify different objects or events in real time effectively.

Researchers have proposed various techniques to prevent mode collapse in GANs, including architectural modifications, regularization methods, and loss functions. Some standard plans include weight normalization, adding noise to the input data, and adding diversity loss terms to the objective function.

One such loss function used to prevent mode collapse is the Wasserstein distance. It is defined as:

$$W(p,q) = \inf E[f(X)]$$

where $X \sim p$ and f is 1- Lipschitz

The Wasserstein distance measures the minimum amount of "work" required to transform one probability distribution, p, into another, q. In the context of GANs, it measures the difference between the real and generated distributions. Adding the Wasserstein distance as a loss term in the objective function encourages the GAN to develop a more diverse range of images and avoid mode collapse. Preventing mode collapse is crucial for ensuring the performance and reliability of GANs in IoT edge devices, where limited computational resources and the need for diverse data can make GANs vulnerable to this problem.

Stability Issues During Training in IoT Resource-Constrained Environments

GANs (Generative Adversarial Networks) has grown in popularity in recent years due to their capacity to produce synthetic data comparable to actual data. However, stability problems might occur while training GANs in IoT resource-constrained settings. Numerous variables, including limited memory, processing capacity, and network bandwidth, may contribute to these problems. The primary issue is the non-convergence of the GAN training process. The objective function in GANs is a minimax game, where the generator tries to maximize the objective function while the discriminator tries to minimize it. The training process terminates when a Nash equilibrium is reached, i.e., when the generator and the discriminator get a balance where neither can further improve the objective function (Mozaffar, M. et al., 2021). However, if the generator or the discriminator becomes too strong, it can result in non-convergence or slow convergence of the training process.

Several techniques have been proposed to address these stability issues in resource-constrained IoT environments. One of these is Wasserstein GAN (WGAN), which replaces the traditional Jensen-Shannon divergence objective function with the Wasserstein distance. This has resulted in improved stability and faster convergence during GAN training.

Adopting conditional GANs, which produce samples contingent upon extra information, is an alternative strategy. This gives the user more control over the created samples, improving convergence and stability during training. While GAN training is still an active area of study, it is vital to remember that additional work is necessary to properly understand and address the stability challenges in resource-constrained IoT systems. However, these approaches have demonstrated promising outcomes.

Lack of Interpretability in IoT Sensor Data

Generative Adversarial Networks (GANs) have successfully generated synthetic data for various applications, including IoT. However, the lack of interpretability of GANs is a significant concern regarding IoT sensor data. Unlike traditional supervised learning algorithms, GANs are complex and opaque models that generate synthetic data without explicitly learning the underlying patterns and relationships in the data. As a result, it isn't easy to understand why a GAN generates the data it does and whether the generated data accurately represents the underlying patterns in the original data (Ganz, F. et al., 2015). This lack of interpretability can have significant implications for IoT applications, particularly in safety-critical domains such as healthcare, where understanding why a model makes specific predictions is essential. In such domains, it is critical to have models that are transparent and can be easily understood by domain experts.

There have been some efforts to address the interpretability issue in GANs, such as using techniques like saliency maps and activation maximization. However, these techniques are still in the early stages of development, and more research is needed to address the interpretability issue in GANs fully.

In conclusion, the need for interpretability in GANs is a significant challenge for their use in IoT applications, particularly in safety-critical domains. Addressing this challenge will require a combination of technical innovations and careful consideration of the ethical implications of using these models in such applications.

Ethical Concerns in IoT Data Privacy and Security

Due to their capacity to produce high-quality synthetic data, Generative Adversarial Networks (GANs) have found extensive applications in several Internet of Things (IoT) applications. However, their use in the IoT has brought up several ethical questions surrounding data privacy and security. One of the main worries is the possibility of using GANs improperly to produce phony data that may be used to influence real-world systems and results. GANs may be used, for instance, to create fake pictures, films, or audio that can be exploited to impersonate people, companies, or even governments. Serious repercussions like identity theft, financial fraud, or political influence may result from this (Hussain, F. et al., 2020).

Another issue to be concerned about is the possibility of GANs revealing sensitive information about persons and organizations. GANs, for instance, may be used to create artificial data that mimics genuine data in terms of its statistical characteristics. This may be used to infer private information about people or businesses, such as their preferences, demographics, or financial situation. It is crucial to carefully analyze the design and deployment of GANs in IoT systems to address these ethical problems. To prevent the synthetic data produced by GANs from being able to be traced back to the original data sources, this may include integrating privacy-preserving techniques like differential privacy.

In addition, it is essential to have clear and transparent policies in place that outline the use of GANs in IoT environments and to ensure that these policies are adhered to by all stakeholders involved. This includes the developers, users, and regulatory bodies, who must work together to ensure that the benefits of GANs are leveraged responsibly and ethically.

Overall, the ethical concerns related to GANs in IoT environments highlight the importance of balancing innovation and technology with social and ethical considerations. This requires a collaborative effort between stakeholders to ensure that GAN use of GANs in IoT environments is aligned with moral principles and protects the privacy and security of individuals and organizations.

Future Directions

Improved Stability and Robustness for GANs in IoT Edge Devices

GANs have been found to exhibit instability difficulties during training, such as mode collapse and oscillation, which can result in a lack of diversity or quality in the produced samples. Regularization methods, including weight and spectral normalization, which can increase the stability of GANs, are one approach to dealing with this problem. Utilizing more sophisticated optimization methods, such as the Wasserstein GAN (WGAN) and its variations, is another strategy that can increase the stability and resilience of GANs.

Another future direction is to address the need for more interpretability in GANs. GANs are known to be "black-box" models, where it can be challenging to understand how the model works and what it is doing. To address this issue, researchers are exploring techniques for making GANs more interpretable, such as using visualization techniques, such as activation maximization, and techniques for explaining the decisions made by the GAN.

Finally, there is also a growing concern about the ethical implications of GANs in IoT data privacy and security. As GANs become increasingly popular and widely used, there is a risk that sensitive data could be generated or manipulated without the user's knowledge or consent. To address these ethical concerns, researchers are exploring techniques for improving the transparency and accountability of GANs, such as using explainable AI techniques and incorporating privacy-preserving mechanisms into GANs.

In conclusion, the future of GANs in IoT edge devices is promising, with researchers exploring several directions for improving the stability, robustness, interpretability, and ethics of GANs. With these efforts, GANs have the potential to play an even more critical role in enabling intelligent applications across industries and revolutionizing the IoT.

Incorporation of Additional Constraints for GANs in IoT Sensor Data

In the future, there is potential for incorporating additional constraints into GANs to handle IoT sensor data better. One area of focus is incorporating domain-specific information, such as physical constraints, into the training process. For example, when generating synthetic images for testing IoT cameras, it may be essential to ensure that the generated images contain objects that follow realistic physical relationships. This can be achieved by incorporating additional constraints, such as object size, depth, and movement, into the GAN's training process.

Another research direction is incorporating additional objectives into the GAN's objective function. For example, adding a reconstruction loss term, which measures the difference between the generated image and the original image, can improve the quality and stability of the generated images. Additionally, incorporating interpretability objectives, such as disentanglement objectives, can help better understand the representations learned by the GAN, which can be useful in IoT sensor data applications where interpretability is important.

Mathematically, the incorporation of additional constraints into the GAN's objective function can be represented as follows:

$J(G,D) = J\_GAN(G,D) + \lambda.J\_Consttraints(G)$

where J (G, D) is the overall objective function for the GAN, J_GAN (G, D) is the standard aim of the GAN function, J_Constraints(G) is the constraint objective function, and λ is a weighting factor that determines the importance of the constraint objective.

In conclusion, the incorporation of additional constraints into GANs can be an essential step towards improving the stability and robustness of GANs in IoT edge devices and better handling IoT sensor data. The future of GANs in IoT is promising, and researchers and engineers will continue to explore new ways to incorporate additional constraints and objectives into GANs to better address the challenges faced in the IoT domain.

GANs in Other Fields Such as Healthcare and Finance for IoT Applications

GANs, particularly in the context of IoT (Internet of Things) applications, have demonstrated tremendous potential in healthcare and finance in recent years. In healthcare, GANs can be employed to create artificial medical pictures and data, which can then be used to train deep learning models, enhance existing data, or share data while maintaining anonymity. For instance, GANs can produce fake X-rays, CT scans, or MRIs that can be used to train deep learning models for prediction, diagnosis, or segmentation tasks in medical image analysis (Lan, L. et al., 2020). Not disclosing actual medical pictures to other parties can protect patient privacy while addressing the limits of limited annotated data in medical imaging.

In finance, GANs can be used to generate synthetic financial data, which can be used for training deep learning models, data augmentation, or privacy-preserving data sharing. For instance, GANs can develop artificial stock prices, credit card transactions, or loan applications, which can be used for training deep learning models for finance tasks such as fraud detection, risk assessment, or portfolio optimization. This can help overcome the limitations of limited annotated data in finance and preserve financial privacy by not sharing accurate financial data with third parties. In healthcare and finance, GANs can be combined with other machine learning techniques and models, such as reinforcement learning, supervised learning, unsupervised learning, or transfer learning, to achieve even better performance and solutions.

However, like any other machine learning technique, GANs also have limitations and challenges in healthcare and finance. For instance, GANs can suffer from mode collapse, where the generator only produces a few limited modes of data distribution, or instability, where the generator and discriminator may not converge during training. Additionally, GANs can also need more interpretability, where it is difficult to understand the generator's or discriminator's decision-making process. Moreover, GANs can also raise ethical concerns, such as privacy and security violations, misuse of synthetic data, or bias and discrimination in the generated data.

Despite these challenges, GANs have shown great potential and impact in healthcare and finance. There is a growing trend and interest in developing more advanced, stable, and interpretable GANs for IoT applications in these domains. This includes improving the stability and robustness of GANs through techniques such as weight normalization, gradient penalty, or cycle consistency, as well as incorporating additional constraints or objectives into the GAN training, such as fairness, diversity, or consistency, to ensure that the generated data is not biased, discriminatory, or unrealistic.

Advancements in Hardware and Infrastructure to Scale GANs for IoT

Advancements in hardware and infrastructure have paved the way for GANs to be used in many industries, including IoT. The development of GPUs, TPUs, and cloud computing have enabled GANs to be trained on more extensive and complex datasets, making it possible to scale them for IoT applications. Significant efforts have been made to improve the efficiency of GANs, including optimization of the network architecture and algorithms and the developing of specialized hardware such as GPUs, TPUs, and dedicated GAN accelerators (Mohammadi, M. et al., 2018). These developments have helped to overcome some of the challenges associated with scaling GANs for IoT applications, such as resource constraints and computational limitations.

In addition, developing new algorithms and techniques for training GANs has also enabled their use in IoT. For example, techniques such as federated learning and transfer learning have been developed to help address data privacy and security issues, making it possible to use GANs in IoT without compromising sensitive information. There have been efforts to create new hardware and infrastructure to support the deployment of GANs in IoT, such as edge devices and fog computing. These advances have enabled the deployment of GANs in resource-constrained environments, such as those found in IoT, making it possible to scale their use for a wide range of applications.

In conclusion, advancements in hardware and infrastructure have played a significant role in enabling the use of GANs in IoT applications. The development of new hardware and algorithms and the optimization of existing infrastructure have helped overcome some of the challenges associated with scaling GANs for IoT, making it possible to use them in a wide range of industries and applications.

CONCLUSION

Summary

- GANs can be used for various tasks in IoT, including image synthesis for object detection in surveillance, security, and industrial inspection.
- GANs can also be used for image-to-image translation for sensor data enhancement and video generation for IoT security surveillance.
- Text-to-image synthesis can also be performed using GANs to create virtual environments for IoT simulations.
- Audio synthesis for voice-controlled IoT devices is another application of GANs.
- Conditional GANs can be used for personalization in IoT devices.
- Wasserstein GANs can be used for anomaly detection in IoT sensor data.
- Style GANs can be used to generate synthetic data for IoT device testing.
- BigGANs can be used for 3D object generation for IoT AR/VR applications.
- GAN mode collapse and stability issues during training in resource-constrained environments are challenges in using GANs in IoT.
- Lack of interpretability, ethical concerns in data privacy, and security are some ethical problems associated with using GANs in IoT.
- The future directions of improved stability and robustness, incorporation of additional constraints, and scaling GANs to handle large data sets are some areas of active research.
- GANs can also be applied in other fields, such as healthcare and finance, for IoT applications.
- Advancements in hardware and infrastructure can help to scale GANs for IoT applications.

REFERENCES

Beaulieu, M., Foucher, S., Haberman, D., & Stewart, C. (2018, July). Deep image-to-image transfer applied to resolution enhancement of sentinel-2 images. In *IGARSS 2018-2018 IEEE International Geoscience and Remote Sensing Symposium* (pp. 2611-2614). IEEE. 10.1109/IGARSS.2018.8517655

Cai, Z., Xiong, Z., Xu, H., Wang, P., Li, W., & Pan, Y. (2021). Generative adversarial networks: A survey toward private and secure applications. *ACM Computing Surveys*, *54*(6), 1–38. doi:10.1145/3459992

Cao, X., Sun, G., Yu, H., & Guizani, M. (2022). PerFED-GAN: Personalized federated learning via generative adversarial networks. *IEEE Internet of Things Journal*. doi:10.1109/JIOT.2022.3172114

Chen, D., Yue, L., Chang, X., Xu, M., & Jia, T. (2021). NM-GAN: Noise-modulated generative adversarial network for video anomaly detection. *Pattern Recognition*, *116*, 107969. doi:10.1016/j.patcog.2021.107969

Dang, X., Liu, G., Tang, X., Wang, S., Wang, T., & Zou, M. (2022). Motor Imagery EEG Recognition Based on Generative and Discriminative Adversarial Learning Framework and Hybrid Scale Convolutional Neural Network. *IAENG International Journal of Applied Mathematics*, *52*(4), 1–9.

Ganz, F., Puschmann, D., Barnaghi, P., & Carrez, F. (2015). A practical evaluation of information processing and abstraction techniques for the internet of things. *IEEE Internet of Things Journal*, *2*(4), 340–354. doi:10.1109/JIOT.2015.2411227

Goodfellow, I., Pouget-Abadie, J., Mirza, M., Xu, B., Warde-Farley, D., Ozair, S., Courville, A., & Bengio, Y. (2020). Generative adversarial networks. *Communications of the ACM*, *63*(11), 139–144. doi:10.1145/3422622

Hussain, F., Hussain, R., Hassan, S. A., & Hossain, E. (2020). Machine learning in IoT security: Current solutions and future challenges. *IEEE Communications Surveys and Tutorials*, *22*(3), 1686–1721. doi:10.1109/COMST.2020.2986444

Jaiswal, A., Sodhi, H. S., Muzamil, H. M., Chandhok, R. S., Oore, S., & Sastry, C. S. (2021). Controlling BigGAN image generation with a segmentation network. In *Discovery Science: 24th International Conference, DS 2021,* (pp. 268-281). Springer International Publishing. https://doi.org/10.1007/978-3-030-88942-5_21

Lan, L., You, L., Zhang, Z., Fan, Z., Zhao, W., Zeng, N., Chen, Y., & Zhou, X. (2020). Generative adversarial networks and its applications in biomedical informatics. *Frontiers in Public Health*, *8*, 164. doi:10.3389/fpubh.2020.00164 PMID:32478029

Li, R., Wang, N., Feng, F., Zhang, G., & Wang, X. (2020). Exploring global and local linguistic representations for text-to-image synthesis. *IEEE Transactions on Multimedia*, *22*(12), 3075–3087. doi:10.1109/TMM.2020.2972856

Liu, J., & Wu, W. (2021). Automatic Image Annotation Using Improved Wasserstein Generative Adversarial Networks. *IAENG International Journal of Computer Science*, *48*(3).

Liu, Y., Xiao, M., Chen, S., Bai, F., Pan, J., & Zhang, D. (2021). An intelligent edge-chain-enabled access control mechanism for IoV. *IEEE Internet of Things Journal*, *8*(15), 12231–12241. doi:10.1109/JIOT.2021.3061467

Lu, H., Du, M., Qian, K., He, X., & Wang, K. (2021). GAN-based data augmentation strategy for sensor anomaly detection in industrial robots. *IEEE Sensors Journal*, *22*(18), 17464–17474. doi:10.1109/JSEN.2021.3069452

Mohammadi, M., Al-Fuqaha, A., Sorour, S., & Guizani, M. (2018). Deep learning for IoT big data and streaming analytics: A survey. *IEEE Communications Surveys and Tutorials*, *20*(4), 2923–2960. doi:10.1109/COMST.2018.2844341

Mozaffar, M., Liao, S., Xie, X., Saha, S., Park, C., Cao, J., & Gan, Z. (2021). Mechanistic artificial intelligence (mechanistic-AI) for modeling, design, and control of advanced manufacturing processes: Current state and perspectives. *Journal of Materials Processing Technology*, *117485*. doi:10.1016/j.jmatprotec.2021.117485

Razavi-Far, R., Ruiz-Garcia, A., Palade, V., & Schmidhuber, J. (2022). *Generative adversarial learning: architectures and applications*. Springer. doi:10.1007/978-3-030-91390-8

Song, S., Chang, K., Yun, K., Jun, C., & Baek, J. G. (2022). Defect Synthesis Using Latent Mapping Adversarial Network for Automated Visual Inspection. *Electronics (Basel)*, *11*(17), 2763. doi:10.3390/electronics11172763

Yang, J., Wang, C., Jiang, B., Song, H., & Meng, Q. (2020). Visual perception enabled industry intelligence: State of the art, challenges and prospects. *IEEE Transactions on Industrial Informatics*, *17*(3), 2204–2219. doi:10.1109/TII.2020.2998818

Chapter 5
Machine Learning–Enabled Internet of Things Solution for Smart Agriculture Operations

Palanivel Kuppusamy

https://orcid.org/0000-0003-1313-9522

Pondicherry University, India

Joseph K. Suresh

Pondicherry University, India

Suganthi Shanmugananthan

Annamalai University, India

ABSTRACT

Internet of things solutions with machine learning capabilities is a hot research area in industries, including agriculture. They can be used for data analysis and further forecasting the big data and intelligent applications in farming. In traditional farming, the main obstacles are disease prediction, automatic irrigation, energy harvesting, and constant monitoring. Today, farmers' cultivation of their crops has changed by introducing automated harvesters, drones, autonomous tractors, sowing, and weeding. Smart farming with ML-enabled IoT systems can improve crop harvesting decisions. The main topic of this chapter is to provide an ML-enabled IoT solution for smart agriculture. The MIoT solution in agriculture allows farmers to use predictive analytics to help them make better harvesting decisions. Designing a MIoT system for smart agriculture can assist farmers in improving yields, planning more effective irrigation, and making harvest forecasts by monitoring essential data like humidity, air temperature, and soil quality via remote sensors.

DOI: 10.4018/978-1-6684-8785-3.ch005

INTRODUCTION

The world's primary source of food is agriculture. Civilizations have depended heavily on it throughout history. The significant complications in traditional agriculture are unpredictable contamination of crops, inclement weather, ineffective pest management regulations, and food safety monitoring. These complications financially impact the agriculture sector, and there are more hazards to the public. Traditional farming techniques are obsolete, and the food supply chain is becoming increasingly complex. The farmer was always out in the field, keeping an eye on the land and the health of the crops (Kodali Ravi et al., 2016).

Motivation

Traditional farming practices (FAO 2015 & FAO 2017) are challenging due to climatic fluctuations and extremities, growing population, urbanization, loss of crops to various diseases, and difficulty detecting weeds. These challenges are motivated to apply a technology-based solution to increase the quality and quantity of the crops.

- Growing crops in populated cities and population hikes could favour increasing urbanization. Therefore, food is needed, and the limited land requires practical solutions to ensure food supply.
- The loss of crops to various diseases is a major issue in farming. Pests and disease control are significant concerns in agriculture. In traditional practice, spraying pesticides over cropping areas produces high environmental costs.
- Weeds are the most critical threats to crop detection. It is challenging to identify weeds and tell them apart from crops. Hence, weed detection is a significant problem in agriculture.
- Crop production can be impacted by extreme weather changes and climates, raising crop costs and degrading crop quality.

Farmers are turning to technologies to automate agricultural production (*Aaron Tan, 2021*), alleviating the need to toil on the land while watching their crops. They change how they cultivate crops using robotic harvesters, drones, self-driving tractors, sowing, and weeding. These operations maximize human labour while raising crop quality and yield.

Advanced Technologies

Advanced technologies can transform inadequate farming methods into smart farming to ensure greater yields and efficiency. Using Information and Communication Technologies (ICT), smart agriculture (or) farming (Dhanaraju et al. 2022) aims to increase agricultural product production and quality with less labor and at the lowest possible cost. Technologies enable them to concentrate on more essential duties to handle boring and repetitive jobs. Cyber-physical systems (CPS), robotics/drones, sensors/IoT devices, agricultural data management, AI, and ML approaches significantly transform process chains in agriculture and forestry. ML and IoT applications contain sensor data quantities, and intelligent processing could bridge the holes between the cyber and physical systems.

In smart agricultural systems, remote sensing technology can decrease waste, increase output, and manage more resources (Anna Triantafyllou et al., 2019). The smart farming system comprised a data collection phase using WSN, a data pre-processing and storage stage, and predictive processing using ML methods. Maduranga & Ruvan Abeysekera (2020) studied the current smart agriculture and farming methods with IoT and ML and anticipated innovative notions of how ML-IoT could be unified in such applications. The smart solutions of smart agriculture (Rajyaguru et al., 2021) are precision agriculture, crop monitoring, livestock monitoring, irrigation management, smart pest control, fertilizer management, and weather forecasting. Numerous studies have demonstrated that data analytics may increase the precision and predictability of smart agricultural systems.

Smart farming consents to examine the plants' growth and to impact the real-time constraints to enhance plant growth and support the farmer's activity. Smart farming manages agricultural activities through technologies that deliver cost-effective farming outputs (Song et al., 2022). It utilizes technologies to gather data, analyze it, and produce actionable insight to manage pre- and post-harvest farming operations. However, the significant challenges of smart agriculture include continuous monitoring, energy harvesting, automatic irrigation, and disease prediction.

The core motivation of this chapter will be on current developments in sensor-based IoT networks that enable ML-based techniques in smart farming. ML-enabled IoT motivates emerging areas to provide effective prominent solutions in agriculture. It comprises current IoT innovation trends that use advanced technologies in smart farming.

- In agriculture, IoT-based solutions enable numerous applications such as crop data analysis, weather forecasts, pest detection, etc. A complete IoT system (IndustryTap, 2022) can integrate various sensors/devices, connectivity, data processing system, and user interface components/ Apps. It produces data for vast and remote agricultural regions, which ML can use to anticipate crops. This ML system can leverage data from IoT to predict yields in huge, remote agricultural areas. IoT in agriculture (using robots, drones, remote sensors, and imagery) can be integrated with ML and analytical tools for crop monitoring, surveying, field mapping, and logical farm management plans to save time and money (Yemeserach Mekonnen et al., 2020).
- Smart farming uses ML algorithms to boost output and lower the possibility of crop damage. ML records irrigation, genetic and environmental variables, seed quality, fertilizer application, and pruning in agriculture. By implementing ML in the agricultural sector, crop fields can produce more crops of higher quality and quantity to meet the rising global need for food (Vishal Meshram et al., 2021).

ML-enabled IoT solutions facilitate the level of computation required in numerous agricultural fields. The combination of ML with IoT sensors can help Big Data analysis and smart farming (Ravesa Akhter et al., 2022) in a shared at device and service levels. Dahane & Benameur (2020) intended an innovative farming system created on an intelligent platform that allowed prediction competencies using AI methods.

Research Objectives

The research objectives of this chapter are below:

a. To explore where MIoT solutions have been hired for smart agriculture.
b. How will smart agriculture benefit from MIoT techniques?
c. To investigate various applications of smart agriculture using the MIoT technique.

d. To propose a MIoT architecture for smart agriculture/farming.
e. What are the main methods used for MIoT-based prediction in smart agriculture?

Goals/Objectives

This chapter aims to facilitate the integration of ML and IoT that focuses on ML-enabled IoT (MIoT), which enhances the data sources' capacity for analysis. With real datasets, this work seeks to demonstrate how to manage heterogeneous information and data. Finding effective strategies to utilize continually recorded and made available data can be the ideal choice to accomplish these aims, as productive agriculture organizations need to increase profitability while reducing costs. Farmers can benefit from deploying remote sensors to monitor critical data such as humidity, air temperature, and soil quality to increase yields, build more effective irrigation systems, and forecast harvests.

Proposed Work

The proposed work in this chapter is presented below:

a. Introduce the MIoT that performs the data exchange between connected devices and various ML algorithms to overcome the challenges discussed in smart farming.
b. Describe how smart agriculture powered by MIoT improves agriculture operations, from sowing seeds to till-crop harvesting.

The MIoT solution in smart agriculture allows predictive analytics to help farmers make better harvesting decisions. The significant research contributions are stated below:

1. The activities and techniques of smart agriculture, including irrigation, soil and seed, disease detection, weed control, pesticides, yield, etc., are considered in this work.
2. This chapter discussed the role of MIoT in smart agriculture and other methods elaborate in smart agricultural operations.
3. This chapter proposed the architectural model for various smart agriculture operations using the MIoT approach.
4. Future works and the challenges and issues in adopting MIoT are discussed.

This chapter is organized as follows. *Section 2* outlines smart agriculture, agricultural technologies, IoT, ML, and MIoT. *Section 3* introduces the literature review, the focus of this chapter, and the proposed methodology. *Section 4* is focused on the MIoT architectural model for smart agriculture, following the challenges and case studies. Finally, *section 5* concludes this chapter.

BACKGROUND

In today's world, agriculture and irrigation must apply ICT to aid agriculturalists and farmers in improving agricultural operations. Digital transformation in smart agriculture involves a human-centered AI method that combines sociological, ethical, and legal issues (*Holzinger et al., 2022*). The evolutions of ML and the IoT allow farmers to improve yield by using effective land utilization, soil fruitfulness, water level, mineral insufficiencies, control pest, trim development, and horticulture. The application of remote sensors provides accuracy and applied agriculture to compact with issues in the field. This improvement might allow agricultural management platforms to organize farm data systematically and boost agribusiness by creating successful plans. Therefore, recent technological development has significantly impacted agriculture, which produces the maximum yield. Hence, this section presents various sources of information required to write this chapter, including agriculture technologies, smart farming and its applications, IoT, ML, and MIoT.

Challenges in Traditional Agriculture System

Agriculture or farming is cultivating the soil, producing crops, and keeping livestock. Adopting conventional farming techniques and the current environmental changes would not sustain such enormous yield creation. Moreover, agriculture needs a consistent expansion sought after with populace development. According to projections from the United Nations (*UN 2015 and 2018*), by 2050, there will be 2 billion more people on Earth, needing a 60 percent increase in food production. The idea has a remarkable impact on agriculture and is designed to use technology in cultivation to enhance farming.

According to *Louis Columbus (2023)*, the perfect challenge for intelligent technology in vast farming is understanding how weather, seasonal sunlight, animal, bird, and insect migration patterns, crop use of specific fertilizers, crop use of pesticides, planting cycles, and irrigation cycles all affect production. Table 1 reports various challenges in the present agricultural and the technological solutions that might be used to overcome these challenges.

Table 1. The challenges in the present agricultural and the solutions

S/N	Problem(s)	Solution(s)	Technology
1	To reduce domestic and wild animals' potential to accidentally destroy crops or burglary at a remote farm location.	Detect animal or human breaches and quickly give out a warning	AI & ML
2	To enhance agricultural yield forecasting	Identify breaches caused by people or animals, then swiftly provide a warning.	AI, ML & Drones
3	Invaluable for crop planning	Yield mapping. (*Tanha et al., 2020*)	3D & ML
4	To improve pest management	Anticipate and recognize insect infestations before they happen. (*United Nations, 2018*).	Drones
5	Shortage of agricultural workers	Use of Agricultural robotics (*Saiz-Rubio & Rovira-Más, 2020*)	Robotics
6	Enhance the food safety and track-and-traceability of agricultural supply chains	Inventory reduction can be reduced by enabling visibility and control across crop/food supply chains.	RFID & IoT
7	Treatment is necessary to increase yields while lowering costs.	Make the optimal blend of biodegradable pesticides, and only permit the use of their products in agricultural areas.	AI, ML, sensor & Drones.
8	To define pricing strategies for a given crop	Forecast the price for crops based on yield rates	ML
9	To improve yield rates are all areas	To determine how successful frequent crop watering is, identify leaks, and improve irrigation systems.	Linear Programming & ML
10	To ensure livestock health	Monitoring livestock's health	AI & ML

To increase agricultural yields and quality, farmers and businesses must adopt data-centric strategies and broaden the breadth and scale of their usage of emerging technologies. Investing in technology can run sustainable and optimal agribusiness. Automation, temperature and climate sensors, detectors, satellite imaging, and Global Positioning Systems (GPS) technologies have resulted in digital transformation in agriculture (*Saurabh Singh, 2022*).

Using technological tools, farmers can better use their limited land, water, and fertilizer resources while increasing crop productivity. These advancements can boost agricultural productivity, boost crop yields, and cut expenses associated with food production while providing algorithms with real-time data. Maximizing fertilizer consumption to increase plant productivity help farmers cut waste and increase output. They improve the farmers' ability to manage their livestock, cultivate crops, and save resources.

Agricultural Technologies (AgTech)

Farmers can be more precise, reduce waste, increase production, and increase profit margins thanks to AgTech and high-tech equipment. *Sivakumar et al. (2021)* provided an overview of the contemporary technology used in agriculture, identified existing and new applications, and talked about the difficulties, potential outcomes, and executions. Drones for crop monitoring, sensors for measuring animals, farm management software, driverless tractors, and many more are examples of emerging farming technologies. The techniques, such as IoT, wireless sensor networks (WSNs), data analytics, and ML in agriculture, were climaxed by *Ravesa & Shabir (2021)*. The trends in agriculture are AI, ML & Data Science, Blockchain, cloud computing, CEA, Data Mining, GIS, IoT, robotics/automation, and regenerative agriculture. According to *Guna Nand Shukla et al. (2022),* the real-time use cases in agriculture included yield and quality assessment, autonomous robots for herding cattle, crop sustainability, weed control by identifying plant/crop species, crop infection, and disease detection, and innovative harvesting and pricing decisions.

- **AI/ML & Data Science.** AI and ML can considerably contribute and profit with their proven data analysis and forecasts (*Khan Rijwan 2021*). The vital agricultural data that IoT devices have collected is evaluated and directed using data science.
- **Blockchain.** Once the crops and food are accessible, farmers have challenges with fair trade, distribution, advertising, and proving the legitimacy of their products. Blockchain enables farmers to protect their crops from theft and fraud, manage the supply chain, and maintain the balance of the food ecology. It also allows farmers to protect their crops from theft and fraud, manage the supply chain, and balance the food ecology (*Singh et al., 2022*). Food traceability, the openness of the food supply chain, crop insurance, e-commerce for agro-trades, and farming subsidies are some real-world uses of blockchain technology (*Khouloud Hwerbi et al., 2022*).
- **Cloud & Edge Computing**. It offers a variety of storage options while handling enormous amounts of data. It oversees real-time data analytics, which includes automating agricultural monitoring and predicting pests and diseases, yields, and the weather (*Uddin et al. 2021*). Edge computing (*Akash 2021*) offers speedier answers and lower latency.
- **Controlled Environment Agriculture (CEA).** As IUFoST (2021) stated, plants are exposed to a carefully regulated amount of light, humidity, nutrients, and water in CEA. Vertical farming, indoor farming, greenhouses, etc., are a few intended indoor environments. Aeroponics, hydro-

ponics, aquaponics, and other innovative methods have been discovered to achieve a balanced plant equilibrium.

- **Data Mining** techniques (*Colombo-Mendoza et al., 2022*) can anticipate production volume from heterogeneous data sources using climate and agricultural production data.
- **Geographic Information System (GIS).** GIS technology uses hardware (satellites, drones, and GPS), software, and data to represent any geographic item spatially. Farmers can use GIS (*Marshet Nigatu, 2019*) to examine a wide range of complex spatial data in the agricultural sector, including the quantity of rainfall, topography, soil elevation, slope aspect, wind direction, flooding, erosion, and much more. GIS applications in agriculture include irrigated landscape imaging, crop health assessment, irrigation amendment analysis, land degradation assessment studies, erosion repair, and suitable drainage elevation models.
- **Internet of Things (IoT).** Various IoT sensors (such as light, humidity, soil moisture, temperature, crop health monitoring, etc.) are inserted into the agricultural farm in smart farming. The IoT can gather information from farm sensors, aerial and ground drones, geofencing, predictive analysis, and greenhouses (*Prajwal & Guru, 2022*). These devices provide real-time object recognition, classification, identification capabilities, crop position data, and soil quality evaluation. IoT is a valuable strategy for increasing agricultural production, especially with agricultural business Apps' introduction.
- **Regenerative Agriculture.** It regenerates the soil in preparation for the upcoming growing season. It regenerates the soil in preparation for the upcoming growing season. (*Livelihoods 2021*). Regenerative agriculture prioritizes soil regeneration, biodiversity, and dispersion. Farmers use reduced tillage, no-till farming, crop rotation, and planting cover crops to maintain soil fertility during regenerative agriculture.
- **Robotics/Automation.** Automation in agriculture can produce vital agricultural goods more rapidly and with less manual effort. Drones, modified tractors, irrigation systems, harvesters, and other advanced technology enable agricultural automation (*Khouloud Hwerbi et al., 2022*). Robots can efficiently complete farming tasks, enabling monitoring and managing natural resources, such as air and water (*OECD, 2021*). They give producers more control over the growing, processing, distribution, and storage of plants and animals, which improves productivity, lowers costs, leads to safer food production and growing environments, and has a more negligible negative environmental and ecological impact.
- **Use Cases of AgTech.** Exclusive use cases of AgTech (*Катерина Зверева 2022*) are Agrobot, the use of robotics, autonomous tractors, and computer vision. Examples include Agrobot picking strawberries from mobile apps while adhering to farmers' requirements, Abundant Robotics harvesting apples with a vacuum, autonomous tractors conducting driverless control, and computer vision-based pesticide sowing and sprinkling (*Biswaranjan Acharya 2022*).

The benefits are greater crop yield and lower water, fertilizer, and pesticide use, which contribute to lower food prices. Other benefits of AgTech include less impact on natural ecosystems, less chemical runoff into rivers and groundwater, and increased worker safety.

Smart Agriculture

Smart agriculture has been adopted due to the rapid growth, development, and fusion of modern technologies with traditional agriculture. Agriculture using IoT, Big Data, GPS, cloud computing, AI, etc., is known as smart agriculture (*Dhanaraju, 2022*). Smart agriculture enhances the caliber and efficacy of farming by utilizing low-cost sensors (such as mechanical, optical, location, and airflow sensors). Amart agriculture applications (*Othmane Friha et al., 2021*) are smart monitoring, water management, pesticide applications, disease control, smart harvesting/reaping, supply chain management, and smart agricultural practices. Innovative agribusiness aims to establish farm management applications, including agricultural weather stations, autonomous irrigation, monitoring (field equipment, animal, and soil), greenhouse automation, insect control, water management, and precision farming (*Reman et al., 2022*). According to the view of Agriculture 5.0 (*Zambon 2019*), farms follow precision agriculture principles and employ machinery that includes unmanned activities and self-directed decision support systems. As a result, Agriculture 5.0 recommends using AI and robots.

Smart Farming. Smart farming (*Sciforce, 2023*) is the management of farms utilizing contemporary ICTs to improve essential human labour while increasing crop quantity and quality. Sensors (for managing soil, water, light, humidity, and temperature), software (specialized software solutions that target specific farm applications), connectivity (cellular and LoRa), location (GPS and satellite), robotics (autonomous tractors and processing facilities), and data analytics are the technologies available to farmers today (analytics solutions, data pipelines for downstream solutions).

As shown in *Figure 1*, smart farming's driving force is IoT sensors/devices that connect machines and sensors integrated into farms to make farming processes data-driven and automated. Farmers do not need to go into the field to monitor field conditions and make strategic decisions for the entire farm or a single plant. The data lies at the heart of IoT-enabled smart farming. Farmers can collect data from objects or sensors and send it over the Internet to AgriCloud. IoT apps installed on a farm can gather and process data in a looped process to improve farming and enable farmers to react quickly to new environmental issues and fluctuations. IoT may improve many facets of agriculture, including grain production and forestry. Precision farming and farming automation can revolutionize agriculture with IoT. The cycle of smart farming is recurring and resembles observation, diagnosis, decisions, and action.

Figure 1. Smart farming model

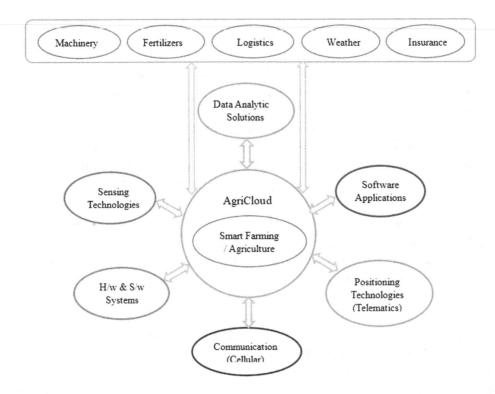

The advantages of smart farming are extensive farmland, autonomous equipment adjustment, and farmland analysis (*Chandan Gaur, 2022*). Farm farming equipment can be adjusted automatically by connecting data about crops, weather, and equipment to modify the temperature, humidity, etc. In large farmlands, an IoT-equipped drone aids in receiving crop condition updates and transmitting real-time field images. Farmers can learn about the state of their fields and crops by studying farmland and utilizing its problems.

Smart Agriculture Applications

Applications for smart agriculture provide promise for enhancing traditional agriculture in underdeveloped areas. Farmers use these applications to gather information on agricultural activities and monitor many aspects of contemporary farms. Some smart agriculture applications (*Utku Kose et al., 2023*) are precision farming, agricultural drones, hopping systems, monitoring livestock and weather conditions, smart greenhouses, AI, and IoT-based computer imagery, etc., are

- **Agricultural Drones.** Drones are employed in agriculture (*Gopal Dutta & Purba Goswami, 2020*) for various functions, such as irrigation, crop monitoring, crop spraying, planting, soil and field studies, and crop health evaluation.
- **Automatic Hydration System.** Farm fields' water usage is effectively controlled by an autonomous hydration system, ensuring the conservation of water resources (*Vunnava et al., 2021*). The

hydration system's sensors identify moisture, temperature, and humidity characteristics and notify farmers to distribute water as needed. Through the use of the Apps, the farmers can remotely activate the pumping motor and begin watering till the moisture level meets the expected range that is defined.

- **Precision Agriculture/Farming.** It introduces IoT-based tactics that raise farming's precision and control (*Rehman et al., 2022*). Farmers can make pesticides and fertilizers more effective or only apply them when necessary.
- **Precision Livestock Farming.** Utilizing smart farming methods, farmers may better track the nutritional requirements of individual animals and change their diets accordingly, lowering illness and boosting herd health (*Rehman et al., 2022*). They can stem the spread of disease by identifying sick animals and separating them from the herd using this knowledge.
- **Smart Greenhouses.** Smart greenhouses may intelligently maintain and monitor the climate without human intervention (*Rayhana et al., 2020*). For subsequent processing and control with the least amount of manual effort, a cloud-based platform stores the data obtained from various sensors in accordance with the needs of the crop.
- **Vertical or Urban Farming.** Vertical farming harvests food in stages and piles it vertically (*Luis Villazon, 2022*). It produces fresher food more rapidly and affordably. Farmers can use their available land more by planting crops that thrive there. With the help of IoT sensors and AI, this method could result in significant agricultural output in a city environment (*Siregar et al., 2022*).

Internet of Things (IoT)

Sensor-based agriculture monitoring systems (*Mathurkar 2014*) have limitations in identifying (or measuring) vegetables from agricultural areas due to conventional color transformation methods. Modern agriculture is made possible by IoT technology, and the agriculture industry is advancing to more significant insights and innovations. The IoT-enabled Agricultural (IoTAg) monitoring system with numerous algorithms, including detection, quantification, ripeness testing, and detection of diseased crops, was proposed by *Kazy Noor-e-Alam Siddiquee et al. (2022)* to get around these drawbacks. IoTAg monitoring (*Quy et al., 2022*) is the fastest-growing technology in the smart and connected agriculture industry.

As shown in *Figure 2*, IoT devices use data from data-gathering sensors to automate and manage farming tasks. The effects of this control could streamline the supply chains and food development at every stage of the farm-to-table process. The IoTAg comprises geofencing, smart greenhouses, livestock tracking, and predictive analytics. As a result, crops produced without pesticides with IoT support yield more for farmers. IoT is expected to help growers and farmers cut waste and increase output by maximizing the use of fertilizers to increase plant efficiency. It allows farmers more control over their animals, crop growth, and cost and resource reduction.

Figure 2. An IoT model of smart agriculture

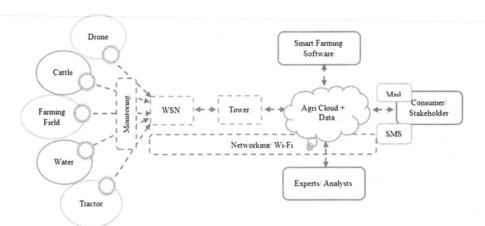

The potential of these devices is significant with the development of 4G/5G wireless communication. Additionally, SCM systems enabled by AI and IoT provide cutting-edge control and visibility into agricultural processes. IoT assistance continuously monitors the land to give farmers valid information. Crop status can be investigated using IoT devices and sensors. IoT tools can track climate change, manage water resources, monitor land use, boost productivity, monitor crops, manage insecticide and pesticide use, manage soil, find plant illnesses, boost crop sales, etc. The rapid development of IoT has enhanced agriculture in several ways, including crop field monitoring, weather forecasting, field control, and other uses.

Machine Learning

The growth of data in agricultural processes has given AI the data it needs to operate and help farmers make better decisions. By enabling farmers to track their crops in real-time and the data produced by IoT, AI technologies can increase the efficiency of the farming process. Different AI techniques (like ANN, CNN, SVM, Decision Tree, Random Forest, etc.) have been utilized in various fields of smart agriculture. For example, smart agriculture is predicted and monitored using image processing with supervised and unsupervised learning techniques (*Supriya Ghavate & Joshi 2021*). With the help of AI and Blockchain, it is possible to construct and run smart agriculture without a centralized authority to provide insights that increase yield and improve performance, ushering in a new era of food efficiency.

AI technology includes ML and DL. ML is a collection of several algorithms that improve the collected data and aid in finding hidden structures (*Sarker, 2021*). Depending on the type of data, task, or issue, it offers a variety of learning methodologies (such as supervised/labeled, unsupervised/unlabelled, semi-supervised, and reinforcement learning). AI applications (Indiana Lee 2022) in agriculture include food recall software, crop visualization, and improvement tools, Blockchain and analytics, and Genomics.

Figure 3 shows a specific ML approach. A learning process is used in the ML technique to complete a task by learning from training datasets (experience). In ML, datasets are collections of examples. After learning, the trained model can categorize, forecast, or fresh cluster instances (testing data) based on the knowledge gained.

Figure 3. A typical machine learning approach

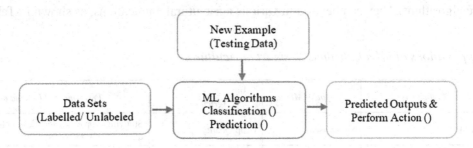

ML algorithms can be used in smart farming to increase crops' productivity, improve crop quality, and reduce the risk of crop damage. Examples of ML algorithms are CNN, circular hough transformation (CHT), color thresholding, and segmentation methods. They can detect, quantify, check ripeness, detect infected crops, and recommend a suitable crop. ML algorithms rely on real-time data to deliver exponential gains for farmers. As shown in Table 2, AI and ML are strong catalysts driving the security of remote facilities, better yields, and pesticide effectiveness.

Table 2. Applications of ML in smart agriculture

S/N	Type of Applications	Goal(s)	Challenges	ML Solution
1	Aerial survey and imaging	To help farmers monitor crops and herds. To boost the precision and efficiency of pesticide spraying.	• to conduct a land survey and monitor livestock and crops	CV & IM
2	Automatic Harvesting Robots	To increase the production rate	• Better ways to gather, sort, and transport products to other locations, • rough picking and handling.	Robot & ML
3	Automatic Weeding	To increase yields at the level of each plant	• Weeding manually can be time-consuming and physically demanding.	ML
4	Crop Management	To produce higher yields	• Higher temperatures, more frequent droughts, cycles of drying, and unexpected wetting.	ML
5	Field Conditions Management	To generate examples of pest management choices, irrigation plans, and fertilization rates based on the prediction.	• Soil and water management	ML
6	Insect detection	To prevent crop lost	• To detect insects in farming	AI, CV, ML & Drones
7	Livestock Management	• Animal welfare assessment • Predicting modeling of animal production • Estimate the environmental impact of livestock operations.	• Help detect and keep track of most livestock health issues, from eating to fertility.	Sensors/ RFID & ML
8	Precision Spraying	To stop the spread of illnesses and pests.	• Precision spraying/replacing farmers in spraying	AI, ML & UAV
9	Price Forecasting for Crops	To predict the priceless assets so that better economic decisions can be made.	• To forecast the prices of crops using climate, government policies, demand, and other factors.	ML
10	Produce Grading and Sorting	It has automated sorting and grading with precision and speed that surpass even those of a skilled expert.	• To distinguish "excellent" produce from the flawed or unattractive.	AI & CV
11	Yield Mapping	• Planning the harvest • Determining what each field can produce	• To draw attention to the soil in diverse agricultural areas • To provide information on moisture content • To allow the farmer to deal with different difficulties on the farm.	ML

CV- Computer Vision, IM - Image Processing.

Various ML models are ANNs, decision trees, support vector machines (SVM), Bayesian networks, and genetic algorithms. They can be used in various agricultural applications, as shown in Table 3.

Table 3. Applications of ML algorithms in smart agriculture

S/N	Algorithms	Type of ML	Application of Smart Agriculture	Performance/ Parameters
1	SVM	Supervised	Crop Quality	Studying crop characteristics, including soil quality, pH, texture, etc.
2	SVM, K-Means, & NN	Supervised	Disease Detection	Detect different types of diseases
3	ANN	Supervised	Field Conditions	Improves the field's usage of irrigation, maintaining temperature and weather occurrences.
4	ANN & SVM	Supervised	Greenhouse Simulation	Calculate the temperature of air, soil, and plants.
5	Deep CNN	Supervised	Plant Seedling Classification	A method for classification and detection that is effective
6	CNN, ANNs, SVM,	Supervised	Species Breeding	High yield, area/field, and climatic conditions
7	CNN, RF, SVM	Supervised & Unsupervised	Species recognition	Consider physical factors - vein morphology, shape, size, and color.
8	NN, ANNs, SVM & DL	Supervised	Species Selection	"
9	ANNs, SVM, DNN & CV	Supervised	Weed Detection	Enhance the detection and removal of weeds
10	ANNs, SVM, M5-Prime RT, kNN, ANFIS, DM, & GP	Supervised	Yield Prediction & Crop Harvest	Weather and climatic variables, including rainfall, temperature, and humidity, as well as soil variables like pH, soil type, and soil fertility

NN - Neural Networks, RT - Regression Trees, kNN - k-Nearest Neighbor, DM - Data Mining,
GP – Genetic Programming, CV – Computer Vision, ANFIS - Adaptive Neuro-fuzzy Interference System

Using ML and data analytics, agricultural areas can make decisions and suggest productive crops. They can improve the accuracy and predictability of smart agriculture.

ML-Based IoT (MIoT) in Smart Agriculture

MIoT-enabled smart farming (active farming) can enhance agricultural operations from seeding to crop harvesting. It increases production and utilizes productive soil, water availability, mineral deficiency, pest management, crop development, and horticulture. In MIoT-enabled smart farming, the remote sensors (such as temperature, humidity, soil moisture, water level, and pH) show accuracy and practical agriculture to deal with field issues (*Sivakumar et al., 2016*). Examples of MIoT applications are an AI-powered drone for field monitoring, an automatic crop watering system, embedded sensors in the field to monitor temperature and humidity, etc. The development of innovative farming methods is gradually increasing crop output, increasing profitability, and decreasing irrigation waste.

The MIoT-enabled smart farming, as depicted in *Figure 4*, consists of remote sensors placed in an agricultural field that transmit data to the Cloud. ML is applied to data collected from the field to forecast outcomes accurately. The decision-making process in agriculture is aided by the findings of the decision tree algorithm, which are provided to farmers through email alert (for example, water supply in advance).

Figure 4. ML-based IoT in smart agriculture

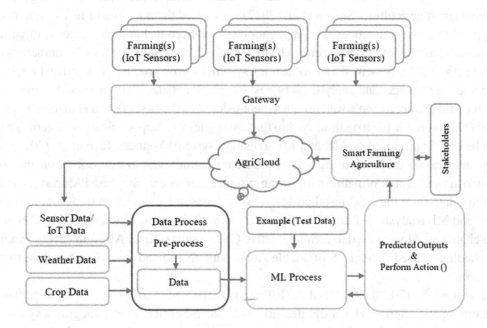

MIoT data analytics in the agricultural sector can boost the amount and quality of crop output to satisfy the rising need for food (*Khan Rijwan et al., 2021*). The advantages are automated business processes, waste reduction, visibility of the supply chain, and security and safety.

Benefits. The benefits of MIoT in smart farming are automation, reduced waste, visibility to the supply chain, and a secure ecosystem. MIoT automates the daily task that occurs in agricultural businesses. It improves the operational efficiency of businesses by reducing waste. MIoT provides more scalability to the business, predicts conflicts that might occur, and warns to manage accordingly. Agriculture industries may control and anticipate risk elements with the support of the safe ecosystem that MIoT offers.

MAIN FOCUS OF THE CHAPTER

Agriculture provides a vital food source for humans with various practical farming techniques for multiple crops. Advanced technologies offer the opportunity to monitor the agricultural environment and ensure the production of high-quality goods. Most studies forfeit sensory data to measure different soil properties, weather, irrigation, and crop properties in smart agriculture. However, due to the fragmentation of agricultural processes, there is still a need for additional research and development in smart agriculture. The perceptions, themes, tools, and developing areas of IoT-based farming, such as Edge of Things, cloud-based farming, IoT-based Greenhouse, mobile and sustainable farming, and big data analytics in agriculture, were covered thoroughly.

Establishing a thriving, sympathetic network for agricultural management is the goal of smart agribusiness research. Samuel et al. (2020) surveyed the different technology methods for crop selection, crop sowing, weed detection, and system monitoring, yielding a productive output using advanced technologies. Tefera et al. (2020) focused on emerging different automation practices as part of the industry's technological evolution. Othmane Friha et al. (2021) offered a complete analysis of evolving technologies for IoT-based smart agriculture. Shaikh et al. (2022) showcased the potential of ICT in traditional agriculture applied to farming practices. The application of advanced technologies successfully transforms conventional agriculture into smart farming (ByM Thilagu & Jayasudha, 2022) with numerous benefits.

Abraham et al. (2022) presented a smart farming system for crop production based on inexpensive IoT sensors, cloud data storage, and analytics services. Kasara Sai et al. (2020) proposed a smart irrigation system that could predict a crop's water requirement using an ML. Hetal Patela et al. (2021) proposed a model and framework for the irrigation system that automates the crop's irrigation system. Thanwamas & Khongdet (2022) proposed an ML and IoT irrigation system. Youness Tace et al. (2022) proposed intelligent and flexible irrigation using ML for smart agriculture with low consumption and cost.

The smart agricultural information monitoring system developed by Saqib and Ahmad (2021) collects the necessary data from the field-installed IoT sensors, measures it, and then transmits it to the Cloud for storage and ML analysis. IoT and CSA data analysis were combined by Rania Ahmed et al. (2022), and Blockchain and AI were used to create effective CSA systems. Eissa Alreshid (2109) examined the IoT/AI technologies used in Smart Sustainable Agriculture (SSA) and IoT/AI architecture to create an SSA platform.

Takudzwa Fadziso (2018) introduced the IoT and a network of Web-connected intelligent devices to enhance farming and improve yields significantly. Farm-as-a-Service (FaaS) integrated system (Kim et al. 2018) enabled high-level application services by managing connected devices, data, and models and operating and monitoring farms. Managing heterogeneous information and data (Balducci et al., 2018) from actual datasets gathers physical, biological, and sensory values. Shi et al. (2019) understood the system architecture and critical technologies for contemporary IoT applications in protected agriculture.

Vadlamudi (2020) examined the functions of IoT in agriculture, traditional and smart agriculture, and recent breakthroughs in IoT. Aman Kumar Dewangan (2020) designed a farm productivity management system with several agriculture IoT devices and a powerful dashboard with analytical capabilities and in-built accounting/reporting features. Vadlamudi (2020) examined recent IoT advancements, contrasted conventional and smart agriculture, and examined the functions of IoT in agriculture.

The domains of AI approaches and datasets utilized in IoT-based prediction in precision agriculture were examined by Fatima et al. (2021). IoT-based Smart Farming (Archana Gupta et al. 2021) enhances the agricultural system by continuously monitoring the field. It provides accurate real-time observation while considering various elements, including humidity, temperature, soil, etc. Alo Sen et al. (2021) concentrated on various case studies in agriculture, such as climate conditions, greenhouse automation, crop management, livestock monitoring, and smart farming management.

Ankita Arun Jagadale (2022) explained how IoT and AI transformed the agricultural industry. The agricultural revolution is due to technological advancements in AI and IoT-based applications (such as drones, UAVs, etc.). Prajwal & Guru (2022) work was about improving current agricultural practices using modern technologies for better yield. Rifat et al. (2022) examined how IoT architectures aid farm fields.

Reddy et al. (2020) introduced a smart irrigation system that predicts a crop's water requirement using an ML with IoT data such as moisture, temperature, and humidity. Tajudin and Shahina Begam (2020) presented an intelligent crop monitoring scheme (ICMS) to support farmers to do agriculture without any complexities. Supriya Ghavate & Joshi H (2020) aimed to use evolving technology for automation.

Jyotir and Vishal (2021) discussed the issues facing the ag industry that MIoT can help with and looked at how these technologies affect the industry. Verma Parul and Kumar Umesh (2021) used ML algorithms to create the IoT architecture for crop predictions. An IoT-enabled ML-trained recommendation system for optimum water use with minimal farmer interaction was proposed by Bhoi et al. (2021). SVM, KNN, and probabilistic neural network classifiers were utilized by Phasinam & Kassanuk (2021) to categorize real-time photos of smart agriculture. Using data analytics and the MIoT system, Ravesa & Shabir (2021) created a prediction model for Apple disease in Kashmir Valley. Palanivel Kuppusmay et al. (2021) studied state-of-the-art technologies and proposed an intelligent agriculture model.

Aliar et al. (2022) thoroughly discussed the structures and strategies used in smart farming and examined various ideas and acceptable solutions. Khongdet & Thanwamas (2022) established a system for real-time image classification in agriculture using IoT-enabled cameras and an ML strategy. The strength and potential of computer tools, such as IoT, wireless sensor networks (WSN), data analytics, and ML in agriculture, were climaxed by Ravesa Akhter & Shabir Ahmad Sofi (2022). In the IoT ecosystem, they projected the Apple disease prediction model in the Apple plantations in Kashmir Valley using data analytics and ML. Table 4 shows the survey of MIoT in smart agriculture applications.

Table 4. Survey of MIoT in smart agriculture applications

S/N	Article/Papers	Applications	App./Model/ Architecture	Technologies/ Techniques	IoT Sensors
1	Reddy et al. (2020)	Smart Irrigation System	Application	MIoT	• Moisture • Temperature & • Humidity.
2	Tajudin & Shahina Begam (2020)	Intelligent Crop Monitoring Scheme (ICMS)	Application	MIoT	High yield
3	Supriya Ghavate & Joshi H (2020)	Farm Yield & Water Management	Application	ML & IoT	Enhance the quality
4	Jyotir & Vishal (2021)	Agro-industry	Application	Technology	Challenges
5	Verma Parul & Kumar Umesh (2021)	Crop Predictions	IoT Architecture	ML & IoT	Prediction
6	Bhoi et al. (2021	Water Management	Recommender System	MIoT	Efficient Water Usage
7	Phasinam & Kassanuk (2021	Real-time Image Processing & Smart Agriculture	Application	SVM, KNN & probabilistic neural network	Real-time operations
8	Ravesa & Shabir (2021)	Data Analytics/ prediction model	Application	MIoT	Case Study Apple disease
9	Sivakumar et al. (2021)	Study of Modern Technologies	Review	ML & IoT	Problems and solutions to smart agriculture
10	Sasi Bhanu et al. (2021)	Handle Water, Environmental hazards, and Fertilizers selection	Review on embedded systems, sensors-based systems	ML & IoT	Automation
11	Aliar et al. (2022)	Smart Agriculture	Designs & Recommendations	Technology	Farming Complexities
12	Khongdet & Thanwamas (2022)	Real-time Image Classification	Framework	MIoT	IoT enabled Camera
13	Ravesa Akhter & Shabir Ahmad Sofi (2022)	Prediction Model	Application	Power of Technologies – IoT with WSN, Analytics, ML	Agriculture Computing

Case Study. IoT infrastructures gather vast data from remote agricultural areas that can be used for ML-based crop predictions. The IoT-based model (Bakthavatchalam et al., 2022) predicts high-yield crops and precision farming. Quy et al. (2022) surveyed many innovative agriculture systems (such as OnFarm, Farmobile, Silent Herdsman Platform, CropX, FormX, Easyfarm, KAA, Farmlogs, etc.) that had been developed and commercialized to raise productivity, diminish human labor, and improve farming efficiency.

Technology scalability should be fostered without compromising the functionality of the current infrastructure (Rifat et al., 2022). The literature analysis did not include agricultural IoT-based services like Software-Defined Networking (SDN)/Network Functions Virtualization (NFV)and fog computing, as well as emerging applications like Agri-voltaic systems and blockchain-based applications.

Issues, Controversies, Problems

From the literature, the main objective of MIoT is to provide complete automation in a smart agriculture that facilitates intelligence through more innovative devices deployed in the fields. ML enables IoT-enabled systems to mimic human-like results, subsequently training from the data and improving their sympathy for the surroundings.

The challenges and issues related to processing IoT data to AgriCloud are data breaches, network latency and failure, power consumption, and a vast volume of data generation.

a. *Data Breaches*. There is an opportunity for a data breach at AgriCloud.
b. *Network Latency*. There is a delay between IoT data generation and processing to the AgriCloud.
c. *Network Failure*. For instance, the network is down, and the digital twin cannot reach the IoT sensors.
d. *Power Consumption*. The energy supplies and battery capacity describe when and how often to send/receive data from the Cloud.
e. *Volume of Data*. It is directly related to the number of sensors producing data.

Proposing a MIoT model for smart farming, MIoT applications in smart industry drives beyond predictable Agribusiness, food supply chain, disease detection, and fertilizer control. These MIoT devices use real-time data to sense patterns and raise an alert in case of uncommon actions. These alarms and alerts can be initiated without human intervention in smart farming.

RESULTS AND DISCUSSIONS

In the modern agricultural era, data produced by MIoT is used to facilitate all interactions between the farmer and the farm process. This section discusses the proposed MIoT model for smart farming.

Smart Farming Operations

According to Alberto Rizzoli (2023), smart farming operations are survey & imaging, automatic weeding, monitoring, disease detection, grading & sorting, and spraying.

- **Survey and Imaging.** Surveying land, keep an eye on crops and livestock. AI, Robotics, image processing, and Robotics with computer vision can be used for surveying, AI can monitor yields and herds, and image processing can help boost the precision and efficiency of pesticide spraying.
- **Smart irrigation.** Smart irrigation uses weather and soil moisture data to determine the irrigation need of the landscape. It can be sprinkler, surface, drip, sub-irrigation, and manual irrigation. Smart irrigation maximizes irrigation efficiency by reducing water waste while maintaining plant health and quality. It can use IoT sensors that calculate the field's special status, including soil temperature, water requirement, weather, etc.
- **Insect and Plant Disease Detection.** Detection of plant diseases and pests is an important activity for better yield. Image processing - classification, detection, and image segmentation methods can be used for plant health.
- **Intelligent Spraying.** Spraying pesticides or fertilizer uniformly across a field can increase the efficiency of agricultural operations. Drones/UAVs and CV techniques can be used in spraying. Pesticides or fertilizer can be evenly sprayed across a field using drones or UAVs in terms of the area and quantity to be sprayed with extreme precision. As a result, there is a considerably lower chance of poisoning water supplies, crops, people, and animals. It can prevent collateral damage to crops or the environment.
- **Automatic Weeding.** Removing weeds physically saves the farmer significant work, moderates the herbicide requirement, and thus makes the whole farming operation much more environmentally friendly and sustainable. Robots with computer vision can spot weeds and eliminate unwanted plants, and AI can find and remove them.
- **Monitoring.** It includes crop, soil, and livestock health.
  a. *Soil Monitoring*. Micro and macronutrients in the soil are critical factors for crop health, yield quantity, and quality. Soil monitoring is essential for the growth and optimization the production.
  b. *Crop Monitoring*. It is essential to track crop health, make accurate yield predictions, and detect crop malnutrition much faster than humans. Drones (or UAVs) can capture aerial image data and train computer vision models, and AI can analyze and interpret this data.
  c. *Animal/Livestock Monitoring*. Determining whether livestock are properly drinking, eating, sleeping, or doing something odd that may indicate disease or behavioral problems is essential. Overhead smart cameras can monitor cattle health and notify the farmers remotely, and CV can count animals, detect infection, identify unusual behavior, and observe significant activities such as giving birth.
- **Grading Process.** After harvesting, the grading process with higher accuracy rates and speed is essential for agricultural operations. A CV can automate the sorting and grading process more accurately than a trained professional.

The above operations assist farmers in understanding critical elements like water, topography, aspect, vegetation, and soil types, as well as in deciding how to utilize limited resources in their production environment and manage them environmentally and sustainably.

Smart Farming Model

Smart farming incorporates technology and sensors in farms to make farming activities data-driven and automated. Farmers can observe field settings and draw tactical decisions for the entire farm without ever setting foot in the field. Smart farming operations collect data from various sensors, enabling more accurate and fast decision-making. IoT devices on a farm can gather and process data in a loop to improve farming and allow farmers to react quickly to new issues and changes in the environment. Farmers can collect data from objects or sensors and send it over the Internet to AgriCloud.

Modern technologies manage smart farming (*Sciforce, 2023*) to boost production while minimizing the need for laborers. Connectivity (cellular and LoRa), data analytics (analytics solutions and data pipelines), location (GPS and satellite), robotics (autonomous tractors and processing facilities), sensors (soil, water, light, humidity, and temperature management), and software are the technologies that are currently available to farmers (specialized software solutions that target specific farm applications). With the help of these technologies, farmers may access information about their crops and their state anywhere, at any time. IoT and other technologies are shown in *Figure 5* as an example of a smart farm.

Figure 5. Smart farming model using MIoT

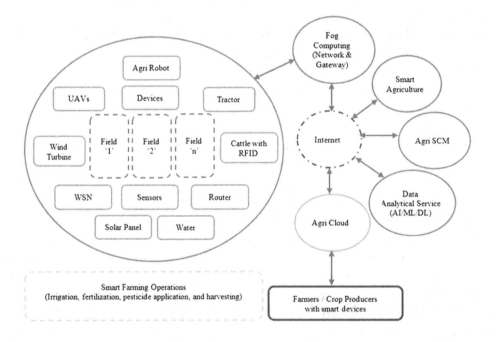

Smart farming comprises sensors and actuators, robots, tractors, RFID, and UAVs to accomplish sensing and control activities. The data collected from the intelligent agricultural devices are forwarded to field gateways via network technologies and then stored in the AgriCloud. Various services and applications can access the cloud data and perform agricultural activities with negligible human communication. Remote monitoring technologies provide farmers access to the farmlands. Hence, farmers can remotely monitor their farms while gathering information on the environmental conditions that affect plants, such as soil fertility, humidity, temperature, and light intensity.

Requirements. *Figure* 5 considers the non-functional requirements (such as accessibility, reactivity, reconfigurability, and extensibility) and functional requirements (such as users' accessibility, processing data, visualizing the data, storing the data, generating alerts) that form the critical features of MIoT architecture model for smart farming.

Data Collection. The smart and interactive agriculture platform offers combined information on traditional agricultural approaches, methods, implements, crop pests, diseases, etc., composed from multiple sources for sustainable agriculture. It allows easy access to the data by users through multiple devices. The data relating to innovative agricultural operations (such as state, telemetry, commands, operational data, metadata, and weather conditions) can be gathered from the smart farm field and forwarded MIoT model for analytics.

- A *state data* can describe the device's current status that can be read/written.
- *Telemetry* data about the environment is collected through sensors.
- *Command* information consists of actions performed by a smart device. The *operational information* includes operating temperature.
- *Metadata* includes device, type, model, date of manufacture, hardware number, etc.
- *Weather data* include temperature, precipitation, humidity, the demand for fertilizers, water usage, etc.

Working Principle. The working principle of MIoT would typically look as follows:

1. IoT sensors integrated with the equipment can measure discrete variables like temperature, vibration, noise, and heat. Then, this data is uploaded to the Agri Cloud for analytics.
2. The ML model is kept on a cloud platform and is fed new data as it comes in.
   a) The processed data is divided into training and verification data by the ML model.
   b) The ML model examines many anomalies, correlations, and projections and then formulates a hypothesis.
   c) The hypothesis must be put to testing and verification.
3. After verification, this is published as an executable endpoint.
4. The trained model may be fed the live streaming data, and based on what it already knows and has been trained to look for, it can infer the condition or health of the machinery.

Figure 6 shows that smart farming follows a cycle of observation, diagnostics, decisions, and actions.

Figure 6. Smart farming life cycle

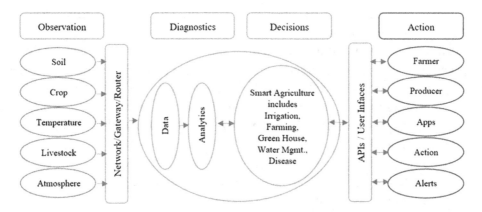

i.   **Observation.** Sensors capture observational data from the atmosphere, soil, livestock, and crops.
ii. **Diagnostics.** The sensor data is supplied into an AgriCloud/hosted IoT platform with established decision rules and models (also known as business logic), which determine the status of the investigated object and pinpoint any shortcomings or demands.
iii. **Decisions.** Following the discovery of problems, the IoT platform's user- and/or ML-driven components decide whether and, if so, which location-specific treatments are required.
iv. **Action.** Following end-user assessment and action, the cycle starts over at the beginning.

MIoT Architecture Model

The MIoT architectural model for smart farming is shown in *Figure 7*. The IoT from smart farming produces large amounts of data from ML. ML creates models that help anticipate future behavior and events by using patterns in past behavior to find and detect trends. Integrating MIoT unlocks insights from otherwise concealed data, enabling quicker automated reactions and better decision-making.

Figure 7. MIoT architectural model for smart farming

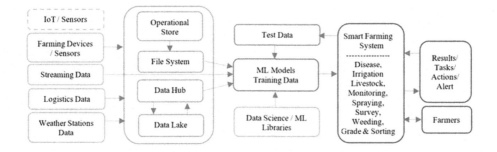

The MIoT platform offers simple access to data kept in operational and historical data storage for training models. This data can be periodically retrieved and sent through an automated pipeline to be transformed and used to train an ML model. Data can be stored locally and retrieved using IoT DataHub connectors or hosted on services like Amazon S3, Microsoft Azure, or Data Lake Storage. Various ML Libraries are Keras, PyTorch, Amazon SageMaker, Tensor Flow, Learn, and R.

Architecture Model. The proposed architecture model includes the sensor & network layer, data storage & processing, analysis, and application layer.

Sensor and Network Layer. This layer gathers and saves large amounts of data about the weather, crop trends, soil quality, harvesting, and satellite photos. The AgriCloud stores the farm's data, making it easily accessible. The saved information can be used to identify a solution and avert significant losses swiftly. The sensor & networking layer includes sensors/actuators, sensor networks (WSNs), agricultural robots/ UAVs/Drones, driverless tractors, and Radio Frequency Identification (RFID) tags to perform sensing and control actions. This layer includes all crop sensors and smart objects for data collection and monitoring.

Sensors perceive and respond to specific inputs, such as illumination, locomotion, pressure, heat, or moisture, and transform them into a representation or signals humans can read for further reading and processing. Various types of sensors used in agriculture enable the need for smart agriculture incorporation. The critical sensors in smart AgriTech are optical, electrochemical, mechanical soil, dielectric soil moisture, location (or agricultural weather stations), electronic, airflow, IoT, GPS, temperature, asset monitoring, accelerometer sensor, and intelligent camera. Sensors can be placed underground, on crops, or UAVs. Underground sensors refer to measurements of moisture, pH, and soil chemical properties such as Sulphur. UAV sensors measure environmental parameters such as humidity, temperature, wind speed, luminosity, or solar radiation. Drones with vision imaging cameras can detect heat from almost all objects and materials, turning them into images and video.

All networking and routing technologies have been established among sensors for data exchange. The smart farming platform uses WSNs to deploy efficient crop, field management, monitoring, and optimization of crop quality. The constant monitoring of environmental parameters helps the grower supervise and maintain optimal conditions to achieve maximum productivity with remarkable energy savings. The sensor and network layer forwards the cultivated field sensory data to the data analytical layer.

Data Storage & Processing Layer. The backbone of ML/predictive analysis is data storage. All the data in cloud-based AgriTech systems is always accessible and may be viewed through gadgets or devices. Additionally, the data storage makes it available to other tools and solutions for additional analysis. The job of this layer is to collect the raw data from the devices or other external services, curate, harmonize, and aggregate it so that it can be presented as context information or given to upstream data processing algorithms or analytics. Agricultural equipment or open-source geo-services are only two examples of other data sources this layer may collect. The data storage & processing layer utilizes the information/ data provided by the sensor & network layer. Various IoT-based protocols perform different agricultural actions with a negligible human interface.

Analytical Layer. Data analysis is the crucial component of intelligent agroecosystems that guarantee efficient pesticide application and disease prevention. The multiple types of data (such as meteorological data, market data, farm data, GIS, and water) available related farms can enable all increased levels of decision-making in smart agriculture. The ideal seed, water, and pesticide requirements for a farm are identified after rigorous examination of all existing data and present-day data formats. Intelligent agriculture systems have an alert mechanism that activates when crop growth irregularities are found. Because of this, it works well when a pest infestation occurs and gives farmers valuable data.

The analytical layer includes the handling and analysis of the collected data. This layer employs efficient data handling and mining techniques to generate accurate predictions and support field operations, including efficient pesticide application, disease diagnostics, and irrigation control. From the data, this layer improves productivity, maximizes crop output, preserves crop quality, and conserves resources. The application layer provides a variety of technologies, such as cloud & fog computing, AI, and big data, which can enhance the system's overall AI/ML/DL capabilities.

Application Layer. The application layer completes the presentation, data processing, forecasting, and data gathering modules. Farmers can start sensor data collecting in a specific field or change existing data in the platform using the data collection module. Farmers can calculate orthophotographs and compute growth indices in the data management module, producing exact data for a field. The visualization module provides graphs and statistical results for each area. The prediction module uses data mining technologies to assess the gathered data properly. The application layer has all appropriate application module interfaces and APIs for implementing alerts on the cultivation process, statistical data visualization, disease and animal detection, fertilizer and water control, and disease and animal detection. It makes it possible for farmers to monitor and control their land efficiently.

Various data visualization techniques enable the simple display of the knowledge obtained through field monitoring, including charts, heatmaps, orthomosaics, and three-dimensional models. The farmer can examine the outcomes generated by the system's operations and take appropriate action. The core server database, housed in the Cloud, contains all system information and data. Farmers can generate statistics about the development of agricultural production based on the collected and processed data, preventing financial loss while enhancing crop quality and achieving a significant increase in production.

Various technology used in each layer is listed in Table 5.

Table 5. Technologies in each layer

S/N	Layers	Functions	Technology Used
1	Sensor & Network Layer	Data collection & routing	Radar, GPS, UAV/Drone, RFID, WSN, 3G/4G/5G, etc.
2	Data Storage & Processing Layer	Data pre-processing & storing	Data Lake and Data Warehouse.
3	Analytical Layer	Data Analysis	Data Mining, AI, ML & DL
4	Application Layer	Results, Alerts & Visualize	Visualization, Dashboard, and Apps.

Web applications can be used to implement intelligent agricultural applications. Applications of smart farming increase extremely transparent farming by focusing on significant agricultural processes, vertical farming, and organic agriculture. The dry farming approach, which includes specialized agricultural practices for non-irrigated production, can benefit from the intelligent farming monitoring system. Moreover, greenhouses can employ this architectural concept to monitor and manage the climate automatically, eliminating the necessity for personal intervention.

Benefits. The MIoT can help in organic & family farming, livestock management, conserving specific species and cultures, and preserving high-quality crops. The predicted outcome is a highly transparent method that supports ecological health as it accomplishes water management and optimizes input usage.

Challenges

The design of a smart farming monitoring system faces several challenges. Some of the challenges are listed below:

a. *Availability.* The sensors must be designed with energy-saving protocols for performance, among other precautions. Hence, they achieve efficient and constant operation during climate change and wildlife interventions.
b. *Interoperability.* With a high-speed WSN, the sensors and actuators from various manufacturers must be connected to and able to communicate.
c. *Computation Limits.* The WSN and IoT have limited computational capabilities.
d. *Routing.* While considering the limited resources, an effective routing strategy can give little delay and deliver efficient services in multiple sensor nodes.
e. *Cost.* This factor is considered during the implementation of networks in smart farming systems. The cost depends on the quality of the materials and the network.

FUTURE RESEARCH DIRECTIONS

Future agriculture systems will go online to improve crop security, crop output, risk mitigation, unmanned farming, greenhouse simulations, etc. Due to the increased scalability, effectiveness, and better resilience of modern approaches like cloud applications and edge intelligence, this research may eventually be expanded to include them. In the twenty-first century, the development of facility agriculture, greenhouse farming, contract farming, and the transition from rural to urban farming practices toward sustainable agriculture with renewable energy and data-driven management may be possible. As a future work, the XAI model (Masahiro Ryo, 2022) can be applied to trust and explainable AI to smart farming for better clarification.

CONCLUSION

Agriculture is considered central to the survival of human beings. Supporting the current practices of traditional agriculture with recent IoT/AI technologies can improve product performance, quality, and volume. This chapter established the importance of employing advanced computing technologies in the agricultural sector, particularly ML and IoT. This study reviewed the IoT/AI technologies discussed in smart agriculture. The significant contribution of this chapter concerns the MIoT technical architecture for smart agriculture. This architecture leads the research and development of a unified platform for smart agriculture to resolve issues in smart agriculture operations. The MIoT technologies for brilliant farming activities were suggested in this chapter. Farmers can collect and evaluate valuable data with the help of innovative farming applications. With intelligent technologies, they can improve agricultural products' sustainability and competitiveness.

This chapter aims to illustrate the IoT and ML-based technologies and applications currently used in agriculture applications. The Cloud of agricultural data, security options, weeding robots, weather forecasts, disaster, and irrigation management, and other IoT services offer an agriculture management system. It also makes crop monitoring, material handling, machine navigation, and harvesting robots easier. The taught skills of machine learning through neural networks, mathematical and statistical modeling for energy saving, and field and animal monitoring systems make IoT technology possible.

Data Availability

The chapter focussed on the qualitative research approach. It studied various datasets available from the corresponding author.

Conflicts of Interest

Conflict of interest is not applicable in this work.

REFERENCES

Зверева, К. (2022). Agribots are coming! How can Robots, Cobots, and Drones Contribute to Farming? (Part II). *East Fruit.* https://east-fruit.com/en/horticultural-business/blogs/agribots-are-coming-how-can-robots-cobots-and-drones-contribute-to-farming-part-ii/

Abraham, A., Dash, S., Rodrigues, J. J. P. C., Acharya, B., & Pani, S. K. (2022). A.I., Edge, and IoT-based Smart Agriculture. Londres (Royaume-Uni): Academic Press. doi:10.1016/C2020-0-00516-5

Abraham, G., & Nithya, M. (2021). Smart Agriculture Based on IoT and Machine Learning. *2021 5th International Conference on Computing Methodologies and Communication*, (pp. 414-419). IEEE. 10.1109/ICCMC51019.2021.9418392

Acharya, B., Garikapati, K. (2022). Internet of things (IoT) and Data Analytics in Smart Agriculture: Benefits and challenges. In A. Abraham, S. Dash, J. Rodrigues, B. Acharya, & S. Pani (eds.) Intelligent Data-Centric Systems, AI, Edge, and IoT-based Smart Agriculture, 3-16. Academic Press. doi:10.1016/B978-0-12-823694-9.00013-X

Akash, S. (2021). Edge Computing: Reshaping the Agricultural Sector with Smart Farming. *Analytical Insight.* https://www.analyticsinsight.net/edge-computing-reshaping-the-agricultural-sector-with-smart-farming/

Akhter, R. & Sofi, R. (2022). Precision agriculture using IoT data analytics and machine learning. *Journal of King Saud University - Computer and Information Sciences, 34* (8). . doi:10.1016/j.jksuci.2021.05.013

Aliar, A. (2022). A Comprehensive Analysis on IoT-based Smart Farming Solutions using Machine Learning Algorithms. *Bulletin of Electrical Engg., and Informatics, 11*(3), 1550-1557. . doi:10.11591/eei.v11i3.3310

Alreshid, E. (2019). Smart Sustainable Agriculture Solution Underpinned by Internet of Things and Artificial Intelligence. *International Journal of Advanced Computer Science and Applications, 10*(5), 93–102. doi:10.14569/IJACSA.2019.0100513

Ashish Choudhary, A. (2021). IoT-based Smart Agriculture Monitoring System. *Circuit Digest.* https://circuitdigest.com/microcontroller-projects/iot-based-smart-agriculture-moniotring-system

Awan, N., Khan, S., Rahmani, M., Tahir, M., Alturki, R., & Ullah, I. (2021). Machine Learning-Enabled Power Scheduling in IoT-Based Smart Cities, Computers, Materials & Continua. *Tech Science Press.* doi:10.32604/cmc.2021.014386

Bakthavatchalam, K., Karthik, B., Thiruvengadam, V., Muthal, S., Jose, D., Kotecha, K., & Varadarajan, V. (2022). IoT Framework for Measurement and Precision Agriculture: Predicting the Crop Using Machine Learning Algorithms. *Technologies, 10*(13), 13. doi:10.3390/technologies10010013

Balducci, F., Impedovo, D., & Pirlo, G. (2018). Machine Learning Applications on Agricultural Datasets for Smart Farm Enhancement. *Machines, 6*(3), 38. doi:10.3390/machines6030038

Bhoi, A., Nayak, R. P., Bhoi, S. K., Sethi, S., Panda, S. K., Sahoo, K. S., & Nayyar, A. (2021). IoT-IIRS: Internet of Things based intelligent-irrigation Recommendation System Using Machine Learning Approach for Efficient Water Usage. *PeerJ. Computer Science, 7*, 578. doi:10.7717/peerj-cs.578 PMID:34239972

Colombo-Mendoza, L. O., Paredes-Valverde, M. A., Salas-Zárate, M. P., & Valencia-García, R. (2022). Internet of Things-Driven Data Mining for Smart Crop Production Prediction in the Peasant Farming Domain. *Applied Sciences (Basel, Switzerland), 12*(4), 1940. doi:10.3390/app12041940

Columbus, L. (2023). 10 Ways AI has The Potential To Improve Agriculture In 2021. *Forbes.* https://www.forbes.com/sites/louiscolumbus/2021/02/17/10-ways-ai-has-the-potential-to-improve-agriculture-in-2021/?sh=68ca39aa7f3b

Dahane, A., Benameur, R., Kechar, B., & Benyamina, A. (2020). An IoT-Based Smart Farming System Using. *Machine Learning,* 1–6. doi:10.1109/ISNCC49221.2020.9297341

Dhanaraju, M., Chenniappan, P., Ramalingam, K., Pazhanivelan, S., & Kaliaperumal, R. (2022). Smart Farming: Internet of Things (IoT)-Based Sustainable Agriculture. *Agriculture, 12*(10), 1745. doi:10.3390/agriculture12101745

Dutta, G., & Goswami, P. (2020). Application of Drone in Agriculture: A Review. *International Journal of Chemical Studies, SP-8*(5), 181–187. doi:10.22271/chemi.2020.v8.i5d.10529

FAO. (2015). Climate Change and Food Security: Risks and Responses. Food and Agriculture Organization of the United Nations Rome.

FAO. (2017). *The future of food and agriculture – Trends and challenges.* Food and Agriculture Organization of the United Nations Rome.

Fatima, N., Memon, K. F., & Ahmed, J. (2021). Precision Agriculture using Internet of thing with Artificial Intelligence: A Systematic Literature Review. *University of Sindh Journal of Information and Communication Technology, 5*(2), 101-110. https://sujo.usindh.edu.pk/index.php/USJICT/article/view/2682

Friha, O., Ferrag, M. A., Shu, L., Maglaras, L., & Wang, X. (2021). Internet of Things for the Future of Smart Agriculture: A Comprehensive Survey of Emerging Technologies. In IEEE/CAA Journal of Automatica Sinica, 8(4), 718-752. doi:10.1109/JAS.2021.1003925

Gebeyehu, M. N. (2019). Remote Sensing and GIS Application in Agriculture and Natural Resource Management. *Int J Environ Sci Nat Res.*, *19*(2), 556009. doi:10.19080/IJESNR.2019.19.556009

Ghavate, S. & Joshi, H. (2021). Smart Farming using IoT and Machine Learning with Image Processing. *International Journal of Current Engineering and Technology*, 15-19.

Gupta, A., Nagda, D., Nikhare, P., & Sandbhor, A. (2021). Smart Crop Prediction using IoT and Machine Learning, *International Journal of Engineering Research & Technology (IJERT), 9*(3), 18-21.

Holzinger, A., Saranti, A., Angerschmid, A., Retzlaff, C. O., Gronauer, A., Pejakovic, V., Medel-Jimenez, F., Krexner, T., Gollob, C., & Stampfer, K. (2022). (2022). Digital Transformation in Smart Farm and Forest Operations Needs Human-Centered AI: Challenges and Future Directions. *Sensors (Basel)*, *22*(8), 3043. doi:10.339022083043 PMID:35459028

Hwerbi, K., Benalaya, N., Amdouni, I. (2022). A Survey on the Opportunities of Blockchain and UAVs in Agriculture. *IEEE 11th IFIP International Conference on Performance Evaluation and Modeling in Wireless and Wired Networks (PEMWN).* IEEE. 10.23919/PEMWN56085.2022.9963871

IndustryTap. (2022). *The Impact of IoT on The Agricultural Sector.* Industry Tap. https://www.industrytap.com/the-impact-of-iot-on-the-agricultural-sector/64411

Jyotir, M. & Vishal, J. (2021). Internet of Things and Machine Learning in Agriculture. *Internet of Things and Machine Learning.* Nova Publishers. doi:10.52305/MTXX5116

Kazy Noor-e-Alam Siddiquee, S. (2022). Development of Algorithms for an IoT-Based Smart Agriculture Monitoring System. Wireless Communications and Mobile Computing. doi:10.1155/2022/7372053

Kodali, R. Jain, V., & Karagwal, S. (2016). *IoT-based Smart Greenhouse.* IEEE. . doi:10.1109/R10-HTC.2016.7906846

Kuppusamy, P., Shanmugananthan, S., & Tomar, P. (2021). Emerging Technological Model to Sustainable Agriculture. In P. Tomar & G. Kaur (Eds.), Artificial Intelligence and IoT-Based Technologies for Sustainable Farming and Smart Agriculture (pp. 101-122). IGI Global. doi:10.4018/978-1-7998-1722-2.ch007

Lee, I. (2022). How the IoT and AI are impacting Agriculture. *Relevant Software.* https://relevant.software/blog/iot-ai-agriculture/

Livelihoods. (2021). Regenerative Agriculture: from the key principles to the practice. *Livelihoods.* https://livelihoods.eu/regenerative-agriculture-from-key-principles-to-practice/

Maduranga, M. W. P., & Abeysekera, R. (2020). Machine Learning Applications in IoT Based Agriculture and Smart Farming: A Review. *International Journal of Engineering Applied Sciences and Technology*, *4*(12), 24–27. doi:10.33564/IJEAST.2020.v04i12.004

Mathurkar, S. S., Patel, N. R., Lanjewar, R. B., & Somkuwar, R. S. (2014). Smart Sensors-based Monitoring System for Agriculture using Field Programmable Gate Array. *2014 International Conference on Circuits, Power and Computing Technologies*, (pp. 339-344). IEEE. 10.1109/ICCPCT.2014.7054914

Mekonnen, Y., Namuduri, S., Burton, L., Sarwat, A., & Bhansali, S. (2020). Review - Machine Learning Techniques in Wireless Sensor Network Based Precision AgricultureJ. *Journal of the Electrochemical Society*, *167*(3), 037522. doi:10.1149/2.0222003JES

Meshram, V., Patil, K., Meshram, V., Hanchate, D., & Ramkteke, S. D. (2021). S.D. Ramkteke, Machine Learning in Agriculture Domain: A state-of-art survey. *Artificial Intelligence in the Life Sciences*, *1*, 100010. doi:10.1016/j.ailsci.2021.100010

OECD. (2021). *Adoption of Technologies for Sustainable Farming Systems, Wageningen Workshop Proc.* Organisation for Economic Co-Operation, and Development.

Oteyo, I. N. (2020). Developing Smart Agriculture Applications: Experiences and Lessons Learnt. *African Conf. on Software Engineering*. Research Gate.

Phasinam, K., & Kassanuk, T. (2022). Machine Learning and Internet of Things (IoT) for Real-Time Image Classification in Smart Agriculture. *ECS Transactions*, *107*(1), 3305–3311. doi:10.1149/10701.3305ecst

Prajwal, V., & Guru, R. (2022). Smart Greenhouse Farming Using IOT and Machine learning. *Journal of Emerging Technologies and Innovative Research*, *9*(8), b708–b718.

Quy, V. K., Hau, N. V., Anh, D. V., Quy, N. M., Ban, N. T., Lanza, S., Randazzo, G., & Muzirafuti, A. (2022). IoT-Enabled Smart Agriculture: Architecture, Applications, and Challenges. *Applied Sciences (Basel, Switzerland)*, *12*(7), 3396. doi:10.3390/app12073396

Rajyaguru, N., Vyas, S., & Vyas, K. (2021). 9 Internet of Things platform for smart Farming. In J. Chatterjee, A. Kumar, P. Rathore, & V. Jain (Eds.), *Internet of Things and Machine Learning in Agriculture: Technological Impacts and Challenges* (pp. 169–202). De Gruyter. doi:10.1515/9783110691276-009

Ravesa, A. & Sofi, S. (2021). Precision agriculture using IoT data analytics and machine learning. *Journal of King Saud University – Computer and Information Sciences*. Science Direct. doi:10.1016/j.jksuci.2021.05.013

Rayhana, R., Xiao, G., & Liu, Z. (2020). Internet of Things Empowered Smart Greenhouse Farming. *IEEE Journal of Radio Frequency Identification*, *4*(3), 195–211. doi:10.1109/JRFID.2020.2984391

Reddy, K. S. P. (2020). IoT-based Smart Agriculture using Machine Learning. *2nd Inter. Conf. on Inventive Research in Computing Applications*, (pp. 130-134). IEEE.

Rehman, A., Saba, T., Kashif, M., Fati, S. M., Bahaj, S. A., & Chaudhry, H. (2022). A Revisit of Internet of Things Technologies for Monitoring and Control Strategies in Smart Agriculture. *Agronomy (Basel)*, *12*(1), 127. doi:10.3390/agronomy12010127

Rifat, A., Patel, P., & Babu, B. S. (2022). The Internet of Things (IoT) in Smart Agriculture Monitoring. *European Journal of Information Technologies and Computer Science*, *2*(1), 14–18. doi:10.24018/compute.2022.2.1.49

Rijwan, K., Indrajeet, K., Jyoti, R., Noor, M., & Shahnawaz, H. (2021). *Opportunities of Artificial Intelligence and Machine Learning in the Food Industry, Journal of Food Quality*. Hindawi. doi:10.1155/2021/4535567

Rizzoli, A. (2023). *8 Practical Applications of AI in Agriculture*. v7Labs. https://www.v7labs.com/blog/ai-in-agriculture

Ryo, M. (2022). Explainable artificial intelligence and interpretable machine learning for agricultural data analysis. *Artificial Intelligence in Agriculture, 6*, 257–265. doi:10.1016/j.aiia.2022.11.003

Saiz-Rubio, V., & Rovira-Más, F. (2020). From Smart Farming towards Agriculture 5.0: A Review on Crop Data Management. *Agronomy (Basel), 10*(2), 207. doi:10.3390/agronomy10020207

Sarker, I. H. (2021). Deep Learning: A Comprehensive Overview on Techniques, Taxonomy, Applications, and Research Directions. *SN Computer Science, 2*(6), 420. doi:10.100742979-021-00815-1 PMID:34426802

Sasi Bhanu, J., Sunitha, D., & Bigul, P. A. (2021). Agricultural Internet of Things using Machine Learning. AIP Conference Proceedings. AIP Publishing. doi:10.1063/5.0058012

Sen, A., Roy, R., & Dash, S. R. (2021). Smart Farming using Machine Learning and IoT. Wiley. doi:10.1002/9781119769231.ch2

Senthil, K. S. D., & Shamili, M. D. (2022). Smart Farming using Machine Learning and Deep Learning Techniques. *Decision Analytics Journal, 3*. doi:10.1016/j.dajour.2022.100041

Shaikh, T. A., Mir, W. A., Rasool, T., & Sofi, S. (2022) Machine Learning for Smart Agriculture and Precision Farming: Towards Making the Fields Talk. In: Archives of Computational Methods in Engineering. doi:10.100711831-022-09761-4

Shetty, S., & Smitha, A. B. (2021). Smart Agriculture Using IoT and Machine Learning. In A. Choudhury, A. Biswas, T. P. Singh, & S. K. Ghosh (Eds.), *Smart Agriculture Automation Using Advanced Technologies. Transactions on Computer Systems and Networks*. Springer. doi:10.1007/978-981-16-6124-2_1

Singh, A., Gutub, A., & Nayyar, A. (2022). *Redefining food safety traceability system through Blockchain: Findings, Challenges, and Open Issues*. Multimed Tools Appl. doi:10.100711042-022-14006-4

Singh, S. (2022). The Definitive Guide to Enterprise Digital Transformation. *App Inventive*. https://appinventiv.com/guide/digital-transformation-for-business/

Siregar, R. R. A., Seminar, K. B., Wahjuni, S., & Santosa, E. (2022). Vertical Farming Perspectives in Support of Precision Agriculture Using Artificial Intelligence: A Review. *Computers, 11*(9), 135. doi:10.3390/computers11090135

Sivakumar, R., & Prabadevi, B. (2022). *Internet of Things and Machine Learning Applications for Smart Precision Agriculture*. IoT Applications Computing. doi:10.5772/intechopen.97679

Song, C., Ma, W., Li, J., Qi, B., & Liu, B. (2022). Development Trends in Precision Agriculture and Its Management in China Based on Data Visualization. *Agronomy (Basel), 12*(11), 2905. doi:10.3390/agronomy12112905

Tajudin, K., & Shahina Begam, I. (2020). Intelligent Crop Growth Management System using Internet of Things. *European Journal of Molecular and Clinical Medicine, 7*(9), 2211–2222.

Talaviya, T., Shah, D., Patel, N., Yagnik, H., & Shah, M. (2020). Implementation of Artificial Intelligence in Agriculture for Optimization of Irrigation and Application of Pesticides and Herbicides. *Artificial Intelligence in Agriculture, 4*, 58–73. doi:10.1016/j.aiia.2020.04.002

Tan, A. (2021). How IoT and Machine Learning are Automating Agriculture, Tech Target. *Computer Weekly.* https://www.computerweekly.com/news/252504285/How-IoT-and-machine-learning-are-automating-agriculture

Tefera, H. A., Huang, D., & Njagi, K. (2020). Implementation of IoT and Machine Learning for Smart Farming Monitoring System [IJSBAR]. *International Journal of Sciences, Basic and Applied Research, 52*(1), 67–77.

Triantafyllou, A., Sarigiannidis, P., & Bibi, S. (2019). Article Precision Agriculture: A Remote Sensing Monitoring System Architecture. *Information (Basel), 10*(11), 348. doi:10.3390/info10110348

Uddin, M. A., Ayaz, M., Mansour, A., Aggoune, H. M., Sharif, Z., & Razzak, I. (2021). Cloud-connected flying Edge Computing for Smart Agriculture. *Peer-to-Peer Networking and Applications, 14*(6), 3405–3415. doi:10.100712083-021-01191-6

UN. (2018). Food & Agriculture Organization of the United Nations; E-Agriculture in Action: DRONES for Agriculture, Bangkok, 2018. UN.

Utku Kose, V. B., Prasath, S., Mondal M., R., Podder P., & Bharati, S. (2023). Artificial Intelligence and Smart Agriculture Applications. Auerbach Publications.

Vadlamudi, S. (2020). Internet of Things (IoT) in Agriculture: The Idea of Making the Fields Talk. *Engineering International, 8*(2), 87–100. doi:10.18034/ei.v8i2.522

Verma, P. & Umesh, K. (2021). "Smart Farming: using IoT and Machine Learning Techniques." Internet of Things and Machine Learning in Agriculture: Technological Impacts and Challenges, Jyotir Moy Chatterjee, Abhishek Kumar, Pramod Singh Rathore, & Vishal Jain, Berlin. De Gruyter. doi:10.1515/9783110691276-001

Villazon, L. (2022). Vertical farming: Why stacking crops high could be the future of agriculture. *Science Focus.* https://www.sciencefocus.com/science/what-is-vertical-farming/

Vunnava, S. L., Yendluri, S. C., & Dhuli, S. (2021). IoT based Novel Hydration System for Smart Agriculture Applications *10th IEEE Inter., Conf., on Communication Systems and Network Technologies.* IEEE. 10.1109/CSNT51715.2021.9509597

Wadhwani, P. (2022). Agriculture Technology Trends: Collaborating Tech with Agriculture. *Bacancy Technology.* https://www.bacancytechnology.com/blog/agriculture-technology-trends

Zambon, I., Cecchini, M., Egidi, G., Saporito, M. G., & Colantoni, A. (2019). Revolution 4.0: Industry vs. Agriculture in a Future Development for SMEs. *Processes (Basel, Switzerland), 7*(1), 36. doi:10.3390/pr7010036

KEY TERMS AND DEFINITIONS

5G Technology: A 5th generation mobile, built on the current 4G network but with increased connection speeds and shorter delays.

Agricultural Robots (Agribots): A robot designed for agriculture to automate tasks for farmers, boosting production efficiency and reducing the industry's reliance on manual labor.

AgTech: Any innovation (including hardware, software, business models, new technologies, and Apps) used across the value chain to improve efficiency, profitability, and sustainability.

Artificial Intelligence (AI): The simulation of human intelligence processes by machines, especially computer systems. AI applications include expert systems, natural language processing, speech recognition, and machine vision.

Autonomy: A technology that can function without being told what to do. Examples are driverless tractors, cultivators, sprayers, and harvesters.

Big data: Technology provides farmers with granular data on rainfall patterns, water cycles, fertilizer requirements, and more. It enables them to make smart decisions (such as what crops to plant for better profitability and when to harvest), ultimately improving farm yields.

Blockchain: Technology is a decentralized, distributed, and public digital ledger used to record transactions across many computers so that the record cannot be altered retroactively without altering all subsequent blocks and the network consensus.

Cloud Computing: Technology is on-demand internet access to computing resources such as applications, servers, data storage, development tools, networking capabilities, and more hosted at a remote Data Center managed by a services provider.

Controlled Environment Agriculture (CEA): The production of plants and their products, such as vegetables and flowers, inside controlled environment structures such as greenhouses, vertical farms, and growth chambers.

Data Mining technology: Sorting through large data sets to identify patterns and relationships that can help solve business problems through data analysis. Data mining techniques and tools enable enterprises to predict future trends and make more-informed business decisions.

Data Science: Technology is an interdisciplinary academic field that uses statistics, scientific computing, scientific methods, processes, algorithms, and systems to extract or extrapolate knowledge and insights from noisy, structured, and unstructured data.

Deep Learning (DL): Technology is a method in AI that teaches computers to process data in a way inspired by the human brain. DL models can recognize the complex picture, text, sounds, and other data patterns to produce accurate insights and predictions.

Edge Computing: A networking focused on bringing computing as close to the source of data as possible to reduce latency and bandwidth use.

Geographic Information System (GIS): A computer system for capturing, storing, checking, and displaying data related to positions on Earth's surface.

Internet of Things (IoT): Technology refers to the collective network of connected devices and the technology that facilitates communication between devices and the cloud system.

IoTAg (IoT-enabled Agriculture): Represents technology wherein agricultural planning and operations connect in previously impossible ways if not for advances in sensors, communications, data analytics, and other areas.

Machine learning (ML): Technology is a type of AI that allows software applications to become more accurate at predicting outcomes without being explicitly programmed to do so. ML algorithms use historical data as input to predict new output values.

ML-enabled IoT (MIoT): A: system consisting of a sensor network that studies lighting and cloud patterns to predict the weather is successfully deployed in agricultural environments.

Precision Agriculture (PA): A farming management strategy based on observing, measuring, and responding to temporal and spatial variability to improve agricultural production sustainability. It is used in both crop and livestock production.

Regenerative Agriculture (RA): An outcome-based food production system that nurtures and restores soil health, protects the climate and water resources and biodiversity, and enhances farms' productivity and profitability. It comprises various innovative technologies to combat climate change challenges, soil health, and protect the land's ecosystem.

Sensor: A device that measures or detects environmental changes to present data for decision-making. This data can deliver benefits through improved crop and livestock yields; reduced wastage and livestock mortality; automation of farm operations; and maintenance or cost savings. Sensors may measure soil moisture/nutrition, weather, and water storage levels.

Smart Farming: This is about using the new technologies that have arisen at the dawn of the Industrial Revolution 4.0 in agriculture and cattle production to increase production quantity and quality by making maximum use of resources and minimizing the environmental impact.

Smart Greenhouse: Combines conventional agricultural systems and new IoT technologies for complete visibility and automation. It helps pinpoint inefficiencies and combat issues that have plagued farming operations to protect crops and maximize yields.

Urban Farming: (urban agriculture or gardening) Cultivating, processing, and distributing food in or around urban areas. It encompasses a complex and diverse mix of city food production activities, including fisheries and forestry.

Vertical Farming: This is the practice of growing crops in vertically stacked layers. It often incorporates controlled-environment agriculture, which aims to optimize plant growth, and soilless farming techniques such as hydroponics, aquaponics, and aeroponics.

Chapter 6
An Investigative Study on Internet of Things in Healthcare

N. A. Natraj
Symbiosis International University, India

Sriya Mitra
Symbiosis International University, India

Giri Gundu Hallur
Symbiosis International University, India

ABSTRACT

The emergence of IoT in the healthcare sector can bring about a surge of innovation and development. This qualitative research paper is aimed to explore the growth, adoption and threats that the emergence of IoT in the healthcare sector holds. IoT has various capabilities that pose an opportunity to provide solutions for a digitised healthcare system that would also aid in battling pandemics such as COVID. IoT has applications in multiple places, such as improving hospital treatment systems, tracking different medical parameters, identifying irregularities. The correct usage of IoT may aid in dealing with a variety of medical issues such as speed, cost, and complexity. It makes the surgeon's job easier by lowering risks and improving overall performance. The growing use of connected devices in healthcare is a major driver for the IoT-Health market.This research identifies the threats and the prospects of IoT in the future of healthcare.

INTRODUCTION

The Internet of Things is a young technology that has just recently emerged. It deals with data exchange with other systems or devices through the Internet or other communication networks. Sensors, connectivity, data processing, and user interface are the four components that make up an IoT system. Sensors/ Devices are the critical components of IoT, which collect data from the environment; few examples include humidity sensors, heart rate monitors and temperature sensors. The following element is con-

DOI: 10.4018/978-1-6684-8785-3.ch006

nection, this describes how the sensors are linked to the cloud. This can consist of various Bluetooth, Wi-Fi or cellular methods, which connect the device to the Internet and enable communication. Once data reaches the cloud, data processing is implemented; this could involve checking whether readings are within the acceptable limit or detecting anomalies and objects as per requirement. Next comes how the data collected and processed is made helpful to the end-user. This can be through a mobile application or alerts. The user may also make judgments using IoT applications depending on the data they have received. Due to its utility, simplicity and cost-effectiveness, the adoption of IoT has seen considerable growth. The basic framework of IoT in health care is shown below in Figure 1.

Figure 1. The basic framework of IoT in healthcare

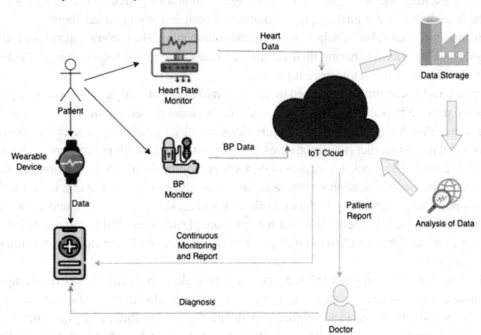

Smart Wearables: Wearables refer to the physical devices worn by individuals to track, analyse and send personal data. These devices can either be wearable or implantable and track physiological (Mathew et al., 2018) data such as heart rate, sleeping pattern, calories burnt. Typically, these wearables take the form of accessories like watches, apparel, or eyewear. The market for such devices is expanding rapidly with a continuous increase every year; According to the most recent projection from Gartner, Inc., worldwide end-user spending on wearable devices would reach $81.5 billion in 2021, up 18.1% from $69 billion in 2020 (Thakor et al., 2021). The healthcare industry is getting the best of wearable technology as vital information like blood pressure, and temperature can be relayed directly to the patients and healthcare professionals. Using this data, healthcare professionals may keep an eye on the patients, follow their movements, and create a personalised strategy.

Smart Hospitals: IoT is also finding its use in Smart Hospitals; using IoT's various tools and capabilities, an integrated healthcare system can be created that provides patients with cost-effective and efficient treatment. The aggregation and analysis of the multiple types of data can help healthcare providers make automated workflows that can be used for non-medical objectives. Hospitals, for instance, employ IoT systems to manage resource utility costs and regulate the temperature in operation rooms. This technology can also give way to the proper administration of hospitals in pandemics such as COVID-19. The information from IoT sensors can help inappropriate distribution and reduction of waste. Additionally, it can lessen the likelihood of accidents and stop medical device theft. (Javaid & Khan, 2021).

Remote Monitoring: Remote monitoring involves collecting data from sensors and then performing different processes based on the data, such as triggering automatic alerts and notifications. IoT enables an organisation to obtain equipment maintenance data and use it to make informed decisions. In the case of healthcare, remote monitoring can be used to track the patients' health continuously and provide timely help when required. In this way, a patient can receive high standard services without even stepping out of the house or consulting doctors routinely. Through remote monitoring, there will also be a lot of patient data available with the doctor, enabling them to make custom and personalised plans.

Remote monitoring can also be helpful for Insurance companies. These companies provide their users with rewards and benefits for sharing their routine activities. The intent behind doing so is to identify fraudulent claims and enable a transparent process

Personalised Medication: Personalised medication makes use of analytics to the data acquired from IoT sensors to provide personalised solutions. It is usually characterised by the 4P's - i.e. personalised, predictive, preventive and participatory. Organisations are using predictive analytics to obtain trends from historical data and predict future activities. The patient's vital health parameters are used to identify the possible risks. The focus lies on providing a system for the detection, prevention and solution to diseases. A personalised medication system, also known as ambient assisted living, has an architecture that requires physical devices wherein data is collected, a processing/storage unit, and a consumer subsystem to receive the personalised health-related solutions.(Datta et al., 2015)The use of Personalised medication ensures a higher standard of quality and reduces a patient's life cycle from enrolment to post medical treatments.

Cloud: A network that supports IoT sensors and apps makes up an IoT cloud. The computers and storage required for real-time operations and processing are included in the cloud components. It also comprises the standards and services required for integrating, controlling, and protecting various IoT applications and devices. For the healthcare sector, patient data can be saved in the cloud. These gadgets are linked to the hospital's network through various locations and operations. By using integration technologies, this data can also be sent to other hospitals or doctors, acting as an inventory of all relevant details of the patient. These cloud-based solutions offer exceptional efficiency, scalability, and reduction in costs while becoming more dynamic and user-focused. Many researchers have suggested the use of cloud computing for Covid 19 as well. On-demand data storage is made possible by leveraging computer system resources, and with the aid of the Internet, patient data can be quickly transferred in urgent situations (Javaid & Khan, 2021). This improves the data's quality while lowering the cost of data storage.

Fog IoT: Fog computing is a decentralised computing architecture that increases the functionality of cloud services by utilising intermediary computing resources (fog nodes) for data processing, storage, and analysis. Coined by Cisco, the term "fog computing" refers to the method of bringing the services of the cloud closer to the 'things' that produce IoT data (Hariri et al., 2019). It is an alternative to the conventional centralised cloud computing architecture for IoT. All of the processing and storage of extremely large volumes of data is handled by cloud servers. This approach turns out to be highly inefficient, slow, and expensive due to the bandwidth limitations of networks catering to more than 30 billion IoT devices worldwide.

Fog computing offers to decentralise things by distributing compute requirements for storage, analysis, and networking—in a way that time-sensitive IoT data is acted on closer to where it is produced. This is achieved through intermediate computing resources (fog nodes) between the devices and the cloud (at network edges) that bring down processing time requirements to milliseconds. This infrastructure provides advantages of low latency due to reduced bandwidth requirements, better security, compliance, connectivity, interoperability, real-time processing, and for all these reasons, better user experience. Moreover, fog computing also reduces energy consumption and minimises time complexity.

Role of IoT in Healthcare

The Internet of Things (IoT) has the potential to revolutionise healthcare by making it smarter and more efficient. These are some examples of how IoT might improve healthcare:

- Remote Monitoring
- Improved Diagnostics
- Enhanced Medication Management
- Efficient Healthcare Delivery

Remote Monitoring

Internet of Things devices may be used to remotely monitor patients' health status in real time. Wearables such as smartwatches, for example, may monitor vital indications such as heart rate, blood pressure, and oxygen levels. This data may be provided to healthcare providers, who can utilise it to make educated patient care decisions.

Improved Diagnostics

IoT devices can aid in the early detection of illnesses by continuously monitoring patients' health. Smart inhalers, for example, can be used to track and analyse patients' medication consumption in order to detect early indicators of asthma attacks.

Enhanced Medication Management

IoT devices can assist patients in better managing their medicine by delivering reminders to take medication at the appropriate time. Furthermore, smart pill dispensers may track medicine consumption and notify healthcare practitioners if individuals forget to take their prescription.

Efficient Healthcare Delivery

IoT can assist healthcare providers in optimising workflows and increasing operational efficiency. Smart hospital beds, for example, may be used to monitor patients' movements and help avoid falls, while IoT-powered asset monitoring can assist hospitals in keeping track of medical equipment.

Ultimately, IoT has the potential to transform the healthcare business by delivering smarter, more efficient treatment and increasing patient outcomes.

Research Objectives

- Explore different applications of IoT
- Identify the threats associated with the applications of IoT
- Identify possible solutions or prospects of the threats

Research Questions

- What are the applications of IoT, and why is it growing?
- What are the concerns with its expansion?
- How can the threats to IoT be tackled effectively?

This paper aims to provide an outlook about the applications and work done on IoT and the drawbacks that need to be taken into account for its growth in the future.

LITERATURE REVIEW

A vast number of papers define what IoT is and the benefits, applications, benefits, and disadvantages associated with it. This section offers a glimpse at the prior IoT for healthcare work that was done. In a 1999 presentation for Procter & Gamble, Kevin Ashton coined the phrase "Internet of things," referring to a technology that used RFID tags to connect various devices in order to control the supply chain. Since then, the Internet of Things has grown and impacted almost every sector of industry. IoT has brought a significant transformation in healthcare by changing the scope of interaction between devices and people in healthcare solutions. Mathew et al. (2018) conducted research on the different healthcare applications of IoT such as physiological Monitoring, Ambulance fitted with sensors, and sensor-enabled pills that can improve the healthcare industry. Technological advancement plays a significant role in the rise of such technologies- sensors, faster internet speeds, and declining equipment costs are all critical drivers of growth in the applications. The solutions prove to be very beneficial for people suffering from chronic diseases and facing issues in mobility. No longer is it necessary to be present in the medical facilities. Medical professionals may keep track of their patients' medication usage, which will help them control expenses. RFID can be used for a drug management system to automate their restocking process, which reduces time, error rate and increases efficiency within hospital pharmacies. The paper mentions privacy reserving algorithms and its requirement but does not discuss the methods related to them. Interoperability is also another essential factor, and the paper does not discuss how it is managed to attain seamless integration with multiple connected devices from different manufacturers.

Chacko et al. further elaborate on the concerns regarding privacy and security in IoT. Such systems involve sending data automatically to the Internet; this poses a risk of being hacked. The system is susceptible to many different forms of assaults, including collision, Denial of Service attacks, and manipulation with nodes. (Chacko & Hayajneh, 2018) The paper gives instances of how devices like pacemakers and insulin pumps can be hacked and used to deliver fatal medications. The paper provides steps to secure IoT in healthcare. The paper suggests methods to improve security through risk assessments, authentication, and a defence depth strategy where all the system layers are protected against specific risks. There are no regulations for m-Health; this way, apps provide users data to many other third-party apps.

Selvaraj et al. provide us with an outlook of the significant areas wherein IoT has found its use, such as Cloud and Big Data. It also provides examples of various architectures involving cloud and IoT, which use machine learning algorithms, temporal fuzzy ant miner tree for early detection of heart diseases and classification of medical records, respectively(Selvaraj & Sundaravaradhan, 2020). The paper suggests using an optimisation algorithm to conserve energy s buneeds toot discuss the method and implementation., cloud storage cost. The paper also states that data can be affected by Noise, and noise removal techniques should be applied but does not elaborate on the methods.

Yannuzzi et al. further elaborate on one of the main applications of IoT, which is the Cloud. Integrating cloud and IoT makes the user experience dynamic and personalised. However, the cloud comes with the limitation of latency.(Yannuzzi et al., 2014) Due to a centralised server, the data requires multi-hops which increases the time taken and may become a critical factor in hospitals. Fog installation can address this issue. Fog computing scales better than the cloud for executing data analytics and making choices in the present. It can aggregate data from the cloud selectively and grow upon geographic expansions.

Narrowband IoT is the more straightforward and leaner version of IoT, which holds a lot of potentials. Routray et al. state that Narrowband IoT is preferred since it is economical and requires fewer resources. The paper discusses the architecture, application and issues in narrowband IoT.(Routray & Anand, 2017) Due to its low power and wide area coverage capabilities, NbIoT is used widely in healthcare. The paper does not mention the delay-tolerant methods which can be used in NbIoT.

In the paper by Javaid et al., we understand how IoT can also prove to be beneficial in pandemics like Covid19. It can be used to keep tabs on employees and patients, cutting down on wait times. (Javaid & Khan, 2021) Applications include networked imaging, clinical operations, drug delivery, and medication management in medical devices. Data analytics can also be used in the data collected through IoT to find solutions to complex decisions.

THEORETICAL FRAMEWORK

The applications of IoT in healthcare are vast and growing every day. However, some particular problems require attention. These connected devices efficiently collect data and provide it to healthcare organisations for analysis and solutions. But this simplicity also contributes to the issue, as patient data security and privacy must be maintained to guarantee privacy. The system must include firewall, anti-malware, and antivirus software to safeguard data from hackers. The data should also be encrypted using efficient, standard tested methods such as AES, RSA and DSA. Manufacturers can also make use of Lightweight cryptography (LWC), a lighter version of cryptography. Lightweight cryptography contains features such as small memory, small processing power, low power consumption, a real-time response that work with resource-constrained devices such as RFID tags, sensors as well as resource-rich devices such as PCs, smartphones.(Umair et al., 2021) A few of the renowned LWC techniques include SPECK, SIMON and PRESENT. Companies can also use authentication methods like biometrics and two-factor authentication to protects users. After implementing such methods, the IoT devices should be tested multiple times before their release in the market.

To guarantee that a patient's safety is not jeopardised, laws and regulations must be adhered to in order to maintain proper privacy. Almost little regulation or control of any type has been applied to the exponential growth of mHealth solutions. Health practitioners need more faith in the existing health applications since so few of them have passed a thorough validation procedure.This needs to be changed; there should be strict and uniform regulations across countries so that companies can follow these and data is kept safe. IoT data is supplied into cognitive automation systems so they can guide and infer information for the user automatically. Figure 2 shows the Mechanism of data flow in IoT in Health Care.

Figure 2. The mechanism of data flow in IoT in healthcare

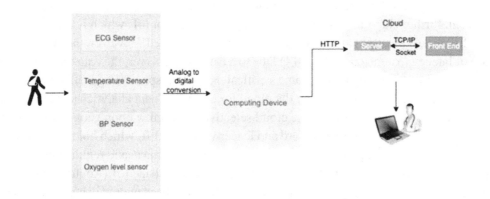

As data is gathered from various sources, there can be some ambiguities; in this case, care must be made to clear out any confusions before the system uses this information. This data can be overwhelming, and many companies are unclear of how to address it. In order to examine and predict future courses of action with high accuracy and advanced decision-making processes, it is necessary to analyse such huge amounts of data using modern analytical techniques. AI techniques for big data analytics, such as ML, NLP, and CI, can be applied to address the issue of ambiguity. Techniques such as active learning, fuzz logic and deep learning help to remove uncertainties from the data collected.

The most critical factors of IoT also involves the fact that it can be used 24/7, and doctors can do remote monitoring for patients who are aged and disabled as well. But the issue in this context was about the conservation of energy. Being available all the time will cause the battery to drain out, so there have to be some mechanisms to reduce this. The energy supply also constrains the performance of the IoT in terms of the quantity of measurements, data analysis, and data exchange. The situation for IoT devices changes with the deployment of energy harvesters. Many of the performance restrictions will disappear or be much diminished. Solar-powered panels, which transform kinetic energy into electrical energy, or the energy produced during vibrations that naturally occur in the manufacturing and transportation sectors, are two potential energy harvesting sources. To save energy, engineers might also employ protocols.

Most cellular technologies turn off their cellular module when a connection to the cellular network is not essential, however power is required to re-establish a connection to the network. In contrast, PSM uses network connection timers to reduce IoT power usage. Even when not in use, IoT devices will stay connected to a network while in the power-saving mode for a certain period of time before being disconnected.

ROLE OF IOT IN HEALTH CARE DURING COVID-19

As a result of COVID-19, there is a greater need than ever for healthcare to be easily accessible, barrier-free, and more efficient in monitoring, testing, and diagnosis. IoT can be helpful in this situation since it prevents the transmission of diseases and makes it convenient to consult with a doctor wherever you are,

- Smart wearables have been in the market for quite some time, but COVID-19 has provided a major boost to this sector. The Global Data research states that the wearables industry will grow from $27 billion in 2019 to $64 billion by 2024 (Kintzlinger & Nissim, 2019)
- Telemedicine saw an increase of 10-fold in the US. Other nations like China and Italy, to mention a couple, have also shown a surge in telemedicine use.

Various companies invested in IoT to devise solutions for COVID-19. WHOOP Inc. created the COVID-19 identification method, and CQUniversity in Australia utilised a WHOOP strap based on respiratory rate. The WHOOP strap collects data and transmits it to a mobile application and thereafter to the WHOOP system. Through their methodology, 20% of people having Coronavirus were identified 2 days before they started showing symptoms for the same and the detection rate increased to 80%.

Philips has developed the Philips Biosensor BX100 to aid in the management of proven and suspected Corona patients in the hospital. The lightweight, disposable biosensor is a single-use, 5-day wearable patch that may be linked to a scalable hub to monitor numerous patients in different rooms. The gadget is designed to integrate with existing clinical procedures for mobile viewing and alerts, and it does not require cleaning or charging.

Researchers from MIT and Harvard University employed AI to test if COVID-19 subjects could be correctly distinguished from a forced-cough mobile phone recording alone. Their research shows that their system correctly discriminates officially tested COVID-19 cases 97.1 per cent of the time, with a 100% asymptomatic detection rate, using cough recordings from more than 5000 people. Cough recordings have previously been used to diagnose conditions, including, pneumonia and asthma, accurately. This shows the value of including such technologies in wearables to offer a non-invasive, real-time solution for sickness diagnosis, pre-screening, and epidemic monitoring.

The DETECT (Digital Engagement & Tracking for Early Control & Treatment) study, conducted by the Scripps Research Translational Institute, gathers information from activity trackers and smart-watches owned by consenting couples as well as self-reported symptoms and test results. They recently discovered that information from wearable technology combined with self-reported symptoms can be used to more accurately detect cases of COVID-19 than symptoms alone. Many other similar initiatives have accelerated their deployment to enable interested parties to openly donate their sensor and clinical data to fight COVID-19.

Standards Related to IoT

IoT devices haven't had a robust security design in the software. This is because security is an added source of expenditure that manufacturers are avoiding. Due to growing privacy and security concerns, this belief has started to alter. There are various types of standards used in addressing security and privacy issues since IoT is used in different fields. These types include Technical, legal, regulatory and social standards.

Healthcare Standards: The International organisation for standardisation has defined ISO ICS 11 with over 1300 standards in health and safety. This applies to a variety of sectors ranging from traditional medicines to medical devices. Though these standards aren't directly associated with IoT, these could be used in the development of new standards (Aravind et al., 2020).

FDA Guidance: The federal and drug administration safeguards public health by keeping a check on food safety, medications, and medical devices. On 15th January 2016, FDA issued guidelines for best practices in cybersecurity management. Medical device manufacturers can follow the directions in the "Draft guidance for industry and food and drug administration staff" to resolve security risks throughout the lifespan of the product. The policy states security standards such as multi-factor authentication, limiting user access and stronger passwords. The healthcare industry should also evaluate its infrastructure and adopt new measures to strengthen cybersecurity (Darshan & Anandakumar, 2015).

Device Communication Standards: Under the CEN ISO/IEEE 11073, there are standards that allow data to flow between medical devices and external systems. The devices that come under this standard include fitness devices, activity monitors and weighing scales. The objective is to achieve an efficient exchange of data between the devices and point of care (Kashani et al., 2021).

IoT Security Standards: There hasn't been a significant effort made to create universal IoT security standards. There are recommendations provided by NIST and ENISA have provided recommendations for IoT device manufacturers and IoT Security respectively.

NIST-Recommended Guidelines for the Manufacture of IoT Devices: In the May 2020 NIST paper "Foundational Cybersecurity Activities for IoT Device Manufacturers," manufacturers of IoT devices were given guidelines on how to make their IoT devices more secure. There are six techniques mentioned in the NIST recommended practices to add security features in IoT devices. Four mentions the thought process involved with regard to security before making the device. The other techniques deal with aspects such as duration of support, end of life of device, updates and vulnerabilities to be communicated with the customer.

European Union Baseline Security Recommendations: The ENISA recommendations were released in November 2017 (Azzawi et al., 2016). There are three main topics for security measures in the paper, "Baseline Security Recommendations for IoT in the context of Critical Information Infrastructures," which was published.

- **Technical Safeguards:** Technical measures include hardware and software security techniques. Hardware root of trust, automated update rollbacks, and difficult-to-crack default passwords that are specific to each device are a few examples of these technical security features.
- **Policies:** Security policies must be comprehensive and well-documented in order to secure an organisation's system. Some of the categories include risk and threat detection and assessment, security by design, and privacy by design. IoT device manufacturers should stay away from cryptographic methods and communication protocols. This increases security because it is simpler for hackers to access IoT devices using recognised protocols and keys. Manufacturers should have procedures in place for looking into and handling security-related events.
- **Organisational, People, and Process Measures:** Employers should be trained in privacy and security procedures as part of organisational, people, and process measures. Data sharing, data processing by third parties, and the development of hardware or software should follow relevant security and legal standards for their services.

CONCLUSION

IoT is improving the healthcare sector in leaps and bounds. It has various applications like patient monitoring, smart wearables and smart ambulances, providing patients with a better and more personalised experience. IoT also proves beneficial for healthcare providers as they can get a lot of data and proceed with analytic data methods to derive optimum solutions while having reduced costs. The use of technologies and techniques such as Cloud, WSN and Fog further expands IoT's utility. Through Cloud storage, essential data like a patient's history can be stored and made available for transmission when required. The use of Integrated cloud and fog technologies will provide all the benefits of the cloud with an additional advantage of higher speed and reduced latency. But with the expansion of IoT, some concerns need to be addressed. To ensure the privacy and security of the patient's data, more effective algorithms and encryptions must be used. While only some companies ask for permission to send patient's data from one app to another, most companies bypass this question due to the lack of regulations. There is a need for strict regulations to ensure that an individual has control over his data and is informed about the people who have access to it. These regulations would also assure healthcare workers that the data is authentic and that their decisions can be based on it. While new applications are coming up very frequently for IoT, the focus has to be shifted to improving the architecture. The battery life and energy consumption needs to be improved.

The amalgamation of various data sources, modern algorithms, and cognitive computing will enhance patient care and help doctors make more accurate and precise clinical decisions. IoT will likely change how healthcare organisations run by enabling new business models based on real-time data acquired by sensors and due to ongoing technical advancement.

REFERENCES

Aravind, A. R., Chakravarthi, R., & Natraj, N. A. (2020, September). Optimal mobility based data gathering scheme for life time enhancement in wireless sensor networks. In *2020 4th International Conference on Computer, Communication and Signal Processing (ICCCSP)* (pp. 1-5). IEEE. 10.1109/ICCCSP49186.2020.9315275

Azzawi, M. A., Hassan, R., & Bakar, K. A. A. (2016). A review on Internet of Things (IoT) in healthcare. *International Journal of Applied Engineering Research: IJAER*, *11*(20), 10216–10221.

Bianchi, C., Tuzovic, S., & Kuppelwieser, V. G. (2022). Investigating the drivers of wearable technology adoption for healthcare in South America. *Information Technology & People*.

Chacko, A., & Hayajneh, T. (2018). Security and privacy issues with IoT in healthcare. *EAI Endorsed Transactions on Pervasive Health and Technology*, *4*(14), 155079. doi:10.4108/eai.13-7-2018.155079

Darshan, K. R., & Anandakumar, K. R. (2015, December). A comprehensive review on usage of Internet of Things (IoT) in healthcare system. In *2015 International Conference on Emerging Research in Electronics, Computer Science and Technology (ICERECT)* (pp. 132-136). IEEE. 10.1109/ERECT.2015.7499001

Datta, S. K., Bonnet, C., Gyrard, A., Da Costa, R. P. F., & Boudaoud, K. (2015, October). Applying Internet of Things for personalized healthcare in smart homes. In *2015 24th Wireless and Optical Communication Conference (WOCC)* (pp. 164-169). IEEE. 10.1109/WOCC.2015.7346198

Hariri, R. H., Fredericks, E. M., & Bowers, K. M. (2019). Uncertainty in big data analytics: Survey, opportunities, and challenges. *Journal of Big Data*, 6(1), 1–16. doi:10.118640537-019-0206-3

Javaid, M., & Khan, I. H. (2021). Internet of Things (IoT) enabled healthcare helps to take the challenges of COVID-19 Pandemic. *Journal of Oral Biology and Craniofacial Research*, 11(2), 209–214. doi:10.1016/j.jobcr.2021.01.015 PMID:33665069

Kashani, M. H., Madanipour, M., Nikravan, M., Asghari, P., & Mahdipour, E. (2021). A systematic review of IoT in healthcare: Applications, techniques, and trends. *Journal of Network and Computer Applications*, 192, 103164. doi:10.1016/j.jnca.2021.103164

Kintzlinger, M., & Nissim, N. (2019). Keep an eye on your personal belongings! The security of personal medical devices and their ecosystems. *Journal of Biomedical Informatics*, 95, 103233. doi:10.1016/j.jbi.2019.103233 PMID:31201966

Mathew, P. S., Pillai, A. S., & Palade, V. (2018). Applications of IoT in healthcare. *Cognitive Computing for Big Data Systems Over IoT: Frameworks, Tools and Applications*, 263-288.

Natraj, N. A., Kamatchi Sundari, V., Ananthi, K., Rathika, S., Indira, G., & Rathish, C. R. (2022, July). Security Enhancement of Fog Nodes in IoT Networks Using the IBF Scheme. In *Third International Conference on Image Processing and Capsule Networks: ICIPCN 2022* (pp. 119-129). Cham: Springer International Publishing. 10.1007/978-3-031-12413-6_10

Perez, A. J., & Zeadally, S. (2017). Privacy issues and solutions for consumer wearables. *IT Professional*, 20(4), 46–56. doi:10.1109/MITP.2017.265105905

Routray, S. K., & Anand, S. (2017, February). Narrowband IoT for healthcare. In *2017 International Conference on Information Communication and Embedded Systems (ICICES)* (pp. 1-4). IEEE.

Selvaraj, S., & Sundaravaradhan, S. (2020). Challenges and opportunities in IoT healthcare systems: A systematic review. *SN Applied Sciences*, 2(1), 139. doi:10.100742452-019-1925-y

Thakor, V. A., Razzaque, M. A., & Khandaker, M. R. (2021). Lightweight cryptography algorithms for resource-constrained IoT devices: A review, comparison and research opportunities. *IEEE Access : Practical Innovations, Open Solutions*, 9, 28177–28193. doi:10.1109/ACCESS.2021.3052867

Umair, M., Cheema, M. A., Cheema, O., Li, H., & Lu, H. (2021). Impact of COVID-19 on IoT adoption in healthcare, smart homes, smart buildings, smart cities, transportation and industrial IoT. *Sensors (Basel)*, 21(11), 3838. doi:10.339021113838 PMID:34206120

Yannuzzi, M., Milito, R., Serral-Gracià, R., Montero, D., & Nemirovsky, M. (2014, December). Key ingredients in an IoT recipe: Fog Computing, Cloud computing, and more Fog Computing. In *2014 IEEE 19th International Workshop on Computer Aided Modeling and Design of Communication Links and Networks (CAMAD)* (pp. 325-329). IEEE.

Yeole, A. S., & Kalbande, D. R. (2016, March). Use of Internet of Things (IoT) in healthcare: A survey. In *Proceedings of the ACM Symposium on Women in Research 2016* (pp. 71-76). 10.1145/2909067.2909079

Chapter 7
Machine Learning–Based Threat Identification Systems:
Machine Learning–Based IDS Using Decision Tree

Jyoti

Department of Engineering and Technology, Guru Nanak Dev University, Regional Campus, Jalandhar, India

Sheetal Kalra

Department of Engineering and Technology, Guru Nanak Dev University, Regional Campus, Jalandhar, India

Amit Chhabra

Department of Computer Engineering and Technology, Guru Nanak Dev University, Amritsar, India

ABSTRACT

The increasing popularity of internet of things (IoT), dissimilar networks, distributed devices, and applications has turned out to be a major call for the identification of novel security threats and tracing malicious network behaviours. An intrusion detection system (IDS) is a self-defense tool for preventing several types of cyberattacks. Latest machine learning (ML) methods are becoming the backbone for constructing intelligent IDS that are highly data driven. This chapter proposes decision tree based IDS for the dataset NSL-KDD. A novel approach has been developed for the ranking of security features. The proposed system has been validated against performance evaluation metrics consisting of recall, precision, accuracy, and F score. The results produced by the proposed system are compared with well-known ML methods including logistic regression, support vector machines and K-nearest neighbor in order to analyse the efficiency.

DOI: 10.4018/978-1-6684-8785-3.ch007

INTRODUCTION

IoT is a communal network of connected devices which facilitates communication between the cloud and the devices, as well as between the connected devices themselves. Distributed Intelligent Systems are becoming popular with the advancements in Artificial Intelligence (AI). A huge amount of data is travelling over millions of distributed heterogeneous networks. IoT and cloud services lead the users to deal with enormous data on daily basis. As the horizon of devices is increasing so is the concern towards security breaches. In today's world cybersecurity services are essential because of massive collection of crucial data over computers and other devices which is used in government sector, business, healthcare, military and financial organizations (Sarker et al., 2021).

In large scale networks, cyber-attacks like computer malware, unauthorized login or denial-of-service (DoS) attack cause huge financial losses (Tariq, 2018). It becomes essential to protect IoT devices against different types of known and unknown attacks in order to gain maximum benefit from emerging technologies. Threat identification systems play a vital role in protecting IoT devices. These systems check whether the traffic, coming in a network is benign or malicious. These are also capable of identifying the type of threat. In this chapter, the researchers have explained the basics of IoT, ML and threat identification system i.e IDS. ML based decision tree approach is used while designing IDS utilizing KDD dataset.

Internet of Things

The Internet of Things (IoT) is a network of connected devices that can sense, act and converse with one another and with the outside world (i.e smart objects). It enables the sharing of information. IoT can develop services either with or without direct human involvementLarge companies, service providers, and sectors like manufacturing, healthcare, smart grids, digital agriculture, and many more are now paying attention to IoT. "Things" are a collection of nodes that have the ability to communicate with one another, either with or without human participation. (HaddadPajouh, 2021). The term things include sensors, smart televisions, smart refrigerators and smart vehicles etc. These things also known as objects or nodes can talk, listen, hear and act in smart manners. The industry 4.0 revolution is beginning to include this usage. There are three layers in the IoT architecture. Several levels of security are needed. Several security issues at each tier of the IoT architecture are depicted in Figure 1.

Application Layer

This layer offers a variety of facilities for use in many IoT applications including smart homes, smart self-driving cars, smart parking systems, etc. This layer ensures the authenticity, privacy and reliability of the data. In order to communicate data over the network, the application layer protocols define how the application interacts with the lower layer protocols. Process-to-process communications are made possible via application layer protocols using ports (Swamy et al., 2017).

Figure 1. Security concerns at different layers of IoT

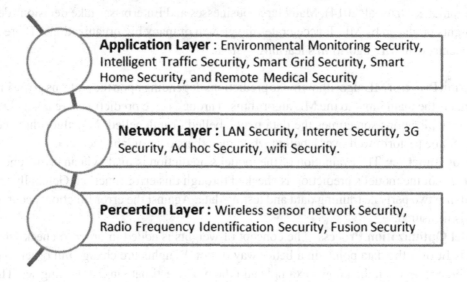

Network Layer

In IoT, the network layer connects all things and enables them to be aware of their environment (Gokhale et al., 2018). Data routing and transmission to various hubs and devices across the Internet is this layer's primary role. At this layer, data transmission, data aggregation, and data filtering are also performed. Yet, the most often used security mechanisms in IoT architectures' network layers are key management and encryption systems, blockchain technology, and intrusion detection systems.

Edge Layer

End-user IoT devices (cloud-edge) can communicate with clients in this layer as well as their working domains, such as sensors, smart metres, or IoT edge layer servers of a gateway (Portilla et al., 2019). Information from the real-time objects is gathered by the Edge or perception layer. The "Sensors" layer of the Internet of things is another name for this layer. This layer's primary objective is to gather data from the environment using sensors and actuators. It gathers information and sends it to the network layer for additional processing. Also, it manages Node collaboration in local and short-range networks for the internet of things. The edge-layer in the IoT architecture is vulnerable to numerous assaults because of its physical accessibility.

Machine Learning

A subfield of computer science and artificial intelligence called "machine learning" focuses on using data and algorithms to mimic how people learn while continuously improving the accuracy of its predictions. The enormous and growing amounts of data, computing capacity and the development of better learning algorithms have all contributed to a recent surge in public interest in ML and AI. Algorithms are trained with the real world examples, so that correct predictions could be made. For accurate predictions statistical methods are well utilized. ML is used in data mining initiatives to unearth important insights (Baştanlar & Özuysal, 2014). Many large businesses and Enterprises take decisions depending upon the insights produced by ML. It accelerates the growth of many big organizations. There are three basic components to an ML algorithm's learning system.

1. **A Decision Process:** ML algorithms make predictions depending upon the patterns given in the data. This data is the main input to the ML algorithms. Through these predictions are made. Depending upon the applications sometimes the data may labelled or unlabelled. ML algorithms come in a variety. Some perform well over labelled while the others over unlabelled data.
2. **An Error Function:** The evaluation of the model's prediction is made via an error function. The correctness of the model's prediction is checked through this error function. Generally the data is divided into two parts i.e training data and testing data. Against the error function, accuracy of the model is measured.
3. **A Model Optimization Process:** The concept of weights is added in order to check whether the model is fit over the data points in a better way or not. Weights are changed in order to decrease the difference between the known example and the model estimate in the training set. The procedure of "evaluate and optimise" will be repeated by the algorithm. The weights keeps updating automatically, until a desired level of accuracy is reached.

Machine Learning Methods

ML techniques (models) can be divided into three groups.

1. **Supervised Machine Learning:** Supervised learning deals with the labelled data. The algorithms are trained through labelled data so that accurate predictions could be made. The model adjusts the value of weights depending upon the training data being provided to it until a well fit is not achieved. This turns out to be a part of the cross validation procedure to make sure if the model does not fit accurately or not. An example where supervised learing is giving benefit to large companies is when an email is classified as a spam or a normal email. The incoming email is filtered automatically. If an email is a spam it is placed in the spam folder automatically. Neural networks, naive bayes, linear regression, logistic regression, random forest, and support vector machines (SVM) are a few techniques used in supervised learning (Badillo et al., 2020).
2. **Unsupervised Machine Learning:** Unsupervised learning deals with unlabelled data. The datasets falls in the category of labelled as well as unlabelled. ML algorithms are used to check and cluster these unlabelled datasets. The motive is to find hidden patterns so that the behaviour could be analysed in order to make correct prediction s(Badillo et al., 2020). There is no need for human intervention. The main application is to make pattern recognition and data analysis. Algorithms used in unsupervised learning include neural networks, k-means clustering, and probabilistic clustering methods.
3. **Semi-Supervised Learning:** In between supervised and unsupervised lies a third domain that is semi supervised Learning. The motive is to extract features from a large unlabelled dataset in order make classification using smaller labelled data. The problem of unavailability of supervised labelled data can be resolved with this type of learning. It also helps if labelling enough data is very expensive.

By utilising efficient ML algorithms, intelligent IDS can be created. In this chapter, researchers have proposed an intelligent Network based IDS on KDD Cup dataset using machine learning based decision Tree approach. ML can help in predicting attacks in the cyberspace. As supervised ML is utilised to address both Regression and Classification issues. Decision trees are thought to be far more effective when creating prediction models (Badillo et al., 2020). Traditional ML techniques used in cybersecurity such as k-nearest neighbour (k-NN) algorithms, Naive Bayes (NB) models, Logistic Regression (LR), Support Vector Machines (SVM) and Decision trees which are crucial in identifying anomalies and are able to solve multiclass problems that classify these attacks into different types (Alzubi et al., 2018). For the purpose of creating data-driven intelligent IDS, developing security models for analysing cyber patterns and detecting risks using cybersecurity data could be quite helpful (Sarker, Kayes, Badsha et al, 2020). The security model makes it possible to detect the malicious activities by analysing the knowledge base. This type of security modelling based on AI driven approach makes cybersecurity computing automated and intelligent as compared with traditional security Systems. In supervised learning, Decision Trees are the famous predictive models which are best for building an IDS based on classification algorithms. A significant benefit of the Decision tree is that it offers a comprehensive set of rules that are simple to combine with real-time technologies (Jijo & Abdulazeez, 2021). After choosing to split the node based on the information gained for each feature, the entire dataset will be separated into nodes that reflect the features. The class is represented by each leaf node (normal or attack). Decisions are made at each node until the leaf node is reached. In this way the decision tree can be used to make a predictive model for the Detection of any potential network Intrusion. Though classification algorithms of machine learning perform well in predictive analytics, but because of the diversity of features of datasets, it becomes challenging to produce accurate results (Umer et al., 2022). The datasets may involve so many features even some features may be of no use at all. It may affect the accuracy of prediction. This thing may also create the problem of overfitting. To provide better security solutions towards the present cloud based services is essential for the maximum success of these services as many people rely over these services for their important tasks.

Intrusion Detection System

With the advent of IoT and cloud computing, digital communication has been increased. With this, more security is needed in order to save various devices and digital assets. Threat Identification systems provide a solution for detecting threats. These systems make the environment aware regarding malicious activities and incoming threats. An IDS makes it easier to identify, locate and control unauthorized System movements including modification, destruction, and unauthorized access (Ahmad et al., 2021). An IDS gathers information from the computer devices and networks, analyse them order to identify the malicious activities which are violating the security policy. It raises an alert whenever it senses an attempt by a malicious user to access the system.

Development of Intrusion Detection models has gained interest of many researchers over time. These models can learn from how the network behaves and forecast hostile activity on the network. A number of researches have been conducted with respect to cybersecurity for detecting Intrusions. IDS can be based on network or can be based on host (Ahmad et al., 2021).

Figure 2. Network-based IDS

There are two types of detection: network-based and host-based. Network-based detection examines and records online network data at the network gateway or server before the assault affects end users, in contrast to host-based detection, which looks at the network data acquired at the attacked System. Network-based IDS watches network traffic and finds suspect activity, whereas Host-based IDS observes a single System and detects suspicious or malicious actions by looking at the operating System files.

Traditional approaches towards security such as firewalls, encryptions etc. are less effective in protecting the system against new intrusion techniques (Ingre et al., 2018). Anti-virus software is capable of protecting the network users from malwares such as viruses, worms, spywares and Trojan horses but these software have certain limitations. They are able to detect only those threats which matches the signatures stored in their database rather than the ones for whom no signature is stored in the database.

IDS falls in two categories (i) Signature Based IDS (ii) Anomaly Based IDS. Intrusion Detection can be signature based (SIDS), i.e., the behaviour of the attack is saved which helps in detecting the similar types of attacks. Signature could be a sequence used by malware. SIDS is capable of detecting attacks based known patterns easily. On the other hand Intrusion Detection can be anomaly based (AIDS) where pattern of the attack is not known. It becomes difficult and challenging to detect such attacks where the pattern is not fixed and no signature is available. AIDS uses data driven approach in order to profile the normal behaviour and detects the malicious activity whenever variations occur. These types of attacks are also known as zero day attacks (Moon et al., 2017). This anomaly-based intrusion disclosing method has the downside of having a high false alarm rate since it interprets unexpected System behaviour as hostile activity or an anomaly. Frameworks for the Detection of these kinds of zero-day attacks are being developed by many academics. The security research community is giving continuous attention towards Detection of malicious activities over the network worldwide.

Signature based Intrusion Detection has a high success rate than anomaly based Intrusion Detection where zero day attacks are considered. Anomaly based Intrusion Detection has a disadvantage that the model may predict the unseen behaviour as anomaly and can produce high false alarm rates. So reduction in high false alarm rate should be the prime consideration of Intrusion Detection approach. Machine Learning and data mining have played a vital role in analysing security patterns and predicting useful decisions with respect to the security of the System.

IDS in IoT Devices

IoT is a fast and upcoming sector with vast applications that are being used in different fields. The users of IoT are increasing day by day as the number of products increase (Remesh et al., 2020). Although users are using IoT devices, but they are unaware regarding the security risks lying underneath. Therefore it is essential that smart devices should be secure in terms of data transmission and privacy of the users. Due to the extensive usage of IoT based networks, attacks to these systems have increased at a rapid rate. IoT devices are the pieces of hardware such as actuators, sensors, appliances or machines which are used for certain application and programmed in a such a way that these perform best in their application areas. All of these IoT devices need to come in contact with the internet or have to become a part of the network in order to transmit data and signals. Communication in IoT networks is incorporated through IoT gateways. Gateway is usually a hardware device which runs application software in order to accomplish certain tasks in the system. IDS can be designed and run in IoT gateway device e.g Arduino or NodeMCU, Mozilla. In this way IDS is implemented in an IoT device thus detecting the malicious traffic. For detecting attacks Machine learning plays a vital role in today's scenario. Machine learning techniques reduce energy consumption, attain better detection rate and reduce false positives.

IDS in Smart Applications of IoT

IoT consists of multiple smart devices. These smart devices make human life very easy. These smart devices are used in diverse spectrum of life. Some of the application areas are Healthcare, industries, E-commerce, banking, Smart homes and Government sector (Slevi & Visalakshi, 2021). In healthcare, psychological information such as temperature, pressure rate, electrocardiograph (ECG), electroencephalograph (EEG), and so on are gathered via embedded or wearable sensors. Additionally, environmental information such as temperature, humidity, date, and time can also be recorded. In Smart homes, the concept of Web of things is incorporated where a central web application monitors and controls various connected IoT devices (Alalade, 2020). Similarly in industries Certain IoT devices are connected and exchange information through internet in order to act smartly. E-commerce, Banking and Government sector uses IoT devices to make quick responses to the queries and for the fast and timely transactions. As the horizon of IoT applications is very vast, so is the security concern. Thus IDS provides a solution for the security concerns of the above IoT applications (Sarwar et al., 2022).

As machine learning is capable of making learning possible on the basis of data, it is a kind of data driven approach that can help in producing successful intelligent security models pertaining to cybersecurity. Defense is needed from both the perspectives (i) physical devices (ii) user's data. Cybersecurity helps to safeguard both devices and services.

Hereby, Researchers have proposed a security model based on decision tree approach of ML.

The following is a summary of the contributions:

- A novel method is proposed for choosing and ranking significant dataset features.
- Decision tree based ML Intrusion Detection system is proposed on selected features which will detect whether the coming network traffic falls into the category of normal or anomaly class.

The rest of the paper is organised as follows. The background and related work on intrusion detection algorithms are presented in Section 2. Section 3 illustrates the suggested decision tree-based intrusion detection methodology. The findings of the tests performed using our suggested Scheme is presented in Section 4.

RELATED WORK

Among the various machine learning methods, tree-based prediction models are considered to perform better in Intrusion Detection and have less false alarm rates comparatively. Denning et al. (Denning, 1987) in the year 1987 proposed the first Intrusion Detection System based on anomaly by providing a general-purpose framework for Intrusion Detection System by keeping an eye out for any unusual patterns of System usage in the audit records of the System. Rai et al. (Rai et al., 2016) proposed an algorithm for signature-based IDS based on decision tree (C4.5) approach by considering Information Gain and Gain ratio on NSL-KDD datasets with an accuracy of 79.52%. AI-Omari et al. (2021) anticipated Tree based IDS model for the prediction of attacks or anomalies in cyberspace considering security feature ranking, Gini Index as the key factors utilizing UNSW-NB 15 dataset showing approximately more accuracy as compared with other classification based models. Sarker et al. (2020) proposed a generalized tree based model named IntruDtree, a security model which is machine learning based considering ranking of security features utilizing popular KDD'99 cup dataset showing more accuracy as compared with other traditional machine learning classification approaches. Ingre et al. (2018) proposed a decision tree-based Intrusion Detection model using Feature Correlation Selection method over NSL-KDD datasets. The proposed method (after feature selection for two classes) gives an accuracy of 98.3%. Moon et al. (2017) proposed an architecture that deals with Advanced persistent threats (APT) using Behavior analysis by experimenting on malicious data installing virtual machine with an accuracy of 84.7%. Malik et al. (2018) proposed a pruned decision tree classifier which monitors the network for the Detection of anomalous connections using Particle swarm optimization using popular KDD'99 cup dataset showing accuracy of 93.53%. Shahri et al. (2016) proposed a hybrid approach centred on Genetic Algorithm (GA) and Support Vector Machine (SVM). GA was used in order to select the relevant features and SVM as the classifier for Intrusion Detection. 97.3% detection rate was given by the proposed scheme but relatively low rate of Detection efficiency is noted for unknown attacks. Ioannou, C., & Vassiliou, V. (2020) used Particle Swarm Optimization Algorithm(PSO) for feature selection and Extreme Gradient Boosting (XGB) model for fitness function. Random Forest machine learning technique has been used for detection. It gives 98% accuracy for binary classification and 83% for multiclass. Sarker, Iqbal H (2020) presented an effective exploration of machine learning modelling for security in order to detect multi-attacks and cyber-anomalies. Correlation technique has been used for relevant feature selection and ranking of important features and ten machine learning classification techniques were employed along with Artificial Neural Network (ANN) based security model. Both UNSW-NB 15 and NSL-KDD datasets are used for the comparison of various security models. Deep learning based Intrusion Detection techniques were presented in (Aslahi-Shahri et al., 2016; Ioannou & Vassiliou, 2020; Javaid et al., 2016; Meng et al., 2018; Shone et al., 2018). DNNs and nonsymmetric deep autoencoder were used for developing the proposed techniques. Related work summary is demonstrated in Table 1 below.

Table 1. Related work summary

Author	Dataset	Method	Results
Ingre et al. (2018)	NSL-KDD	Decision Tree with pruning technique	Binary Classification Accuracy = 90.3% Five class Classification Accuracy = 83.7%
Rai et al. (2016)	NSL-KDD	C4.5 decision Tree	Accuracy = 79.5%
Al-Omari, Mohammad, et al. (2021)	UNSW-NB 15	Decision Tree	Training Accuracy = 96.7% Testing Accuracy = 96.7% AUC = 96.7%
Sarker, Iqbal H., et al. (2020)	Not Mentioned	Decision Tree	Accuracy = 98% P = 98% Recall = 98% Fscore = 98%
Dong et al. (2020)	NSL-KDD	Bagging Algorithm, Multiple C4.5 decision trees	True positive rate = 99.4%
Li, Yinhui, et al. (2012)	KDDcup99	Clustering, colony algorithm, Support vector machines	Accuracy = 98.6%
Javed Asharf, Nour Moustafa, et al.[42]	KDD99	Recurrent Neural Networks (RNNs)	Accuracy = 96.8%
Abebe Diro, Abdun Mahmood, et al. (2021)	NSL-KDD LSTM	Naïve-bayes classifier	Accuracy NSL-KDD=99.78% LSTM=99.91%
Samir Fenanir, Fouzi Semchedine, etal(2021)	UNSW-NB 15, KDD dataset	Decision Tree, Support Vector Machine, Logistic Regression	Accuracy rate 95.76%
Thien Duc Nguyen, Samuel Marchal, etal(2019)	TPR and FPR	Federated learning approach	Accuracy=95.6%
Ansam Khraisat, Iqbal Gondal, etal(2021)	NSL - KDD	Complex pattern-matching	Accuracy = 97%
Mohamed Faisal Elrawy, Ali Ismail Awad, etal(2021)	True Positive Rate (TPR), False Positive Rate (FPR)	Hybrid intrusion detection techniques IPv6 over low – wireless personal area networks (6LoWPAN)	Accuracy = 80%

Apart from decision tree approach of machine learning, there are certain other techniques available which performs good with respect to implementing IDS. These techniques are Random Forest, Support vector machine, Gradient descent, Naïve Bayes classifier, Recurrent neural network, Deep neural network.

PROPOSED WORK

This section contains proposed model which is decision tree based IDS model. It is made up of various modules. First module contains data exploration which is performed on the popular security dataset NSL-KDD which consist of 41 features. Different security features of NSL-KDD are explored in detail. In the second module data pre-processing is performed as there are repetitions in the dataset so data cleansing is an important aspect towards the success of machine learning centred Intrusion Exposure model. In this phase data encoding, data normalization and Feature ranking and selection is done over

NSL-KDD. Feature ranking is performed on the basis of Mutual Information. Impurity of the respective nodes is calculated and on the basis of this, importance score be assigned to the features. Then the features having high importance score are considered. In the third module dataset has been split into training and testing parts. Decision tree based model is trained over training data. The proposed model is tested in the final module using testing data that either returns the final class as normal or attack. Figure 3 illustrates the proposed model.

Figure 3. Proposed decision tree based intrusion tree model

Data Exploration

A newer version of the KDD '99 dataset is called NSL-KDD (Kaggle, n.d.). The information is acquired through the simulation of a number of invasions or intrusions in a military network environment. This Intrusion dataset divides the network traffic into two categories- anomaly and normal. Dataset is publically available on Kaggle (Mohammed & Ahmed, 2019) – the largest data science and machine learning community in the world. The original KDD '99 consists of a lot redundant data that's why we use NSL-KDD security dataset in our experiments. There are numerous datasets for NSL-KDD in the dataset repositories, among these "KDDTarin" and "KDDTest" are taken for training and testing correspondingly. It has 125973 records in the training dataset and 22544 records in the test dataset. The NSL-KDD dataset contains four types of attacks and is made up of 41 characteristics and one class attribute. Attacks can be denial of service (DoS), User to Root (U2R), Remote to Local (R2L) or Probe attacks (Dong et al., 2020). The dataset consists of a total of 41 features. Among these 41 features, 3 features i.e Service, Flag and Protocol type are nominal. Rest of the features are of either Integer type or float type which classify them as quantitative.

Table 2. Dataset features with types

Feature Type	Feature Name
Integer	Duration, Dst_host_count, Src_bytes,, Dst_host_srv_count, Dst_bytes, Count, Land, Is_guest_login, Wrong_fragment, Is_host_login, Hot, Num_failed_logins, Num_shells, Num_compromised, Root_shell, Su_attempted, Num_root, Num_file_creations, Logged_in, Num_access_files, Num_outbound_cmds, Srv_count, Urgent
Nominal	Protocol_type, Flag, Service
Float	Dst_host_diff_srv_rate, Serror_rate, Dst_host_srv_rerror_rate Srv_serror_rate, Rerror_rate, Dst_host_srv_rerror_rate, Same_srv_rate, Dst_host_serror_rate, Diff_srv_rate, Dst_host_srv_serror_rate, Srv_diff_host_rate, Dst_host_rerror_rate, Dst_host_same_src_port_rate, Srv_rerror_rate, Dst_host_same_srv_rate

The above features need to be processed for effectively analysing the network behaviours which will lead to developing intelligent Intrusion Detection System.

Data Preprocessing

The data requires preprocessing before actually being processed. This preprocessing may lead to several steps as the various features may differ with each other on the basis of certain parameters such as data distribution, density, type etc. So it becomes essential to prepare the dataset as per requirements. KDDTrain+ and KDDTest+ from the NSL-KDD dataset have been utilised in our research. The table below shows the detailed overview of such instances.

Table 3. The two sets of statistics from the NSL-KDD dataset

NSL-KDD		
Class	**KDDTrain+**	**KDDTest+**
Normal	67,343	9711
U2R	52	200
DoS	45,927	7458
R2L	995	2754
PRB	11,656	2421
Attacks	58,630	12,833
Total	125,973	22,544

Our proposed scheme pre-processes the data as mentioned in the following sections.

Feature Numericalization

The Security dataset consists of numerical as well as categorical or nominal data values. In NSL-KDD protocol_type, service and flag along with the class values are nominal. These nominal values need to be converted into vector so that can be applied to Intrusion Detection model (Kaggle, n.d.). Another name for feature encoding is feature numericalization. Many nominal and ordinal encoding techniques are available such as one hot encoding, mean encoding, label encoding, target guided ordinal encoding etc. In our case label encoding is the best one as there may be significant increase in feature dimensions with the use of other techniques. Label encoding easily converts all the nominal values into numeric values. Feature Encoding is performed in Python using sklearn library.

Data Normalization

The data must be normalised during the data pre-processing phase. The values corresponding to different features differ significantly and distributed differently. For some data points values are much higher and for some data points values are much lower. To fill this gap it is important to rescale the values (Tavallaee et al., 2009). So to apply the data to the security model, it becomes necessary to firstly scale the data equally. The features which have substantial difference in data scales are rescaled using following formula:

$$Y` = \frac{Y - Y_{min}}{Y_{max} - Y_{min}} \tag{1}$$

The feature's maximum and minimum values are represented by Y_{max} and Y_{min}, respectively. If the value of Y is extreme in the column, the numerator is identical to the denominator consequently the value of Y' will be 1. If the value of Y is minimum in the column, the numerator will be 0, thus Y' will be 0. Value of Y' will range from 0 to 1 for Y values between minimum and maximum. In this way normalization will be applied to those features which are highly distributed.

After normalization all the features are encoded so that it becomes ready for further feature selection which is discussed in the next section.

Figure 4. Different density plots for duration, src_bytes, srv_count, and serror_rate features

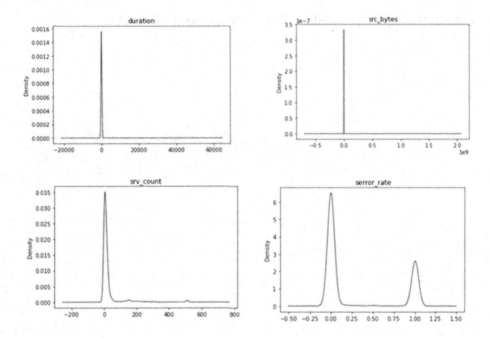

Feature Importance and Selection

The chosen dataset consists of 41 features. As dataset comprises too many features but all of them are not equally important for our proposed model. We have calculated the importance of each feature and taken only those features which are significant. In data mining the aspect of dimensionality of the dataset is a concept of consideration which influences the processing time and storage space (Ali et al., 2014). Ambiguity in the dataset may lead to wrong classification of attacks (Albulayhi et al., 2022). In our approach, to select the most significant features the concept of mutual Information has been used. A non-negative quantity called "mutual information" illustrates the interdependence of random variables. Higher values indicate high dependence, while zero indicates that the random variables are independent. Mutual Information calculation is based on entropy estimation.

Mutual Information can be calculated among two random variables M and N. It measures the amount of knowledge on N brought by M or we can say that the amount of knowledge imposed by N on M. If there is no dependency among M and N then their mutual information is zero.

$$I(E;F) = H(E) - H(E|F) \qquad (2)$$

Here H(.) is entropy, conditional entropies are represented by H(E|F) and H(F|E) and joint entropy is shown via H(E;F) (Tariq, 2018).

The following is the formula of Entropy.

$$\text{Entropy}(X) = \sum_{i=1}^{c} P(x_i) log_b P(x_i) \tag{3}$$

Where 'P' is simply the highest probability of an element/class 'i' in our data.

Table 4. Mutual information of different features

Feature Name	Mutual Information
ssrc_bytes	0.928013
Service	0.761364
Flag	0.674956
diff_srv_rate	0.655961
same_srv_rate	0.616609
dst_host_diff_srv_rate	0.612346
Count	0.564668
dst_host_same_srv_rate	0.554909
dst_host_srv_count	0.534711
dst_bytes	0.499432
dst_host_serror_rate	0.498888
serror_rate	0.473665
dst_host_srv_serror_rate	0.442665
srv_serror_rate	0.419883
dst_host_same_src_port_rate	0.385203
logged_in	0.349955
dst_host_srv_diff_host_rate	0.327261
srv_count	0.296387
dst_host_count	0.288737
protocol_type	0.263309
dst_host_rerror_rate	0.201517
srv_diff_host_rate	0.189649
dst_host_srv_rerror_rate	0.154058
rerror_rate	0.145776
srv_rerror_rate	0.099879
Duration	0.074366
wrong_fragment	0.059868
Hot	0.059459
num_compromised	0.040852
is_guest_login	0.009442
root_shell	0.005095
num_failed_logins	0.003263
num_access_files	0.003218
num_shells	0.002443
num_root	0.001953
Urgent	0.001945
su_attempted	0.001520
num_file_creations	0.001280
is_host_login	0.000080
Land	0.000000
num_outbound_cmds	0.000000

Here Table 4 represents the values of mutual information when implemented using sklearn library of python and importing mutual_info_classif. It is evident from the table that not all features are equally important. Features are arranged in descending order on the basis of mutual information gain showing least important features at the bottom.

Figure 5. Features mutual information gain

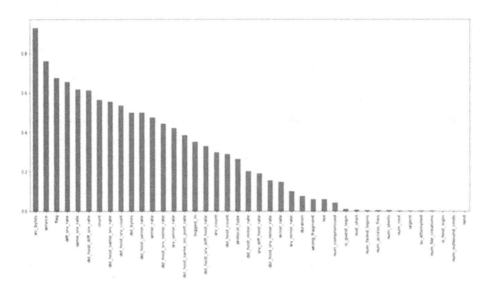

DESIGNING INTRUSION DETECTION DECISION TREE (INTRUSION TREE)

As soon as all necessary features have been selected based upon their weightage, we design tree-based model which follows data driven approach for intelligent Intrusion detection. Rather than taking all the features available in the dataset, we have taken top 16 features to build our Intrusion Decision tree. The design of the tree begins with root node. The decision for choosing root node depends upon information gain. When developing a tree, the feature with the most information gain serves as the root. Root node splits the given dataset into subsets and creates the tree incrementally. Information gain, sometimes known as IG, is a statistical parameter that determines how well a certain feature distinguishes training instances from their targeted classification.

Mathematically, IG is denoted as:

Information Gain(L,M) = Entropy(L) – Entropy(L,M) (4)

Where L® Current state and M ® Selected attribute

To effectively separate the nodes, CART (classification and Regression Trees) methods use the gini impurity (Osanaiye et al., 2016). The Gini impurity decides which node is to be chosen for binary split. The node with the lowest Gini is considered to be highly impure. Gini impurity is also known as an "impurity" metric as it shows how the model diverges from a pure division. The Gini impurity is calculated as shown in the formula

$$Gini = 1 - \sum_{i=1}^{n} \left(p_i \right)^2 \tag{5}$$

where total number of classes is represented by n and p_i is the likelihood of selecting a data point with class i. Thus how much a particular feature decreases the impurity is calculated using the above mentioned formula. The importance of a feature depends upon how much it decreases the impurity.

The entire process of building Intrusion Detection tree is represented in Algorithm 1. An Intrusion training dataset D is given, D={X_1, X_2, ..., X_s} where s is the data size. Diverse cyber-attack classes Labels = {attack, normal} have been represented in the training data. N-dimensional features represent each instance. The outcome is IntrusionTree, a classification rule based tree related to D.

Algorithm 1: IntrusionTree

```
Data: Dataset: D  // Dataset is made up of several instances with various
attributes and corresponding attack class labels.
Result: IntrusionTree
1.         Procedure IntrusionTree (D, Feature_list, LABELS)
2.         Calculate feature weightage
Feature_weight ¬ calculateWeight(feature_list)
// using formula    I(E ;F) = H(E) - H(E | F)
3.         Relevant_Feature_List ¬ SelectFeatures (feature_list, Feature_weight, n)
//Extract relevant features
4.         GenerateTree (D, Relevant_Feature_List, LABELS)
5.         R ¬ Create_Root_Node()
6.         If all instances in D are members of the same class, LABELS, then
           Return R as a leaf node having respective label from LABELS
7.         end
8.         Find the decision attribute A_best_split for the purpose of split-
ting the Tree and assign A_best_split to the node R.
9.         foreach attribute value v ∈ A_best_split do
10.            Create subset D_sub of D containing v.
11.               If D_sub in not empty then
12.            Attach the note returned by GenerateTree (D_sub, {Relevant_Feature_
List -  A_best_split },    LABELS) to node R;
13.               end
14.               assign a leaf labelled with common class in D to R;
15.         end
16.         If Relevant_Feature_List is blank then
17.            Return R as leaf node having a label from LABELS
18.         end
19.         Return R
```

EXPERIMENTAL RESULTS AND PERFORMANCE EVALUATION

This section comprises experiments which are précised over NSL –KDD that a cyber security dataset. The results of this work are briefly analysed and reported. Firstly, experiment is setup to evaluate our proposed IntrusionTree model. The suggested model is also assessed using Evaluation Metrics (Accuracy, Recall, Fscore, Precision), and the outcomes are contrasted with those of other conventional ML models.

Experiment Design

Firstly, in our experiment the processes mentioned in section 3 were implemented. Then we considered training and test data splitting. In our dataset, we have two separate files for training and testing data as KDDTrain+ and KDDTest+. KDDTrain+ has been utilised to train the model we propose. The proposed model has been examined and validated using KDDTest+. The statistics of NSL KDD is represented in Table 3. here were two experiments run. The initial experiment was using the proposed model on the features that were chosen based on the section 3 pre-processing. In a subsequent experiment, several conventional ML models, including Logistic Regression (LR) etc. were used. All experiments are implemented on Python.

The proposed Intrusion Detection model named IntrusionTree is implemented using Jupyter Notebook. Mutual Information based method is used and implemented in Jupyter Notebook for the reduction of dimensionally. The complete system is driven by an Intel Core i5 processor running at 2.20 GHz and 4 GB of RAM. The success of the suggested Intrusion Detection model is estimated using the publicly accessible dataset NSL-KDD.

Table 5. Implementation environment

Environment	Description
Personal Computer	Windows 10, 2.20GHz processor speed, 4 GB RAM
Programming Language	Version 3.8 of Python with ML libraries
Implementation Platform	The proposed model is developed and put into practise using Jupyter Notebook. The development of AI-based applications also uses Jupyter.
Database management System	MS Excel 2016

Evaluation Matrix

The efficiency of IDS is evaluated using significant metrics named Accuracy, Recall, Precision, and Fscore. The values depend upon the following terms (Gomes et al., 2020):

TP: It represents "True positives," or the number of rows that were actually classified as intrusion.
TN: It denotes true negatives that means the number of rows truly identified normally.
FP: It represents the amount of rows that were safe but were mistakenly identified as intrusions, or false positives.

FN: t represents the amount of rows that are dangerous but were mistakenly identified as normal as a false negative.

$$\text{Accuracy} = \frac{TN + TP}{TN + FP + TP + FN} \tag{6}$$

$$\text{Precision} = \frac{TP}{TP + FP} \tag{7}$$

$$\text{Recall} = \frac{TP}{TP + FN} \tag{8}$$

$$\text{F1 Score} = 2 * \frac{Precision * Recall}{Precision + Recall} \tag{9}$$

Experiment 1

Outcome findings are provided in this subsection to demonstrate that our suggested IntrusionTree, which is based on machine learning technique, is capable of effectively detecting cyber intrusions and is able to produce results for unknown test situations. Training and test subsets have been made for the model's testing in order to determine the outcomes. To report true positives, false positives, true negatives, and false negatives, a normalised confusion matrix is created. The confusion matrix is shown in Figure 6. The numbers are 15468, 12108, 1380, and 144, respectively, for true positives, true negatives, false positives, and false negatives. Table 6 displays the prediction results for the classes "attack" and "normal" based on these values in terms of precision, accuracy recall, and fscore.

Table 6. Results of each class's specific proposed intrusion tree model

Class	Accuracy	Precision	Recall	Fscore
Normal	97%	96.7%	97%	96%
Attack	97%	97%	96.7%	97%

Figure 6. Normalized confusion matrix

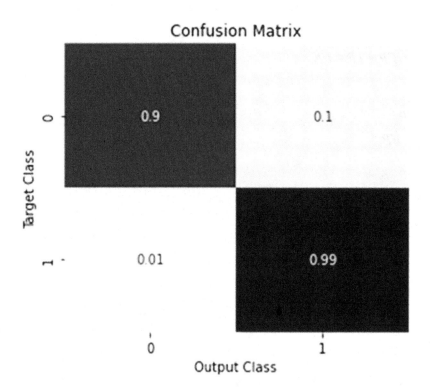

Experiment 2

Through this experiment, we can assess how well our intrusion detection model for cyber security stacks up against widely used machine learning-based techniques. First, we selected baseline techniques like k-Nearest Neighbor (KNN), well known Logistic Regression (LR), and Support Vector Machines(SVM) to compare our suggested model with conventional techniques. We have used same dataset for computing the results with these traditional methods, so that there will be fair comparison. Feature Encoding and scaling are also performed for all these models upon the same dataset. Same set of training and testing data is used for each of the machine learning-based model.

Results for each method (technique) are shown in Figure 7 along with comparisons to our suggested model using the evaluation metrics. Figure 7 demonstrates that our suggested model is superior to the established techniques. The causes of the improved performance are data pre-processing which removes the redundant data, feature encoding which converts the nominal data into numerical, scaling which helps in uniform distribution of the data and selection of best features on the basis of mutual information of all the features. It also minimizes the over-fitting issues. The model is capable to reduce the computational complexity as well as able to enhance the prediction rate for unseen test cases. y examining the data and doing comparisons, it is clear that our IntrusionTree model outperforms other conventional machine learning categorization techniques.

Figure 7. Analysis of the precision, accuracy, recall, and fscore of various classical machine learning models

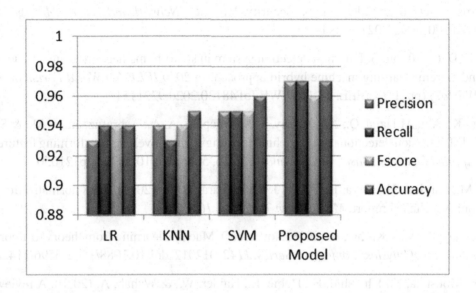

CONCLUSION AND FUTURE WORK

In this research, we present a security model for intelligent Intrusion detection that can competently and successfully identify assaults and unknown test scenarios. The suggested model uses binary classification for its purpose (normal and attack). The key machine learning approach steps, encoding and rescaling, have also been applied. The Gini index is used as a splitting criterion in the suggested system (IntrusionTree). In order to choose the best characteristics that have the most control on classification, feature selection uses the mutual information of each feature. Using independent testing data files from the benchmark NSL-KDD dataset, the suggested model is examined. The overall accuracy of binary categorization is excellent. The choice of features is crucial for improving accuracy. The proposed model's effectiveness has been demonstrated in the experiments in terms of Accuracy, Recall, Precision, and F Score.

Our next effort will involve evaluating our proposed work with more extents (dimensions) of IoT security characteristics. The focus of subsequent research is on machine learning-based security modelling while also working to make security models lightweight.

Funding: No financial support was provided to the authors for this research.
Conflicts of Interest: The authors of the above research declare no conflicts of interest.

REFERENCES

Ahmad, Z., Shahid Khan, A., Wai Shiang, C., Abdullah, J., & Ahmad, F. (2021). Network intrusion detection system: A systematic study of machine learning and deep learning approaches. *Transactions on Emerging Telecommunications Technologies*, *32*(1), e4150. doi:10.1002/ett.4150

Al-Omari, M., Rawashdeh, M., Qutaishat, F., Alshira'H, M., & Ababneh, N. (2021). An intelligent tree-based intrusion detection model for cyber security. *Journal of Network and Systems Management*, *29*(2), 1–18. doi:10.100710922-021-09591-y

Alalade, E. D. (2020, June). Intrusion detection system in smart home network using artificial immune system and extreme learning machine hybrid approach. In *2020 IEEE 6th World Forum on Internet of Things (WF-IoT)* (pp. 1-2). IEEE. 10.1109/WF-IoT48130.2020.9221151

Albulayhi, K., Abu Al-Haija, Q., Alsuhibany, S. A., Jillepalli, A. A., Ashrafuzzaman, M., & Sheldon, F. T. (2022). IoT intrusion detection using machine learning with a novel high performing feature selection method. *Applied Sciences (Basel, Switzerland)*, *12*(10), 5015. doi:10.3390/app12105015

Ali, P. J. M., Faraj, R. H., Koya, E., Ali, P. J. M., & Faraj, R. H. (2014). Data normalization and standardization: A technical report. *Mach Learn Tech Rep*, *1*(1), 1–6.

Alzubi, J., Nayyar, A., & Kumar, A. (2018, November). Machine learning from theory to algorithms: An overview. *Journal of Physics: Conference Series*, *1142*, 012012. doi:10.1088/1742-6596/1142/1/012012

Asharf, J., Moustafa, N., Khurshid, H., Debie, E., Haider, W., & Wahab, A. (2020). A review of intrusion detection systems using machine and deep learning in internet of things: Challenges, solutions and future directions. *Electronics (Basel)*, *9*(7), 1177. doi:10.3390/electronics9071177

Aslahi-Shahri, B. M., Rahmani, R., Chizari, M., Maralani, A., Eslami, M., Golkar, M. J., & Ebrahimi, A. (2016). A hybrid method consisting of GA and SVM for intrusion detection system. *Neural Computing & Applications*, *27*(6), 1669–1676. doi:10.100700521-015-1964-2

Badillo, S., Banfai, B., Birzele, F., Davydov, I. I., Hutchinson, L., Kam-Thong, T., & Zhang, J. D. (2020). An introduction to machine learning. *Clinical Pharmacology and Therapeutics*, *107*(4), 871–885. doi:10.1002/cpt.1796 PMID:32128792

Baştanlar, Y., & Özuysal, M. (2014). Introduction to machine learning. *miRNomics: MicroRNA biology and computational analysis*, 105-128.

Denning, D. E. (1987). An intrusion-detection model. *IEEE Transactions on Software Engineering*, *SE-13*(2), 222–232. doi:10.1109/TSE.1987.232894

Diro, A., Mahmood, A., & Chilamkurti, N. (2021). Collaborative Intrusion Detection Schemes in Fog-to-Things Computing. *Fog/Edge Computing For Security, Privacy, and Applications*, 93-119.

Dong, R. H., Yan, H. H., & Zhang, Q. Y. (2020). An Intrusion Detection Model for Wireless Sensor Network Based on Information Gain Ratio and Bagging Algorithm. *International Journal of Network Security*, *22*(2), 218–230.

Gokhale, P., Bhat, O., & Bhat, S. (2018). Introduction to IOT. *International Advanced Research Journal in Science. Engineering and Technology*, *5*(1), 41–44.

Gomes, C. M. A., Lemos, G. C., & Jelihovschi, E. G. (2020). Comparing the predictive power of the CART and CTREE algorithms. *Avaliação Psicológica*, *19*(1), 87–96.

HaddadPajouh, H., Dehghantanha, A., Parizi, R. M., Aledhari, M., & Karimipour, H. (2021). A survey on internet of things security: Requirements, challenges, and solutions. *Internet of Things*, *14*, 100129. doi:10.1016/j.iot.2019.100129

Ingre, B., Yadav, A., & Soni, A. K. (2018). Decision tree based intrusion detection system for NSL-KDD dataset. In Information and Communication Technology for Intelligent Systems (ICTIS 2017)-Volume 2 2 (pp. 207-218). Springer International Publishing. doi:10.1007/978-3-319-63645-0_23

Ioannou, C., & Vassiliou, V. (2020, May). Experimentation with local intrusion detection in IoT networks using supervised learning. In *2020 16th International Conference on Distributed Computing in Sensor Systems (DCOSS)* (pp. 423-428). IEEE. 10.1109/DCOSS49796.2020.00073

Javaid, A., Niyaz, Q., Sun, W., & Alam, M. (2016, May). A deep learning approach for network intrusion detection system. In *Proceedings of the 9th EAI International Conference on Bio-inspired Information and Communications Technologies (formerly BIONETICS)* (pp. 21-26). 10.4108/eai.3-12-2015.2262516

Jijo, B. T., & Abdulazeez, A. M. (2021). Classification based on decision tree algorithm for machine learning. *Evaluation, 6*, 7.

Li, Y., Xia, J., Zhang, S., Yan, J., Ai, X., & Dai, K. (2012). An efficient intrusion detection system based on support vector machines and gradually feature removal method. *Expert Systems with Applications*, *39*(1), 424–430. doi:10.1016/j.eswa.2011.07.032

Malik, A. J., & Khan, F. A. (2018). A hybrid technique using binary particle swarm optimization and decision tree pruning for network intrusion detection. *Cluster Computing*, *21*(1), 667–680. doi:10.100710586-017-0971-8

Meng, W., Tischhauser, E. W., Wang, Q., Wang, Y., & Han, J. (2018). When intrusion detection meets blockchain technology: A review. *IEEE Access : Practical Innovations, Open Solutions*, *6*, 10179–10188. doi:10.1109/ACCESS.2018.2799854

Mohammed, M. N., & Ahmed, M. M. (2019). Data Preparation and Reduction Technique in Intrusion Detection Systems: ANOVA-PCA. *International Journal of Computer Science and Security*, *13*(5), 167–183.

Moon, D., Im, H., Kim, I., & Park, J. H. (2017). DTB-IDS: An intrusion detection system based on decision tree using behavior analysis for preventing APT attacks. *The Journal of Supercomputing*, *73*(7), 2881–2895. doi:10.100711227-015-1604-8

Nguyen, T. D., Marchal, S., Miettinen, M., Fereidooni, H., Asokan, N., & Sadeghi, A. R. (2019, July). DÏoT: A federated self-learning anomaly detection system for IoT. In *2019 IEEE 39th International conference on distributed computing systems (ICDCS)* (pp. 756-767). IEEE. 10.1109/ICDCS.2019.00080

Kaggle. (n.d.). *Home*. Kaggle. https://www.kaggle.com/

Osanaiye, O., Cai, H., Choo, K. K. R., Dehghantanha, A., Xu, Z., & Dlodlo, M. (2016). Ensemble-based multi-filter feature selection method for DDoS detection in cloud computing. *EURASIP Journal on Wireless Communications and Networking, 2016*(1), 1–10. doi:10.118613638-016-0623-3

Portilla, J., Mujica, G., Lee, J. S., & Riesgo, T. (2019). The extreme edge at the bottom of the Internet of Things: A review. *IEEE Sensors Journal, 19*(9), 3179–3190. doi:10.1109/JSEN.2019.2891911

Rai, K., Devi, M. S., & Guleria, A. (2016). Decision tree based algorithm for intrusion detection. *International Journal of Advanced Networking and Applications, 7*(4), 2828.

Remesh, A., Muralidharan, D., Raj, N., Gopika, J., & Binu, P. K. (2020). Intrusion Detection System for IoT Devices. *2020 International Conference on Electronics and Sustainable Communication Systems (ICESC)*, Coimbatore, India. 10.1109/ICESC48915.2020.9155999

Sarker, I. H., Abushark, Y. B., Alsolami, F., & Khan, A. I. (2020). Intrudtree: A machine learning based cyber security intrusion detection model. *Symmetry, 12*(5), 754. doi:10.3390ym12050754

Sarker, I. H., Furhad, M. H., & Nowrozy, R. (2021). Ai-driven cybersecurity: An overview, security intelligence modeling and research directions. *SN Computer Science, 2*(3), 1–18. doi:10.100742979-021-00557-0

Sarker, I. H., Kayes, A. S. M., Badsha, S., Alqahtani, H., Watters, P., & Ng, A. (2020). Cybersecurity data science: An overview from machine learning perspective. *Journal of Big Data, 7*(1), 1–29. doi:10.118640537-020-00318-5

Sarwar, A., Hasan, S., Khan, W. U., Ahmed, S., & Marwat, S. N. K. (2022, March). Design of an advance intrusion detection system for IoT networks. In *2022 2nd International Conference on Artificial Intelligence (ICAI)* (pp. 46-51). IEEE. 10.1109/ICAI55435.2022.9773747

Shone, N., Ngoc, T. N., Phai, V. D., & Shi, Q. (2018). A deep learning approach to network intrusion detection. *IEEE Transactions on Emerging Topics in Computational Intelligence, 2*(1), 41–50. doi:10.1109/TETCI.2017.2772792

Slevi, S. T., & Visalakshi, P. (2021, November). A survey on Deep Learning based Intrusion Detection Systems on Internet of Things. In *2021 Fifth International Conference on I-SMAC (IoT in Social, Mobile, Analytics and Cloud)(I-SMAC)* (pp. 1488-1496). IEEE. 10.1109/I-SMAC52330.2021.9641050

Swamy, S. N., Jadhav, D., & Kulkarni, N. (2017, February). Security threats in the application layer in IOT applications. In *2017 International conference on i-SMAC (iot in social, mobile, analytics and cloud)(i-SMAC)* (pp. 477-480). IEEE. 10.1109/I-SMAC.2017.8058395

Tariq, N. (2018). Impact of cyberattacks on financial institutions. *Journal of Internet Banking and Commerce, 23*(2), 1–11.

Tavallaee, M., Bagheri, E., Lu, W., & Ghorbani, A. A. (2009, July). *A detailed analysis of the KDD CUP 99 data set. In 2009 IEEE symposium on computational intelligence for security and defense applications*. IEEE.

Umer, M. A., Junejo, K. N., Jilani, M. T., & Mathur, A. P. (2022). Machine learning for intrusion detection in industrial control systems: Applications, challenges, and recommendations. *International Journal of Critical Infrastructure Protection*, *38*, 100516. doi:10.1016/j.ijcip.2022.100516

Vinayakumar, R., Alazab, M., Soman, K. P., Poornachandran, P., Al-Nemrat, A., & Venkatraman, S. (2019). Deep learning approach for intelligent intrusion detection system. *IEEE Access : Practical Innovations, Open Solutions*, *7*, 41525–41550. doi:10.1109/ACCESS.2019.2895334

Vujović, Ž. (2021). Classification model evaluation metrics. *International Journal of Advanced Computer Science and Applications*, *12*(6), 599–606. doi:10.14569/IJACSA.2021.0120670

Yin, C., Zhu, Y., Fei, J., & He, X. (2017). A deep learning approach for intrusion detection using recurrent neural networks. *IEEE Access : Practical Innovations, Open Solutions*, *5*, 21954–21961. doi:10.1109/ACCESS.2017.2762418

Chapter 8
Future Outlier Detection Algorithm for Smarter Industry Application Using ML and AI:
Explainable AI and ML for Smart Industry Evolution Using ML/AI Algorithms and Implementations

Kunal Dhibar

Bengal College of Engineering and Technology, India

Prasenjit Maji

ⓘ https://orcid.org/0000-0001-8057-6963

Dept of CSD, Dr. B.C. Roy Engineering College, Durgapur, India

ABSTRACT

Throughout many real-world investigations, outliers are prevalent. Even a few aberrant data points can cause modeling misspecification, biased parameter estimate, and poor forecasting. Outliers in a time series are typically created at unknown moments in time by dynamic intervention models. As a result, recognizing outliers is the starting point for every statistical investigation. Outlier detection has attracted significant attention in a variety of domains, most notably machine learning and artificial intelligence. Anomalies are classified as strong outliers into point, contextual, and collective outliers. The most significant difficulties in outlier detection include the narrow boundary between remote sites and natural areas, the propensity of fresh data and noise to resemble genuine data, unlabeled datasets, and varying interpretations of outliers in different applications.

DOI: 10.4018/978-1-6684-8785-3.ch008

INTRODUCTION

Most real-world datasets contain data observations that are unlikely to correspond to the particular framework and/or characteristics of such dataset. Outliers are assertions that differ markedly from most of the data points found in the datagram. Outlier identification is a problem that must be addressed in a variety of usage, which include fake prevention and detection (e.g., potentially malicious utilization credit and debit cards or even different kinds of monetary transactions), healthcare information analyzation (e.g., capable of recognizing dynamics are changing to therapeutic interventions among patient populations), fault detection in production processes, and detection of network intrusions, among others. Additionally, the presence of outliers effects several data processing tasks, necessitating the limitation or elimination of outlier observations. Recognizing outliers in multidimensional data is a tough task that gets increasingly difficult when dealing with high-dimensional datasets.

Outlier identification is a significant topic in data mining which has been widely investigated for many years by different scholars. The definition of an anomaly is "an occurrence in a dataset that seems to be incongruous with the balance of that the collection of data," according to. Mining outliers' methodologies may be utilized in a range of areas, notably identifying credit card fraud, detection of network intrusions, and environmental monitoring. In feature extraction, there are two primary missions. To begin, we should define what data are considered outliers in a given set. Secondly, a productive technique for calculating these outliers is required. The outlier problem was initially investigated by the statistical community. They assume that the presented dataset is normally distributed, and an entity is referred to as an outlier if it differs significantly considerably from these dividends. In contrast, finding a suitable allocation for responsible for considerable is practically difficult. To solve the above noted shortcoming, the information management industry has proposed a number of model-free solutions. Outliers based on distance and outliers based on density are two examples.

Outlier detection is a prominent subject in the data mining field since it intends to find trends that happen seldom when compared to other data mining methodologies. An outlier is a finding that differs substantially from and then contradicting the main section of a datagram, as if though created by a separate methodology. Outlier detection is critical because outliers may provide both raw patterns and actionable insights about something like a dataset. Outlier detection is used in a broad variety of situations, as well as criminal identification, credit card fraud analysis, detection of network intrusions, medical diagnostics, defective detection in essential safety systems, and image processing abnormalities detection. There has been an awful lot of effort recently in outlier detection research, with several approaches described. Current research on outlier detection may be classified into three types based on whether or not label evidence is accessible or can be used to create outlier detection approach: unstructured, supervised, and semi-supervised strategies.

Figure 1. Different outlier approach

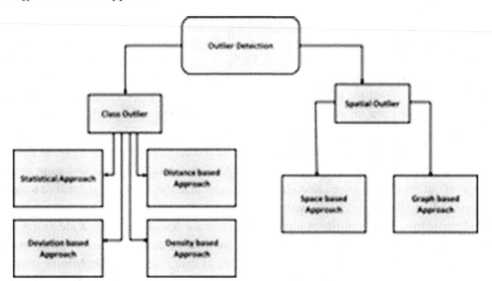

LITERATURE REVIEW

Hawkins established an accepted definition of outlier, stating that outliers are measurements that differ considerably from the rest of the data. For something like the interpretation of significant, substantial, and meaningful data, outlier identification is essential. Outlier detection techniques are used to find anomalies in information contexts including such higher dimensional data, unpredictable data, real - time processing, network data, and time series data.

- Outlier recognition is achievable through initial data set creation examination and model analysis.
- By just using the permitted operations. This characteristic relies on algorithmic statistics that favors one controller over another.

Christy proposed employing spatial outlier recognition and distance-based background subtraction algorithms to locate oddities. Their solution has three constructed biomedical datasets. In regards of precision, their exploratory outcomes show that perhaps the cluster-based outlier classification strategy surpasses the distance-based outlier detection approach. Across geographical and temporal datasets, the results of R. P.S. Manikandan's cluster-based K-medoids methodology are evaluated using the ROSE (Rough Outlier Set Extraction) approach. To form clusters, they used the K-means algorithm. His novel strategy, as contrasted to the ROSE method, consumes less calculating effort and has a lower error rate. In addition, the KSVM approach raises the total amount of outliers.

Bin Wang produced the very first paper on Distance-based outlier identification on incomplete data, realizing the need to discover outliers on incomplete data because existing algorithms only dealt with deterministic data. He introduced a dynamic programming approach and a grid-based reduction methodology for spotting anomalies in data sets. Outlier detection issues in time series data streams were identified by Salman Ahmed Shaikh and Hiroyuki Kitagawa, who proposed an approach that relies on a continuously distance-based outlier detection algorithm. This approach involves progressive processing of state-changing items. The proposed state change cell-based technique is used to identify the most likely outliers, as well as the algorithm's effectiveness is evaluated at the conclusion of processing.

Ville Hautamaki identified flaws in the prior k-nearest neighbor graph methodology and developed a graph-based outlier identification method based on kNN. Experiments are done using both real and fictitious datasets. This method works well with synthetic datasets but not so well with real ones. This model recognizes the causes of poor performance on real-world datasets. Sourajit Behera and Rinkle Rani examined numerous density-based algorithms, including LOF, OPTICS, DBSCAN, and DENCLUE, using Breast Cancer Data. Single cluster Hadoop, noise precision identification level, frequency of abnormal occurrences identified on high-dimensional information, the ability to handle fluctuating density, input variables and sophistication, and so forth are the elements utilized to draw the comparison. The execution time to build clusters was used to compare the results with other algorithms, and the comparative analysis shows that the OPTIC algorithms outperform other approaches.

Charu C. Agrawal created a fuzzy inference system on dubious datasets using a density-based outlier classification method. In this proposed technique, they used various probabilistic computations in addition to density estimation and sampling. They have also used micro clustering properties to efficiently examine large datasets. They compared their strategy to deterministic strategies and found that perhaps the probability - based model outperformed them all. Seung Kim suggested using KD tree encoding to minimize calculation costs on very large log data sets. Because the computation time of a density-based approach is rather long, their suggested solution lowers the detection accuracy of a widely used density-based LOF algorithm. Jihyn Ha discovered two weaknesses in previous distance-based and density-based methods. The local density challenge, low-density pattern challenge, and mathematical model susceptibility are among them. To address these factors, two new observations have indeed been recommended: local identification of outliers and universal identification of anomalies.

Armin Danshpazhouh and Ashkan Sami developed an outlier detection methodology based on entropy. Before disclosing the top N outliers, they rescued unambiguous negative samples from formerly labelled positive data. The bulk of outlier detection methods make use of either or both supervised and unsupervised learning. This recommended part employs semi-supervised learning. Researchers also compared its application to other learning methodologies. Chang-Tien Lu devised a method for identifying geographic outliers that solves the shortcomings of the spatial outlier associated set of processes. Their proposed method is effective in identifying topographical outliers. The research was conducted utilizing a real-world demographic dataset. Their proposed method detects false spatial outliers but outperforms in finding true spatial outliers. This article presents an overview of three frequently used outlier identification strategies based on its research in diverse fields: Cluster-Driven, Concentration-Centered, and Distance-Based detection algorithm.

DISCUSSION

Current strategies for generic outlier identification may be generally classified into four basic approaches: statistical distribution-based approaches, distance-based approaches, density-based approaches, and subspace-learning approaches.

The quantitative dispersion techniques use a "discordance test" with respect to some known properties of the database, which might include the mean, variance, and/or an anticipated distribution of the data, to identify any outlier observations in relation to the selected model. The majority of methodologies in this classification are intended for univariate datasets, or data sources with a single characteristic; but nevertheless, several problems concerning outlier detection in multidimensional datasets tried to present the COPOD outlier detection method, which was inspired by quantitative methodology to model multiple regression distribution of the data. COPOD first builds the experimental copula and afterwards uses the fitted model to determine the tail frequency of each integrative approach in order to determine whether it constitutes a conventional or outlier occurrence. The probabilistic dispersion of the information and the accompanying parameters for the dataset framework are not always established from the beginning, which is a key difficulty for approaches that rely on it. Moreover, the statistical features of the dataset may influence outlier detection owing to the masking or swamping effect.

Outliers are discovered using distance-based approaches, which measure the disparities among data points. Outliers are information observations that are missing a sufficient number of nearby observations within a given distance threshold. The Functional Outlier Map (FOM) approach was the first attempt to use graphical tools to locate outliers in multidimensional data obtained. These techniques employ statistical depths calculations and dynamic topological derived from them to find outliers. Prykhodko et al. found outliers in multidimensional kinds of behaviors data using both univariate and multivariate normalization processes. They used the squared Mahalanobis distance as well as a quantile of the Chi-Square distribution to accomplish this. Caberoa et al. proposed an outlier detection approach that merges projections into relevant subspaces using archetype analysis and a nearest-neighbor algorithm. A key problem with distance-based algorithms is their rely on global dataset information, as their effectiveness is dependent on the neighborhood width of observation, particularly addition to their dependency on statistical dataset features (e.g., mean values).

Density-based approaches rely on the local outlier factors of data points, which are derived by taking the local density of their neighborhoods into consideration. While techniques in this category achieve adequate accuracy without relying on any presumptions about the dataset's dispersion, they have a substantial computational cost, especially for high dimensional huge datasets. As one of the subspace-learning-based outlier detection techniques, Zhao et al. introduced LOMA, a local outlier recognition approach for massive high-dimensional datasets. LOMA uses attribute relevance analysis to decrease data. Additionally, it employs particle swarm optimization to efficiently explore sparse subset of features with low data density, i.e., a limited handful of the data set's observations. Our proposed approach is similar in that it employs subspace-learning, however unlike property significance analysis, we employ the analysis of principal components to minimize dimension. The situation in which the training dataset already knows whether each occurrence is typical or unusual is alluded to as supervised outlier detection. One type of support vector machine (OCSVM) or support vector data description (SVDD) creates a hypersphere around normal data and utilizes it to identify an unknown sample as an inlier or outlier. The supervised outlier detection problem is troublesome in many real-world applications because gathering labelled data across the entire training sample is generally expensive, time demanding, and subjective.

Unsupervised outlier identification algorithms without previous knowledge of the class distribution are broadly characterized as distribution-based, distance-based, density-based, and clustering-based. Outliers do not obey the distribution-based method, which implies that all data points are created by a certain statistical model. Unfortunately, in many real-world applications, the assumption of an underlying distribution of data points is not always accessible. Distance-based approach was firstly investigated by Knox and Ng. An object o in a dataset X is an outlier if at least $p\%$ of objects in X are further than the distance D from o. The global parameters p and D are not suitable when the local information of the dataset varies greatly. Representatives of this type of approaches include *K*-nearest neighbor (*KNN*) algorithm and its variants.

No approach outperformed the others across all four datasets. PCA fared best for the astrophysics and F-MNIST datasets, although negatively selection and exclusion forest become quite for the Earth science and planetary datasets, correspondingly. This highlights the need of providing a diverse selection of algorithms and ways for fast comparing them in DORA, because the best strategy will vary based on the dataset. The purpose of this study was to demonstrate how DORA could be utilized to simplify outlier identifying trials and compare results among datasets from different geographies. As a result, hyperparameter tweaking, which might improve outcomes for every dataset, is not possible; we will investigate this more in the future. Prior research has emphasized the challenges of defining uniform outlier identification criteria that effectively portray how models will perform in real-world scenarios while also permitting cross-dataset comparability (Campos et al., 2016). With this in mind, we chose two complementary metrics: accuracy at N = n outliers, which measures the percentage of known outliers when the number of options equals the number of outliers, and Mean Discovery Rate, which measures the average fraction of known outliers. It is very difficult to develop tests to evaluate outlier detection algorithms for real-world use cases since acquiring labelled samples of outliers, inliers, or both for evaluation is difficult, if not impossible. Moreover, labelling might be subjective or inaccurate, especially in the case of scientific datasets. In prior work on the astrophysics dataset, for example, a dataset of known outliers was accessible through human annotation, but the remaining samples in the dataset used for evaluation were not known to be inliers or outliers. Unlabeled samples that are actually outliers (as was revealed to be common in prior work (Wagstaff et al., 2020a)) are reported as false positives, resulting in artificially low evaluation metrics.

Table 1. Big data stream with outlier detection

(Maiorana, Li, 2017) **HSDE-based outlier detection method**
Input: *a // The time series* *Max p // Top of the hierarchy* *k // the k in k-distance* *λ // local outlier factor limit LOF* *Output*: **LOF** *(a)*
Step 1*: HSS = HSDE (a)* **Step 2***: **foreach** aj in HSS* **do** *{* *Calculate k-distance neighborhood of a***j**, *Calculating the object's routing range a***j**; *Determining the ρ***k** **(a)**; *Assessing the LOF* **(aj)**; *if (LOF(aj) ≥ λ)* *Label aj as outlier;* *}* *Step 3:* **R***eturn LOF(a)* *// Where ρ(x)* **is a** *level of the local density of the object x. ρ(x) is the ratio of the number of the reachability distances of the object x and the total sum of the reachability distance of the objects.* *LOF (aj) =* **1 -** *ρk* **(a)**
(Mei, Derong, 2017) **UELM based outlier algorithm**
Input: *The training data: A ∈ RN×d.* *Output*: **T***he label vector of cluster yi* **corresponding** *to xi*
Step 1*: Laplacian* **L** *of* **A** *Graph is created* **Step 2***: {a*i*, b*i*}* **ra***ndom values are generated for each hidden neuron,* *and H ∈ RN×L output matrix is calculated;* *Step 3:* **if** *L ≤ N* **then** **F***ind the generalized Eigen vectors k2, . . .,* **kM+1 of Equa***tion* **11.** **Let** *β = [k̃2, k̃3, . . ., k̃M+1].* **else** **Fin***d the generalized Eigen vectors u2, . . ., uM+1 of Equation 13.* **Let** *β = H[ũ2, ũ3, . . ., ũM+1].* **Step 4: Calcul***ate th***e** *embedding matrix: E = Hβ;* *// E a***s a poi***nt a***re** *treated as each row of points, and cluster the N points into K* **C***lusters using* **t***he k-means algorithm.* *Step 5: Let P be* **t***he* **lab***el vector of cluster index for all the points.* *Step 6: Return P;*
(Mei, Derong, 2017) **CB outlier detection on each slave node**
Input. *The cluster set SETS* **res***erved on slave node S,* **int***egers k, n,* **t***he* **t***hreshold L B* *Output.* **T***he* **CB** *outliers in P*

Continued on following page

Table 1. Continued

Step 1: *foreach cluster Ci in SETS* **do**
St*ep 2:* **In** *Ci increasing order, points are ordered based on their distances from the centroid point;*
Step 3: **Points** *are scanned in reverse* **order;**
Step 4: **foreac***h scanned point p*
 do
Initialize a heap nn; // To reserve the current kNNs of p
k distemp = ∞; // The largest distance from the points in nn to p
boolean potential_outlier = true;
Visit the points from p to the both sides to search p's kNNs;
Step 5: foreach visited point q
 do
if *q is before p and*
            dis(q, C.centr) < dis(p, C.centr) − k distemp then
Before q from the list visited points are erased;
else if q is behind p and
            dis(q, C.centr) > dis(p, C.centr) + k distemp then
Erase the points behind q from the visited list;
 else
Update nn and k distemp;
Step 6: if nn meets the **con***dition of Corollary 5 then*
potential_outlier = false;
 break
Step 7: **if** *potential_outlier then*
*Send p to the master node to updat***e** *LB;*

APPLICATOIN / USUAGE

ML-enable IoT applications: Machine learning (ML) and artificial intelligence (AI) are playing an increasingly important role in the development of outlier detection algorithms in IoT applications. Outlier detection is an essential component of IoT systems as it can help identify anomalous events or behavior that may indicate a problem or a threat. These algorithms can learn from historical data and identify patterns and anomalies that may not be immediately apparent to human operators. ML-based outlier detection algorithms can be supervised or unsupervised, depending on whether labeled training data is available. Supervised outlier detection algorithms use labeled data to learn what is normal and what is anomalous, allowing them to identify outliers in real-time data streams. Unsupervised outlier detection algorithms, on the other hand, use statistical techniques such as clustering and density estimation to identify data points that fall outside of the normal range. The use of machine learning and artificial intelligence in outlier detection algorithms is becoming increasingly important in a wide range of IoT applications such as healthcare, manufacturing, transportation, and many others. However, implementing these algorithms can be challenging due to issues related to data quality, lack of standardization in data formats, and privacy concerns. Nevertheless, the benefits of using ML-based outlier detection algorithms in IoT applications are substantial and are expected to grow as the field of IoT continues to evolve.

ISOLATION FOREST

(Zhong-Min, Guo-Hao, 2019) The research proposes the Isolation Forest (iForest), a novel isolation approach. Anomalies are expected to be few and varied, and to be more subject to isolation than ordinary measures. Outliers are geographically isolated and have a sparse geographical distribution. Outliers are more likely to occur in sparse areas of a feature space. As a result, readings in these locations may be considered outliers. Measurements in thick portions are difficult to extract when a dataset is partitioned recursively and randomly, necessitating multiple iterations. Measurements in sparse regions, on the other hand, are readily separated and take fewer rounds to divide.

To overcome this issue, iForest splits a dataset using a binary tree, specifically an isolation tree (iTree). In an isolation tree, measurements with shorter pathways are more likely to be outliers. The following is an explanation of iTree. Outlier detection performance of a single isolation tree is often low. Thus, the iForest consists of multiple isolation trees. The construction of the iForest is shown in Algorithm 1.

(Zhong-Min, Guo-Hao, 2019) Algorithm 1- *iForest* (X, t, W)	(Zhong-Min, Guo-Hao, 2019) Algorithm 2- *iNNE generation* (X, t, W)
1: *iForest* ← ∅ 2: $l = ceiling(\log_2 W)$ 3: **for** $i = 1$ **to** t **do** 4: X_i = RandomSubset(X,W) 5: *ConstructTree$_i$* 6: *iForest* = *iForest* ∪ {*iTree$_i$*} 7: **end for** 8: **return** *iForest*	1: *iNNE* ← ∅ 2: **for** $i = 1$ **to** t **do** 4: S_i = *RandomSubset*(D, W) 5: B_i ← ∅ 6: **for** $i = 1$ **to** $\|S_i\|$ **do** 7: $B_i = B_i$ ∪ {B(c)} 8: **end for** 9: iNNE = iNNE ∪ {B_i} 9: **end for** 10: **return** *iNNE*

There are three parameters: the input measurements X, the number of iTrees t, and the number of measurements contained in each iTree W. In addition, the upper bound of the height of the tree is calculated as l *ceiling(*$\log2_w$*)*. The measurements in X are divided into t subsets and the number of measurements in each subset is W. Then, the corresponding iTree of each subset is constructed. Finally, the algorithm returns the collection of iTrees, namely iForest.

ISOLATION USING NEAREST NEIGHBOR ENSEMBLES

(Zhong-Min, Guo-Hao, 2019) The main idea of the isolation using nearest neighbor ensemble (iNNE) is as follows. For an instance x in the training set, the isolation of x is implemented by building a hypersphere that only covers x. The radius of the hypersphere is determined by the distance between x and its nearest neighbor (NN) in the training set. Therefore, if x locates in a sparse area, the corresponding hypersphere is large. On the contrary, a random selected subset S of size 9 D 8. if x locates in a dense area, the corresponding hypersphere is small. In general, outlier instances appear in sparse areas, while normal instances appear in dense areas. Thus, the radius of the hypersphere can be used to detect outliers. The details of the iNNE are as follows.

Table 2. Comparison of different outlier detection algorithm

Variables	Algorithms / Techniques		
	Cluster	**Distance**	**Density**
Computation expense	Less	Less	Moderately High
Effectiveness	Very productive	Moderately useful	Effective
Big density data	reliable	pertinent	beneficial
sophistication	Less intricate	Moderately nuanced	Highly nuanced

The isolation using nearest neighbor ensembles contains two stages: training stage and evaluation stage. In the training stage, t sets of hyperspheres are built from t subsets. The number of elements of each subset is W. Details of the training stage can be found in Algorithm 2. There are three input parameters In Algorithm 2. D is a training dataset, W is the size of a subset, and t is the number of subsets. Similar to the iForest algorithm, the instances in D are divided into t subsets. Then, the corresponding hyperspheres are constructed for each instance in a subset by definition 8. Finally, the algorithm returns the set of these hyperspheres, namely iNNE. As we need to compute the distance between two instances, the time complexity of the training stage is $O\ (tW^2)$. The memory complexity in the training stage is dominated by t sets of hyperspheres, so the memory requirement is $O\ (tW)$, where t and W are constants (Zhong-Min, Guo-Hao, 2019). Insurance companies often face the problem of client moral risk in client management, including information asymmetry, defaulting insurance premium and fraud after accidents. Such problems not only dramatically cut down the profit and damage the development of insurance companies but also threat the property and safety of policyholders and cause the unnecessary waste of social resources. Thus, one of the key tasks' insurance companies have to accomplish is to reduce the potential risk through effective fraud detection, that is to say, to find out exceptional policyholders who may deserve a further scrutiny. Some classifier based on data mining can tell whether an instance is a normal claim or abnormal claim. Outlier detection being able to rank top suspicious instances is more accurate and saves lots of energy and time. The term "outlier" can refer to any single data point of dubious origin or disproportionate influence.

The appearance of noisy data, which could also impair system reliability and readability, may highlight important information. Policyholder documentation may now be saved electronically thanks to advances in data storage technologies. Some policyholder factors, including as age, employment, and credit history, are closely associated to future activities and may show the insured's morality risk. To identify outliers, several risk assessment methodologies have been developed. Traditional statistical approaches are inapplicable since they are based on specific statistical distributions. Employing these methods to scan massive datasets containing millions upon millions of records is neither economical nor practical.

Prominent mining techniques include the K-means algorithm, neural network models, the wave cluster approach, and the density-based outlier mining algorithm (DB-algorithm), which can deal with clusters of any shape and performs well on huge databases. Nevertheless, they require unknown domain-dependent characteristics as inputs, such as the number of clusters. These parameters are not only difficult to get, but their changes have a considerable influence on the outcomes. Furthermore, if a significant number of solitary points or loosely packed clusters arise, a DB approach that concentrates on global context may overlook certain critical outliers that diverge from their local clusters. As a result, the nonparametric outlier mining algorithm is recommended.

Figure 3. Multiple Big data streams using outlier detection

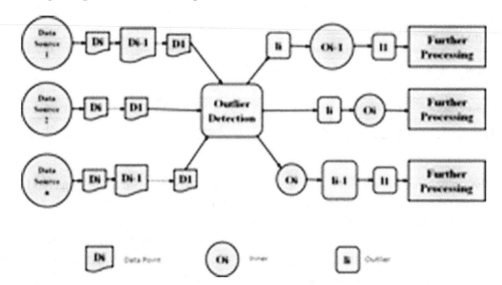

Table 3. Table of pros and cons of outlier detection techniques

Technique	Pros	Cons
Statistical Based	Outliers can be found via temporal correlation.	Due to no prior understanding of data distribution in real life, parametric models are useless.
Clustering Based	Need the selection of a threshold measure to describe data similarity.	Higher processing cost for multivariate data with outlier detection.
Nearest neighbour Based	Easy to use in IoT scenarios with many forms of data.	A multivariate dataset's distance calculation has a high processing cost.
Classification Based	No explicit statistical model or predicted parameters are required.	While more computationally intensive than statistical and clustering approaches, good recognition accuracy.
Artificial Intelligent Based	When adding new data or rules, the system doesn't require any fresh training.	High memory use as the number of rules increases.
PCA Based	It may be applied to large and multivariate datasets.	Higher cost of calculation.
Hybrid Based	Benefit from the advantages of one or more techniques to implement an optimal solution for a specific application.	Using several strategies could raise the cost of calculation and use up the few sensor resources.

(Samara, M.A.; Bennis, I.; 2022)

Key Uses and Challenges

The specific qualities of huge data streams, such as transiency, uncertainty, multiple perspectives, dynamic connection, and dynamic data distribution, make outlier detection more challenging. All off-line store-and-process algorithms for outlier identification are inapplicable to massive data streams since huge data streams are initially infinite and data points are transient. Outlier detection systems should recognize outliers as soon as they come and compare data points gradually to the summary. Furthermore, each data point in a massive data stream is associated with a timestamp. Large data streams are susceptible to variations in rate and accuracy; the timestamp information reveals the relational state. Approaches for identifying outliers typically compare the piece of evidence to the remainder of the data points in the setting. Additionally, as the environment and patterns change, the distribution of data in massive data streams changes over time. A movement in distribution of the data is referred to as concept drift. Outlier detection differs based upon that entity and realm. Approaches to outlier detection should be applicable to related entities and effective of coping with concept drift.

Finally, data points in a big data stream are prone to uncertainty, and in many applications, the arrival rate is high but not constant. External events can jeopardize data sources like sensors. Large data stream outlier detection systems should be able to determine the consistency of questionable data points and adjust processing schedule appropriately. Outlier detection in enormous data streams differs significantly from detection in static data, periodic data, or tiny data streams. Finding entities in data streams that act differently than usual objects is an intriguing study subject for big data mining. Most existing outlier detection methodologies and tactics for enormous data streams, in general, cover just a fraction of the problem's conceptualization and sophistication.

The Drawbacks of Manual Outlier Detection Businesses today manage thousands of pieces of information that may be relevant to their key performance indicators (KPIs), as well as a wide range of outliers to review and evaluate. Whether an anomaly represents a problem or an opportunity, timely intervention is critical to achieving positive outcomes in the context of unpredictability. A single API downtime might cost a corporation money per second. A firm may gain critical time to roll back an improvement and restore income stream by decreasing the period between incidence and detection. Yet, for more than a few dozen measures, individually checking each sign for unusual data points is difficult. Besides which, manually monitoring dashboards and alerts in traditional BI solutions fails to supply organizations with the real-time insight they are seeking to stay ahead of situations before they become excessively costly and damaging. Manual detection is insufficient even when the anomaly presents an opportunity rather than a problem. An extraordinary spike in users or transactions from a certain geographic location, for example, might be the consequence of a successful social media advertising campaign that made headlines in that region. Given the brief length of such surges, your organization has a little window of opportunity to capitalize on and transform that engagement in to other logins and transactions.

CONCLUSION

This research focuses on methods for identifying outliers from the standpoint of data mining. We will first go through the basics of outlier detection. After that, we'll delve into the topic's history and inspirations. We also examine related research on outlier detection. After that, we examine six types of outlier detection techniques: multivariate statistical techniques, depth-based approaches, deviation-based perspectives, distance-based perspectives, high-dimensional frameworks, and density-based methodologies. The fifth section examines three of the most often used outlier detection methods: cluster-based, distance-based, and density-based outlier detection. A few strategies for optimum Outlier detection application in everyday life are indeed offered.

The future of anomaly detection will be proactive rather than reactive, changes produce disruptions, which can have serious consequences for your organization. But, if you can detect the changes in your environment in near real-time, you may avoid these interruptions and hence outages. Being reactive to outages is no longer an option in today's digital business world, as apps power a large percentage of the organization.

REFERENCES

Cassisi, C., Ferro, A., Giugno, R., Pigola, G., & Pulvirenti, A. (2013). Enhancing densitybased clustering: Parameter reduction and outlier detection. *Information Systems*, *38*(3), 317–330. doi:10.1016/j. is.2012.09.001

Shiblee Sadik, M, & Gruenwald, L. (2010). DBOD-DS: Distance-based outlier detection for data streams. *Database Expert Syst Appl.*, *62*(61), 122–136.

Wu, K, Zhang, K, Fan, W, & ... (2014). RS-Forest: A rapid density estimator for streaming anomaly detection. *IEEE Int Conf Data Min.*, *2014*, 600–609. PMID:25685112

Chugh, N, Chugh, M, & Agarwal, A. (2014). Outlier detection in streaming data: a research perspective. *2014 Int Conf Parall Distrib Grid Comput.* (pp. 429–4). IEEE.

Schubert, E, Zimek, A, & Kriegel, HP. (2015). Fast and scalable outlier detection with approximate nearest neighbor ensembles. *Database Syst Adv Appl.*, *90*(50), 19–36.

Yuan, Y, Cao, H, Zhang, Y, & ... (2016). Outlier mining based on neighbor-density-deviation with minimum hypersphere. *Information Technology and Control*, *45*(3), 267–277.

Paul, S. KP Bhaumik,. (2016). *AIDCOR: artificial immunity inspired density based clustering with outlier removal.* Int J Mach Learn Cybernet.

Ding, S., Wu, F., Qian, J., Jia, H., & Jin, F. (2015). Research on data stream clustering algorithms. *Artificial Intelligence Review*, *43*(4), 593–600. doi:10.100710462-013-9398-7

Gandomi, A., & Haider, M. (2015). Beyond the hype: Big data concepts, methods, and analytics. *International Journal of Information Management*, *35*(2), 137–144. doi:10.1016/j.ijinfomgt.2014.10.007

Hido, S., Tsuboi, Y., Kashima, H., Sugiyama, M., & Kanamori, T. (2011). Statistical outlier detection using direct density ratio estimation. *Knowledge and Information Systems*, *26*(2), 309–336. doi:10.100710115-010-0283-2

Namiot, D. (2015). On big data stream processing. *Int J Open Inform Technol.*, *3*(8), 48–51.

Nori, F., Deypir, M., & Sadreddini, M. H. (2013). A sliding window-based algorithm for frequent closed itemset mining over data streams. *Journal of Systems and Software*, *86*(3), 615–623. doi:10.1016/j. jss.2012.10.011

Samara, M. A., Bennis, I., Abouaissa, A., & Lorenz, P. (2022). A Survey of Outlier Detection Techniques in IoT: Review and Classification. *J. Sens. Actuator Netw.*, *11*(1), 4. doi:10.3390/jsan11010004

Silva, A., & Antunes, C. (2015). Multi-relational pattern mining over data streams. *Data Mining and Knowledge Discovery*, *29*(6), 1783–1814. doi:10.100710618-014-0394-6

Sun, D. W., Zhang, G. Y., & Zheng, W. M. (2014). Big data stream computing: Technologies and instances. *Journal of Software, 25*(4), 839–862.

Yeh, M-Y, Dai, B-R, & Chen, M-S. (2007). Clustering over multiple evolving streams by events and correlations. *IEEE Transactions on Knowledge and Data Engineering*, 1349–1362.

Ren, J, Wu, Q, Zhang, J, & ... (2009). *Efficient outlier detection algorithm for heterogeneous data streams*. Six Int Conf Fuzzy Syst Knowl Discov.

- Tan SC, Ting KM, Liu TF. (2011). Fast anomaly detection for streaming data. Proc Twenty-Second Int Jt Conf Artif Intell. (pp. 1511–1516). Springer.

Singh, K, & Upadhyaya, S. (2012). Outlier detection: Applications and techniques. *IJCSI Int J Comp Sci Iss., 9*(3), 307–323.

Yogitaa, D.T. (2012). A framework for outlier detection in evolving data streams by weighting attributes in clustering. *2nd Int Conf Commun Comput Secur.* (pp. 214–222).

Sadik S. & Gruenwald L. (2013). Research issues in outlier detection for data streams. *ACM SIGKDD Explor Newslett arch, 15*(1), 33–40.

Koupaie, HM, Ibrahim, S, & Hosseinkhani, J. (2013). *Outlier detection in stream data by clustering method*. Int J Adv Comput Sci Inf Technol.

Vijayarani, S, & Jothi, P. (2013). An efficient clustering algorithm for outlier detection in data streams. *International Journal of Advanced Research in Computer and Communication Engineering, 2*(9), 3657–3665.

Niennattrakul V, Keogh E, Chotirat AR. (2013). Data editing techniques to allow the application of distance-based outlier detection to streams. *IEEE 13th Int Conf Data Min.* (pp. 947–952). IEEE.

Thakkar P, Vala J, & Prajapati V. (2016). Survey on outlier detection in data stream. *Int J Comput Appl., 136*(2), 0975– 8887.

Almusallam, NY, Tari, Z, Bertok, P, & ... (2016). *Dimensionality reduction for intrusion detection systems in multi-data streams-A review and proposal of unsupervised feature selection scheme*. Emerg Comput.

Xhafa, F., & Naranjo, V. (2015). Processing and analytics of big data streams with Yahoo! *S4. IEEE 29th Int Conf Adv Inform Network Appl.* (pp. 263–270). IEEE. 10.1109/AINA.2015.194

Zhang, Y., Hamm, N. A. S., Meratnia, N., Stein, A., van de Voort, M., & Havinga, P. J. M. (2012). Statistics-based outlier detection for wireless sensor networks. *International Journal of Geographical Information Science, 26*(8), 1373–1392. doi:10.1080/13658816.2012.654493

Pang, LX, Chawla, S, Wei, L, & ... (2013). On detection of emerging anomalous traffic patterns using GPS data. *Data & Knowledge Engineering, 87*(9), 357–373.

Hossein, MK, Ibrahim, S, & Hosseinkhani, J. (2013). Outlier detection in stream data by clustering method. *Int J Adv Comp Sci Inform Technol., 2*(3), 25–34.

Smrithy, GS, Munirathinam, S, & Balakrishnan, R. (2016). Online anomaly detection using non-parametric technique for big data streams in cloud collaborative environment. [Big Data]. *IEEE Int Conf Big Data*, *2016*, 1950–1955.

Zhou, X.-Y. (2007). A fast outlier detection algorithm for high dimensional categorical data streams. *Journal of Software*, *18*(4), 933–942. doi:10.1360/jos180933

Elahi, M. (2008). Efficient clustering-based outlier detection algorithm for dynamic data stream. *IEEE Fifth Int Conf Fuzzy Syst Knowl Discov*. (pp. 298–304). IEEE.

Bhosale, SV. (2014). Outlier detection in straming data using clustering approached. *International Journal of Computer Science and Information Technologies*, *5*(5), 6050–6053.

Sreevidya, SS. (2015). Detection of outliers in data stream using clustering method. *Int J Sci Eng Technol Res.*, *4*(3), 559–563.

Kumar, D, Bezdek, JC, Rajasegarar, S, & ... (2016). Adaptive cluster tendency visualization and anomaly detection for streaming data. *ACM Transactions on Knowledge Discovery from Data*, *11*(2), 24–40.

Chapter 9
Edge Computing:
Optimizing Performance and Enhancing User Experience

Kavita Srivastava

Institute of Information Technology and Management, GGSIP University, India

ABSTRACT

This chapter discusses the emerging paradigm of edge computing and its potential to optimize performance and enhance user experience in modern computing systems. The chapter begins by introducing the concept of edge computing, its definition, and its various applications. It then highlights the key benefits of edge computing. The chapter also delves into the various architectures and models of edge computing. It examines the challenges associated with edge computing, such as resource constraints, security, and privacy issues. The authors then provide an overview of the technologies and tools used in edge computing. They discuss how these technologies can be used to optimize performance and enhance user experience in edge computing systems. The chapter also presents several use cases and real-world applications of edge computing, including smart homes, autonomous vehicles, and healthcare systems. The authors examine the benefits and challenges of using edge computing in these domains and provide insights into how to optimize performance and enhance user experience.

INTRODUCTION

Edge computing involves processing and analyzing data closer to where the data is actually originated. This is in contrast to using data from a cloud server. Hence, with edge computing, we basically deal with the distributed data rather than centralized one. It makes it a distributed computing paradigm. The term "edge" refers to the outer layer of an organization's network where data is generated, and the processing and analysis of that data occurs. Edge computing aims to reduce data processing time and bandwidth requirements by moving data processing closer to where the data is generated.

Traditional cloud computing models rely on data being sent to centralized data centers for processing and analysis, which can create latency and bandwidth issues, particularly in the case of large-scale data sets or real-time data streams. Edge computing addresses these challenges by performing data process-

DOI: 10.4018/978-1-6684-8785-3.ch009

ing and analysis at the edge of the network, where the data is generated, without the need for data to be sent to a centralized location.

Edge computing has emerged as a result of the increasing demand for real-time, high-performance, and low-latency processing of data generated by Internet of Things (IoT) devices, autonomous systems, and other emerging technologies. By processing data closer to the source of data generation, edge computing offers several advantages over traditional cloud computing models, including reduced latency, improved data security, and enhanced reliability.

In fact, edge computing represents a significant shift in the way organizations process and analyze data, and it has the potential to transform industries ranging from healthcare to manufacturing to transportation.

Literature Survey

Mittal, S. et al., explored the integration of edge computing with cloud computing in order to improve the performance and efficiency of computing systems. The authors first provide an overview of the two computing paradigms, highlighting their respective advantages and disadvantages.

They then propose a hybrid approach that combines the benefits of both edge computing and cloud computing, called the Edge-Cloud architecture. In this architecture, the authors propose that computation is distributed across both edge devices (i.e., devices located closer to the end-users) and cloud servers, depending on the specific requirements of the application.

The paper also discusses various challenges associated with integrating edge and cloud computing, including issues related to network latency, security, and resource allocation. The authors propose solutions to these challenges, such as the use of machine learning algorithms for resource allocation and the use of blockchain technology for enhancing security.

Srivastava, K. et al., presented a solution for implementing edge computing in an Internet of Things (IoT)-based weather monitoring application. The authors propose a system architecture that includes a network of sensors, a gateway, and an edge server, which are designed to collect, process, and transmit weather data to a cloud-based application.

The authors describe the design and implementation of the system, including the hardware and software components used. They also provide experimental results to show the performance of the system, such as the latency and throughput of data transmission between the sensors, gateway, edge server, and cloud.

Cao, K. et al., provided a comprehensive overview of the state-of-the-art research in edge computing. It begins by introducing the concept of edge computing and highlighting its potential benefits in terms of reducing network latency, improving application performance, and enabling real-time processing of data. The authors also discuss the challenges of implementing edge computing, such as the limited computing resources and the need for efficient data management.

The paper then presents a detailed review of the existing literature on edge computing, including various architectures, frameworks, and techniques for deploying and managing edge computing systems. The authors cover topics such as edge intelligence, security and privacy, resource allocation, and network optimization.

Talebkhah, M. et al., provided a comprehensive overview of edge computing, its architecture, applications, and future perspectives. The authors start by introducing the concept of edge computing and its potential benefits in reducing latency, improving QoS, and enabling new applications in different domains. They then provide a detailed analysis of the architecture and components of edge computing, including edge devices, edge servers, and edge gateways. The authors also discuss the different technologies and protocols used in edge computing, such as fog computing, software-defined networking, and virtualization.

The paper then focuses on the applications of edge computing in different domains, including smart cities, healthcare, transportation, and industry 4.0. The authors analyze the key requirements and challenges of edge computing in these domains, such as real-time processing, data security, and energy efficiency. They also provide a detailed analysis of the different use cases and solutions proposed in the literature to address these challenges.

Finally, the authors discuss the future perspectives and challenges of edge computing, including the integration with other emerging technologies such as 5G, AI, and blockchain, and the need for standardization and interoperability.

Xue, H. et al., presented a comprehensive survey of the current state-of-the-art in edge computing for the Internet of Things (IoT). The authors first introduce the concept of edge computing and its potential benefits in IoT, such as reducing latency, improving reliability, and reducing network traffic. They then provide an overview of the key technologies and components of edge computing, including edge devices, edge servers, and edge middleware.

The authors then survey the existing literature on edge computing for IoT, focusing on the key issues and challenges in this area, such as resource management, security and privacy, and service orchestration. They analyze the different approaches and solutions proposed in the literature, such as edge caching, edge analytics, and edge intelligence.

The authors also identify and discuss several open research issues and future directions for edge computing in IoT, such as edge intelligence and machine learning, edge orchestration and automation, and edge security and privacy.

Fan, Z. et al., proposed an edge computing service architecture based on Information-Centric Networking (ICN), which aims to address the challenges of providing efficient and scalable services at the edge. The proposed architecture leverages the ICN paradigm to enable content-based routing, caching, and delivery, which can improve the performance and reliability of edge computing services.

The authors first provide an overview of the proposed architecture and its key components, such as the Content-Centric Service Function (CCSF) and the Service Function Chain (SFC). They then present a detailed analysis of the design and implementation of the proposed architecture, highlighting its benefits and limitations.

The authors also conduct a set of experiments to evaluate the performance of the proposed architecture, comparing it with traditional cloud-based and edge-based service architectures. The results show that the proposed architecture can significantly improve the response time and throughput of edge computing services, especially for content-centric applications.

Wang, Y. et al., provided a survey of the literature on edge computing, with a focus on the mainstream dimensions of edge computing. The authors first introduce the concept of edge computing and its potential benefits, such as reduced latency, improved scalability, and enhanced security. They then identify the mainstream dimensions of edge computing, which include architecture, resource management, security and privacy, and applications.

The authors survey the literature on edge computing, covering various aspects such as architectures, resource management, security and privacy, and applications. They classify the literature based on these mainstream dimensions and provide a critical analysis of the key findings.

The paper provides a valuable contribution to the field of edge computing by providing a comprehensive survey of the literature and highlighting the main trends and challenges in the field. The authors also provide recommendations for future research directions, such as developing new architectures and resource management techniques for edge computing, enhancing the security and privacy of edge computing systems, and exploring new applications and use cases.

Ramya, R. et al., presented a survey of the literature on edge intelligence in IoT-based computing platforms. The authors first provide an overview of the IoT and its various components, such as sensors, gateways, and cloud computing. They then introduce the concept of edge intelligence and its potential benefits, such as reduced latency, improved reliability, and enhanced security.

The authors survey the literature on edge intelligence in IoT-based computing platforms, covering various aspects such as architectures, machine learning algorithms, resource management, and applications. They classify the literature based on these aspects and provide a critical analysis of the key findings.

The paper provides a valuable contribution to the field of edge intelligence in IoT-based computing platforms by providing a comprehensive survey of the literature and highlighting the main trends and challenges in the field. The authors also provide recommendations for future research directions, such as developing new machine learning algorithms and techniques for edge intelligence in IoT-based computing platforms, enhancing the security and privacy of edge intelligence systems, and exploring new applications and use cases.

Srivastava, K. et al., discussed the potential of combining deep learning techniques with edge computing to improve the efficiency and accuracy of data analysis. The author provides an overview of the challenges associated with traditional centralized data processing approaches, such as latency, bandwidth limitations, and security risks.

The paper then describes the concept of edge computing, which involves processing data closer to the source of generation, and highlights the benefits of this approach in terms of improved response times, reduced network traffic, and enhanced data privacy.

The author then discusses the potential of deep learning techniques in edge computing, particularly in the context of analytics and data processing. The paper provides several case studies of deep learning-based edge computing systems, highlighting the benefits of these systems in terms of improved efficiency, accuracy, and privacy.

The paper also discusses the challenges associated with implementing deep learning on edge devices, such as limited processing power, memory, and energy resources. The author provides several solutions to these challenges, such as model compression and hardware optimization.

Kong, L. et al., presented a comprehensive survey of edge computing for the Internet of Things (IoT). The authors first provide an overview of the IoT and its potential benefits, as well as the challenges it faces, such as limited computing resources, energy constraints, and security and privacy concerns.

The authors then present a detailed survey of the literature in the field of edge computing for IoT, covering various aspects such as architectures, communication protocols, resource management, security and privacy, and applications. They classify the literature based on these aspects and provide a critical analysis of the key findings.

Sánchez, J.M.G. et al., presented a systematic mapping study of edge computing for cyber-physical systems (CPS), with a particular focus on trustworthiness. The authors first provide an overview of edge computing and its potential benefits for CPS, including reduced latency, improved reliability, and enhanced security.

The authors then present a systematic mapping of the literature in the field, identifying and analyzing 80 relevant papers. They classify these papers according to various criteria, such as the application domain, the type of edge computing architecture, and the level of trustworthiness.

The paper provides a detailed analysis of the results, highlighting the main trends and challenges in the field. The authors conclude that trustworthiness is a critical factor in the design and implementation of edge computing for CPS, and that there is a need for more research in this area.

B. Li et al. proposed a robustness-oriented approach to deploying edge applications in edge computing environments. The authors propose a system called READ (Robustness-Oriented Edge Application Deployment), which uses reinforcement learning to optimize the deployment of edge applications based on robustness metrics. The proposed approach is expected to improve the robustness and resilience of edge applications.

Saeed, E. et al., investigated the performance of various edge computing models for Internet of Things (IoT) applications. The authors first provide an overview of edge computing and its potential benefits for IoT, including reduced latency, improved reliability, and enhanced privacy and security.

The authors then present three edge computing models for IoT: fog computing, mobile edge computing, and cloudlet computing. They evaluate the performance of these models using various metrics, including response time, energy consumption, and throughput. The evaluation is conducted using simulations and experiments.

The paper presents a detailed analysis of the results, comparing the performance of the three edge computing models under different scenarios. The authors conclude that the choice of edge computing model depends on the specific requirements of the IoT application, such as the size of data, the processing capability of devices, and the network conditions.

Hua, H. et al., provided a comprehensive overview of the integration of edge computing and artificial intelligence (AI), specifically from a machine learning perspective. The authors first introduce the concept of edge computing and its benefits, including reduced latency, improved scalability, and enhanced privacy and security.

The authors then discuss the integration of AI with edge computing, highlighting the challenges associated with this integration, such as the need for efficient resource allocation and data management. The paper also discusses various machine learning techniques that can be applied to edge computing, such as deep learning, reinforcement learning, and transfer learning.

In addition, the authors provide a survey of the current state of research in the field, including various applications of edge computing with AI, such as smart homes, autonomous vehicles, and healthcare. The paper also includes a discussion of the future prospects for edge computing with AI, including potential challenges and opportunities.

How Does Edge Computing Work?

Edge computing is a distributed computing paradigm that involves processing data on devices closer to the source of data instead of in a centralized cloud or data center. Here is a general overview of how edge computing takes place from its starting point till the end:

Data Generation. Edge computing begins with the generation of data from devices such as sensors, IoT devices, and other connected devices.

Edge Nodes. The data is then processed and analyzed at the edge nodes, which are devices such as routers, gateways, and servers that are located closer to the data source.

Data Filtering and Processing. The data is then filtered and processed at the edge nodes to extract relevant information and insights. This helps to reduce the amount of data that needs to be transmitted to the cloud or data center, which saves time and reduces latency.

Local Storage. Once the data is processed at the edge nodes, it may be stored locally for further analysis or sent to the cloud or data center for storage and analysis.

Communication. Edge computing requires communication between the edge nodes and other devices, as well as with the cloud or data center. This is typically done through wireless or wired networks.

Cloud or Data Center. Finally, the processed data may be sent to the cloud or data center for additional analysis, storage, and processing. This allows for more advanced analysis and insights that may not be possible at the edge.

Therefore, edge computing provides a way to process and analyze data in real-time, closer to the source of the data, which can help organizations to make faster and more informed decisions. In fact, Edge Computing offers several benefits to organizations as well as end users. For instance, processing data at the edge results in reduced latency so the response time of applications is reduced. Since, there is less transfer of data from cloud to edge servers, so the overall bandwidth consumption is reduced. Hence, the users experience reduction in cost and performance improvement. To reap these benefits, it is essential to optimize the processes of edge computing at various stages.

Optimization in Edge Computing

Optimization in Edge Computing refers to the process of improving the efficiency, performance, and reliability of edge computing systems through various optimization techniques. Edge computing optimization involves optimizing various aspects of the edge computing system such as resource allocation, task offloading, network topology, security, and privacy, latency, and energy consumption.

The goal of edge computing optimization is to ensure that the system can provide computing and storage services to devices and applications at the edge of the network in a reliable, efficient, and secure manner. By optimizing various aspects of the system, edge computing optimization can help to reduce latency, improve energy efficiency, enhance user experience, and reduce the overall cost of operation.

Optimization techniques that can be used in edge computing optimization include machine learning, artificial intelligence, mathematical optimization, and heuristic algorithms. These techniques can be used to make intelligent decisions about resource allocation, task offloading, network topology, and other aspects of the system.

Performance optimization is critical in edge computing to ensure that applications and services run smoothly and provide a high-quality user experience.

In short, edge computing optimization is a critical process that can help to ensure the effectiveness and efficiency of edge computing systems, leading to better performance, reliability, and security, and ultimately, better user experience. The following strategies can be adopted for enhancing user experience.

Edge Computing can provide numerous benefits, including faster processing times, reduced network latency, and improved reliability. However, to ensure a positive user experience, there are several strategies that can be employed. Here are some strategies for enhancing user experience in edge computing:

Minimize latency. One of the primary benefits of edge computing is reduced latency. To maximize this benefit, it's important to ensure that data processing and analysis happens as close to the source as possible. This can be achieved through the use of edge devices and gateways, which are designed to bring compute and networking capabilities closer to the end-user.

Optimize network performance. To ensure a smooth user experience, it's important to optimize network performance. This can be achieved through the use of load balancing, caching, and other techniques to ensure that network traffic is distributed efficiently and that data is transmitted quickly and reliably.

Use intelligent caching. Caching can be an effective way to improve performance in edge computing environments. By storing frequently accessed data locally on edge devices, you can reduce the need to access data over the network, which can help to minimize latency and improve overall performance.

Employ predictive analytics. Predictive analytics can be used to anticipate user needs and provide a more personalized experience. By analyzing data in real-time and making recommendations based on user behavior, you can create a more engaging and satisfying user experience.

Focus on security. Security is always a top concern in any computing environment, but it's especially important in edge computing, where data is processed and analyzed closer to the source. To enhance security, it's important to use encryption, authentication, and other techniques to protect data and prevent unauthorized access.

By implementing these strategies, you can help to ensure a positive user experience in edge computing environments, which can lead to increased engagement and loyalty among users.

What Aspects of Edge Computing are Amenable to Optimization?

Several aspects of edge computing can be optimized to improve its overall performance and efficiency. Some of the key aspects that can be optimized include:

Resource allocation: Edge computing involves the allocation of computing and storage resources to devices and applications at the edge of the network. Optimizing resource allocation can help to ensure that the available resources are used efficiently and effectively, and that devices and applications have the resources they need to perform their tasks.

Task offloading: Task offloading involves transferring computing tasks from devices to edge servers or cloud data centers to reduce the computation time and energy consumption of the devices. Optimizing task offloading decisions can help to ensure that tasks are offloaded to the most appropriate location, based on factors such as the computation complexity of the task, the energy level of the device, and the latency requirements.

Edge caching: Edge caching involves storing frequently accessed data at edge servers to reduce the data retrieval time and energy consumption of devices. Optimizing edge caching can help to ensure that the most frequently accessed data is stored in the cache, and that data is stored for an appropriate length of time.

Network topology: The topology of the network can also be optimized to improve the performance of edge computing. Optimizing the network topology involves ensuring that devices are connected to the most appropriate edge servers and that data is routed efficiently between devices and servers.

Security and privacy: Edge computing involves processing and storing sensitive data at the edge of the network, making security and privacy a critical aspect of edge computing. Optimizing security and privacy measures can help to ensure that data is protected from unauthorized access and that privacy is maintained for users.

Latency and energy consumption: Optimizing latency and energy consumption is another key aspect of edge computing. Techniques such as compression, dynamic resource allocation, and optimization algorithms can be used to minimize latency and energy consumption and improve the overall performance of the network.

Optimizing these aspects of edge computing can help to ensure that the network is efficient, reliable, and secure, and that users have a high-quality experience.

Resource Allocation

Resource allocation in edge computing involves allocating computing and storage resources to devices and applications at the edge of the network. The goal of resource allocation is to ensure that the available resources are used efficiently and effectively, and that devices and applications have the resources they need to perform their tasks. In order to perform resource allocation, the following steps may be followed.

Determine the resource requirements. The first step is to determine the resource requirements of each device and application. This includes identifying the computing and storage resources required to perform the tasks, as well as the latency and energy requirements.

Analyze the available resources. The next step is to analyze the available resources at the edge of the network, including computing and storage resources, network bandwidth, and energy levels. This information is used to determine the total available resources.

Develop a resource allocation model. Based on the resource requirements and available resources, a resource allocation model is developed to determine the best allocation of resources to devices and applications.

Allocate resources. The final step is to allocate resources to devices and applications based on the resource allocation model. This may involve dynamically allocating resources based on the changing demands of devices and applications.

Optimization techniques such as mathematical optimization, heuristic algorithms, and machine learning can be used to develop and implement the resource allocation model. These techniques can help to ensure that resources are allocated efficiently and effectively, and that devices and applications have the resources they need to perform their tasks.

Basically, resource allocation in edge computing is a critical process that helps to ensure that devices and applications at the edge of the network have the resources they need to perform their tasks, leading to better performance, reliability, and security.

Yang. Z. et al. (2020), proposed an approach that focuses on optimizing the allocation of wireless resources and the partitioning of tasks in mobile edge computing environments. This is an important area of research, as effective resource allocation and task partitioning are critical to achieving efficient and reliable mobile edge computing systems.

Sun D. et al., (2020) proposed an approach that focuses on a demand-side resources collaborative optimization system, that enables efficient resource management in distributed computing systems. The use of cloud and edge computing can provide additional computational resources and enhance system performance.

Y. Song et al., (2017) proposed an approach to QoS-based task distribution in edge computing networks for IoT applications. This approach considers both the QoS requirements of tasks and the computing resources of edge servers to allocate tasks to appropriate servers. The proposed approach improves the utilization of computing resources and achieves high Quality of Service(QoS).

Wei X. et al., (2023) proposed a joint optimization framework for resource placement and task dispatching in edge clouds across multiple timescales. The authors consider both the short-term performance of individual tasks and the long-term resource utilization of edge servers. They formulate the problem as a mixed-integer programming problem and propose an efficient heuristic algorithm to solve it. With their approach they were able to achieve better resource utilization and QoS than baseline methods. In fact, both short-term performance and long-term resource utilization improved with this method.

J. Lu et al., (2020)proposed an optimization strategy for scheduling tasks in a Storm-based edge computing environment. The authors aim to improve the efficiency of task scheduling and reduce the time for task completion by optimizing the allocation of tasks to edge nodes. The proposed strategy results in improving task completion time and reducing resource consumption.

H. Badri et al., (2018) proposed a risk-based optimization framework for resource provisioning in mobile edge computing (MEC) networks. The authors formulate the resource provisioning problem as a two-stage stochastic optimization problem, where the goal is to minimize the expected cost while ensuring a specified quality of service (QoS) for the users. They use a risk-neutral approach and propose a linear programming formulation for solving the optimization problem. Simulation results show that their proposed approach can improve the QoS and reduce the cost compared to other baseline approaches. Hence this method can be used for resource provisioning in MEC networks that considers the uncertainties in the network and user demands.

F. Costanzo et al. (2022), proposed a dynamic resource optimization and altitude selection algorithm for UAV-based multi-access edge computing (MEC) networks. The proposed algorithm considers the resource requirements of tasks and the energy consumption of UAVs to jointly optimize the resource allocation and altitude selection of UAVs. The results show that the algorithm can improve the performance of UAV-based MEC networks. The results indicate that the proposed algorithm is a novel algorithm for optimizing the resource allocation and altitude selection of UAVs in MEC networks and it can achieve better performance compared to baseline methods.

Z. Su et al (2022)., proposed a grasshopper optimization algorithm (GOA) to solve the 0-1 knapsack computation resources allocation problem in mobile edge computing (MEC) networks. The authors formulate the problem as an integer programming problem and use GOA to find the optimal solution.

D. Gao et al., (2021) proposed a resource optimization framework for multi-user MIMO systems in edge cloud computing environments. The proposed framework jointly optimizes the transmit beamforming and resource allocation to maximize the system throughput while satisfying the quality of service (QoS) constraints. It is a comprehensive framework for resource optimization in multi-user MIMO systems in edge cloud computing environments. The authors have formulated the optimization problem and provided a solution that takes into account QoS constraints.

D. Song et al., (2021) proposed a cloud-edge computing resource collaborative optimization method for power distribution fault analysis services. The proposed method optimizes the resource allocation and scheduling for tasks in the cloud and edge computing nodes to minimize the energy consumption and improve the fault analysis efficiency. The authors have considered both energy consumption and fault analysis efficiency in their optimization model.

C. Yin et al. (2020), presented an optimization method for resource allocation in fog computing, aiming to maximize the number of completed tasks with limited computing and communication resources. The proposed method formulates the optimization problem as a mixed-integer programming (MIP) problem and then uses a genetic algorithm to find an approximate solution. The results show that it can effectively allocate resources and achieve high task completion rates. The proposed method is a well-designed optimization method for resource allocation in fog computing, and the proposed approach provides a good balance between resource utilization and task completion rates.

Z. Yang et al. (2020), proposed a joint optimization approach for wireless resource allocation and task partition in mobile edge computing (MEC) systems. The proposed approach aims to maximize the total system utility, which is defined as the sum of the task completion rates and the energy efficiency. The authors formulate the optimization problem as a mixed-integer non-linear programming (MINLP) problem and then solve it using a sequential convex programming (SCP) algorithm. The results show that it can achieve a higher total system utility compared to some baseline approaches. It provides a good tradeoff between task completion rates and energy efficiency.

Z. h. Huo et al., (2020) proposed a two-step multi-objective optimization framework for the microgrid scheduling problem based on cloud-edge computing. The first step involves optimizing the scheduling of the microgrid using a genetic algorithm (GA), and the second step involves optimizing the communication and computing resources allocation for the microgrid using a particle swarm optimization (PSO) algorithm. The proposed framework demonstrated its effectiveness in achieving a trade-off between the objectives of minimizing the cost and maximizing the reliability of the microgrid.

Task Offloading

Task offloading in edge computing refers to the process of transferring computing tasks from devices to edge servers or cloud data centers to reduce the computation time and energy consumption of the devices. In edge computing, task offloading is often used to reduce latency, improve energy efficiency, and enhance the overall performance of the network. The following steps may be followed to perform task offloading.

Determine the offloading decision factors. The first step is to determine the factors that should be considered when making offloading decisions. This may include the computation complexity of the task, the energy level of the device, the latency requirements, the available network bandwidth, and the reliability of the network connection.

Analyze the task and device characteristics. The next step is to analyze the characteristics of the task and device, including the computation requirements, memory requirements, and energy consumption. This information is used to determine whether task offloading is feasible and to identify the most appropriate offloading strategy.

Select an offloading strategy. Based on the offloading decision factors and task and device characteristics, an offloading strategy is selected. This may involve offloading the task to an edge server or cloud data center, or it may involve performing the task locally on the device.

Execute the offloading strategy. The final step is to execute the offloading strategy. This may involve transferring the task data to the edge server or cloud data center, executing the task on the remote server, and transferring the results back to the device.

Optimization techniques must ensure that tasks are offloaded to the most appropriate location, based on the offloading decision factors and task and device characteristics.

X. Duan et al., (2020) proposed an offloading strategy for edge computing in the Internet of Things (IoT) that considers the quality of service (QoS) and energy efficiency. The authors present a mathematical model to minimize energy consumption and delay in offloading computation tasks to edge servers.

J. Fan et al., (2022) proposed a task caching and computation offloading scheme for a multi-user mobile edge computing network. The authors present a mathematical model to minimize the total latency and energy consumption in the network. They also propose a heuristic algorithm to solve the problem and conduct simulations to evaluate the performance of the proposed scheme.

H. Zhou et al., proposed a joint optimization algorithm for computing offloading and service caching in edge computing-based smart grids. The algorithm is designed to minimize the total energy consumption of the system by jointly optimizing computing offloading and service caching. The proposed approach is a valuable contribution to the field of edge computing and smart grids. The joint optimization algorithm proposed in the paper could potentially lead to significant energy savings in smart grid systems

D. Ye et al., (2022) proposed an edge computing offloading algorithm based on second-order oscillatory particle swarm optimization (SOPS). The algorithm is designed to minimize the total energy consumption of edge devices by optimizing the offloading decision for each task. Basically, the proposed method can lead to significant energy savings.

T. Gao et al., (2022) proposed a particle swarm optimization (PSO) with Lévy flight for service caching and task offloading in edge-cloud computing. The proposed PSO algorithm optimizes the resource allocation problem by considering the energy consumption and service delay.

J. Xiong et al., (2019) proposed a joint bit allocation and trajectory optimization approach for task offloading in UAV-aided edge computing. The authors formulate the problem as a nonlinear integer programming problem and propose a solution based on Lagrange relaxation and subgradient optimization.

X. Q. Pham et al. (2021), proposed **a** QoE-based optimization approach to computation offloading in vehicle-assisted MEC. The approach involves two stages: (i) QoE estimation using a deep neural network (DNN) model, and (ii) QoE-based computation offloading optimization using a genetic algorithm (GA).

Edge Catching

Edge caching is a technique used in edge computing to store frequently accessed data closer to the end-users, reducing latency and improving performance. In edge caching, data is stored in cache memory located at the edge of the network, such as on an edge server or a device itself. This data is then quickly accessible to end-users when requested, reducing the need to fetch data from a distant data center. The following are the advantages of caching data at the edge.

Reduced latency. Edge caching reduces the latency of accessing data, as the data is stored closer to the end-users. This can lead to better user experience and faster application performance.

Improved scalability. Edge caching can improve the scalability of applications, as it reduces the load on the central data center and allows for more efficient use of network resources.

Reduced network congestion. By storing frequently accessed data at the edge, edge caching reduces the amount of data that needs to be transmitted over the network, reducing network congestion and improving overall network performance.

The following steps specify how edge caching is done.

Identify frequently accessed data. The first step in edge caching is to identify the data that is frequently accessed by end-users. This may include static content such as images and videos, as well as dynamic content such as user profiles and social media updates.

Cache data at the edge. Once the frequently accessed data has been identified, it is stored in cache memory located at the edge of the network, such as on an edge server or a device itself. The data is cached in a way that optimizes retrieval time, such as using a least-recently-used (LRU) algorithm.

Serve cached data to end-users. When an end-user requests data that has been cached at the edge, the data is quickly served from the cache memory, reducing latency and improving performance. If the requested data is not available in the cache memory, it is fetched from the central data center.

Optimization techniques must ensure that data is cached in the most efficient and effective way, leading to improved performance, scalability, and reliability of the network.

RH A. et al., proposed an approach that focuses on collaborative edge data caching, which is an important area of research given the increasing demand for edge computing and the need for efficient data management in distributed environments. Collaborative caching has the potential to improve system performance and reduce network congestion by leveraging the caching capabilities of multiple nodes in a network.

Network Topology

Network topology optimization in edge computing refers to the process of designing and optimizing the communication network between edge devices, edge servers, and cloud data centers to improve performance, reduce latency, and increase energy efficiency. Some techniques for network topology optimization in edge computing are given below.

Decentralized topology. In a decentralized topology, edge devices communicate directly with other nearby devices, reducing the need for communication with a central hub. This reduces latency and improves reliability.

Hierarchical topology. A hierarchical topology involves dividing the network into multiple layers, with edge devices communicating with edge servers, and edge servers communicating with cloud data centers. This reduces the load on edge devices and improves scalability.

Fog computing. Fog computing involves deploying computing resources and data storage closer to the edge, reducing the need for communication with a central data center. This reduces latency and improves performance.

Load balancing. Load balancing involves distributing the workload across multiple edge servers to ensure that no single server is overloaded. This improves performance and reduces the risk of failure.

Network slicing. Network slicing involves dividing the network into multiple virtual networks, each optimized for a specific type of application. This improves performance and reliability, as each network can be optimized for the specific demands of the application.

Security and Privacy

Security and privacy are critical concerns in edge computing, as data is often stored and processed closer to the end-users. Some ways to optimize security and privacy in edge computing are given below.

Secure communication. Secure communication protocols such as TLS/SSL can be used to encrypt communication between edge devices, edge servers, and cloud data centers. This can prevent eavesdropping and data tampering.

Authentication and access control: Strong authentication mechanisms such as multi-factor authentication can be used to ensure that only authorized users and devices can access the network and data. Access control mechanisms such as role-based access control can be used to ensure that users and devices have access only to the data and resources they need.

Data encryption. Data encryption can be used to protect sensitive data stored in edge devices, edge servers, and cloud data centers. This can prevent data breaches and unauthorized access to sensitive data.

Threat detection and prevention. Threat detection and prevention mechanisms such as firewalls and intrusion detection systems can be used to detect and prevent cyber attacks on the network and devices.

Privacy-preserving data processing. Techniques such as differential privacy and homomorphic encryption can be used to ensure that data is processed in a privacy-preserving manner, without revealing sensitive information. Differential privacy is a technique that provides statistical guarantees for protecting the privacy of individuals in a dataset while allowing for useful analysis to be performed on the data.

Security monitoring and auditing. Regular security monitoring and auditing can help to identify and address security vulnerabilities and threats in the network and devices.

Latency and Energy Consumption

To monitor and optimize latency and energy consumption in edge computing, you can use the following techniques.

Profiling and monitoring. Profiling and monitoring techniques can be used to gather data on the performance and energy consumption of edge devices, edge servers, and cloud data centers. This data can be used to identify bottlenecks and optimize the performance and energy consumption of the system.

Resource allocation. Resource allocation techniques can be used to allocate computing resources, such as CPU, memory, and network bandwidth, based on the demands of the applications and devices. This can help to optimize the performance and energy consumption of the system.

Load balancing. Load balancing techniques can be used to distribute the workload across multiple edge servers to ensure that no single server is overloaded. This can help to optimize the performance and energy consumption of the system.

Power management: Power management techniques, such as dynamic voltage and frequency scaling (DVFS) and sleep states, can be used to reduce the energy consumption of edge devices when they are idle or underutilized.

Task offloading: Task offloading techniques can be used to offload computationally intensive tasks to edge servers or cloud data centers, reducing the energy consumption and latency of edge devices.

Edge caching: Edge caching can be used to reduce the latency of frequently accessed data by storing it closer to the end-users, reducing the need for communication with cloud data centers.

Optimization techniques are needed to ensure that the system is designed and optimized to provide the best possible performance and energy efficiency in edge computing.

Yang Y. et al. (2021), proposed an approach that focuses on improving the energy efficiency of mobile edge computing, which is an important area of research given the increasing demand for mobile devices and the growing importance of edge computing in mobile networks. The use of deep reinforcement learning and optimization techniques could potentially yield significant improvements in energy efficiency, as these approaches are well-suited to handling complex, dynamic environments.

Abe M. et al. (2020), explored the use of distributed edge computing to optimize video communication. The authors propose a system that uses edge devices to perform video processing, such as encoding and decoding, rather than relying solely on central servers. The proposed approach is expected to reduce latency and improve the quality of video communication.

Y. Li and S. Wang (2018) proposed an energy-aware algorithm for placing edge servers in mobile edge computing (MEC) environments. The proposed algorithm considers the energy consumption of edge servers and mobile devices to determine the optimal placement of servers to minimize energy consumption while meeting the demand for computation and communication resources.

T. Zhao et al. (2022), proposed a joint optimization approach for reducing latency and energy consumption in mobile edge computing (MEC)-based proximity detection systems in road networks. The proposed approach considers the trade-off between latency and energy consumption to determine the optimal placement of edge servers and the allocation of computation and communication resources.

Xiao, K. et al., (2018) proposed a heuristic algorithm for server placement in edge computing, based on resource requirements forecasting. The algorithm aims to minimize the energy consumption and latency while ensuring the reliability of the service. The authors used a simulation environment to evaluate the performance of the proposed algorithm and compared it with other state-of-the-art approaches. The results showed that the proposed algorithm achieved better performance in terms of energy consumption, latency, and reliability.

Takeda, A. et al., (2020) proposed a joint optimization approach for edge server and virtual machine placement in edge computing environments. This approach aims to minimize the energy consumption and latency while ensuring the reliability of the service.

X. Chen et al. (2020), proposed an edge server placement algorithm for edge computing environments to minimize energy consumption and network delay. The proposed algorithm takes into consideration the traffic demand, energy consumption, and delay constraints to place the edge servers. The proposed approach can significantly impact the overall performance of the system. It takes into account multiple objectives, including energy consumption and network delay, and considers the traffic demand as a crucial factor in determining the edge server placement.

P. Chen et al. (2022) proposed a multi-slot dynamic computing resource optimization algorithm for edge computing systems. The authors use a Markov decision process (MDP) model to capture the dynamics of the computing resource allocation problem over time. They propose a heuristic algorithm for solving the MDP problem and show that it can achieve near-optimal performance. It can significantly reduce the energy consumption and improve the resource utilization compared to other baseline approaches.

T. Ogino (2021) proposed a simplified multi-objective optimization algorithm for flexible IoT edge computing. The proposed algorithm is designed to optimize energy consumption, response time, and throughput for a given workload. The algorithm uses a heuristic approach to reduce the search space and selects the optimal solution based on a Pareto front.

P. Lai et al., (2022) proposed a dynamic user allocation method for stochastic mobile edge computing (MEC) systems. The goal is to allocate users to edge servers to minimize the total energy consumption and maximize the number of served users. The authors use a Markov decision process (MDP) model to formulate the problem and propose a dynamic programming algorithm to solve it. The proposed method is evaluated through simulations and compared with two other baseline methods. It is a novel solution using Markov decision process and dynamic programming.

X. Diao et al. (2019), proposed a fair data allocation and trajectory optimization scheme for UAV-assisted mobile edge computing. The authors formulate the problem as a mixed-integer nonlinear programming problem and propose a heuristic algorithm to solve it. The proposed scheme can improve the fairness of data allocation and reduce the computation latency.

Z. Wu et al. (2022), proposed a joint deployment and trajectory optimization scheme for unmanned aerial vehicle (UAV)-assisted vehicular edge computing networks. The authors formulate the problem as a mixed-integer nonlinear programming problem and propose a heuristic algorithm to solve it. The proposed scheme leads to the improvement in the network throughput and reduces the computation latency.

There are some strategies for performance optimization in edge computing which are given below.

Edge node placement. One of the key strategies for performance optimization in edge computing is to ensure that edge nodes are placed strategically to minimize latency and network congestion. By placing edge nodes closer to the user, service providers can reduce latency and improve the performance of real-time applications.

Content caching. Content caching involves storing frequently accessed content closer to the user, which can reduce latency and network congestion. By caching content at edge nodes, service providers can improve the performance of applications and services and provide a smoother user experience.

Load balancing. Load balancing involves distributing network traffic across multiple edge nodes to ensure that no single node is overloaded. By load balancing network traffic, service providers can improve the performance of applications and services and prevent downtime due to overload.

Network optimization. Network optimization involves configuring the network to ensure that it is performing at its best. This may involve optimizing network protocols, using compression techniques to reduce data transfer size, or using quality-of-service (QoS) techniques to prioritize network traffic.

Predictive analytics. Predictive analytics can help service providers anticipate user needs and optimize the user experience accordingly. By analyzing historical data on user behavior, network performance, and other relevant factors, predictive analytics can help service providers optimize their applications and services for performance.

Containerization. Containerization involves packaging an application and its dependencies into a container that can be deployed anywhere. By using containerization, service providers can improve the performance of applications by making them more scalable and easier to deploy across multiple edge nodes.

In general, performance optimization is critical in edge computing to ensure that applications and services run smoothly and provide a high-quality user experience. By focusing on edge node placement, content caching, load balancing, network optimization, predictive analytics, and containerization, service providers can optimize their applications and services for performance and deliver a seamless user experience.

CONCLUSION

In conclusion, edge computing is a transformative technology that is poised to revolutionize the way data is processed, analyzed, and stored. By bringing processing power and analytics capabilities closer to the source of data, edge computing enables faster, more efficient decision-making and unlocks new possibilities for innovation in industries ranging from healthcare to transportation to manufacturing. While there are challenges and limitations associated with edge computing, such as security and scalability concerns, there are also promising solutions and emerging trends that suggest a bright future for this exciting technology. As edge computing continues to evolve and mature, it is likely to play an increasingly critical role in the development of emerging technologies such as 5G networks and augmented reality, and to shape the future of technology in ways that we have yet to fully imagine. Optimization of processes at different stages of Edge Computing is necessary to enhance the user experience. There are many approaches that can be adopted for optimization including optimal resource allocation, caching at the edge, optimized offloading, latency reduction, and energy preservation. In addition, for optimization, the techniques such as Integer Programming, Machine Learning, Heuristic Techniques, Reinforcement Learning, and Particle Swarm Optimization can be used.

REFERENCES

A., R., S., K., N., N., & P., P. (2022). *Collaborative Edge Data Caching In Online*. 2022 Second International Conference on Advanced Technologies in Intelligent Control, Environment, Computing & Communication Engineering (ICATIECE), Bangalore, India. 10.1109/ICATIECE56365.2022.10047075

Badri, H., Bahreini, T., Grosu, D., & Yang, K. (2018). Risk-Based Optimization of Resource Provisioning in Mobile Edge Computing. *2018 IEEE/ACM Symposium on Edge Computing (SEC)*, Seattle, WA, USA. 10.1109/SEC.2018.00033

Chen, P., Xu, H., Fan, X., Hu, J., & Song, T. (2022). Multi-Slot Dynamic Computing Resource Optimization in Edge Computing. *2022 14th International Conference on Wireless Communications and Signal Processing (WCSP)*, Nanjing, China. 10.1109/WCSP55476.2022.10039042

Chen, X., Liu, W., Chen, J., & Zhou, J. (2020). An Edge Server Placement Algorithm in Edge Computing Environment. 2020 12th International Conference on Advanced Infocomm Technology (ICAIT), Macao, China. 10.1109/ICAIT51223.2020.9315526

Costanzo, F., Lorenzo, P. D., & Barbarossa, S. (2020). *Dynamic Resource Optimization and Altitude Selection in Uav-Based Multi-Access Edge Computing.* ICASSP 2020 - 2020 IEEE International Conference on Acoustics, Speech and Signal Processing (ICASSP), Barcelona, Spain. 10.1109/ICASSP40776.2020.9053594

Diao, X., Zheng, J., Cai, Y., Wu, Y., & Anpalagan, A. (2019, December). Fair Data Allocation and Trajectory Optimization for UAV-Assisted Mobile Edge Computing. *IEEE Communications Letters*, *23*(12), 2357–2361. doi:10.1109/LCOMM.2019.2943461

Duan, X., Xu, F., & Sun, Y. (2020). Research on Offloading Strategy in Edge Computing of Internet of Things. *2020 International Conference on Computer Network, Electronic and Automation (ICCNEA)*, Xi'an, China. 10.1109/ICCNEA50255.2020.00050

Fan, J., Lan, W., Geng, S., & Zhao, X. (2022). *Task Caching and Computation Offloading for Muti-User Mobile Edge Computing Network.* 2022 4th International Conference on Communications, Information System and Computer Engineering (CISCE), Shenzhen, China. 10.1109/CISCE55963.2022.9851119

Gao, D., Cheng, H., Han, Z., & Yang, S. (2021). *Resource Optimization for the Multi-user MIMO Systems Assisted Edge Cloud Computing.* 2021 IEEE 6th International Conference on Signal and Image Processing (ICSIP), Nanjing, China. 10.1109/ICSIP52628.2021.9688949

Gao, T., Tang, Q., Li, J., Zhang, Y., Li, Y., & Zhang, J. (2022). A Particle Swarm Optimization With Lévy Flight for Service Caching and Task Offloading in Edge-Cloud Computing. *IEEE Access : Practical Innovations, Open Solutions*, *10*, 76636–76647. doi:10.1109/ACCESS.2022.3192846

Genda, K., Abe, M., & Kamamura, S. (2020). *Video Communication Optimization Using Distributed Edge Computing.* 2020 21st Asia-Pacific Network Operations and Management Symposium (APNOMS), Daegu, Korea (South). 10.23919/APNOMS50412.2020.9237044

Huo, Z. h., Wang, P., Zhang, S.-j., Wang, D., & Kong, Z. (2020). *A Two-Step Multi-objective Optimization Frame-work for Microgrid Scheduling Problem Based on Cloud-edge Computing.* 2020 IEEE 4th Conference on Energy Internet and Energy System Integration (EI2), Wuhan, China. 10.1109/EI250167.2020.9347092

Lai, P. (2022). Dynamic User Allocation in Stochastic Mobile Edge Computing Systems. *2022 IEEE World Congress on Services (SERVICES)*, Barcelona, Spain. 10.1109/SERVICES55459.2022.00030

Li B. (2022). READ: Robustness-Oriented Edge Application Deployment in Edge Computing Environment. *IEEE Transactions on Services Computing, 15*(3), 1746-1759. . doi:10.1109/TSC.2020.3015316

Li, Y., & Wang, S. (2018). An Energy-Aware Edge Server Placement Algorithm in Mobile Edge Computing. *2018 IEEE International Conference on Edge Computing (EDGE)*, San Francisco, CA, USA. 10.1109/EDGE.2018.00016

Lu, J., Liu, X., Yue, T., Wang, W., & Zhao, L. (2020). *Research on Edge Computing-oriented Storm Edge Node Scheduling Optimization Strategy.* 2020 IEEE Sustainable Power and Energy Conference (iSPEC), Chengdu, China. 10.1109/iSPEC50848.2020.9350984

Ogino, T. (2021). Simplified Multi-objective Optimization for Flexible IoT Edge Computing. 2021 4th International Conference on Information and Computer Technologies (ICICT), HI, USA. 10.1109/ICICT52872.2021.00035

Pham, X. Q., Huynh-The, T., & Kim, D.-S. (2021). *A QoE-based Optimization Approach to Computation Offloading in Vehicle-assisted Multi-access Edge Computing.* International Conference on Information and Communication Technology Convergence (ICTC), Jeju Island, Korea. 10.1109/ICTC52510.2021.9621000

Song, D., Yuan, L., Zhao, W., Yu, Q., Du, J., & Pan, K. (2021). Cloud-Edge Computing Resource Collaborative Optimization Method for Power Distribution Fault Analysis Service. *2021 China International Conference on Electricity Distribution (CICED)*, Shanghai, China. 10.1109/CICED50259.2021.9556733

Song, Y., Yau, S. S., Yu, R., Zhang, X., & Xue, G. (2017). An Approach to QoS-based Task Distribution in Edge Computing Networks for IoT Applications. *2017 IEEE International Conference on Edge Computing (EDGE)*, Honolulu, HI, USA. 10.1109/IEEE.EDGE.2017.50

Su, Z., He, G., & Li, Z. (2022). *Using Grasshopper Optimization Algorithm to Solve 0-1 Knapsack Computation Resources Allocation Problem in Mobile Edge Computing.* 2022 34th Chinese Control and Decision Conference (CCDC), Hefei, China. 10.1109/CCDC55256.2022.10034236

Sun, D., Li, X., Zhao, L., Wang, Y., Cao, X., & Yang, J. (2020). *Study of Demand-Side Resources Collaborative Optimization System Based on the Cloud & Edge Computing.* 2020 IEEE 5th International Conference on Cloud Computing and Big Data Analytics (ICCCBDA), Chengdu, China. 10.1109/ICCCBDA49378.2020.9095622

Takeda, A., Kimura, T., & Hirata, K. (2020). *Joint optimization of edge server and virtual machine placement in edge computing environments.* 2020 Asia-Pacific Signal and Information Processing Association Annual Summit and Conference (APSIPA ASC), Auckland, New Zealand.

Wang, T., Zou, Y., Zhang, X., Liu, J., & Wu, J. (2023). *Automotive Mixed Criticality DAG Function Scheduling Optimization Based on Edge Computing.* 2023 6th World Conference on Computing and Communication Technologies (WCCCT), Chengdu, China. 10.1109/WCCCT56755.2023.10052398

X. Wei, A. B. M. M. Rahman, D. Cheng D., & Wang, Y. (2023). Joint Optimization Across Timescales: Resource Placement and Task Dispatching in Edge Clouds. *IEEE Transactions on Cloud Computing, 11*(1), 730-744. . doi:10.1109/TCC.2021.3113605

Wu, Z., Yang, Z., Yang, C., Lin, J., Liu, Y., & Chen, X. (2022, February). Joint deployment and trajectory optimization in UAV-assisted vehicular edge computing networks. *Journal of Communications and Networks (Seoul)*, *24*(1), 47–58. doi:10.23919/JCN.2021.000026

Xiao, K., Gao, Z., Wang, Q., & Yang, Y. (2018). A Heuristic Algorithm Based on Resource Requirements Forecasting for Server Placement in Edge Computing. *2018 IEEE/ACM Symposium on Edge Computing (SEC)*, Seattle, WA, USA. 10.1109/SEC.2018.00043

Xiong, J., Guo, H., & Liu, J. (2019, March). Task Offloading in UAV-Aided Edge Computing: Bit Allocation and Trajectory Optimization. *IEEE Communications Letters*, *23*(3), 538–541. doi:10.1109/LCOMM.2019.2891662

Yang, Y., Hu, Y., & Gursoy, M. C. (2021). Deep Reinforcement Learning and Optimization Based Green Mobile Edge Computing, 2021 IEEE 18th Annual Consumer Communications & Networking Conference. CCNC. doi:10.1109/CCNC49032.2021.9369566

Yang, Z., Xie, J., Gao, J., Chen, Z., & Jia, Y. (2020). Joint Optimization of Wireless Resource Allocation and Task Partition for Mobile Edge Computing. *2020 IEEE/CIC International Conference on Communications in China (ICCC)*, Chongqing, China. 10.1109/ICCC49849.2020.9238820

Ye, D., Wang, X., & Hou, J. (2022). An Edge Computing Offloading Algorithm Based on Second-Order Oscillatory Particle Swarm Optimization. *2022 3rd Information Communication Technologies Conference (ICTC)*, Nanjing, China. 10.1109/ICTC55111.2022.9778592

Yin, C., Li, T., Qu, X., & Yuan, S. (2020). *An optimization method for resource allocation in fog computing.* 2020 International Conferences on Internet of Things (iThings) and IEEE Green Computing and Communications (GreenCom) and IEEE Cyber, Physical and Social Computing (CPSCom) and IEEE Smart Data (SmartData) and IEEE Congress on Cybermatics (Cybermatics), Rhodes, Greece. 10.1109/iThings-GreenCom-CPSCom-SmartData-Cybermatics50389.2020.00139

Zhao, T., Liu, Y., Shou, G., & Yao, X. (2022, April). Joint optimization of latency and energy consumption for mobile edge computing based proximity detection in road networks. *China Communications*, *19*(4), 274–290. doi:10.23919/JCC.2022.04.020

Zhou, H., Zhang, Z., Li, D., & Su, Z. (2023, April 1). Joint Optimization of Computing Offloading and Service Caching in Edge Computing-based Smart Grid. *IEEE Transactions on Cloud Computing*, *11*(2), 1122–1132. doi:10.1109/TCC.2022.3163750

Chapter 10
The Role of Wireless Body Area Networks in Smart Healthcare System in the Context of Big Data and AI

Shahad P.
Indian Institute of Information Technology, Kottayam, India

Ebin Deni Raj
Indian Institute of Information Technology, Kottayam, India

ABSTRACT

This chapter covers various technical features available for building a smart healthcare system. Its different components are included as separate sections along with introduction and conclusion parts. There are many other sections incorporated here, namely infrastructure development, streaming data management, cloud computing and machine learning techniques for data analysis. The advancements in sensor technology and internet of things could enable real-time monitoring of several body parameters. The values from these sensors connected to the body can be Integrated for analysis. Which can enhance our health monitoring system. The way of harvesting information from these fast-moving data is also explored here in this chapter. Further, the role of machine learning techniques to build intelligence to the system is discussed in detail. This chapter is organized as a systematic study to form a smart healthcare system which are accepted by the people because of its analytical accuracy.

INTRODUCTION

Traditional industries transform to smart industries with advancements in information and communication technologies. The growth of artificial intelligence together with the advancement of Rapid increase in computation power, availability of health data from various devices and health applications and the support of transforms the healthcare industry into a smart healthcare industry (Chui et al., 2017). To-

DOI: 10.4018/978-1-6684-8785-3.ch010

day's healthcare system recognizes the impact of ICT to improve the quality of healthcare and turns into a smart healthcare system. Prevention of diseases, impact diseases on the body and their treat are together means "healthcare". The smartness of healthcare system is achieved through the introduction of advanced diagnostic tools, better treatment and usage of devices that improve the quality of life of anyone to everyone. Health improvement through internet resources, mobile resources, record management by electronic means, health management with smart device etc. are treated as smart health services. All these services will reduce efforts of patients to get service of experts, quicker and accurate responses for monitored conditions and reduce cost of treatment.

A Smart healthcare system follows a three-layered architecture. First layer is called collection layer which constitutes sensors which are part of wireless body area network. This layer is responsible for collecting sensed data and sends either to the edge devices or clouds for storage or processing. Edge systems can be any device which is capable to do basic processing with sensed data. The point where data reaches first from sensors are sometimes called Base Station (BS) The transmission layer, second layer, consists of a gateway and base stations along with internet support. Third layer, Analysis Layer, is formed by incorporating cloud facilities with data mining algorithms and machine learning technologies (Chen et al., 2018). Health status of a user and corresponding medical treatment measures can be obtained from such systems with Artificial intelligence.

Figure 1. Artificial intelligence

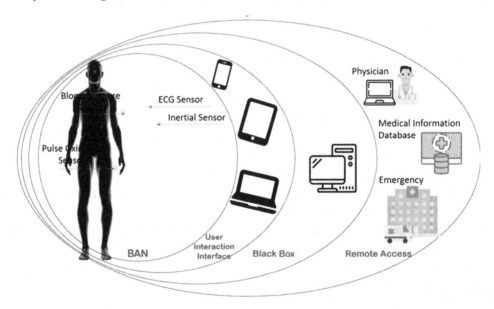

The core component of smart healthcare systems are sensors. Sensors are devices which used to detect changes or variation in its environment. Based on the environment from which changes are detected, method and purpose of these devices, they are classified into different categories. Deployment of sensors in an environment are the means to collect various data related to the environment (Aggarwal et al., 2013). Group of such sensor nodes together with other networking devices forms a Wireless Sensor Network (WSN). These small devices are communicating with short range signals. These networks face challenges in terms of battery life, short communication range, modest processing power, little storage capacity etc. Here we are considering one such network call Wireless Body Area Network and is considered as part of a smart healthcare system.

Sensors connected in proximity to the human body, combined with wireless communication technologies form a Wireless Body Area Network (WBAN). Wearing smart gadgets or devices, implanting communicating devices to the human body brings connection and a hence a network among themselves around patients' body (Sethi & Anand, 2019). We can connect all the sensors within two meters range and form a body centered system by using the Personal Area Network (PAN). This body centered system supports computer assisted rehabilitation or early detection of medical conditions with the help of a number of physiological sensor values participating in Wireless Body Area Network. A real time update of medical records is possible with the data from sensors of WBAN and remote monitoring of these updates plays a big role in smart health systems.

Data is new fuel which runs many of the current businesses. Introducing new sensors to determine new parameters triggers the growth of such businesses. Incorporating different sensors and sensor connected objects contributing much to the whole data which helps in automation and hence smartness in the business. The wireless sensor networks together with Internet forms internet of things which are great contributors of streaming data to the industry. A large collection of sensors can be used to individually gather data and send data through a coordinator node to the internet in an IoT system. This generated big data requires computers, mobile phones, computer networks, and more software and computational intelligence for sending, storing, data analysis and retrieving information (Dey et al., 2018). IoT helps to connect physicians, patients and managers through smart devices and provide facilities for monitoring sensed data after processing and enables communication with patients with their physiological data. All the data collected from these IoT devices needs to be uploaded to the shared cloud storage and it could be analyzed using big data technology.

A huge volume of data with high velocity of data generation with more varieties are usually called "Big Data". Traditional methods or platforms are unable to deal with such big data so there are many new techniques are introduced (Manogaran et al., 2018). Tremendous growth in big data analysis in the past few years leads to the generation of intelligence for big data analysis in smart systems. Massive amount of structured or unstructured data are generated by public and private healthcare organizations on every moment. These data include streaming data from sensors, readings taken by various equipment's, variety of images for diagnosis like magnetic resonance imaging (MRI), computed tomography (CT) and radiography. There are lots of social media contents research findings and other sources. Here we focus on the sensor data connected in Wireless Body Area Network. The proper extraction of data from these sources will give useful and crucial information for healthcare system. Determining association or correlation among the values from these varieties of sources leads to the development of a good model that can save millions of people with reduced cost for treatment.

The best smart healthcare system not just analyses body parameter but it also cares about surrounding environment. Atmospheric parameters like humidity, temperature and quality of air have great impact in health of a person. So, it should be considered while analyzing the reason of health. These sensor values are good support for wearable and carry-on sensors. So it should be considered while analyzing the reason of health. These sensor values are good support for wearable and carry-on sensors. It should be utilized at its maximum for better accurate results. Data received in the processing centers either from dedicated or non-dedicated sensors are inputs for processing which are used for decision making processes. The data collected from the healthcare systems needs to be modeled for building recommendation engines. To scale system resources cloud service can be utilized.

Analysing or processing fast moving data is new challenge in machine learning which are required to explore. The machine learning branch of computer science deals with theory for constructing a learning system, its performance measure, properties and algorithms. On the basis of the way of learning process there are three types of Machine Learning. They are supervised learning, unsupervised learning, and reinforcement learning. In supervised learning a mapping functions is required to mark labels to the instances. These functions maps input to outputs by training the function with labelled data. Collected data will split into training set and test set. They are split with 90 10 rules where 90 percentage of data will use for training and 10 percentage for testing in contrast with the supervised learning, unsupervised learning draw inferences from the data set without labelling them. The collective feedback from interactions with corresponding environment makes learning more advanced. This sort of learning is called reinforcement learning.

The statistical impact of WBAN in the near future depends on factors like improvements in technology, acceptance by the users, regulatory rules etc. Since more WBAN devices are connected to the people its adoption is increasing day by day. The capability of these WBAN devices to generate continuous data and its real time processing improves healthcare. This can even help in personalized medical support. At the same time users have major concern on their privacy since it is related to their personal data. So, for getting the real picture of impact of WBAN in healthcare we need continuous monitoring.

This segment of knowledge is organized as five sections, with basic concepts in Wireless Body Area Network as first session. Its formation with participating sensors is also covered in this session along with the concepts of internet of things. Section 2 details about various data generating from WBAN devices and its management. This section covers challenges in processing of these streaming data. Reliable architecture for processing Bigdata is discussed in fourth section. Services from the cloud environment and other advancements in bigdata processing tools are covered. Artificial Intelligence through Machine Learning algorithms which help data modelling for easy analytics of data were discussed in Section 5, summary of the chapter along with conclusions is added in the last section.

INFRASTRUCTURE WITH WBAN AND IOT

The sensors concerned with measuring different parameters in the proximity of human body have key roles in screening, diagnostics and treatment. Advancements of internet infrastructure together with sensors enables different things to communicate these things include human body and convert it to smart healthcare system. Incorporating technical growth with traditional systems ensures rehabilitation, instantaneous responses, quick and accurate decision making, better consultation etc. Most important and currently connected sensors in WBAN are capable of reading parameters temperature of human body,

E-health services (Zulkernine et al., 2013) grows beyond the limits with sensor devices and through the small edge processing systems like Raspberry pi. E- health systems with biometric sensors and medical equipment are developed by Arduino and Raspberry Pi experts. These sensors have different sensing, computation and storage capabilities with a limitation of energy sources (Fortino et al., 2012). The values from these sensors should be monitored in real time and need to be kept in remote storage for offline and real time data analysis. A prolonged health monitoring without restricting day to day activities is always a challenge along with the radiation effects (Agarwal & Guo, 2015) (Al Islam & Arifin,

2016) of these devices.This could be attained to an extent through the development of intelligent and affordable smart healthcare systems (Ghamari et al., 2016). Values from these sensors are monitored in real-time for the diagnostic purposes through better data analysis tools.

Figure 2. Few smart devices participating in the formation of wireless body area networks

Various environments need different kinds sensors to record changes. Dedicated to fields like medical, entertainment, fitness, life style etc. there are hundreds of sensors (Alam & Hamida, 2014). To form WBAN it requires some or many of the above sensors. Data from them are collected in edge processing device for filtering and hence reducing network traffic. There are challenges in defining pre-processing and filtering rules. Technological support in terms of processing, storage, tools and interactions for these streaming data are still in infancy. Lack of experts in dedicated tools like Apache Kafka, Apache Storm, Apache flink etc. are also an important challenge. These devices are capable of alleviating processing load, reduce storage spaces and increases scalability. Learning from maximum records while keeping least instances is the best feature of an efficient streaming analysis system.

Connected sensors form a WBAN which follows a four layered architecture (Ghamari et al., 2016). In Layer- 1 communication among nodes is through wireless network or with different light weight protocols using in the IoT network. The Layer-2 consists of interaction devices like mobile phones, laptops, Raspberry pi or any other edge processing devices. Access point is organized as layer-2 which will connect to wired network through Wi-Fi or other standards. Layer-3 is responsible for carrying data from access point to destination where data analysis is performed. This layer contains a core component called Decision Measuring Unit (DMU) responsible for data collection, filtration and analysis. Healthcare services to the patients are provided by the Layer-4. Decisions are stored in remote systems which are accessible to the hospitals. Doctors or other professionals could comment or label required data for improving performance of machine learning systems. This added knowledge could use as an emergency service or as health service with long term intension.

Standardization of products will enhance its acceptability. Smart healthcare system development also needs standardization of sensors, connecting technology, data transfer methods and its processing. Achieving energy efficiency through light-weight protocols and. End devices, usually sensor, in

WBAN are connecting to coordinator nodes through specific lower power protocols like Zigbee. This is standardized as IEEE 802.15.4. These standards are capable of communicating with Wi-Fi or TC/IP protocol standards. This enables transfer of data to long distances like cloud servers. Interference of different radio signals like Zigbee, LTE, Wi-Fi are to be considered before deployment of new architecture.

In the formation WBAN wired network technologies are also taking part. Wearable sensors can detect physical signs of different parts of the human body. Quality of life could drastically improve with Internet of Things in the smart healthcare industry. Real time processing of steaming data will generate valuable insights. It generates a huge volume of data streams which call for efficient real time processing. Real time processing of these big data requires designing and implementation of real time processing which in turn demands knowledge concept drift detections, methods for model adaptation etc. (Cort'es et al., 2015). Diagnostics, medical assistance, delivery of medications and services are the important domains where IoT technology can provide support. The low-cost point of service is another advantage of IoT technology. Incorporating many sensing and monitoring devices for smart healthcare systems aims at making the feeling that those living at home are not abandoned.

IoT enabled smart health is characterized by number of interconnected devices, their types, interacting processes and applications, data generating devices, number of data entry points to the system, decision points, etc. These IoT devices are categorized into two major classes; one includes smartphones, tablets and laptops and the other is a set of interconnected sensors. These sensors will generate data in different volumes, velocities and varieties. Based on the timeframe of data generation, they are classified into real time streaming sensors, discrete data sources and one time data sources. IoT devices in WBAN are also classified based on the connection with patients as implantable and wearables. There are certain devices which get connected as and when required. Based on their connectivity with the network, they are categorized to wired, wireless or non-connected devices.

Smart devices connected to the is much more than users on internet (Katal et al., 2013). Communication between these connected objects is possible through internet. These physical objects or things on internet are representing Internet of Things. This term is first used by Ashton to include the physical objects on the internet. Connections like GPRS, RFID, LAN, DSL Wi-Fi, and 3G made communication possible between objects, people and objects and also people among themselves. Major big data sources are such kinds of things in the communication endpoints. Sensor technologies, Internet of Things and Bigdata created new learning horizon and hence new challenges. As they are not following generic structure it is more preferable to use NoSQL than SQL.

Through the IoT gateway or other edge devices different IoT devices can communicate with or without intervention of human beings. Data collection process can make it easier and more dynamic by using Artificial Intelligence and Machine Learning. IoT technology has advantages like ubiquity, electronic communication between connected devices made easy, saving time and money for transferring data packets over the network, reducing human intervention and automating tasks. At the same time, it has to face disadvantages like lack of international standard of compatibility for IoT. A single bug may lead to corruption of a large system which necessitates dealing with millions of IoT devices. There are challenges in collection and management of data from all those devices, more possibility for being hacked.

STREAMING SENSOR DATA MANAGEMENT

Storing all sensor data in the backing store for long term will not make sense. These sensors generate a variety of data with the characteristics of heterogeneity, variety, noise, high redundancy and unstructured features (Skourletopoulos et al., 2017). Abundant data generating from sensor nodes could be structured, semi-structured or unstructured. Characteristics of such data includes volume, velocity, and variety and are termed as big data. More number of v's are adding to the definition of Bigdata to deal with value and veracity properties (Aydin et al., 2015). Extracting information from such data faces challenges in their mining, management and privacy. But are useful in preventive medicines, monitoring of patients, follow up care, permanent care, disease management etc.

Streaming data from connected sensors in WBAN phases challenges in real-time processing their storage and memory. Large scale system to overcome challenges are not feasible for general researchers (Pramanik et al., 2018). Cloud computing services, and Distributed computing are contributing much for processing and storage of fast-moving data. Development of new technologies like Fog Computing Cloud Computing and Internet of things could scale this system. Bigdata management involves new theoretical contents with algorithms and specific frameworks and tools. Traditional technologies (Nivash et al., 2014) cannot store, manage or perform analytics on these kinds of data. Management of data is possible with understanding of properties (5 V's) of Bigdata. Installation of millions of cameras, usage of smart devices, number of users in internet etc. resulted in generation of huge volume of data. Their velocity of generation matters while performing analytics. Application determines structure of data, it can be images, videos, text message or numerical data. This property is known by variety term in 5 V's. Among such data determining best data is a different challenge and is defined with fourth 'V' veracity. All these are to be taken into consideration for inferring some knowledge and value from it. This is coming under 5th V called value (Miler et al., 2011).

Date we need to manage in Bigdata is either comes in predefined format, structured, or native formats, unstructured. Smart healthcare system should have capability to process Electronic Health Record and Electronics Medical Records. These records are in structured form with details of patient, date of birth, phone number, address etc. To convert existing system to smart system it is necessary to use old documents in digital format. This type of data is to be analysed in semi-structured format (Sakr & Elgammal, 2016). A good smart healthcare system will integrate data from different sources which are in different format and extract value from them. Present healthcare organizations are in are trying to build efficient way automate all existing documents to analysis format and generate meaning full data.

Depending on data collection mode and environment Bigdata is categorized as (1) Medical data from the hospitals, which are kept in different formats and are to be converted to the electronic form for analysis (2) Expert opinions in the form of texts, audios or videos in the internet of social media (3) Trusted and authentic health data providing by government and medical institutions for studies (4) Data collected with devices connected to a particular person. These data are generated from individual wearable or implanted sensors. Such devices can supply pressure, heart beat, blood sugar, breath, sleep, temperature and movement in a streaming fashion. All the categories of together builds a Bigdata based smart healthcare system.

Smart healthcare promises Right Living, Right Care, Right Provider, Right Innovation and Right Value. Data scientists determines patterns and correlations exist among wide variety of fast-moving data. Data scientists are using modern tools to transform high dimensional and large volume of data. Valuable data collected from WBAN sensors contain hidden information that can be used for diagnosis or fault detection. A scalable and distributed storage will enable analytics more accurate. This needs to keep data for particular period of time. Filtering and dimensionality reductions with prototypes for determining best samples will help efficient utilization of storage spaces. Database management system Cassandra will be suitable option for large volume of critical sensor data while for small or medium and non-critical data MongoDB is right option.

High dimensional data will be received as input to the Bigdata management system. Determining principal components from high dimensional data will consume more time. Determining relationship between different attributes is also time consuming. All these will affect model building in terms of response time and size. This smart era increases availability of personal data through different gadgets. Mobile phones are greatest source among them. smart cameras, smart watches, different sensor implanted clothes, intestinal gas capsules and many more are supporting data generation part of smart healthcare systems. This sensor stream data is good indicator for detection of diseases in early stages, prognosis becomes simpler and patient specific management is possible. Data from these devises will help to create Maintaining a Human Activity recognition (HAR) which is very important for health monitoring.

Human activities can recognize with two major approaches, one through capturing streaming video and other through smart devices connected to the internet. Because of privacy issues it is preferable to choose wearable device-based methods to maintain the HAR than video-based approaches. To obtain HAR, we require data to be collected from a various range of sources. So, it is required to fuse data from these multiple sources. Since data generated from the sensors connected from WBAN are of heterogeneous nature, it is so hard to analyse. So, health data gathered will be stored in an edge system, like personal computer, mobile phone, etc. it comes for the long term, storage will be in RDBMS MySQL database server. It doesn't mean that data from all sensors in WBAN are integrated to a single format in a single system. Different WBAN applications will keep data from various sensors and are stored in separate systems. Most sensor data are in the form of tables with different table formats and data types. So, as a solution to the storage and retrieval of heterogeneous Bigdata management, we need to decide whether SQL or NoSQL databases are apt for the purpose. Real bottleneck lies in analysis of the big data volume and extracting the useful information.

Choosing right database from a set of different database management systems available in the NoSQL (Al Rasyid et al., 2016) category is another task in analysis. Processing data on the fly and keeping few most relevant records is best strategy for long term tasks. There are open-source database management systems with different approaches of data storage. Few of them are key-value pair type, document storage type and wide column stores (Al Rasyid et al., 2015). Each of these categories have several database systems. Database Management Systems like Redis, Riak and Tokyo Cabinet etc are coming under Key value pair category. While Amazon SimpleDB, MongoDB, and CouchDB are following document types to organize data by assigning a unique key for each document. Wide Column Stores database store tables of records on multiple systems after partitioning them in column wise and raw wise. Databases like HBase, Cassandra and Hyper Table are falling under this category. All such databases provide support for managing Bigdata.

As a result of unprecedented data rates in terms of volume and velocity it is necessary to utilize exclusive technologies for analysis on Bigdata (Zulkernine et al., 2013). After deciding storage mechanism and retrieval method of Bigdata next stage is choosing method to integrate sensor data from any WBAN application. To integrate the sensor data knowledge on latest a large-scale data processing system is required. This in turn utilizes new programming models like MapReduce (van der Veen et al., 2012) and Hadoop like large scale data integration and processing technologies. The term "Hadoop" is commonly used for all ecosystems which run on HDFS file systems. There are many different tools integrated to the Hadoop ecosystems. Apache Sqoop is one among them to load data from RDBMS(MySQL) and convert to HDFS. Sensor data in HDFS format will be processed and backed up with Apache Hive and MapReduce will do querying and processing of data.

Integration of Hadoop architecture with Open stack clouds could be used for many types of critical monitoring (Dean & Ghemawat, 2008). All three categories of cloud computing services, infrastructure as a service (IaaS), platform as a service (PaaS) and software as a service (SaaS) have services to support Bigdata. The most flexible and basic cloud computing model open stack is a free open standard cloud computing platform falling under the IaaS category. IaaS provides the access and management to computing hardware, storage, networking and operating systems with a configurable virtual server. There are various deployment models available with Open stack. They are i) On premise distribution ii) Open stack based public cloud iii) Hosted open stack private cloud iv) Appliance based open stack and v) OpenStack as a Service.

CLOUD COMPUTING AND BIG DATA

The personalized smart healthcare system has to analyse data from smart devices and other connected sensor reading in the form stream. Data from the near body sensors are analysed with artificial intelligence and will turn into information for physicians. Fitting of such data to the workflow is made easy through various cloud computing services (Raj et al., 2014). Large sets of collected data are hard to manage by traditional systems and methods. Cloud computing plays a critical role here in providing an environment for a common user to perform big data processing. Cloud computing infrastructure and services reduces scalability issue and attracts more researchers to the field. High-powered servers from different vendors with limited free access unlimited paid access are treated as "cloud" for our purpose. Cloud services enables d querying of large data sets faster than a standard computer. There are different categories of cloud computing services provided by various providers. In the cloud computing stack, the Software as a Service (SaaS) arranged on top, Platform as a Service (PaaS) in the middle, and Infrastructure as a Service (IaaS) at the bottom. Leading cloud services providers like Google Cloud Platform, Microsoft Azure, IBM Cloud Services, Amazon Web Service (AWS) are providing free services for practising. All these vendors provide different types of cloud Application Programming Interfaces. Provisioning of cloud hardware, software and platforms for the developing numerous applications or services are made possible and easy with these interfaces.

Bigdata management, processing and analysis is difficult with basic computers. But advanced computing architecture, parallel and distributed systems, various database management system can together help in scaling system performance. Cost of this system is very high and it is reduced by the development of cloud computing facilities. Cloud computing service uses resource allocation models and algorithms to improve autoscaling and auto provisioning capability. In this era there are exclusive cloud support for Bigdata mining platforms to provide various statistical and analytical functions (Ta et al., 2016). A good collaboration between patients, developers and scientists will help to manage and develop bigdata analytics algorithms which eventually enhance results. The term "Bigdata" and "Cloud Computing" are entirely different but are correlated. "Bigdata" refers collected data with 5 V's properties while "Cloud Computing" mean mechanisms of actions performed on this data. There are different products available with cloud packages which include database management systems, machine learning capabilities, identity management systems, cloud based virtual machines and containers and more.

There are three main sources which generate major portion of Bigdata which are machine data, social data and transactional data. They can either unstructured or structured. For the unstructured data it is difficult to model and analyse. Social data mainly come from video uploads, general media, likes, comments, tweets & retweets that are uploaded via various social media platforms. Consumer behavior and sentiments are the valuable insights from such data which influence market analysis. The second main contributor of big data is machines which are installed with sensors. As the Internet of things was growing, this category of data grew even more. In the near future, sensors in road cameras, satellites, games, medical devices and smart devices will deliver more data in high volume, velocity and variety with values. The third category of data are from payment orders, delivery receipts, storage records and Invoices which are categorized as transactional data. Despite having enough data, organizations struggle to make sense from these and to put them for good use.

The data harnessed through these different media can utilize cloud computing platforms for storage, searching, editing and to obtain future insights. The free service like Google Cloud Messaging (GCM) can be used by developers to send messages to multiple platforms (Rahamathulla, 2020). Sensor data in non-standard format could be transformed to standard format through artificial intelligence services of cloud computing environment. Real time data processing and analytics of Bigdata are possible in fraction of seconds with cloud infrastructure. State of the art infrastructure of cloud computing enables faster execution with paying money for the time and power we are using (Ifrim et al., 2017). There are specific platforms for Bigdata analytics in the cloud environment and are named as Cloud Based Analytics as a Service (CLAaaS) (Zulkernine et al., 2013).

Big data related researches were facing lots of issues including the support in terms of resources which include both software and hardware for versatile analytical tools, analytic workflow management system, enabling customized and role-based service delivery, ubiquitous access, knowledge delivery, etc. Cloud technology is advanced to provide ubiquitous and scalable computing resources required for the development of projects related to Bigdata. CLAaaS provides resources to work with wide varieties of analytic tools.

Applications that demand high performance computing systems can be provided with cloud services. Smart health monitoring needs cloud support for pre-processing of data in the absence of edge processing systems, storage facility for storing selected instances, for analytic task on stored and newly generating data and for visualizations of results. All these functions have to be performed on Big Data and are done with the support from IoT cloud. To make these tasks even simpler, techniques like data reduction are applied.

Storage and Analysis of data from different locations increased relevance of cloud-based hardware and software. To provide services for Bigdata with reduced cost and increased efficiency it is important to have Big Data as a Service (BDaaS). Organizations can understand and gain insights from this large information set with the help of Big Data as a Service (BDaaS). In general, services that involves managing or running Bigdata on the cloud is usually referred to as BDaaS. As a part of BDaaS, the cloud provides statistical analysis tools to generate insights from large dataset. BDaaS uses different application software for different purposes. These include application software for querying on the collected set of data, mining of useful data and for analytical processing.

Core BDaaS, Performance BDaaS, Feature BDaaS and Integrated BDaaS are variants of available BDaaS. Core DBaaS is solely implemented on PaaS. NoSQL and Hadoop are integrated to the virtual machines to provide BDaaS. Even without improved capabilities, it has the advantage of quick implementation. Limitations of Core BDaaS overcomes by Performance BDaaS, by implementing it on PaaS and IaaS. To provide abstraction and to improve productivity in bigdata services Service model Feature BDaaS is implemented on PaaS and SaaS. Integrated BDaaS integrates all good features of all the three models by implementing them in three layers of the cloud namely, IaaS, SaaS and PaaS.

Sensors in smart healthcare systems require support from different cloud services for different purposes like data ingestion, storage and analytics. As a first step we need to ingest data into the cloud. Tools are available for various phases of Bigdata analysis with cloud services. They are providing in many names by different service providers. Range of services includes services for capturing streaming data from connected devices in the IoT environment. Storage of data their computations and processing for knowledge extractions. In order to connect IoT devices and to receive data from the respect environment through MQTT protocol we can use Amazon IoT service. But for continuous capturing and storing of terabytes of data per hour we can use another AWS service called Kinesis. Ingestion of streaming data from healthcare environment (Taher et al., 2019) can do with this Kinesis service. Similarly, many other clouds service providers also handle such cases with different service names. The performance analysis of such systems could be performed by considering some parameters like response time with varying velocity and volume of data to analyses.

DATA ANALYSIS AND MACHINE LEARNING

The diverse range of physiological and functional data are increasingly available with increasing popularity of smart wearable devices. These smart devices could be a part of internet of things and hence of smart healthcare system. These devices reduce human intervention in detecting health related parameters. Data generating by these smart devices can be interpreted with machine learning and data analysis techniques to support clinical decision making. Early medication or prognostics were possible with these health data. Learning from such streaming data or from Bigdata is done with recent and promising learning techniques. Distributed and Parallel learning, Transfer Learning, Federated Learning, Reinforcement learning are few of them.

Machine Learning algorithms analyses and interprets data to recognize underlying patterns and trends in available dataset. The models built with available dataset needs updations in real-time. Selection of right learning method is necessary to process massive amounts of changing diverse and real-time data. This will offer quick and effective decision making at a reduced cost. Application of efficient methods and techniques like machine learning and deep learning on Bigdata will produce wealth of information. Machine learning algorithms can be used to design the classification methods of time series data and for their analysis (Obinikpo & Kantarci, 2017). Personalized health care service is popularizing than generalized methods with smart healthcare systems.

A smart health system can be built by passing through different phases. First phase is acquiring and accessing recorded data. Second phase is data extraction which is more difficult than the first phase. The acquired data are usually heterogeneous in nature because they usually contain free text data, structured data, image data and sensor data. These extracted data further require expert assistance to remove errors and inconsistencies and this phase is called data cleaning. Third phase is aggregation of instances related to a particular person from all sources of data. In last phase models are developed and displayed with modern visualization tools for easy understanding.

Human Activity Record (HAR) created with data collected by different means will use for exploiting data and for performing actions. Continuously collecting physiological data exhibits new behaviors or concepts which are not pre-defined in history. This is called concept drift and is to be handled with streaming data analysis tools. Such data analysis Retraining task on such fast-moving stream is difficult to achieve. So, an on-the-fly analysis with limited window size will give real-time responses. This approach could avoid centralized storage and analysis with high end systems and reduce many other resource consumptions.

Machine Learning algorithms are applied after preprocessing of collected data and are passed through feature extraction and feature selection before going for learning with a data analytics algorithm. These learning methods can be any of supervised, unsupervised, semi supervised learning and reinforced methods. The algorithms like K-Nearest Neighbors (KNN), Linear Discriminant Analysis (LDA), Support Vector Machines (SVM), Random Forest, Neural Networks (NN) and Deep Learning fall under the category of supervised learning where data are labeled before applying these algorithms. All machine learning tasks are aiming to form regression equations, build classifier or clustering. Target variables value is predicting with generalized equation or finding label of a particular instance are done with decision tree or random forest like basic algorithms and with modern approaches.

Supervised learning generally goes through five different phases (i) Collection of Dataset for Training and Testing (ii) Extraction of Features (iii) Choosing Suitable Machine Learning Algorithm (iv) Constructing Model (v) Evaluate and Compare the Performance of Algorithm with other Algorithms. Many programming languages like MATLAB, Python, Java, C++, etc. provide packages for supervised learning algorithms. It is required to overcome the issues like multidimensional feature vectors, bias, noise exist in input data, scaling of tasks, data redundancy and heterogeneity among etc.

Another important type of learning method is unsupervised learning where prediction of structure or pattern from input data is done by algorithms, without labeling data sets. Unsupervised learning algorithms mainly deal with the design of models to form hidden patterns and relationships within unlabeled data using machine learning algorithms. Clustering, association rule mining and anomaly detection are important categories of algorithms in unsupervised learning. Anomaly detection primarily deals with noise, deviations, exceptions and recognition of outliers prevailing in the unlabeled data. The faults in sensor, monitoring systems, medical errors, abnormal signals, etc. are examples of anomalies. A semi supervised learning method applies in situations where number labeled data is very less as compared to unlabeled data. During real time streaming data generation, data mislabeling happens mainly due to expensive labeling process, network delay, communication error and other reasons.

Information can extract from streaming data with different modern learning approaches without storing all data instances. There are Federated learning, Reinforcement learning, Online learning, Incremental learning etc. to perform analytics with few parts of entire data set without storing them and without multiple iterations. To deploy machine learning algorithms, it is required to ensure that preprocessing pipelines and transformations are done properly. Representation learning is good approach to set rules for selecting representatives from huge dataset. This probabilistic model that captures posterior probability distribution and determine attributes that has best impact on the output. During supervised learning these representative rules of input can use at the ingestion end to filter out unnecessary records. Representation learning can contribute in feature selection and feature extraction to reduce dimension of dataset. Target of training will decide rules in this learning process and it can increase effectiveness of dataset.

Decision making from streaming sensor data can achieved with deep learning techniques. Non-linear relationship between the attributes can determine by adding more layers in deep learning architecture.

Predictions in complicated relations could also do with this learning approach. Hidden perceptron layers in conventional Artificial Neural Networks (ANN's) helps in identifying patterns (##NO_NAME##, 2022). Deep Belief Networks (DBN's) and Convolutional Neural Networks (CNN's) are two most basic approaches of deep learning. Analytics solutions on large Datasets are possible with advanced graphic processors with increased processing power Using mobile sensor data, deep learning can produce highly accurate results, generate intrinsic features and learn from unlabeled data (Cort'es et al., 2015).

Modeling of data by deep learning is performed by two main stages, learning algorithms with different parameters and updating model's parameters. Learning algorithm calculates the gradient of the model's parameters through iterating different batches independently and the average of these computed values will be used to update the model's parameters. Number of instances and dimension of Dataset will decide of time to get outputs in deep learning models. Parallel or distributed approaches on sliced dataset can improve models' performance in terms of response time. Open-source platform Apache Spark (Al Rasyid et al., 2016) with components spark master and spark workers can use for analysis on such data.

Deep learning network receives data from sensors or other devices in smart healthcare systems. These data are iteratively passes through different layers in neural network. Number of layers will improve perfection but consume more time. So iterative process of evaluation is not suitable for streaming data. Hence new approach like reinforcement learning should be adopted for time to time or instance-based changes in the models. A reward-based system will make model parameters adjustment with each instance or small batch of data. Before applying any machine learning approach dimension of input data should be reduced through feature selection methods. Filters and wrappers are two most important methods for feature selections. Features are ranked on the basis of their relationship with model parameters. On the other hand, wrapper methods consider relationships between features. Feature extraction is the process following feature selection and is of two categories based on their transforming methods. They are linear and nonlinear transformations. These extracted features related to the activities will specify some patterns and these patterns are used for classification and modeling.

Learning process could be made faster by allocating several workstations in distributed fashion for parallel processing. This avoids centralized processing of data and hence saves time and energy. Building decision rules with pattern of transactions are stacked for future purposes. This learning can apply for further learning process in streaming data. Recent and exclusive technologies should be integrated to deal with fast flowing smart healthcare data from the origin point to storage destination. Electronic records including physiological signal can store with Apache Hadoop acritudes. Similarly, we can prefer a cloud-based framework employing a Lucene based distributed search cluster to deal with issues of privacy protection, to ensure highly concurrent and scalable medical record retrieval and to facilitate data analysis in a self-caring setting.

Devices are becoming self-aware, proactive, self-adaptive and prescriptive with artificial intelligence and machine learning tools. New facts in the form of patterns are not easy to identify by human eyes. They can be identified by using machine learning techniques which add more intelligence to the system. This intelligence and cognitive behavior can also be added to deep learning and AI which will enhance decision making capability. A framework with cognition or intelligence can consider all health indicators and determine whether emergency response or specialized medical care is required. So, the systems with artificial intelligence can support various applications in healthcare, like, patient monitoring, smart health record, smart pill dispensing, intelligent disease detection, mobile healthcare, emergency response, smart alerts, remote medical equipment operation and control, etc.

CONCLUSION

Experts from different fields like medical science, computer science, social science, and business together with advanced technology Could make healthcare industry smart. Automation will bring quicker responses to the requests. Industrial collaborations could reduce investment. The opportunity to provide centralized service will enhance coverage area of service by experts. Researches in processing of huge volume of high velocity heterogeneous data will help to processes more data with existing resources. Continuous monitoring of patient's body parameters helps in reducing hospitalized patients count. Time to get acceptance of this smart industry highly depends on cost of smart devices and supporting infrastructure. Huge volume of continuously generating heterogeneous data handling by big data analyst with support of machine learning and artificial intelligence determines accuracy of result and hence trust of patients in this industry. Necessity of developing drift-based learning methods without storing data for long time will improve scalability of model. Infrastructure developments like cloud services has role in collaboration from various countries around the world. Achieving artificial intelligence through machine learning within the constraints should be the focus of researchers.

REFERENCES

Agarwal, K., & Guo, Y. X. (2015), May. Interaction of electromagnetic waves with humans in wearable and biomedical implant antennas. In *2015 Asia-Pacific Symposium on Electromagnetic Compatibility (APEMC)* (pp. 154-157). IEEE. 10.1109/APEMC.2015.7175377

Aggarwal, C. C., Ashish, N., & Sheth, A. (2013). *The internet of things: A survey from the data-centric perspective. InManaging and mining sensor data.* Springer.

Al Islam, N., & Arifin, F. (2016). Performance analysis of a miniaturized implantable PIFA antenna for WBAN at ISM band. In *2016 3rd International Conference on Electrical Engineering and Information Communication Technology (ICEEICT)* (pp. 1-5). IEEE. 10.1109/CEEICT.2016.7873145

Al Rasyid, M. U., Lee, B. H., & Sudarsono, A. (2015, January 1). Implementation of body temperature and pulseoximeter sensors for wireless body area network. *Sensors and Materials, 27*(8), 727–732.

Al Rasyid, M. U., Yuwono, W., Al Muharom, S., & Alasiry, A. H. (2016). Building platform application big sensor data for e-health wireless body area network. In *2016 International Electronics Symposium (IES)*, (pp. 409-413). IEEE. 10.1109/ELECSYM.2016.7861041

Alam, M. M., & Hamida, E. B. (2014, May). Surveying wearable human assistive technology for life and safety critical applications: Standards, challenges and opportunities. *Sensors (Basel), 14*(5), 9153–9209. doi:10.3390140509153 PMID:24859024

Alsheikh, M. A., Niyato, D., Lin, S., Tan, H. P., & Han, Z. (2016, May 20). Mobile big data analytics using deep learning and apache spark. *IEEE Network, 30*(3), 22–29. doi:10.1109/MNET.2016.7474340

Aydin, G., Hallac, I.R., & Karakus, B. (2015). Architecture and implementation of a scalable sensor data storage and analysis system using cloud computing and big data technologies. *Journal of Sensors*.

Chen, M., Li, W., Hao, Y., Qian, Y., & Humar, I. (2018, September 1). Edge cognitive computing based smart healthcare system. *Future Generation Computer Systems*, *86*, 403–411. doi:10.1016/j. future.2018.03.054

Chui, K. T., Alhalabi, W., Pang, S. S., Pablos, P. O., Liu, R. W., & Zhao, M. (2017, December). Disease diagnosis in smart healthcare: Innovation, technologies, and applications. *Sustainability (Basel)*, *9*(12), 2309. doi:10.3390u9122309

Cort'es, R., Bonnaire, X., Marin, O., & Sens, P. (2015, January 1). Stream processing of healthcare sensor data: Studying user traces to identify challenges from a big data perspective. *Procedia Computer Science*, *52*, 1004–1009. doi:10.1016/j.procs.2015.05.093

Dean, J., & Ghemawat, S. (2008). MapReduce: Simplified data processing on large clusters. *Communications of the ACM*, *51*(1), 107–113. doi:10.1145/1327452.1327492

Dey, N., Hassanien, A. E., Bhatt, C., Ashour, A., & Satapathy, S. C. (Eds.). (2018). *Internet of things and big data analytics toward next-generation intelligence*. Springer. doi:10.1007/978-3-319-60435-0

Fortino, G., Giannantonio, R., Gravina, R., Kuryloski, P., & Jafari, R. (2012, December 24). Enabling effective programming and flexible management of efficient body sensor network applications. *IEEE Transactions on Human-Machine Systems*, *43*(1), 115–133. doi:10.1109/TSMCC.2012.2215852

Ghamari, M., Janko, B., Sherratt, R. S., Harwin, W., Piechockic, R., & Soltanpur, C. (2016, June). A survey on wireless body area networks for healthcare systems in residential environments. *Sensors (Basel)*, *16*(6), 831. doi:10.339016060831 PMID:27338377

Ifrim, C., Pintilie, A. M., Apostol, E., Dobre, C., & Pop, F. (2017). The art of advanced healthcare applications in big data and IoT systems. In *Advances in mobile cloud computing and big data in the 5G Era* (pp. 133–149). Springer. doi:10.1007/978-3-319-45145-9_6

Gillis, A. (2022). Internet of Things. *TechTarget*. https://internetofthingsagenda.techtarget.com/definition/Internet-of-Things-IoT

Katal, A., Wazid, M., & Goudar, R. H. (2013). Big data: issues, challenges, tools and good practices. In *Proceedings of the 6th International Conference on Contemporary Computing* (IC3'13), (pp. 404–409). IEEE. 10.1109/IC3.2013.6612229

Khan, N., Yaqoob, I., Hashem, I.A., Inayat, Z., Ali, M., Kamaleldin, W., Alam, M., Shiraz, M., & Gani, A. (2014). Big data: survey, technologies, opportunities, and challenges. *The Scientific World Journal*.

Kibria, M. G., Nguyen, K., Villardi, G. P., Zhao, O., Ishizu, K., & Kojima, F. (2018, May 17). Big data analytics, machine learning, and artificial intelligence in next-generation wireless networks. *IEEE Access : Practical Innovations, Open Solutions*, *6*, 32328–32338. doi:10.1109/ACCESS.2018.2837692

Manogaran, G., Varatharajan, R., Lopez, D., Kumar, P. M., Sundarasekar, R., & Thota, C. (2018, May 1). A new architecture of the Internet of Things and big data ecosystem for securing smart healthcare monitoring and alerting systems. *Future Generation Computer Systems*, *82*, 375–387. doi:10.1016/j. future.2017.10.045

Miler, M., Medak, D., & Odobasic, D. (2011). Two-tier architecture for web mapping with NoSQL database couch DB. *Geospatial Crossroads GI Forum, 11*, 62–71.

Nivash, J. P., Raj, E. D., Babu, L. D., Nirmala, M., & Manoj, K. V. (2014). Analysis on enhancing storm to efficiently process big data in real time. In *Fifth International Conference on Computing, Communications and Networking Technologies (ICCCNT),* (pp. 1-5). IEEE. 10.1109/ICCCNT.2014.7093076

Obinikpo, A. A., & Kantarci, B. (2017, December). Big sensed data meets deep learning for smarter healthcare in smart cities. *Journal of Sensor and Actuator Networks.*, *6*(4), 26. doi:10.3390/jsan6040026

Pike, M., Mustafa, N. M., Towey, D., & Brusic, V. (2019). Sensor Networks and Data Management in Healthcare: Emerging Technologies and New Challenges. In *2019 IEEE 43rd Annual Computer Software and Applications Conference (COMPSAC),* (pp. 834-839). IEEE.

Pramanik, M. I., Lau, R. Y., Demirkan, H., & Azad, M. A. (2018, September 1). Smart health: Big data enables a health paradigm within smart cities. Expert Systems with Applications. 2017 Nov 30;87:370-83..

Chen, M., Li, W., Hao, Y., Qian, Y., & Humar, I. (2022). Edge cognitive computing based smart healthcare system. *Future Generation Computer Systems*, *86*, 403–411.

Rahamathulla, M. P. (2020, March 1). Cloud-based Healthcare data management Framework. *KSII Transactions on Internet and Information Systems*, *14*(3).

Raj, E. D., Nivash, J. P., Nirmala, M., & Babu, L. D. (2014). A scalable cloud computing deployment framework for efficient MapReduce operations using Apache YARN. In *International Conference on Information Communication and Embedded Systems (ICICES2014),* (pp. 1- 6). IEEE.

Ravi, D., Wong, C., Lo, B., & Yang, G. Z. (2016, December 23). A deep learning approach to on-node sensor data analytics for mobile or wearable devices. *IEEE Journal of Biomedical and Health Informatics*, *21*(1), 56–64. doi:10.1109/JBHI.2016.2633287 PMID:28026792

Sakr, S., & Elgammal, A. (2016, June 1). Towards a comprehensive data analytics framework for smart healthcare services. *Big Data Research.*, *4*, 44–58. doi:10.1016/j.bdr.2016.05.002

Salem, O., Serhrouchni, A., Mehaoua, A., & Boutaba, R. (2018, May 31). Event detection in wireless body area networks using Kalman filter and power divergence. *IEEE Transactions on Network and Service Management*, *15*(3), 1018–1034. doi:10.1109/TNSM.2018.2842195

Sethi, D., & Anand, J. (2019, September 12). Big data and WBAN: Prediction and analysis of the patient health condition in a remote area. *Engineering and Applied Science Research.*, *46*(3), 248–255.

Shahad, P. (2021). Challenges in Streaming Data Analysis for Building an Adaptive Model for Handling Concept Drifts. In *2021 International Conference on System, Computation, Automation and Networking (ICSCAN),* (pp. 1-6). IEEE.

Skourletopoulos, G., Mavromoustakis, C. X., Mastorakis, G., Batalla, J. M., Dobre, C., Panagiotakis, S., & Pallis, E. (2017). Data and cloud computing: a survey of the state-of-the-art and research challenges. In Advances in mobile cloud computing and big data in the 5G Era, (pp. 23-41). Springer.

Ta, V. D., Liu, C. M., & Nkabinde, G. W. (2016). Big data stream computing in healthcare real-time analytics. In *2016 IEEE International Conference on Cloud Computing and Big Data Analysis (ICCCBDA)*, (pp. 37-42). IEEE.

Taher, N. C., Mallat, I., Agoulmine, N., & El-Mawass, N. (2019). An IoT-Cloud Based Solution for Real-Time and Batch Processing of Big Data: Application in Healthcare. In *2019 3rd International Conference on Bio-engineering for Smart Technologies (BioSMART)*, (pp. 1-8). IEEE.

van der Veen, J. S., van derWaaij, B., & Meijer, R. J. (2012). Sensor data storage performance: SQL or NoSQL, physical or virtual. In *Proceedings of the IEEE 5th International Conference on Cloud Computing (CLOUD '12)*, (pp. 431–438). IEEE. 10.1109/CLOUD.2012.18

Zulkernine, F., Martin, P., Zou, Y., Bauer, M., Gwadry-Sridhar, F., & Aboulnaga, A. (2013). Towards Cloud -Based Analytics-as-a-Service (CLAaaS) for Big Data Analytics in the Cloud. *2013 IEEE International Congress on Big Data*, Santa Clara, CA. 10.1109/BigData.Congress.2013.18

Chapter 11
Significance of Fog Computing to Machine Learning–Enabled IoT for Smart Applications Across Industries

Mohan Raj C. S.
Hindusthan College of Arts and Science, India

A. V. Senthil Kumar
🆔 https://orcid.org/0000-0002-8587-7017
Hindusthan College of Arts and Science, India

Meenakshi Sharma
🆔 https://orcid.org/0000-0002-6958-8741
University of Petroleum and Energy Studies, India

Ibrahiem M. M. El Emary
King Abdulaziz University, Saudi Arabia

Rohaya Latip
Universiti Putra Malaysia, Malaysia

Saifullah Khalid
🆔 https://orcid.org/0000-0001-9484-8608
Civil Aviation Research Organisation, India

Chandrashekar D. V.
🆔 https://orcid.org/0000-0001-9798-0305
T.J.P.S. College, India

ABSTRACT

Industry 4.0 refers to the phase of transition that is taking place, enabling automation and data interchange in industrial technologies and processes. Fog computing architecture can provide real-time processing, nearby storage, extremely low latency, dependability, large data rates, and other requirements for industrial Internet of Things (IIoT) applications. In the context of IoT applications, fog infrastructure and protocols are the main areas of interest. The phrase "fog computing," sometimes known "edge cloud," is a new paradigm. Between edge devices and Cloud Core, it adds another layer. Along with providing computing, storage, and networking capabilities, it also fills a need left by the cloud. The main features of fog computing are covered in this chapter, along with current research on the subject and a focus on the difficulties encountered when creating its architectural design.

DOI: 10.4018/978-1-6684-8785-3.ch011

1. INTRODUCTION

In reference to the industrial revolution, fog computing, often known as "fogging," is an expanded type of cloud computing that provides applications and services (low latency and high processing) to autonomous heterogeneous devices inside an industry. The purpose is to place intelligence control, processing, and storage close to data devices. Real-time services with high data processing, maximum capacity, and scalability are crucial for industry 4.0. Due to its many advantages over cloud computing, fog computing offers the finest options for this kind of setting. By introducing the notion of network edge computing, the extension of cloud computing seeks to reduce the load on the cloud.Low latency and improved cache memory are necessary for real-time services and decision-making processesin industrial automation(J. Li, F. R. Yu, G. Deng, 2020)

On the basis of geographic distribution, consider real-time applications, low latency, location awareness, the number of nodes, and edge devices with cache support. In between the internet's cloud of devices and end users' devices are virtualized nodes, also known as cloudlets or fog nodes. With superior QoS parameters performance and coverage of crucial IoT criteria, fog computing offers services and applications similar to those offered by clouds. The following are significant benefits of fog computing that affect IoT adoption .When data is stored on network edge nodes, there is no longer a need to retrieve it from distant clouds, which ends transmission delays.Fog computing helps IoT applications process and analyse data more quickly.The processing and computation time will be reduced by data storage on edge nodes.Cache-enabled nodes will stop transmitting pointless data(A. V. Serbanescu, S. Obermeier, 2020)

Industrial cyber-physical systems make it possible for physical objects and processes to be closely associated with computing, communication, and control systems online. Transmissions are facilitated through cyberphysical interfaces that connect the two worlds. Employingwireless sensors, mobile devices like smartphones and tablets, and others. These cyber-physical interfaces conceptually display "cyber twins," which are virtual representations of real-world physical things in the cyberspace. To provide operational insights and guide decision-making, these virtual objects may then be individually and/or collectively analysed, questioned, or simulated (A. Banga,et al., 2019)

The ML models and associated IoT data operations include feature extraction, grouping, and classification. Analytics, such as k-means, k-nearest neighbours (k-NN), support vector machines (SVM), liner regression, and DNN, have been thoroughly examined in the design of ML algorithms faces additional challenges because of the resource-constrained context. There are certain universally applicable algorithms that adapt to resource and distribution limits, such as k-NN and some specific neural network techniques. The second, however, necessitates often time-consuming

Figure 1. Fog computing

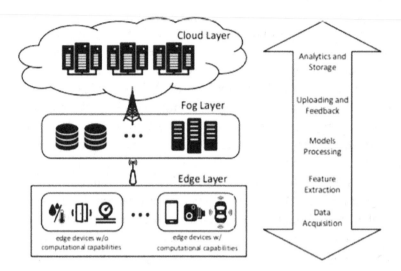

fine-tuning of the granularity of the environment sensor data and the environment model. One possibility is early sensor fusion utilising a 3D voxel environment model (F. Daryabar,et al., 2021).

In addition to using edge intelligence to lessen the quantity of data transmission and storage required on the cloud, the fog-level IoT system produces a faster and more effective knowledge transfer to the cloudAccording to the authors in, in order to allow regional IoT networks to execute edge analytics, future IoT systems need incorporate intelligence at the fog layer. By using local storage, fog-layer analytics, and decision-making, data management may also be enhanced. We need algorithms that are resource-efficient, using little memory and energy, to make this happen. Microsoft recently demonstrated ML models that can be used by small IoT devices.The created algorithm has been disclosed, and it may be used with an Arduino Uno board.(A. Trombetta, 2021).

The organisation and the end user are liberated from having to specify numerous specifics thanks to cloud computing. For latency-sensitive applications that depend on nearby nodes to achieve their delay needs, this joy turns into a problem. In addition to location awareness and low latency, a new generation of Internet installations, most notably the Internet of Things (IoTs), also needs mobility support and geo-distribution. We contend that a new platform—which we've dubbed Fog Computing, or simply Fog—is required to satisfy these demands since fog is a low-lying cloud. Additionally, we contend that there is a productive interaction between the Cloud and the Fog, with Fog Computing enabling a new generation of applications and services as opposed to can-nibalising Cloud Computing.(A. Carcano et al., 2021).

In figure 1,The structure of this chapter is as follows. We introduce the Fog Computing para-digm and outline its features as well as those of the platform that supports it in the second section services. In the section that follows, we'll take a closer look at a few noteworthy applications and services that support our claim that the Fog is an essential part of the platform needed to enable the Internet of Things. In the fourth session, we look at big data and analytics in the context of interesting applications (T. H. Morris.et al., 2021)

The understanding that some of these applications require both long-term global data mining and real-time analytics exemplifies how the two technologies interact and play complimentary roles.

Figure 2. Fog ruling areas

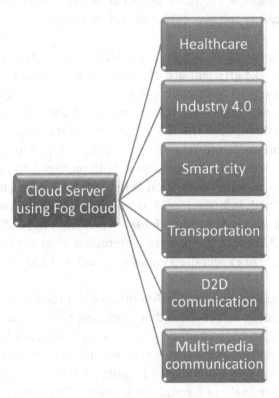

We wrap off by discussing the current state of fog computing and discussion of future work(D. Miorandi et al., 2021)

1. **Low Latency Requirement**: Many IoT applications, particularly those for industrial use cases and the Internet of Vehicles, have stringent service latency requirements.

    Particularly, millisecond-scale end-to-end latencies are required for smart manufacturing systems. Latencies under a few tens of milliseconds are often needed for the drone flying controls and vehicle-to-vehicle communications. Using merely the far clouds won't be sufficient to address any of these situations(M. A. Muna and N. Moustafa, 2020).

2. **Limited Link Bandwidth**: As the number of IoT devices linked wirelessly increases, the link bandwidth is likewise becoming increasingly crowded. It becomes clear that we shouldn't try to send all the data, such as the information gathered at the sensors, to the distant clouds for additional processing. On one side, there are restrictions on the wireless spectrum. Moreover, we lack the bandwidth necessary to upload all of the data to the cloud. However, studies show that the majority of the data produced by the endpoints may be handled locally without negatively affecting the performance of the entire system(T. H. Morris et al., 2021)

3. **Restricted Computing Power**: Because of their high cost and limited energy supply, IoT devices often have a limited computing capacity. Offloading the computational work to distant

clouds is the current solution. However, the offloaded activities will undoubtedly experience additional communication and processing delays as a result.

In addition to offloading to distant clouds, we should strive to offload the computing workload to nearby edge nodes and make efficient use of the network's available computing capacity.

Forecasts IN Mist Processing The following move is to perform computerized examination and prognostics for fabricating machines and cycles using cloud-based AI in the wake of get-together huge volumes of rough data through IoT frameworks and collaboration processes (F. Bonomi., 2020).

The procedures of AI normally fall into three wide classes, including managed learning, solo learning, and cooperative separating that utilization both managed and solo learning.

As a contribution to a preparation calculation in light of datasets and marked data, different elements or traits will be acquired. Informational indexes for the preparation of a learning calculation are likewise utilized. To assess the learning calculation, testing informational indexes are utilized. Upon an assessment of an AI model, the potential aftereffect of an occasion can be anticipated. AI can be utilized in assembling to expect the wear of gear and to decide whether support is required (S.K Datta et al., 2021).

Since AI on large preparation information might take impressive memory and computer chip centers, it assists with speeding up prescient demonstrating to carry out cloud AI calculations. A few business answers for decentralized cloud preparing have been made.

The most prominent benefit of AI is that pretrained models can anticipate results with raised accuracy nearly immediately. The reproduction in independence that goes on for a couple of days is presently diminished extensively to a couple of moments (Manimurugan & Narmatha, 2020).

In figure 2 refer, Our AI and anticipating routine incorporates arrangement, preparing, approval, forecast and, case by case, frame recognition. It ought to be noticed, regardless of whether the essential routine is direct prior to introducing each move toward detail, that it is perceived as an iterative strategy. Any step prior to delivering prepared examples can be reviewed by the need to work on estimate accuracy. For instance, an alternate AI calculation could be picked for approval on a test set that implies an unacceptable model. On the off chance that a preparation set is considered insufficient, extra data ought to be acquired. We characterize our showing process and our standard estimate bit by bit beneath. Regarding the heterogeneity figuring setting, the recommended procedure analysed the wellspring of energy utilization.(C. Narmatha et al., 2020)

Furthermore, the procedure proposed utilized a technique in view of the fractal methodology for AI model to help the calculation for assigning assets. Four modules present the proposed system to booking as displayed in In the first place, the cognisance model assembles asset jobs for all actual gadgets in all bunches like circle use, memory use and computer chip use. To assess asset use, the observation model will communicate these data into the estimate model (M. Viju Prakash et al., 2020). The model estimate investigates use reports and sends the results to the model scheduler. On the foundation of the estimate model results, the scheduler model is responsible for assigning assets to explicit hosts. At last, the task model allots have assets. A suitable virtual machine is utilized for approaching requests of cloud clients. The essential job of the task model. Numerous ongoing calculations have found that the asset heap of the central processor has an compelling impact and strong review relationship. Enormous information presents different deterrents to cloud research (M.Shanmuganathan et al., 2020). The principal factors that make distributed computing important for information handlingare execution, sensibility and minimal expense. Security and security are by the by essential issues for touchy information capacity. We really want secure handling particularly for wellbeing geoinformatics applications with touchy information (K. Batri, and S. Anbu Karuppusamy, 2019).

To relieve information assurance and security dangers, the product ought to be utilized for restricted information access inside the framework, in consistence with the application scope. The data ought to be moved for the finish of the information investigation to the following stage after capacity. Information assurance and security will help through this (S. Manimurugan et al., 2019).

1.1. Fog to Boost IoT Application Security and Privacy

Fog computing, in contrast to Cloud computing, has its own traits, including a dispersed nature, resource limitations, and distant operation. In the context of security and privacy for IoT applications, these features have produced a special environment. The Fog can deploy and provide more computational power to secure and safeguard IoT devices, but because of these qualities, it also creates extra security and privacy issues. In other words, because it offers specific local security and privacy services like local monitoring and threat detection, it may improve the overall security of the IoT layer (Ericsson, 2021).

In figure 3 shows, according to the available research, the majority of studies found that fog reduces the security and privacy of IoT systems. Nevertheless, some experts come to the conclusion that when compared to the Cloud, Time-sensitive data processing and local data storage are made easier by fog computing. This lessens the volume and distance of data sent to the Cloud, which in turn lessens the effect of IoT applications' security and privacy problems. Fog computing speeds up decision-making and decreases response times, enabling quicker, more localised responses to security and privacy issues.

Figure 3. Fog server

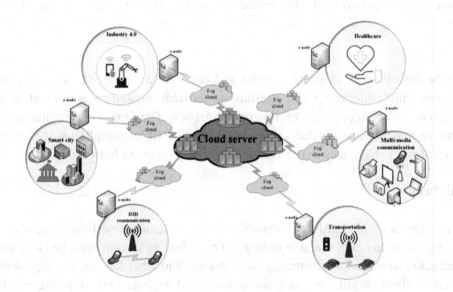

Fog to boost IoT application security and privacy: show in 4 Additionally, fog nodes can act as IoT device proxies. The necessity to connect with the Cloud to update authorisation procedures is removed by these proxies, which can handle and update these processes. Fog computing will improve the security and privacy of IoT applications, claim Alrewas Cloud data security. In addition to the IoT application security and privacy benefits, Gai et al50 argue that Fog computing characteristics can be used to enhance the security and privacy of IoT users. The authors proposed a model called Fog-based Multi-Layer Access Control that utilizes the computation and storage facilities in the Fog server.

This model is responsible for deploying the access control strategies for each application. It also helps in avoiding using limited computing resources of IoT devices and achieves a changeable access control strategy that enables meeting the requirements of multiple manipulation environments

When paying as you go for consumers using Web applications and batch processing, the cloud computing concept is a cost-effective substitute for owning and operating private data centres (DCs). There are several elements that support the economy of Using homogenous computing, storage, and networking components results in cheaper OPEX at the size of mega DCs. Better predictability of huge aggregation allows for higher utilisation without performance degradation.

1.2 Fogcomputing Architecture

There hasn't been a unified architecture for fog computing research that can be used to many applications. Based on the literature that is currently accessible, the architecture in this article offers a comprehensive understanding of the Fog computing-IoT architecture(McKinsey, 2021).

In figure4, Each user's device in this architecture, as seen in may be connected to a single Fog node through wired or wireless access media like WiFi, Bluetooth, 5G, or ZigBee. Fog nodes can also connect with one another wirelessly or through wired networks. The IP core network connects each Fog node to the Cloud. Utilizing virtualization technologies like software-defined networks and network functions virtualization, network virtualization and traffic engineering are possible. Three levels in this design may be distinguished: the device layer, the fog layer, and the cloud layer.

1.2.1. Security

Security can be described in a variety of ways and can have a variety of characteristics, including privacy, anonymity, integrity, trust, attestation, verification, and measurement. which are the OpenFog reference architecture's essential characteristics. The OpenFog architecture will make it possible to create environments that are adaptable and flexible enough to solve a wide range of security issues, from IoT devices to the cloud and the fog layers in between.

1.2.2. Scalability

Scalability includes scalable hardware, scalable software, scalable performance, scalable capacity, scalable dependability, and scalable security. The OpenFog design must be flexible enough to adjust to changes in workload, performance, system cost, and other factors. The OpenFog architecture should allow elastic scaling of small mission-critical deployments up to big mission-critical deployments dependent on demand due to the range of application scenarios for fog computing.

1.2.3. Openness

Composability, interoperability, open communication, location transparency, etc. are all examples of openness. It is crucial for the development of a widespread fog computing ecosystem for Internet of Things applications. The OpenFogarchitecture enables the flexible discovery, gathering, and redistribution of computing, network, and storage resources anywhere in the network. It can also allow the portability and fluidity of applications and services at the point of instantiation(Worldwide internet of things predictions for 2020.).

Figure 4. Fog steps

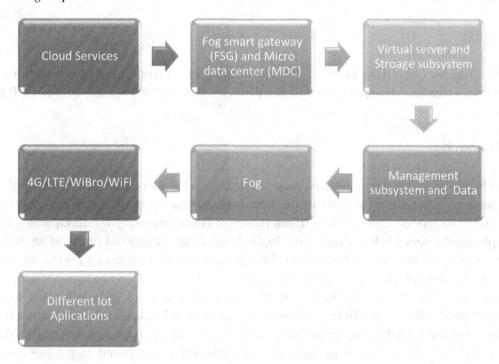

1.2.4. Autonomy

Autonomy promotes the independence of discovery, orchestration and administration, security, operation, and cost-saving, among other things. It is supported across the hierarchy and is necessary for the continued supply of functionality. Decisions may be made at any level of a deployment's structure, including those closer to the device or at higher order layers, according to the OpenFog design.

1.2.5. Programmability

Programmability helps with resource-efficient deployments, adaptable infrastructure, cost-effective operations, improved security, and more. It is the cornerstone of flexibility, liberty, and openness.

Highly adaptive programming is supported by the OpenFog architecture at the hardware and software levels.

1.2.6. Reliability

Availability, and Serviceability (RAS): RAS has three main areas: Hardware, software, and operations make up the three key components of reliability, availability, and serviceability (RAS). It is present in all effective system designs and assumes a significant role in the OpenFog architecture. Under both favourable and unfavourable operating situations, OpenFog architecture will continue to provide the intended functionality, guaranteeing accurate operation, continuous management, and orchestration.

1.2.7. Agility

Agility focuses on converting massive amounts of data into insightful actions. Additionally, it addresses the extremely dynamic nature of fog deployments and the requirement for swift change response.

1.2.8. Hierarchy

A hierarchy has many different aspects, such as devices in the hierarchy, monitoring and control in the hierarchy, operational support in the hierarchy, surrogacy in the hierarchy, and business support in the hierarchy, among others. Even though it is not necessary for all OpenFog designs, most deployments nevertheless express it. On the basis of the functional needs of an end-to-end IoT system, the resources of the OpenFog reference may be understood as a logical hierarchy(D. G. INFSO, "Internet of Things in 2020).

Device layer: In general, the devices in this layer are dispersed geographically, have poor calculation speeds, and have a finite amount of storage space. These gadgets gather raw data and deliver it to the top layer (the fog layer) for processing and storage(] D. Guinard et al., 2021).

Fog layer: It is located close to the device layer and it is composed of a large distributed number of Fog nodes. The Fog nodes consist of network equipment (mobile or static) with higher storage and computation resources than the device layer capabilities such as gateways, routers, access points, switches, access points, and so on. This layer tends to extend Cloud computing to the device layer. The Fog layer provides many services and real-time analysis to IoT devices data. In fact, the Fog layer has more capabilities than IoT devices in terms of computing, temporarily storing, and transmitting a summary of the IoT devices data to the Cloud layer. Fog data centre contributes toward achieving multi-tenant virtualization, enhance computation, storage, and other resource sharing requirements to meet user demands. It also isolates the data and IoT applications. The computation power in this layer reduces the processing load on the limited resources IoT devices and helps in computation offloading to the Cloud servers. Since Fog nodes are connected to the Cloud data centre, more powerful computing and storage capabilities can be provided to IoT devices.(Yazici V, Kozat UC, 2020)

Figure 5. Architecture of fog computing

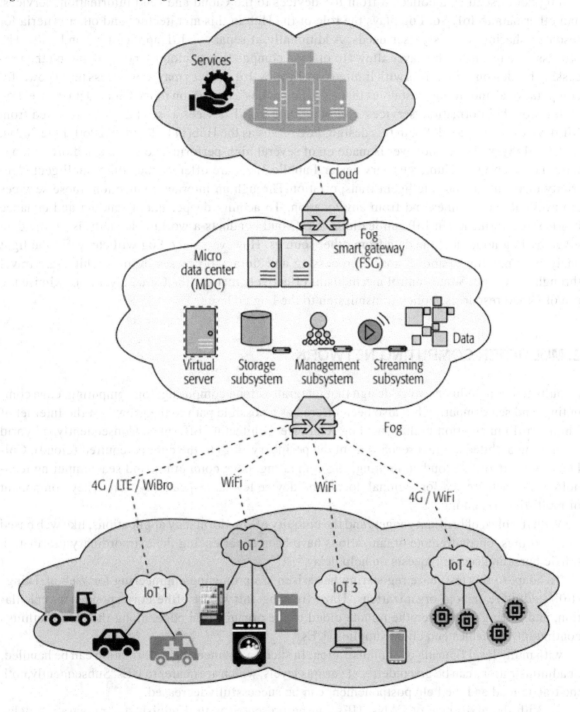

Fog nodes send data collected from IoT devices to the Cloud and send information, services, and other data to IoT. So, Fog plays the role of the Hub in this architecture and other criteria for resource sharing to satisfy user needs. Additionally, it separates IoT applications and data. The processing capacity at this layer allows to offload computation to cloud servers, reducing the processing burden on IoT devices with limited capabilities. IoT devices may now access more powerful computational and storage resources thanks to Fog nodes' connection to the Cloud data centre. Fog nodes provide information, services, and other data to IoT devices as well as data collected from IoT devices to the Cloud. So, in this design, Fog serves as the Hub(ITU-T Std. Y.2060, June 2021).

Cloud layer: The cloud layer is made up of several high-performance servers and offers long-term data archiving. Numerous services and applications are offered, including intelligent electricity distribution and intelligent transportation.Through an internet connection, these services are available at all times and from any location. To acquire deeper understanding and enhance business performance in IoT applications, the Cloud conducts a worldwide analysis on the data given by Fog nodes and the data from other sources. However, since Fog will carry out the light analysis as mentioned above, not all processing and storage processes in this architecture travel through the Cloud. Some control mechanisms effectively manage the Cloud layer to maximise the use of Cloud resources. Policy transmission to the Fog nodes.

2. MULTI-TIER COMPUTING NETWORK

A multi-tier computing network design that integrates cloud computing, fog computing, edge computing, and sea computing has also been offered as a possible path to the growth of the Internet of Things (IoT) in addition to the OpenFog reference architecture offerings. Consequently, a hybrid computing architecture that reaches from the periphery (fog) to the core is required (cloud). Collaborations amongst cloud computing, fog computing, edge computing, and sea computing technologies—each created for regional, local, and device levels, respectively—are a key component of multi-tier computing.

With the blast of brilliant gadgets and the ubiquity oflow-dormancy applications, like web based of recordings, currentremote organizations have been experiencing the extraordinaryinformation traffic burst and severe requests on help delay.

To adapt to this test, haze registering has arisen asa promising engineering for Web of Things (IoT) andfuture remote organizations. Haze figuring shifts part ofthe correspondence, calculation, and storing assets fromthe remote cloud to the organization edge, along the cloud-tothings continuum. It enables end client supplies (UEs)

with multi-level figuring or administration. In such an engineering, information can be handled, or administrations can be given,deftly at various levels, which are nearer to UEs. Subsequently,both the traffic load and the help postponement can be successfullydecreased.

1. With the assistance of FANs, UEs can be presented with diminishedadministration delay and improved nature of administration (QoS). Formodel, delay-lenient administrations can be given by remoteFCN, while delay-touchy applications can be handled atadjoining FANs. Through viable client booking, thetraffic burden and administration postponement can be extraordinarily

decreased.Various exploration endeavours have been devoted to theclient booking issue in multi-level haze figuring organizations. Shah-Mansouri et al. fostered a client planningcalculation to boost every UE's nature of involvement, basedon possible game hypothesis. Liu et al. inferred a curved improvement based joint client planning and asset portioncalculation to limit the complete framework cost for a cloud-mistfiguring network with non-symmetrical numerous entrances. Zhaoet al. used the Lyapunov advancement procedures to

plan a web-based joint client booking and asset distributioncalculation to boost the typical organization throughput for amist empowered content conveyance organization. Liu et al. proposed a disseminated document situation and client planning calculation for aversatile haze reserving administration organization, using matching hypothesis.

Albeit various parts of client booking in multi-levelhaze registering networks have been examined in written works, Powerful client planning plan actually faces difficulties, particularly when the expense model is thought of. By and large, the FCNis worked by a telecom administrator, who signs a help contractwith UEs, while the FANs are have a place with various people.

To all the more likely propel the FANs to share assets and expectin storing, the expense model, particularly for FANs, ought to bethought about.

In this chapter, a bound together multi-level expense model, including theadministration delay and a straight reverse interest dynamic instalmentconspire, and the came about cost-minimization client bookingissue, are researched, in a multi-level haze figuring network comprising of one FCN, different FANs and UEs. Thefundamental commitments of this chapter are summed up as follows.• A bound together multi-level expense model, including the helpdelay and a direct converse interest dynamic instalmentplot, is proposed for a multi-level haze figuring network comprising of one FCN, various FANs and UEs.Also, the came about cost-minimization client planningissue is figured out.• A potential game is figured out to demonstrate the cost minimization client planning issue and the presenceofnash harmony (NE) is demonstrated. Likewise, the comparing conveyed calculation, specifically Cost-Situated Client2019 IEEE INFOCOM WKSHPS: ECOFEC 2019: Financial aspects of Mist, Edge and Distributed computing

• Broad recreations are directed to show the execution of the COUS calculation and the highlights of the proposed cost model. The hypothetical confirmations and reproduction results demonstrate the way that the COUS calculation would be able offer close ideal execution concerning the in general cost, and both the FCN and FANs can profit from the unique instalment plot, contrasted and the fixed instalment conspire. Notwithstanding, the powerful instalment conspire will bring about uncalled for responsibility conveyance among FANs.

In figure 6, The remainder of this chapter is coordinated as follows. The framework model of multi-level haze figuring networks is given in Segment II, along with the numerical detailing of the expense model and the came about cost-minimization client booking issue. In view of potential game, the client planning game is created and broke down in Area III. The NE of this game is demonstrated to exist and the comparing circulated client planning calculation, i.e., COUS calculation, is proposed.

Then, at that point, Area IV assesses the exhibition of created calculation and the highlights of proposed cost model through reproduction.

IoT applications are becoming more intelligent as a result of relevant data expansion, strong processing, and advanced algorithms. Additionally, common IoT applications are moving away from straightforward data sensing, gathering, and representation activities and toward more involved data extraction and analysis. However, the applications often adhere to the standards and guidelines established by a particular industrial area. When combined with environment cognition, big data, and AI technologies, multi-tier computing resources have the potential to create a user-centric strategy in which various IoT services are automatically adapted in accordance with the needs of certain applications and end users.

A large company with a top-down, multi-tier organisational structure can be compared to intelligent IoT networks and services with integrated, multi-tier computing resources. Managers and employees at various levels within the company have various resources, capabilities, and responsibilities in terms of data access and processing, task assignment, customer development, and decision making.

In figure 7, Cloud computing is comparable to the highest rung in a company's hierarchy, with the most information sources, the sharpest analytical skills, the largest storage capacity, and the most power over decisions. Consequently, cloud computing is anticipated to manage difficult global tasks like cross-domain data analysis and processing, abnormal behaviour diagnosis and tracing, hidden problem prediction and searching, new knowledge discovery and creation, and long-term strategic planning. On the other side, edge computing is comparable to front-line employees, who have the fewest resources and skills but may work directly with clients across a range of application areas. Edge computing is hence effective at handling local delay-sensitive operations including data collecting, data compression, information extraction, and event monitoring.

Fog computing, which is akin to middle management in a firm, exists between the cloud and the edge inside the IoT network. like a productive.

Fog computing is a hierarchy of shared computing, communication, storage, and AI algorithms that can cooperatively handle difficult tasks at the regional level, such as cross-domain data analysis, multi-source information processing, and on-site decision making for a wide service coverage. Fog computing is a management system with many levels of resources, duties, and responsibilities. Fog computing can give a flexible way to combine dispersed resources at various physical or logical locations in the IoT network since user requirements are typically dynamic in terms of time and geography, providing timely and efficient services to any consumers.

The multi-tier's devices, or objects, are comparable to the company's clients, who have a wide range of requests and needs for intelligent applications and services. Processing, communication, storage, and power resources are all finite for each device. However, taken as a whole, they support the idea of marine computing at the device level, which allows real-time data sensing, environment cognition, mobility control, and other fundamental tasks for individual objects.

It will be essential to offer a control and orchestration framework to allocate compute, communication, and storage resources in order to make the principles above a reality.

As a result, the interoperability and integration of multi-tier computing sources in the IoT networks may be used to construct intelligent on-demand services. In addition to distributed fog and edge computing with shared resources, accessible environments, and straightforward algorithms for real-time decision making, centralised cloud computing with enormous resources, a secure environment, and powerful algorithms are required to best meet the needs of any user.

Figure 6. Multi-tier computing network

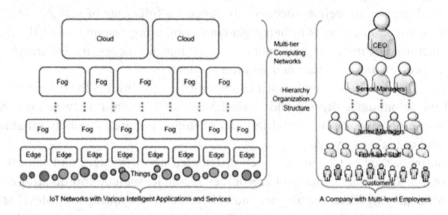

Figure 7. Mul-tier computing networks

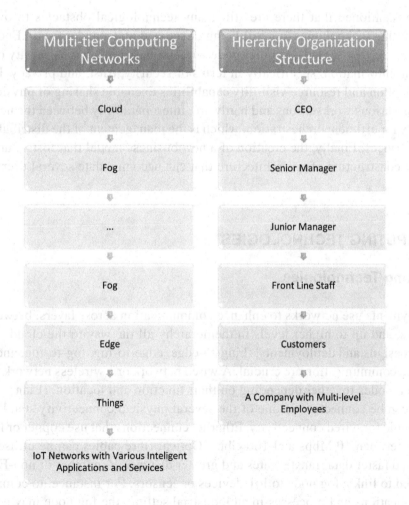

Future multi-tier computing networks will be able to provide densely distributed points for computation and storage, as well as successfully support a full range of services in various environments and applications, thanks to heterogeneous computing resources and the collaborative service architecture. For instance, low-latency connections are necessary for smart automobiles and drones to receive control signals and monitoring data.

Broad connection bandwidths and robust computation capabilities at the device and edge levels are required for autonomous driving and three-dimensional virtual reality gaming. Systems for managing smart cities and the industrial internet of things require extremely reliable networks, secure data, and available services.

Fog computing serves as the link between distributed network edges and centralised clouds, making it essential for managing multi-tier resources.Examples include coordinating cross-domain collaborations, deploying hierarchical rules and policies, and integrating multi-level services. It is also in charge of vertical and horizontal interoperations. Fog computing, when combined with the edge, improves the availability, adaptability, effectiveness, and cost-effectiveness of computing resources and intelligent services in IoT network(J. Bruner, Industrial Internet. O'Reilly, March 2021.).

It should be mentioned that there are still many technological obstacles to overcome since multi-tier computing technologies are still being developed and standardised ..End users should be able to utilise a nearby shared computer node without jeopardising the security of their identities and personal information, specifically in terms of security, trust, and privacy. It is necessary to build virtualization and resource visibility capabilities to enable sharing of physical computing resources across various workstations and hardware. Interoperability between the network's nodes is also necessary for efficient orchestration, which is the management of the distribution of diverse resources and services. Finally, the creation of a new business model that fosters an ecosystem is required for the construction of an architecture that can accommodate several user groups and is multi-tenant.

3. FOG COMPUTING TECHNOLOGIES

3.1. Networking Technologies

Most fog deployments use networks to enhance communication across layers, between fog nodes, between sensors, and up to higher levels in thehierarchy all the way to the cloud. Therefore, for fog computing designs and deployments, thing-to-edge, edge-to-fog, fog-to-fog, and fog-to-cloud connectivity and communication are crucial. A wired network or a wireless network can be used to connect different nodes together depending on their function and location. (Liang K et al., 2020)

A fog node can be connected via one of the several physical connectivity standards, kinds, or interfaces available for wired connectivity. Ethernet connections that use copper or fibre generally provide speeds between 10 Mbps and 100 Gbps. Optical fibre cables can be utilised for connections that demand faster data transfer rates and greater distance. A number of non-Ethernet protocols may be used to link a fog node to IoT devices or sensors. For instance, to communicate with lower layer applications and processes in an industrial setting, the fog node may need to support a Controller Area Network (CAN) bus or other fieldbus protocols. Guaranteed data transmission is essential for industrial automation applications.Time Sensitive Networking (TSN), also known as Deterministic Ethernet, is this kind of networking (which often uses Ethernet). TSN prioritises control traffic in a typical Ethernet context using standards-based time synchronisation technology (such as IEEE 1588) and bandwidth reservation (class-based QoS).

There are three main categories of wireless connectivity: wireless WAN (WWAN), wireless LAN (WLAN), and wireless personal area networks.(WPAN). When extensive geographic area coverage is necessary, WWAN technologies are deployed. Cellular technologies (such 3G, 4G LTE, and 5G), narrow band IoT (NB-IoT), low-power wide area networks (LPWAN), and other technologies all have their own protocols and standards. WLANs use a variety of topologies and protocols, but Wi-Fi has come to be recognised as the standard for WLANs. For smaller geographic regions, such as those found inside of a building or campus, WLANs are a useful option. WPAN is distinguished by its limited communication range, low power consumption, and affordable price.

WPANs can be utilised with home automation systems and wearable technology. Bluetooth, IR, ZigBee, ZWave, and IEEE 802.15.4 are the most popular WPAN technologies (Low Rate WPAN). Additionally, near field communication (NFC) may be employed when fog nodes support equipment that requires close proximity connectivity.

3.2. Computing Technologies

The computing needs of equipment, especially edge devices, will rise as more and more data is created and processed close to end users under strict latency restrictions. Additionally, general purpose compute at the edge will remain crucial as AI approaches are increasingly used to realise the autonomy and agility of fog computing. Typically, one or more multi-core Central Processing Units will be used to implement computation (CPU).Some fog nodes demand CPU throughput above what can be economically (power or processing efficiently) given by ordinary contemporary server and corporate CPU chips. This is especially true for those fog nodes using improved analytics. In such situations, accelerator modules will be arranged adjacent to the processor modules (or closely connected) to offer additional computing throughput. The most popular ones are Field Programmable Gate Arrays (FPGAs) and Graphics Processing Units (GPUs).

Numerous basic cores may be found in thousands in GPUs. They can be a factor of ten quicker for applications that can effectively take use of their huge parallelism and offer considerable power and space savings. Each common CPU may support many GPUs. However, in order to attain this capacity, the node's physical and electrical connections would need to be improved, which might result in an increase in the node's overall power consumption. Large groups of hardware resources that can be programmed at the gate level make up FPGAs. They may be set up with unique logic designs to effectively handle extremely particular challenges. The increased power savings compared to other accelerators, however, may need more lower level understanding depending on implementation (e.g., VHDL). Oftentimes, FPGAs are more energy-efficient than discrete GPU(] M. Hajibaba and S. Gorgin., 2020).

3.3. Storage Technologies

Fog nodes will develop storage tiers that are generally only seen in data centres as they gather and analyse data from throughout the hierarchy. As a result, fog nodes will need a variety of storage.

The most popular ones are fixed spinning discs, solid state drives, and RAM arrays. The node will need to process sensor-generated data in a manner that is almost real-time.

While compared to added delay when accessing non-volatile storage, RAM arrays meet this criteria. Additionally, many fog nodes will have on package memory to accommodate the latency requirements of certain situations. Solid for the majority of fog applications, state drives or flash-based storage may be employed due to their dependability, IOPs, low power needs, and environmental robustness. These include SSDs with SATA and PCIe connections. New programming paradigms and new solid state media classes are also starting to appear. These include NVDIMM-P and 3DXpoint.

Storage devices should provide encryption, key management, and authentication for fog computing by supporting protocols like AES-256 and TCG Opal, among others. Additionally, the storage devices in virtualized fog computing systems should provide ID-based performance allocation by offering programmable storage resources (IOPS or bandwidth) to certain applications. The majority of fog installations also place a high priority on enabling data encryption, real-time information, early alerts about the state of the media, and self-healing capabilities.(Bonomi F et al., 2020)

3.4. Virtualization Technologies

The above-mentioned fog computing systems are examples of highly virtualized architectures. Processing, accelerators, storage, and networking operations should all be virtualized in such designs to increase the fog system's effectiveness, adaptability, and management. Depending on the software layers that operate on the hardware, virtualization may also involve elements of containerization. Consequently, the infrastructure requires both hardware virtualization technologies and software virtualization technologies.

Nearly all processor hardware that would be utilised to construct fog platforms has hardware-based virtualization techniques. It could be crucial for system security as well. A physical system may be shared by several entities thanks to hardware virtualization for I/O and computing. In order to prevent VMs from using instructions or system components that they are not intended to use, virtualization is also highly helpful. Fog computing also uses containerization technologies to aid with isolation in addition to hardware-based virtualization approaches. In a fog computing environment, containers may offer a reduced weight isolation technique. The OS alone provides the isolation assurances; silicon does not entirely provide them. As a result, the criteria for isolation are now relevant to the programme executing on the silicon.Software virtualization solutions are required to allow hardware-based virtualization and OS-based separation for operating software and application microservices. These "appliances" have become more prevalent as Software Defined Networking (SDN) solutions have replaced discrete devices are being used more often as software solutions in Linux containers and virtual machines. In figure 8,Applications and microservices operating on the software backplane can be discretely separated using software containers. Unlike a VM, a software container frequently does not need or include a separate OS.To segregate the application's view of the operating system, a container utilises resource isolation for the CPU, memory, block I/O, network, etc. as well as different namespaces.

The kernel is shared by several containers, but each container can be restricted to only utilise a certain amounts of resources, such as CPU, memory, or I/O.High-distributed systems are facilitated by containers because they allow numerous applications torun on one physical compute node, several virtual machines, and many physical compute nodes.

Artificial Intelligence

In the majority of fog computing implementations, advanced analytics jobs are handled and system performance is optimised using AI approaches. As an illustration, consider how each floor, wing, or even each room might have its own fog node and create a hierarchical control system in a smart building. Each fog node may give occupants of the building with compute and storage infrastructure to support smartphones, tablets, and desktop computers, as well as emergency monitoring, security duties, climate and lighting management, and computing and storage services. For extensive analytics, locally saved operating history may be compiled and uploaded to the cloud. These insights may be used with AI to produce improved models, which are then downloaded to the nearby fog infrastructure and run.

AI research is now at the forefront. Generally speaking, supervised learning, unsupervised learning, and reinforcement learning are the three categories into which AI approaches or machine learning methods may be divided. In a supervised learning system, both the input data and the intended output data are given. To offer a learning foundation for upcoming data processing, input and output data are labelled. In supervised learning, classification and regression are the two major techniques. Unsupervised learning, on the other hand, is a technique that enables computers to categorise both physical and ethereal items without giving them any prior knowledge about the objects. Clustering is the most typical kind of unsupervised learning. To improve its performance, reinforcement learning enables the machine or software agent to adapt its behaviour depending on feedback from the environment.

The processing power of traditional machine learning methods is constrained when dealing with unprocessed natural data. Designing a feature extractor for a machine learning system often involves extensive engineering knowledge. Geoffrey Hinton, YoshuaBengio, and Yann LeCun worked to promote deep learning as a way to overcome the limitations of traditional machine learning methods and achieve considerable success.

With the help of the technique known as "deep learning," computer models made up of many processing layers (also known as "neural networks") may learn representations of data with various degrees of abstraction. These techniques have significantly raised the bar in a variety of fields, including voice recognition, object identification, and visual object recognition.Convolutional neural networks (CNN) are built to handle data that is presented as several arrays, such as a colour picture made up of three 2D arrays representing the pixel intensities for the three colour channels. Recurrent neural networks are frequently preferable for tasks that need sequential inputs, such as voice and language (RNNs).

The long short-term memory (LSTM) networks, which employ unique hidden units to recall inputs for a long period, are a VARIATION OF RNN. Long-term time series data are ideally suited for LSTM networks.(Moysen J, Giupponi L (2020))

Figure 8. Virtualization techniques

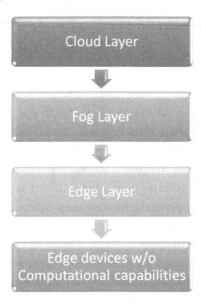

4. A FOG-ENABLED PLATFORM FOR INDUSTRIAL IOT

In the near future, the Internet of Things (IoT) will be used by everything. In 2020, 20 billion IoT/ M2M devices are expected to be online, according to Gartner, Inc. It is Predictably, the enormous number of IoT/M2M devices will cause a data tsunami to hit IoT/M2M systems typically set up on the Cloud. This chapter suggests the use of "Fog Computing" as an alternative to using more cloud resources to address such overloading concerns in order to increase the scalability of IoT/ M2M platforms.

The goal of fog computing is to extend the continuum from the cloud to physical objects (compute, network, and storage) . When compared to cloud computing, fog computing can offer reduced latency and faster reaction times due to its closeness and proximity. As a result, fog may be thought of as a cloud's extension. The Fog network's architecture may be hierarchical, allowing data from end users or devices to be processed at many levels, each of which performs specialised data filtering and analytics before determining whether to pass the data on to the next level for more Fog/ Cloud processing.Our earlier chapter work examined the scalability of an IoT/M2M platform in the cloud, where the resources in the cloud may be scaled up or down in response to the demand of IoT traffic. However, resolving all of the IoT traffic issues in the cloud won't entirely address the scalability issues. The IoT traffic has to be controlled before it enters the cloud in order to increase the scalability of the IoT/M2M platform. This means that the Fog network, which acts as a bridge between the Cloud and physical objects, must have scalable capacity, i.e., the ability to flexibly and dynamically scale serving instances among Fog nodes in response to incoming traffic.

Fog scalability may be used to Industry 4.0, a use case in which a significant number of sensors and actuators have been installed on the manufacturing floor. The manufacturers may simply create industrial IoT (or IIoT) applications on their platforms using these linked sensor and actuators [5].

The majority of data gathered by industrial sensors is still transferred to the cloud to be processed for analytics. However, this initial deployment has shown that this cloud-centric architecture has to be improved in terms of real-time responsiveness and system scalability.A worldwide IoT/M2M platform called oneM2M offers service layer functionality for M2M communications. It offers a framework for the integration of various IoT/M2M technologies in addition to managing heterogeneous IoT/M2M devices.

However, the oneM2M platform still struggles with scalability, which is a problem. Our chapter focuses on extending oneM2M's high scalability from the Cloud to the Fog by moving it to a Fog Computing architecture.In order to further the idea of fog computing and specify a set of system-level architectural frameworks, the OpenFog Consortium was established in 2015. A key element of fog computing offered by OpenFog is the fog node, which gives end devices access to compute, networking, storage, and acceleration resources. In order to support the OpenFog architecture, the OpenFog reference architecture has emphasised eight key pillars: security, scalability, openness, autonomy, RAS (Reliability, Availability, and Serviceability), agility, hierarchy, and programmability. In this study, we concentrate on the Fog architecture's scalability pillar. nodes for service processing.

1) First Level MN-CSE: This level of MN-CSEs registers to the second level MN-CSE and constructs the resource tree corresponding to the underlying sensor devices. at order to be informed of the events that interest them, the second level MN-AE would subscribe to certain resources at the first level MN-CSE.

2) First Level MN-AE: The job of this AE is to keep an eye on the sensor input. It will therefore continuously conduct resource discovery and subscribe recently found resources of interest to the first level MN-CSE. Additionally, it will continue to keep an eye on the sensor data in the resource tree of the First Level MN-CSE. This AE may identify anomalous occurrences using data analytics and alerts the second level MN-AE.

4.1. Inspiration

Equipment developers gather the performance data necessary to comprehend the statistical characteristics of any potential flaws a product may have throughout the course of its expected lifetime. This information facilitates the creation of a failure model unique to each product that may be used to calculate the likelihood of failures and/or malfunctions. Unavoidably, such a failure model is based on a number of presumptions about the operating environment in which the relevant product is used.The operating envelope in the product's specifications and the product's blueprint both take these presumptions into account.However, it is impossible to fully predict a product's real operational circumstances at the time its failure model is created. The likelihood of a malfunction and/or failure before a planned maintenance, however, increases when a production is operated in conditions that go beyond those specified in its specifications. Because of this, using the failure model alone is insufficient to account for the impact of the operational situation(] F. Bonomi et al., 2021).

The resolution of this mismatch and provision of the most accurate information of upcoming malfunctions and failures would necessitate consideration of additional information about the operational context. However, because such information is controlled by the operational management of each specific instance of the product, the equipment maker does not have easy access to it. A better modelling match may be obtained by integrating the manufacturer's failure model's predicted envelope with the administrator's infrastructure's extensive operating history.

The importance of operational data is further highlighted by the diversity of client demands that are placed on contemporary automation systems. The latter, initially intended to function under central supervision for the manufacturing of huge numbers of a relatively small number of product kinds, are increasingly being charged with the small-scale production of customised goods.

4.2. Overview of APO

In Figure 9 Fog ModellingTo increase the efficacy of applying equipment failure models, we suggest the establishment of a so-called Adaptive Operations Platform (AOP), which would take into account the deployment and configuration of each unique infrastructure.The platform connects to equipment makers and the operational managers of Supervisory Control and Data Acquisition (SCADA) infrastructures from the viewpoint of the value chain, providing the following capabilities:

Enabling the equipment maker to: • Manage the release of the failure model linked to each specific kind of equipment (i.e., publish, update, and withdraw).

- Control each managed (e.g., published) failure model's access permissions.
- Keep track of how each managed (e.g., published) failure model is accessed and used.

Figure 9. Fog modelling

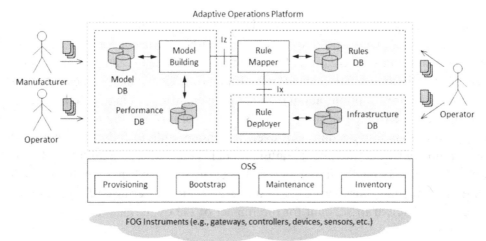

The management of the SCADA infrastructure will be made possible by this, allowing them to:
• Manage the subscription (i.e., subscribe and unsubscribe) to the failure model related to each specific kind of equipment.

• Search a failure model database for models that fit a certain set of criteria.

AOP is built upon the service capabilities of the following levels, as shown in Fig. 3:

• The Operational Support System (OSS), which uses the Fog Infrastructure to provide the customary asset management and business support functions (e.g., inventory, maintenance, provisioning, etc.). The Fog Infrastructure consists of networking equipment with specific Fog capabilities and offers end-to-end communication services.

AOP offers a number of functional aspects in order to effectively use important Fog Infrastructure characteristics(] P. Shvaiko and J. Euzenat.2005).

The Model Building (MB) functional element combines static data on the failure models of the various equipment types found in the industrial site with dynamic data gathered during the

operation of the latter. The failure model database (shown as Model DB in Fig. 9) offers an interface for equipment makers to register and update this information. The former data are obtained from this database.

In Figure 10, The latter information is obtained via the equipment performance database (shown as Performance DB in Fig. 9), whose information is filled with information on equipment performance under the supervision of the administration of the industrial infrastructure.

The SCADA historical database is frequently found to store operational data (such as measurements of operating conditions, etc.) for the equipment that makes up the infrastructure. The MB will create a "fused model" for each kind of equipment by fusing information from the Model DB and the Performance DB. The fused model, or a variant of it that better matches the operational environment, may be thought of as a refined version of the failure model for the specific industrial infrastructure.

In The Software Defined Networking (SDN) infrastructure's rule mapper (RM) functional element is tasked with translating the fused model into a set of traffic handling regulations. The RM communicates with the MB through the Iz interface (Fig. 3) for each specific piece of equipment in order to access the correct fused model. Under the management of the RM element, traffic handling rules (also known as DMo rules) are stored in the rules database (shown as Rules DB in Fig. 9).

The Rule Deployer (RD) functional element computes the deployment plan to apply this collection of traffic handling rules to the relevant elements after receiving a description of the SDN infrastructure's capabilities and a set of traffic handling rules.The devices that make up the infrastructure are described in a database (shown as Infrastructure DB in Fig. 9), together with their deployment configuration and connections (such as networking topology, etc.), as well as the related configuration and provisioning settings. Access to traffic management rules is made possible by the RM element via the Iz interface depicted in Fig. 9

5. INDUSTRIAL CYBER-PHYSICAL SYSTEM

5.1. Fog Computing, Its Security, and Privacy

Utilizing cyberphysical interactions, real-time embedded machine learning is delivered via the fog computing infrastructure. For various engineering applications (such as equipment prognostics), the cloud platform holds production-ready machine learning models encoded as Predictive Modelling Markup Language (PMML), which are distributed and performed by fog nodes placed within the facility's local network. Factory-to-cloud connections rely on the facility's current security rules and services managing internet communications, even while these local operations provide data security and privacy.

After the cloud receives communications from the factory, the request is authenticated using the 128-bit Global Unique Identifier of the fog node (GUID). The GUID is looked up, the engineering applications handled by the node are identified, and the necessary PMML models are returned for download or synchronisation from a cloud database of registered devices. The fog is used to store the downloaded PMML models, allowing for their execution inside the factory's physical walls and the delivery of real-time forecasts and decision-making (such as control modifications) without the need for ongoing connections to the cloud.

The IoT is creating vast amounts of data quickly and in a wide range. Since IoT devices are unable to process this volume of data, it is sent to the cloud for analysis and decision-making. Instead of sending data to distant clouds, time-sensitive applications need rapid judgement and action based on this data. A novel computer paradigm known as fog or edge computing is presented to evaluate this sort of data close to the network edge (] F. Bonomi et al., 2021). Fog computing is becoming more popular as a cloud computing extension to enable computation closer to the end devices. Fog nodes in fog computing offer a variety of resources for the processing required at the networkedge. Commonplace PCs, routers, access points, and base stations might be these fog nodes.

Some of the security challenges that exist in cloud computing are carried over into fog computing and can be resolved utilising similar procedure similar to those employed in other areas of the current IT environment (] P. Shvaiko and J. Euzenat.2005), control methods and processes. The enemies can take over a portion of the private or personal networks in networks with conventional security measures, nevertheless, and launch assaults to jeopardise security and privacy.

Fog computing architecture is open by design to facilitate heterogeneity and interoperability, allowing users to instal and plug in their own devices to function as full-fledged fog nodes. As a result, attackers can connect their specially constructed devices to the target network and launch an MITM attack to eavesdrop on and/or alter the data. BYOD (Bring Your Own Device) is becoming the top issue for network administrators since they have little to no control over such devices due to the emergence of mobile malware. These reliable gadgets could interact with unsecured surroundings beyond business hours and could be hacked. The security concerns brought forth by BYOD have not received enough attention from researchers, and it is likely that as time goes on, this issue will worsen.By utilising a device authentication technique, the related security concerns may be reduced.

Because fog environments involve several service providers, users, and remote resources, enforcing security in them is a significant issue. It is difficult to design and implement authentication and authorisation methods that can coexist with several fog nodes with various computational capabilities. Potential solutions include trustworthy execution environments and public-key infrastructures.

Users of fog deployments also must plan for the failure of individual sensors, networks, service platforms, and applications. To help with this, they could apply standards, such as the Stream Control Transmission Protocol, that deal with packet and event reliability in wireless sensor network

5.2. Use of Electricity

There are a lot of nodes in a fog environment. Due to the distributed nature of the processing, it may be less energy-efficient than in centralised cloud systems. Utilizing effective filtering and sampling techniques, combined computer and network resources, and efficient communications protocols like CoAP.

Figure 10. Using normal router

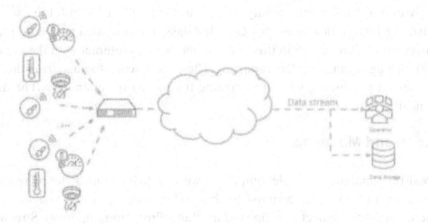

6. FOG-COMPUTING APPLICATIONS

Fog computing might be useful for many different purposes.

6.1. Monitoring of Activities and Health

Healthcare, where real-time processing and event response are essential, may benefit from fog computing. Fog computing is one method that has been suggested for using to identify, forecast, and stop stroke victims from falling. 6 The learning algorithms for fall detection are dynamically distributed across edge hardware and cloud resources. The results of the experiments showed that this system had a faster reaction time and used less energy than cloud-only methods.

A smart healthcare system based on suggested fog computing allows low latency, mobility assistance, and location and privacy awareness.

6.2. Intelligent Utility Services

It is possible to utilise fog computing with smart utility services8, which aim to improve energy generation, distribution, and invoicing. In these situations, edge devices have the ability to provide customers' mobile devices with more granular energy usage information than standard smart utility services (for instance, hourly and daily readings rather than monthly measurements). These cutting-edge technologies may also estimate daily energy costs, recommend the most cost-effective energy source at any given moment, and advise when to use household appliances to cut down on utility use.

6.3. Games, Cognitive Systems, and Augmented Reality

Applications that use augmented reality and are sensitive to latency heavily rely on fog computing. For instance, the EEG Tractor Beam enhanced multiplayer online brain-computer interaction game continuously classifies brain states in real-time on fog devices and then fine-tunes classifier models on cloud servers based on data gathered via electroencephalogram sensors. 9

Individuals with impaired mental acuity may do a variety of tasks with the use of a wearable cognitive-assistance system that leverages Google Glass devices, such as learning the names of people they meet but can't recall.10 In this application, devices interact with the cloud to do tasks that can wait, such logging and problem reporting.The system sends footage from the Glass camera to the fog devices for processing while performing time-sensitive operations.The method shows how utilising nearby fog[20].

6.4. Simulation and Modeling

To make fog real-time analytics possiblecomputing, we need to look into different resource management and scheduling methods.ApplicationsCloud and edge resourcesSoftwaredeedsnetworkingIoT sensors and actuators that connect machines to machinesProgrammingmodelsStream processing, sense-process-actuateManagement of multitenant resourcesResource scheduling, operator placement, and ow placementhandling of raw databoth prolong and monitoringadministration of services and APIsdiscovering API Identification and verification API structureImage Architecture for fog computing. End devices, like as sensors and actuators, are present in the bottom layer together with applications that improve their functioning. These components communicate with edge devices, such gateways, and subsequently cloud services using the network at the following layer. The infrastructure is managed via the resource-management layer, which also makes quality-of-service enforcement possible. Finally, apps use programming paradigms for fog computing to give intelligent services to users (Cisco 819 integration services router, 2021).

In Figure 11. Scenario using a router supporting DMoSmart cities, intelligent transportation systems, smart healthcare, public safety, smart grid, industry 4.0, smart homes, and smart buildings are just a few of the domains where applications have been developed and supported by a variety of fog computing technologies (D. G. INFSO, "Internet of Things in 2020). Fog computing is emerging as a contemporary and competitive environment in a number of different areas. These innovations and applications frequently have a profound effect on how people live their daily lives. But not all of them have matured to a suitable degree and need more thought. Existing methods mostly rely on modelling and simulation to attempt to respond to new queries, resolve issues, or make decisions in a particular subject. Today, a variety of simulation tools for fog computing environments are available, each with a particular method and set of features (Cisco connected grid router, 2020).

Designing fog computing simulators that satisfy the requirements for generality, scalability, efficiency, mobility support, and low-latency is a challenging issue. Finding the components of the simulator is one of the process's most difficult difficulties. Every choice is based on the real cost of setting up and running the fog computing system, including all of its supporting apps, computer resources, and infrastructure. Next, we list a number of deployment-related difficulties.

Figure 11. Scenario using a router supporting DMo

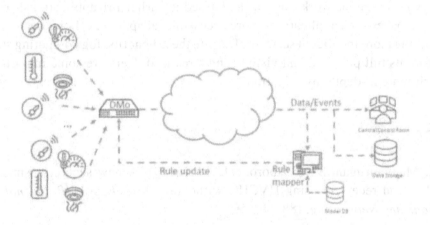

7. CONCLUSION

In our contemporary world, a wide range of IoT devices have been widely used for environmental monitoring, infrastructure management, intelligent manufacturing,The effective resource sharing, data processing, information extraction, and real-time decision making are necessary for operations optimization, safety and surveillance, remote healthcare, and other applications that create increasing amounts of data.Tens of billions of smartphones and vehicles connecting to communication networks for various mobile applications and interactive services, such as online gaming, high-resolution video streaming, augmented/virtual reality, and autonomous driving, make this problem much more difficult in mobile environments.

Despite varying industrial sectors and service scenarios, there is a clear trend towards more intelligent and sophisticated tasks, such as cross-domain data analysis, pattern recognition, and behaviour prediction, in addition to more conventional, straightforward tasks like data sensing, collection, and representation. We should create a user-centric strategy to enable diverse IoT applications to be automatically set and adjusted in accordance with particular user preferences, service situations, and performance needs in various industrial sectors in order to support this trend.

Multi-tier computing resources and adaptive algorithms must be widely used in this strategy at the global, regional, local, and device levels. However, cloud computing alone cannot successfully support intelligent transportation as a centralised solution.

Fog computing has recently been suggested and used as an extension of cloud computing along the continuum from the cloud to things [1, 2], enabling the fusion and cooperation of various computing resources at the cloud, in the network, at the edge, and on the things (devices). In other words, fog computing offers a fresh architecture for efficiently pooling dispersed computing resources at various levels and connecting complex service requirements with the finest local resources. By doing this, it makes sure that scenario analysis, fast data processing, and decision-making occur at the places that are most convenient for where the data is created and has to be used. Fog computing can effectively handle some of the aforementioned infrastructure issues, including network connection, communication capacity, and service latency for supporting moreintelligent IoT applications and services.

Fog computing maintains and makes use of the multi-tier processing, communication, and storage capabilities inside the networks, as was addressed in earlier IoT applicationsbetween objects and the cloud. It makes such applications more accessible, adaptable, efficient, and cost-effective while also enabling more intelligent services. Despite these benefits, fog computing still has a long way to go before its full potential and vision can be realised. Here are some important scientific areas that merit more in-depth investigation.

REFERENCES

Almutairi, S., Manimurugan, S., & Aborokbah, M. (2019). A new securetransmission scheme between senders and receivers using HVCHC withoutany loss. *EURASIP Journal on Wireless Communications and Networking*, (88), 1–15.

Asuncion, A., & Newman, D. (2020). *UCI machine learning repository.*

Gao, W., Morris, T., Reaves, B., & Richey, D. (2020). On SCADA controlsystem command and response injection and intrusion detection. 2010 eCrime Researchers Summit, Dallas, TX.

Banga, A., Gupta, D., & Bathla, R. (2019). Towards A Taxonomy of CyberAttacks on ScadaSystem. *International Conference on Intelligent Computing and Control Systems (ICCS),* (pp. 343-347). IEEE.

Bank, R. K., Priyadarshini, R., Dubey, H., Kumar, V., & Mankodiya, K. (2019). FogLearn: leveraging fog-based machine learning for smart system bigdata analytics. In *Geospatial Intelligence: Concepts, Methodologies,Tools, and Applications* (pp. 1225–1241). IGI Global.

Batri, K., & Anbukaruppusamy, S. (2019). Improving TCP Performance in AdHoc Networks. *European Journal of Scientific Research, 65*(2), 237–245.

Beaver, J. M., Borges-Hink, R. C., & Buckner, M.A. (2020). An Evaluationof Machine Learning Methods to Detect Malicious SCADA Communications. *International Conferenceon Machine Learning & Applications.* IEEE.

Bonomi, F. (2021). Connected vehicles, the internet of things, and fogcomputing. In *The eighth ACM international workshop on vehicularinter-networking (VANET),* (pp. 13-15). ACM.

Carcano, A., Coletta, A., Guglielmi, M., Masera, M., Nai Fovino, I., & Trombetta, A. (2021). andA. Trombetta, "A Multidimensional Critical StateAnalysis for Detecting Intrusions in SCADA Systems,". *IEEE Transactions on Industrial Informatics,* 7(2), 179–186. doi:10.1109/TII.2010.2099234

Dakheel, A. H., Dakheel, A. H., & Abbas, H. H. "Intrusion detectionsystem in gas-pipeline industry using machine learning," Periodicals ofEngineering and Natural Sciences, vol. 7, pp. 1030-1040, 2019.

Daryabar, F., Dehghantanha, A., & Udzir, N. I. (2021). Towards secure modelfor SCADA systems. International Conferenceon Cyber Security, CyberWarfare and Digital Forensic. CyberSec.

Datta, S. K., Bonnet, C., & Haem, J. (2020). Fog computing architecture to enableconsumer centric internet of things services. In *In 2020InternationalSymposium on Consumer Electronics* (pp. 1–2). ISCE.

Fournier-Viger, P., Nkambou, R., & Tseng, S. M. (2021). RuleGrowth: Miningsequentialrulescommon to several sequences by pattern-growth. Symposium on Applied Computing. ACM.

Hoffmann, H. (2020). Kernel PCA for novelty detection. *Pattern Recognition,* 40(3), 863–874. doi:10.1016/j.patcog.2006.07.009

Khazai, S., Homayouni, S., Safari, A., & Mojaradi, B. (2021). Anomalydetection in hyperspectral images based on an adaptive support vectormethod. *IEEE Geoscience and Remote Sensing Letters,* 8(4), 646–650. doi:10.1109/LGRS.2010.2098842

Kim, J., Ha, T., Yoo, W., & Chung, J. M. (2019). Task Popularity based EnergyMinimized Computation Offloading for Fog Computing WirelessNetworks. *IEEE Wireless Communications Letters,* 8(4), 1200–1203. doi:10.1109/LWC.2019.2911521

Li, J., Yu, F. R., Deng, G., Luo, C., Ming, Z., & Yan, Q. (2020). IndustrialInternet: A Survey on the Enabling Technologies, Applications, andChallenges. *IEEE Communications Surveys and Tutorials,* 1504–1526.

Li, Q., Zhao, J., Gong, Y., & Zhang, Q. (2019). Energy-efficient computationoffloading and resource allocation in fog computing for Internet ofEverything. *China Communications,* 16(3), 32–41.

Liu, J., Xiong, K., Fan, P., Zhong, Z., & Letaief, K. B. (2019). Optimal Design ofSWIPT-Aware Fog Computing Networks. arXiv preprintarXiv:1901.08997.

Manimurugan, S. & Al- Mutari, S. (2020). A Novel Secret Image HidingTechnique for Secure Transmission. *J. Theo. Appl. Inf. Tech,* 95 (1), 166-176.

Manimurugan, S. Al-Mutairi, S., Aborokbah, M., Chilamkurti, N., Ganesan, S., & Patan, R. (2020). Effective Attack Detection in Internet of MedicalThings Smart Environment Using a Deep Belief Neural Network. IEEE.

Manimurugan, S., & Narmatha, C. (2020). Secure and Efficient Medical ImageTransmission by New Tailored Visual Cryptography Scheme with LSCompressions. *International Journal of Digital Crime and Forensics*, *7*(1), 26–50. doi:10.4018/IJDCF.2015010102

Manimurugan, S., & Narmatha, C. (2020). A New Approach for IRIS Imageidentification using modified contour segmentation. IEEE international conference on green computing, communication and electricalengineering. IEEE.

Manimurugan, S., Porkumaran, K., & Narmatha, C. (2020). The New Block PixelSort Algorithm for TVC Encrypted Medical Image. *Imaging Science Journal*, *62*(8), 403–414. doi:10.1179/174 3131X14Y.0000000078

Morris, T. H., Jones, B. A., Vaughn, R. B., & Dandass, Y. S. (2021). Deterministic Intrusion Detection Rules for MODBUS Protocols. *Hawaii International Conference on System Sciences*. IEEE.

Muna, M. A., Moustafa, N., & Sitnikova, E. (2020). Identification of maliciousactivities in industrial internet of things based on deep learning models. *Journal of Information Security and Applications*, *41*, 1–11.

Nakhkash, M. R., Gia, T. N., Azimi, I., Anzanpour, A., Rahmani, A. M., & Liljeberg, P. Analysis of Performance and Energy Consumption ofWearable Devices and Mobile Gateways in IoT Applications. *COINS '19: Proceedings of the International Conference on Omni-Layer IntelligentSystems*, (pp.68-73). ACM. 10.1145/3312614.3312632

Narmatha, C., Manimegalai, P., & Manimurugan, S. (2020, October). A LosslessCompression Scheme for Grayscale Medical Images Using a P2-Bit ShortTechnique. *Journal of Medical Imaging and Health Informatics*, *7*(6), 1196–1204. doi:10.1166/jmihi.2017.2212

Saravana Balaji, B., Salih Mohammed, A., & Al-Atroshi, C. (2020). Adaptability ofSOA in IoT Services-An Empirical Survey. *Int. J. Comput. Appl*, *182*(31), 25–28.

Scholkopf, B., Smola, A., & Müller, K.-R. (1998). Nonlinear Component Analysis as a Kernel Eigenvalue Problem. *Neural Computation*, *2*(5), 1299–1319. doi:10.1162/089976698300017467

Serbanescu, A. V., Obermeier, S., & Yu, D. Y. (2020). ICS threat analysisusing a large-scale honeynet. *3rd International Symposium for ICS &SCADA Cyber Security Research 2020 (ICS-CSR 2020) 3*, (pp. 20-30). ACM.

Shanmuganathan, M., Almutairi, S., Aborokbah, M. M., & Ganesan, S., & Ramachandran, V. (2020, August). Review of advanced computational approaches onmultiple sclerosis segmentation and classification. *IET Signal Processing*, *14*(6), 333–341. doi:10.1049/iet-spr.2019.0543

Siddavatam, I. A., Satish, S., Mahesh, W., & Kazi, F. (2020). An ensemblelearning for anomaly identification in SCADA system. *2020 7th International Conference on Power Systems (ICPS)*, (pp. 457-462). ACM.

Tax, D. M. J., & Duin, R. P. W. (2020). Data domain description using supportvectors. European Symposium on Artificial Neural Networks. IEEE.

Yi, S., Hao, Z., Qin, Z., & Li, Q. (2020). Fog computing: Platform andapplications. In 2020 Third IEEE Workshop on Hot Topics in WebSystems and Technologies (HotWeb), (pp. 73-78). IEEE.

Zhang, Y., Niyato, D., Wang, P., & Kim, D. I. (2020). Optimal energymanagement policy of mobile energy gateway. IEEE T Veh. *Technol*, *65*(5), 3685–3699.

Zhu, B., Joseph, A., & Sastry, S. (2021). A Taxonomy of Cyber Attackson SCADA Systems. 2021 *International Conference on Internet of Things and 4th International Conference on Cyber, Physical and Social Computing, Dalian*. ACM.

Zolanvari, M., Teixeira, M. A., & Jain, R. (2019). Effect of ImbalancedDatasets on Security of Industrial IoT Using. *Machine Learning*.

Chapter 12
New Cloud Computing–Based Strategy for Coordinating Multi–Robot Systems

Claudio Urrea

University of Santiago of Chile, Chile

ABSTRACT

Construction and manufacturing are performed in environments where automation is highly appreciated as it improves security and efficiency. This article proposes a new cloud computing-based strategy for coordinating multi-robot systems aimed to transport and manipulate heavy objects like a beam. The proposed solution is applied to a team of dynamically modeled selectively compliance assembly robot arms (SCARAs) with four degrees-of-freedom (4-DoF), simulated in virtual reality modeling language (VRML) and coordinated by a cloud server. The SCARAs are organized on a leader/follower configuration, so that the leader performs a computed trajectory, constantly reporting its position to the remote server and this, in turn, resends this information to the followers. The SCARAs are modeled using MatLab and they communicate through a transmission control protocol (TCP) with a Python process running in the cloud server to demonstrate the feasibility of incorporating cloud resources in a multi-robot system, testing their performance through simulation.

1. INTRODUCTION

The implementation of robots in construction has reduced the risk of accidents and improve productivity. Manipulating or transporting heavy structures like steel beams has been one of the main goals in the field (Cai et al., 2019; Chen et al., 2021b; Khetre et al., 2021). Commonly, human operators perform these tasks by grabbing the ends of the object and moving them cooperatively. However, for robotic manipulators to conduct these tasks, complex coordination strategies are required, which has been considered a major challenge since the 1980s (Cena, et al., 2013; Cheng, et al., 2021; Urrea et al., 2022).

A Multi-Agent System (MAS) is composed of autonomous problem-solving entities –each one with their own information and interests– that share an environment and work together to find answers that

DOI: 10.4018/978-1-6684-8785-3.ch012

are beyond individual capabilities (Francisco et al., 2019; Maleš et al., 2019; Gielis et al., 2022). A Multi-Robot System (MRS) is formed by a group of robots organized to maintain communication and share an assigned task in a synergic manner, so each member uses its resources to achieve a common goal (De Almeida et al., 2019; Yan et al., 2023). Therefore, MRSs are more efficient, robust and cost effective at manipulating objects (Hichri et al., 2019; Wang et al., 2023).

MRSs are widely studied in a number of different situations, such as space monitoring (Teruel et al., 2019; Tu et al., 2021), task allocation algorithms by auction (Sullivan et al., 2019; Cao et al., 2021), collaboration between assembly and aerial robots (Mathews et al., 2019; Zhang et al., 2021), odor source localization (Xing Chen et al., 2019; Wei et al., 2021), decentralized control for cooperative construction (Deng et al., 2019; Wu et al., 2021), and collaborative object manipulation with a human operator (Yousefizadeh et al., 2019; Semeraro et al., 2023).

To achieve collective coordination between robots, path computing metaheuristic models inspired in biological communities such as bat swarms (Suárez et al., 2019; Yilmaz et al., 2020), or ant colonies (Jian et al., 2019; Alarcon et al., 2021) have been employed, as well as other models based on physical systems governed by Newton Laws such as gravitational search or a Swarm Intelligence Inspired Framework (Wang et al., 2019; Goswami et al., 2021). Nevertheless, searching for a path from origin to destination is a time-consuming and resource-demanding process that often cannot be managed by the computer machines embedded in the robot, since both its data processing and storage capacities are limited (Li et al., 2023).

Remote servers technologies integrated with MRSs have given rise to Cloud Robotics (CR) (Du et al., 2019; Kumari et al., 2020). In CR, robots share information through a middleware, and therefore tasks that consume high processing resources are assisted by more powerful computing systems (Wang et al., 2022).

Recent studies have developed control strategies for manufacturing devices (Lu et al., 2019; Chen et al., 2021a), creating technologies such as the Internet of Things (IoT) integrated in industrial scenarios (Nord et al., 2019; Abbas et al., 2020), especially in applications such as construction and smart buildings control. The inclusion of cloud computing is suggested in several studies on the subject, in which robots delegate the high burden of information processing required by some tasks to cloud services (Tang et al., 2019; Craveiro et al., 2019; Jia et al., 2019; Zhou et al., 2019; Pires et al., 2020). Even data structures such as images, maps and videos can be processed thanks to Big Data systems (Varlamov, 2021; Hang 2022).

In this line, in construction tasks such as manipulating heavy beams or similar objects, the proposed MRS solution should meet physical stability and efficient computable coordination strategy requirements. This paper deals with the development of a strategy for coordinating Selective Compliance Articulated Robot Arms (SCARAs) supported by the Google Cloud platform, which aim to complete the task of piling up blocks. The derivation of the coordination task to a cloud server provides a powerful computing resource without the need of having it embedded in the robot or in a local data center, hence preventing the hardware from exposure to the hazardous environment of the construction field. The objective of this application is to demonstrate the feasibility of modeling a communication and coordination service for SCARA with construction purposes, in a real cloud platform and with a leader/follower mode. To present the results, a 3D simulation is provided in Virtual Reality Modeling Language (VRML). The simulation is applied to the dynamic model of SCARA robots with one redundant prismatic joint on its base (Urrea et al., 2022a).

2. MATERIALS AND METHODS

The solution described is applied to a pair of SCARA-type manipulator robots with one redundant prismatic joint, which cooperate with each other to lift a beam that has been randomly placed on the floor and move it to the top of a tall structure, as shown in Figure 1. The robots are organized in a leader/ follower configuration in which the leader grabs a beam end and starts moving along a calculated path, while the follower modifies its own position to keep distance from the leader.

These robot arms have 4 Degrees-of-Freedom (4-DoF) to place their end effector. The first and forth joints are prismatic and establish the vertical position of the end effector, while the second and third joints are rotational and determine position in the horizontal plane. In this way, final location depends on the general position of the joints as well as the length of the links. In addition, prismatic joints are dynamically uncoupled from rotational joints. For the simulations, robots are modeled on MatLab®, Simulink® and Simscape®.

Figure 1. Pair of SCARA-type manipulator robots with a redundant prismatic joint

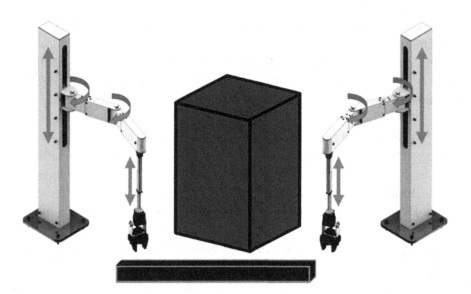

Both leader and follower execute a Java coded Application Program Interface (API) like the one developed in Urrea et al., 2020. The difference between both APIs is that the former reports the robot position to the server, while the latter waits for this information from the server to then transmit it to the follower.

The server corresponds to a Unix based Google Cloud computing instance whose operating system is CentOS 7.0, which also runs a Python programmed API to maintain communication with the robots.

Residual Mean Square (RMS) and Residual Standard Deviation (RSD) are two statistical indicators used to compare actual results with expected ones (Urrea et al., 2021). The distance between end effectors ideally equals the length of the beam moved all along the moving trajectory. The RMS and RSD indicators are employed to assess the accuracy of the real trajectory:

$$\text{RMS} = \sqrt{\frac{\sum_{i=1}^{n}(d_i - l)^2}{n}}, \tag{1a}$$

$$\text{RSD} = \sqrt{\frac{\sum_{i=1}^{n}(d_i - l)^2}{\sum_{i=1}^{n}d_i^2}}. \tag{1b}$$

where d_i is the distance between end effectors, l is the length of the beam and n is the total of movement trajectory points.

For simplicity reasons, the beam is considered a long rectangular object lying somewhere on the floor of the workspace. Its detection procedure is beyond the scope of this paper, and therefore it will not be covered further. However, the position of the beam center and its ends, from it will be gripped by the end effectors of the robots, will be assumed.

Furthermore, the grippers of the robots will be modeled since the end effectors make solidary contact with the beam as they touch it. Since no specific action is performed by the end effectors, they have no actuators.

2.1. Communication and Server

The scheme described in Figure 2 shows how the leader reports its current position to the server, and the server sends the information to the follower. Communication is performed through Transmission Control Protocol (TCP), where a socket is defined with the server's Internet Protocol (IP) address that the robots connect to in order to send a data stream.

In this architecture, the robots act as clients to the server. This implies that they send TCP commands to the server while the latter keeps monitoring its socket. The sockets in the server are developed in Python language. The server is permanently running, expecting to receive the command from the leader robot.

Figure 2. Client/Server communication scheme

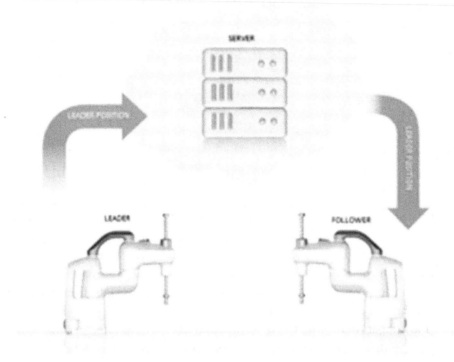

2.2. Redundant Manipulator With 4-DoF

The SCARA architecture used in this paper is the one displayed in Figure 3. It is composed of four links and four joints, which translates into four joint position variables, namely d_1, $\theta 2$, $\theta 3$ and d4. The reason for selecting this architecture is that the combination of the first and fourth joints make it possible to grip objects from the floor and lift them over tall structures, which is the goal of this work. Furthermore, as mentioned above, the second and third joints are mechanically uncoupled from the prismatic ones, which makes control less complex.

For simulation purposes, the dynamic and kinematic models developed in Urrea et al., 2021b will be used with an extension to an additional prismatic joint.

Figure 3. Representation of the 4-DoF manipulator robot

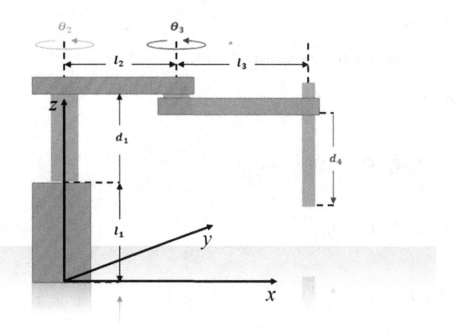

2.2.1. Kinematics

Considering the reference system and the variables specified in Figure 3, and using the sine theorem and the cosine theorem, SCARA kinematics are defined by equation 2.

$$d_1 = z - l_1, \tag{2a}$$

$$\theta_3 = sin^{-1}\left[\pm\sqrt{1 - \left(\frac{x^2 + y^2 - l_2^2 - l_3^2}{2l_2 l_3}\right)^2}\right], \tag{2b}$$

$$\theta_2 = tan^{-1}\left(\frac{y}{x}\right) - tan^{-1}\left(\frac{l_3 sin\theta_3}{l_2 + l_3 cos\theta_3}\right), \tag{2c}$$

$$d_4 = z - l_1 - d_1. \tag{2d}$$

The above provides the relative positions of each joint, d_1, $\theta 2$, $\theta 3$ and d4, so that the end effector can reach a space position x, y and z, depending on the lengths of the links l1, l2 and l3.

2.2.2. Dynamics

The SCARA dynamics is defined by equations 3 and 4.

$$F_1 = M_{11}\ddot{d}_1 + G_1, \tag{3a}$$

$$\tau_2 = M_{22}\ddot{\theta}_2 + M_{23}\ddot{\theta}_3 - C\dot{\theta}_3^2 - 2C\dot{\theta}_2\dot{\theta}_3 + G_2, \tag{3b}$$

$$\tau_3 = M_{23}\ddot{\theta}_2 + M_{33}\ddot{\theta}_3 - C\dot{\theta}_2^2 + G_3, \tag{3c}$$

$$F_4 = M_{44}\ddot{d}_4 + G_4. \tag{3d}$$

where the parameters M_{11}, M_{22}, M_{23}, M_{33}, M_{44}, C, G_1, G_2, G_3, and G_4 are calculated as follows:

$$M_{11} = m_2 + m_3 + m_4, \tag{4a}$$

$$M_{22} = m_2 l_{C_2}^2 + m_3 \left[l_2^2 + l_{C_3}^2 + 2l_2 l_{c_3} \cos\theta_3 \right] + m_4 \left[l_2^2 + l_3^2 + 2l_2 l_3 \cos\theta_3 \right] + I_2 + I_3, \tag{4b}$$

$$M_{33} = m_3 l_{C_3}^2 + m_4 l_3^2 + I_3, \tag{4c}$$

$$M_{23} = m_3 l_{C_3} \left[l_{C_3} + l_2 \cos\theta_3 \right] + m_4 l_3 \left[l_3 + l_2 \cos\theta_3 \right] + I_3, \tag{4d}$$

$$M_{44} = m_4, \tag{4e}$$

$$C = m_3 l_2 l_{C_3} \sin\theta_3, \tag{4f}$$

$$G_1 = -M_{11}g, \tag{4g}$$

$$G_2 = m_2 l_{C_2} g cos\theta_2 + m_3 g \left[l_{C_3} cos(\theta_2 + \theta_3) + l_2 cos\theta_2 \right], \tag{4h}$$

$$G_3 = m_3 g l_{C_3} cos(\theta_2 + \theta_3), \tag{4i}$$

$$G_4 = -M_{44} g. \tag{4j}$$

where F_1, d_1, F_4, and d_4 are the forces and linear positions for the prismatic joints 1 and 4, respectively; $\tau 2$, $\theta 2$, $\tau 3$, and $\theta 3$ are the torques and angular positions for joints 1, 2, and 3, respectively, and; mi, l_i, and l_{C_i}, are the masses, length and distance to centroids of the i-*th* link, with i=2, 3, 4; g corresponds to gravity acceleration and $I2$ and $I3$ are the inertia tensors of links 2 and 3, respectively.

2.2.3. Actuators

All four joints in the SCARA are actioned by DC motors (Urrea et al., 2021c). Therefore, the model requires the DC motor armature voltage as input, and responds with a joint position, as illustrated in Figure 4.

Figure 4. Block diagram of the actuator model

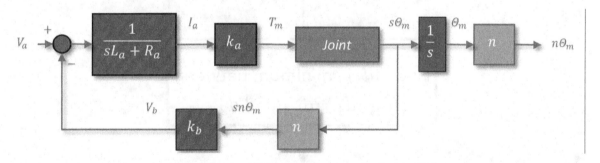

In Figure 4, L_a is the armature inductance and R_a the armature resistance. The torque, T_m, applied to the SCARA joint is proportional to the armature current, I_a, by the constant k_a, causing an angular speed $s\Theta_m$ output, which, in turn, causes a proportional back electro-motive voltage V_b by a constant k_b.

Additionally, the motor is connected to the joint through a gear train, which increases torque and reduces speed in an n factor.

2.3. Coordination Algorithm

The SCARA robots coordinate with each other in a leader/follower configuration. In other words, one of them assumes the task of calculating and moving along a trajectory from the initial to the final position of the beam, leading the follower SCARA robot through the process. The latter is not aware of the destination position since it only follows the leader.

However, the leader and follower roles are dynamically assigned depending on their position with respect to the beam they are aiming to move. This implies that each SCARA will be nominated leader or follower every time a beam needs to be moved.

In the situation proposed in this paper, the SCARA robot that is closer to any of the ends of the beam will be picked as a leader, as explained in Figure 5.

In Figure 5, each $d_{j,k}$ is the distance from the base of the j-th SCARA and the k-th end of the beam. Additionally, $md_{j,k}$ is the minimum $d_{j,k}$. Therefore, the j-th SCARA, indicated in $md_{j,k}$ is assigned as leader.

2.4. Trajectory Generation

Figure 5. Leader/Follower coordination diagram

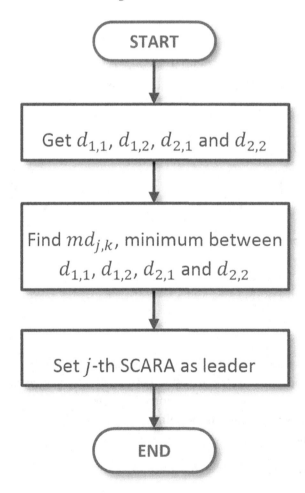

The cooperative trajectory is based on the trajectory desired for the transported beam. Basically, the beam path is composed of three segments, each one aimed to: (i) lift the beam from the floor, (ii) displace the beam to its horizontal end position, (iii) move both end effectors down to release the beam on the platform. During all three segments, the leader and the follower maintain continuous communication.

It is assumed that each robot places its end effector on the closest end of the beam and starts its trajectory from there. The ends of the beam are named P_L and P_F and are manipulated by the leader and the follower, respectively.

The path is calculated from a Cartesian origin placed on the floor, so that the base of the leader and the follower are located in position X_L^B and X_F^B, respectively, the center of the beam follows the path $X_C(t)$, and P_L and P_F ends rotate along the trajectory $X_L^E(t)$ and $X_F^E(t)$, respectively, around $X_C(t)$.

As mentioned above, the leader frequently reports its end effector position to the follower. Hence, the latter calculates the point it should move towards as a function of the leader's position. According to that, the paths of the end effector of each robot, $X_L(t)$ and $X_F(t)$, with reference to their respective bases, are:

$$X_L\left(t\right) = X_C\left(t\right) + X_L^E\left(t\right) - X_L^B, \tag{5a}$$

$$X_F\left(t\right) = X_L\left(t\right) - 2 \cdot X_L^E\left(t\right) + X_L^B - X_F^B, \tag{5b}$$

where, considering a $2r$ length beam:

$$X_L^E\left(t\right) = R\left[\varphi(t)\right] \cdot \begin{bmatrix} r \\ 0 \\ 0 \end{bmatrix}, \tag{5c}$$

and $R[\varphi(t)]$ is a rotational transformation matrix defined as:

$$R\left[\varphi(t)\right] = \begin{bmatrix} \cos\varphi(t) & -\sin\varphi(t) & 0 \\ \sin\varphi(t) & \cos\varphi(t) & 0 \\ 0 & 0 & 1 \end{bmatrix}, \tag{5d}$$

Also, both $X_C(t)$ and $\varphi(t)$ are modeled as a cubic polynomic path as follows:

$$X_C\left(t\right) = A_0 + A_1 t + A_2 t^2 + A_3 t^3, \tag{6a}$$

$$\varphi\left(t\right) = \Phi_0 + \Phi_1 t + \Phi_2 t^2 + \Phi_3 t^3. \tag{6b}$$

In equation 5, $A^k = \left(A_x^k, A_y^k, A_z^k\right)$ are three-component vectors and Φ_k are scalar coefficients, where $k=0,1,2,3$. They are determined as follows:

$$A_0 = X_C(0), \tag{7a}$$

$$A_1 = (0,0,0), \tag{7b}$$

$$A_2 = \frac{3}{t_f^2} \left[X_C(t_f) - X_C(0) \right], \tag{7c}$$

$$A_3 = -\frac{2}{t_f^3} \left[X_C(t_f) - X_C(0) \right], \tag{7d}$$

$$\Phi_0 = \varphi(0), \tag{7e}$$

$$\Phi_1 = 0, \tag{7f}$$

$$\Phi_2 = \frac{3}{t_f^2} \left[\varphi(t_f) - \varphi(0) \right], \tag{7g}$$

$$\Phi_3 = -\frac{2}{t_f^3} \left[\varphi(t_f) - \varphi(0) \right], \tag{7h}$$

where t_f is the time required to complete the trajectory, $X_C(0)$ is the initial beam position and $\varphi(0)$ its initial rotation angle against the horizontal axis.

As explained above, the complete trajectory is divided into three segments, which determine how $X_C(0)$, $X_C(t_f)$, $\varphi(0)$ and $\varphi(tf)$ are selected. On segment (i) and (iii) the movement is purely vertical, so initial and final positions vectors only differ in the vertical component, and the rotation angle is constant. On (ii) the beam only moves horizontally, so the vertical component remains constant and the rotation varies from the initial to the final angle.

2.5. Server Development

The robots communicate with the server through TCP. While they are moving the beam, the leader informs its position to the server. Then, the server provides this information to the follower, which modifies its own position to move along with the leader and keep distance from it in order to prevent damage.

Figure 6 shows a flow chart of the communication process between the leader and the follower through the server. It shows the leader's *n*-points trajectory $P_L(t_k)$, where $k=1,2,3,\ldots,n$, that starts on P_{L_i} and ends on P_{L_F}. Additionally, it presents the follower's *n*-points trajectory $p_F(t_k)$, where $k=1,2,3,\ldots,n$. The initial positions of the leader and the follower are equal to the initial positions of each end of the beam, P_{E_1} and P_{E_2}, respectively.

Furthermore, Figure 5 shows a highlighted block that represents the leader-follower communication procedure that is running on the remote server. As explained previously, the communication is carried out by TCP, with port Prt_L assigned to the leader and Prt_F to the follower. Likewise, the sockets $Sckt_L$ and $Sckt_F$ for leader and follower, respectively, open through those ports; the algorithm is detailed in Figure 7. At the beginning, and empty string *DT* is defined and every chunk of data *Dt* received from $Sckt_L$ is

Figure 6. Communication process flow chart

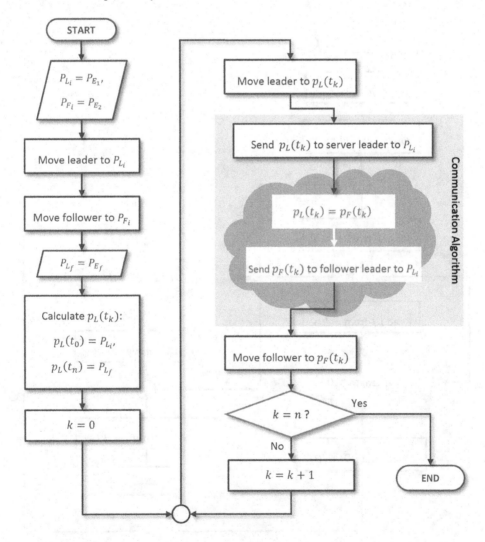

concatenated to it until it becomes bigger than the maximum buffer length B. After that, the data stored in DT is sent through $Sckt_F$. Finally, the algorithm empties DT and returns to listening $Sckt_L$.

2.6. Simulations Parameters

As explained above, the solution proposed is simulated over a 3D model designed in VRML. The parameters are set as shown below.

Figure 7. Server listening to leader port and replying through follower port

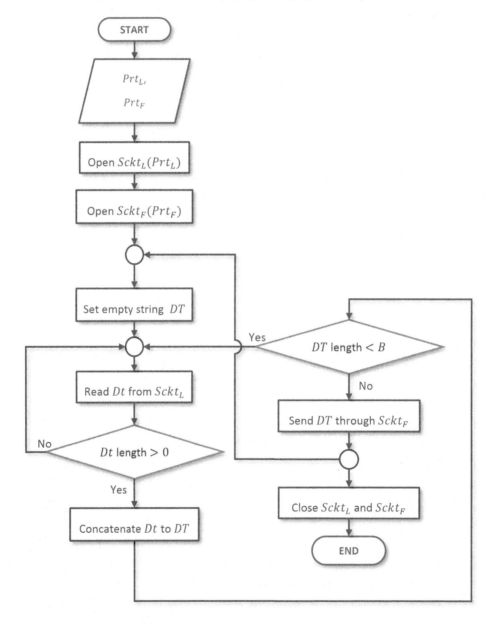

The length of the links and mass of robot 1 and robot 2 are detailed in the tables below:

The conversion factors of gear trains are assumed equal in both robots. Their values are listed in the table below for each of the four joints:

In addition, the actuator model parameters are considered the same in each joint for both robots:

All the robot joints are considered to initiate their paths from rest, so their initial velocity is zero.

Table 1. Simulation parameters in robot 1

Parameter	Value
l_1	0.8 m
l_2	1.2 m
l_3	1.2 m
l_{C_2}	0.6 m
l_{C_3}	0.6 m
m_2	12.0 kg
m_3	6.0 kg
m_4	2.0 kg
I_2	0.44 kg·m²
I_3	0.08 kg·m²

Table 2. Simulation parameters in robot 2

Parameter	Value
l_1	0.8 m
l_2	1.4 m
l_3	1.4 m
l_{C_2}	0.7 m
l_{C_3}	0.7 m
m_2	12.0 kg
m_3	6.0 kg
m_4	2.0 kg
I_2	0.44 kg·m²
I_3	0.08 kg·m²

Their initial positions are listed in the table below:

where X_1^B and X_2^B are the base positions of robot 1 and 2, and X_1^i and X_2^i are the end effectors

Table 3. Conversion factor of gear trains

Conversion Factor	Value
n_1	561.6
n_2	600
n_3	561.6
n_4	561.6

initial positions of robot 1 and robot 2, respectively.

Table 4. Parameters of the actuators model

Parameter	Value
R_a	1.6 Ω
L_a	4.8 mH
k_a	0.35 N·m/A
k_b	0.4 V·s/rad

The beam is considered a parallelepiped randomly placed on the floor, hence $z=0$ always at the initial position, and the objective is to move it to a destination over a platform at a specific height. These

Table 5. Initial position of the robots

Parameter	Value
X_1^B	(1.6, 0, 0) m
X_2^B	(-1.8, 0, 0) m
X_1^i	(2.0, 1.0, 2.4) m
X_2^i	(-2.0, 1.2, 0.8) m

parameters are listed below:

The results of the simulation will be provided in the next section.

3. RESULTS

The robots, the beam and the platform are assigned coordinate systems after which the cooperative

Table 6. Parameters of the beam and the environment

Parameter	Value
Mass	80 kg
Initial Position of the end 1	(-0.53242, -0.27371, 0) m
Initial Position of the end 2	(0.45820, 0.43567, 0) m
Central destination position	(0.00000, -0.40000, 1.8) m
Destination rotation	0 rad
g	9.8 m/s^2

work algorithm described above automatically chose robot 1 as the leader and robot 2 as the follower; this decision was made based on the shortest end-effector distance from the robot selected as leader to the closest end of the beam. Then, for the trajectory generation process and the consequent gripping movement of the beam ends by the end-effector of each robot, the kinematic and dynamic equations of these SCARA robots are used. In this way, once the beam is grabbed by the robots, it is transported to the surface of the platform, where it is released.

The SCARA robots and their workspaces are modeled in VRML. Some images along the simulation are shown in Figure 8.

According to the path generation algorithm, inverse kinematics and the actuators model, the path, torque, current and voltage graphs of the joints are shown from figures 9 to 16. The results of these simulations show the excellent dynamic performance of each of the measured variables.

These results indicate that the use of a cloud server is a good option for robot-to-robot communication in construction work, as it enables more efficient handling of heavy materials.

Figure 8. Result of beam manipulation and transportation

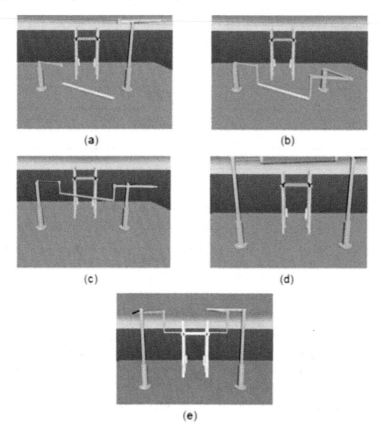

(a) (b) (c) (d) (e)

4. CONCLUSION

In this article, a new cloud computing-based strategy for coordinating multi-robot systems aimed to

Figure 9. Path, torque, current, and voltage of actuator 1 on leader

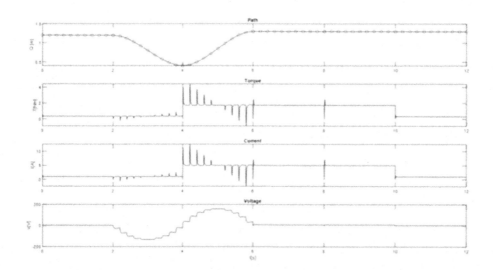

Figure 10. Path, torque, current, and voltage of actuator 1 on follower

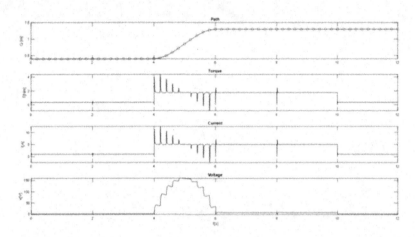

Figure 11. Path, torque, current, and voltage of actuator 2 on leader

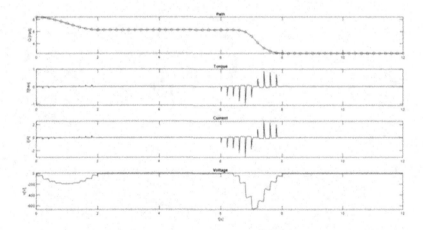

Figure 12. Path, torque, current, and voltage of actuator 2 on follower

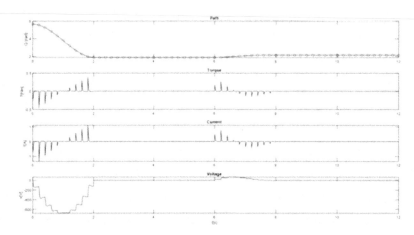

Figure 13. Path, torque, current, and voltage of actuator 3 on leader

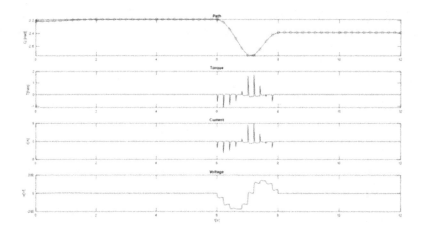

Figure 14. Path, torque, current, and voltage of actuator 3 on follower

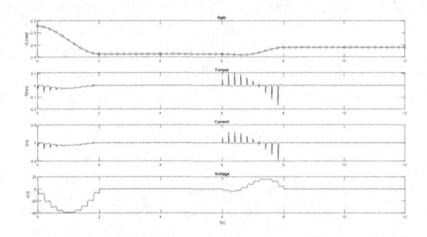

Figure 15. Path, torque, current, and voltage of actuator 4 on leader

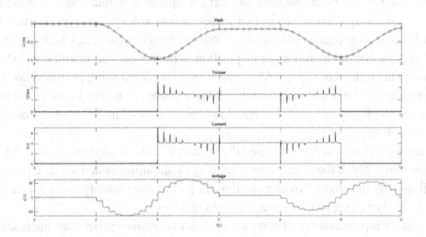

Figure 16. Path, torque, current, and voltage of actuator 4 on follower

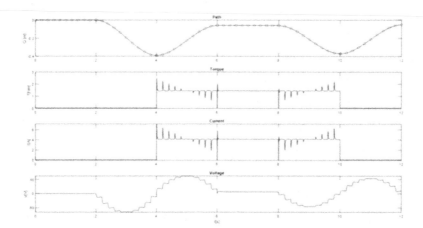

transport and manipulate heavy objects like a beam was developed. The proposed solution was applied to a team of dynamically modeled Selectively Compliance Assembly Robot Arms with 4-DoF, simulated in Virtual Reality Modeling Language and coordinated by a cloud server.

The SCARA robots have been modeled on their kinematic, dynamic and actuators; therefore, mechanical variables like torque, and electrical variables such as voltage and current were obtained.

The SCARA robots were organized in a leader/follower configuration, in which the leader performed a computed trajectory, constantly reporting its position to the remote server and this, in turn, forwarded this information to the followers. The SCARA robots were modeled using MatLab® and communicated through a Transmission Control Protocol with a Python process running in the cloud server, in order to demonstrate the feasibility of incorporating cloud resources in a multi-robot system by testing their performance through simulation.

The integration of a cloud service allowed the system to use a high complexity leader/follower coordination algorithm without the need to integrate large computing systems, since the server could be as robust as allowed by the cloud provider and thus all processing capacity and information is available for the SCARA robots as long as they have access to Internet.

The virtual reality representation allowed for a correct coordination between the leader and follower during the movement of all the objects.

The remote service was programmed in Python, and consisted of a TCP socket communication where data was received from the leader through a port and replicated through another port to the follower. Nevertheless, this service can be developed with any other programming language in which TCP communication can be modelled. Therefore, compatibility would not be an issue. Furthermore, the efficiency of the remote service depends on the speed and robustness of the network.

It is well known that construction and manufacturing are carried out in environments where automation is highly appreciated, as it improves safety and efficiency; therefore, the results obtained in this article are an initial proposal for the handling of heavy beams, which can be easily extended to other heavy materials.

REFERENCES

Abbas, H., Al-Fuqaha, A., Guizani, M., Rayes, A., & Mohammadi, M. (2020). IoT Applications in Industrial Scenarios: A Comprehensive Survey. *IEEE Communications Surveys and Tutorials*, *22*, 2595–2620. doi:10.1109/COMST.2020.2994304

Alarcon, F., & Tovar, E. (2021). Ant Colony Optimization for Multi-Robot Systems. *IEEE Robotics and Automation Letters*, *6*, 7163–7169. doi:10.1109/LRA.2021.3093659

Cai, S., Ma, Z., Skibniewski, M. J., & Bao, S. (2019). Construction automation and robotics for high-rise buildings over the past decades: A comprehensive review. *Advanced Engineering Informatics*, *42*, 100989. doi:10.1016/j.aei.2019.100989

Cao, C., Shi, Y., & Su, Y. (2021). A Distributed Task Allocation Algorithm for Heterogeneous Robot Teams. *IEEE Transactions on Automation Science and Engineering*, *18*, 666–680. doi:10.1109/TASE.2020.2991522

Cena, C. G., Cardenas, P. F., Pazmino, R. S., Puglisi, L., & Santonja, R. A. (2013). A cooperative multi-agent robotics system: Design and modelling. *Expert Systems with Applications*, *40*(12), 4737–4748. doi:10.1016/j.eswa.2013.01.048

Chen, H., Li, X., Zhang, S., & Li, Y. (2021a). Intelligent Control Strategy for Smart Manufacturing Systems: A Comprehensive Review. *IEEE Transactions on Industrial Informatics*, *17*, 1973–1990. doi:10.1109/TII.2020.3020355

Chen, J., Zhao, X., & Yan, G. (2021b). Robotic manipulation of large-scale structures in civil engineering. *Automation in Construction*, *121*, 103471. doi:10.1016/j.autcon.2020.103471

Chen, X., & Huang, J. (2019). Odor source localization algorithms on mobile robots: A review and future outlook. *Robotics and Autonomous Systems*, *112*, 123–136. doi:10.1016/j.robot.2018.11.014

Cheng, B., Dong, M., Zheng, Y., Zhang, X., & Wu, X. (2021). An intelligent robot system for steel beam transportation and assembly. *Journal of Intelligent & Robotic Systems*, *103*, 249–259. doi:10.100710846-020-01322-3

Craveiro, F., Duarte, J. P., Bartolo, H., & Bartolo, P. J. (2019). Additive manufacturing as an enabling technology for digital construction: A perspective on Construction 4.0. *Automation in Construction*, *103*, 251–267. doi:10.1016/j.autcon.2019.03.011

De Almeida, J. P. L. S., Nakashima, R. T., Neves-Jr, F., & de Arruda, L. V. R. (2019). Bio-inspired on-line path planner for cooperative exploration of unknown environment by a Multi-Robot System. *Robotics and Autonomous Systems*, *112*, 32–48. doi:10.1016/j.robot.2018.11.005

Deng, Y., Hua, Y., Napp, N., & Petersen, K. (2019). A Compiler for Scalable Construction by the TERMES Robot Collective. *Robotics and Autonomous Systems*, *121*, 103240. doi:10.1016/j.robot.2019.07.010

Du, H., Xu, W., Yao, B., Zhou, Z., & Hu, Y. (2019). Collaborative optimization of service scheduling for industrial cloud robotics based on knowledge sharing. *Procedia CIRP*, *83*, 132–138. doi:10.1016/j.procir.2019.03.142

Francisco, M., Mezquita, Y., Revollar, S., Vega, P., & de Paz, J. F. (2019). Multi-agent distributed model predictive control with fuzzy negotiation. *Expert Systems with Applications*, *129*, 68–83. doi:10.1016/j.eswa.2019.03.056

Gielis, J., Shankar, A., & Prorok, A. (2022). A Critical Review of Communications in Multi-robot Systems. *Current Robotics Reports*, *3*(4), 213–225. doi:10.100743154-022-00090-9 PMID:36404913

Goswami, A., & Banerjee, B. (2021). A Swarm Intelligence Inspired Framework for Cooperative Multi-Robot Systems. *IEEE Transactions on Systems, Man, and Cybernetics. Systems*, *51*, 1697–1709. doi:10.1109/TSMC.2019.2952262

Han, Z. (2022). *Research on Big Data Mining Application of Internet of Things Based on Artificial Intelligence Technology. 2022 International Conference on Computing, Robotics and System Sciences (ICRSS)*, Macau, China. 10.1109/ICRSS57469.2022.00025

Hichri, B., Fauroux, J. C., Adouane, L., Doroftei, I., & Mezouar, Y. (2019). Design of cooperative mobile robots for co-manipulation and transportation tasks. *Robotics and Computer-integrated Manufacturing*, *57*, 412–421. doi:10.1016/j.rcim.2019.01.002

Jia, M., Komeily, A., Wang, Y., & Srinivasan, R. S. (2019). Adopting Internet of Things for the development of smart buildings: A review of enabling technologies and applications. *Automation in Construction*, *101*, 111–126. doi:10.1016/j.autcon.2019.01.023

Jian, Y., & Li, Y. (2019). Research on intelligent cognitive function enhancement of intelligent robot based on ant colony algorithm. *Cognitive Systems Research*, *56*, 203–212. doi:10.1016/j.cogsys.2018.12.014

Khetre, A., & Gupta, P. (2021). Path planning of mobile robots for efficient transportation of heavy loads using a novel modified differential evolution algorithm. *Applied Soft Computing*, *104*, 107194. doi:10.1016/j.asoc.2021.107194

Kumari, S., & Kaur, A. (2020). Towards Cloud Robotics: Survey of Trends and Technologies. *IEEE Access : Practical Innovations, Open Solutions*, *8*, 156800–156819. doi:10.1109/ACCESS.2020.3019383

Li, G., Liu, H., Huang, T., Han, J., & Xiao, J. (2023). An effective approach for non-singular trajectory generation of a 5-DOF hybrid machining robot. *Robotics and Computer-integrated Manufacturing*, *80*, 102477. doi:10.1016/j.rcim.2022.102477

Lu, Y., & Xu, X. (2019). Cloud-based manufacturing equipment and big data analytics to enable on-demand manufacturing services. *Robotics and Computer-integrated Manufacturing*, *57*, 92–102. doi:10.1016/j.rcim.2018.11.006

Maleš, L., Marčetić, D., & Ribarić, S. (2019). A multi-agent dynamic system for robust multi-face tracking. *Expert Systems with Applications*, *126*, 246–264. doi:10.1016/j.eswa.2019.02.008

Mathews, N., Christensen, A. L., Stranieri, A., Scheidler, A., & Dorigo, M. (2019). Supervised morphogenesis: Exploiting morphological flexibility of self-assembling multirobot systems through cooperation with aerial robots. *Robotics and Autonomous Systems*, *112*, 154–167. doi:10.1016/j.robot.2018.11.007

Nord, J. H., Koohang, A., & Paliszkiewicz, J. (2019). The Internet of Things: Review and theoretical framework. *Expert Systems with Applications*, *133*, 97–108. doi:10.1016/j.eswa.2019.05.014

Pires, J. N., Faria, B. M., & Neto, P. (2020). Cloud robotics: A survey on current trends and future challenges. *Journal of Intelligent & Robotic Systems*, *100*, 1–19. doi:10.100710846-019-01077-4

Semeraro, F., Griffiths, A., & Cangelosi, A. (2023). Human–robot collaboration and machine learning: A systematic review of recent research. *Robotics and Computer-integrated Manufacturing*, *79*, 102432. doi:10.1016/j.rcim.2022.102432

Suárez, P., Iglesias, A., & Gálvez, A. (2019). Make robots be bats: Specializing robotic swarms to the Bat algorithm. *Swarm and Evolutionary Computation*, *44*, 113–129. doi:10.1016/j.swevo.2018.01.005

Sullivan, N., Grainger, S., & Cazzolato, B. (2019). Sequential single-item auction improvements for heterogeneous multi-robot routing. *Robotics and Autonomous Systems*, *115*, 130–142. doi:10.1016/j.robot.2019.02.016

Tang, S., Shelden, D. R., Eastman, C. M., Pishdad-Bozorgi, P., & Gao, X. (2019). A review of building information modeling (BIM) and the internet of things (IoT) devices integration: Present status and future trends. *Automation in Construction*, *101*, 127–139. doi:10.1016/j.autcon.2019.01.020

Teruel, E., Aragues, R., & López-Nicolás, G. (2019). A distributed robot swarm control for dynamic region coverage. *Robotics and Autonomous Systems*, *119*, 51–63. doi:10.1016/j.robot.2019.06.002

Tu, X., Ma, O., Zhang, L., Yang, J., Wang, Y., & Cai, H. (2021). Distributed Multi-Robot System for Planetary Surface Exploration. *IEEE Access : Practical Innovations, Open Solutions*, *9*, 12987–12996. doi:10.1109/ACCESS.2021.3053461

Urrea, C., & Agramonte, R. (2022a). Evaluation of Parameter Identification of a Real Manipulator Robot. *Symmetry*, *14*(7), 1446. https://www.mdpi.com/2073-8994/14/7/1446. doi:10.3390ym14071446

Urrea, C., Kern, J., & Álvarez, E. (2022b). Design of a generalized dynamic model and a trajectory control and position strategy for n-link underactuated revolute planar robots. *Control Engineering Practice*, *128*, 1–13. doi:10.1016/j.conengprac.2022.105316

Urrea, C., & Matteoda, R. (2020). Development of a virtual reality simulator for a strategy for coordinating cooperative manipulator robots using cloud computing. *Robotics and Autonomous Systems*, *126*, 103447. doi:10.1016/j.robot.2020.103447

Urrea, C., & Páez, F. (2021a). Design and Comparison of Strategies for Level Control in a Nonlinear Tank. *Processes (Basel, Switzerland)*, *9*(5), 735. doi:10.3390/pr9050735

Urrea, C., & Pascal, J. (2021b). Design and validation of a dynamic parameter identification model for industrial manipulator robots. *Archive of Applied Mechanics*, *91*(5), 1981–2007. doi:10.100700419-020-01865-2

Urrea, C., & Pascal, J. (2021c). Dynamic Parameter Identification based on Lagrangian Formulation and Servomotor-type Actuators for Industrial Robots. *International Journal of Control, Automation, and Systems*, *19*(8), 2902–2909. doi:10.100712555-020-0476-8

Varlamov, O. (2021). "Brains" for Robots: Application of the Mivar Expert Systems for Implementation of Autonomous Intelligent Robots. *Big Data Research*, *25*, 100241. doi:10.1016/j.bdr.2021.100241

Wang, F., Laili, Y., & Zhang, L. (2022). Multi-granularity service composition in industrial cloud robotics. *Robotics and Computer-integrated Manufacturing*, *78*, 102414. doi:10.1016/j.rcim.2022.102414

Wang, S., Wang, Y., Deying, L., & Zhao, Q. (2023). Distributed Relative Localization Algorithms for Multi-Robot Networks: A Survey. *Sensors (Basel)*, *23*(5), 2399. doi:10.339023052399 PMID:36904602

Wang, Y., Yu, Y., Gao, S., Pan, H., & Yang, G. (2019). A hierarchical gravitational search algorithm with an effective gravitational constant. *Swarm and Evolutionary Computation*, *46*, 118–139. doi:10.1016/j.swevo.2019.02.004

Wei, C., Ma, C., & Duan, H. (2021). Multi-robot olfactory system for odor mapping and plume tracking. *Measurement*, *178*, 109249. doi:10.1016/j.measurement.2021.109249

Wu, C., Luo, S., Zuo, Y., & Chen, C. (2021). Decentralized multi-robot system for cooperative assembly. *Robotics and Autonomous Systems*, *133*, 103739. doi:10.1016/j.robot.2020.103739

Yan, F., Feng, S., Liu, X., & Feng, T. (2023). Parametric Dynamic Distributed Containment Control of Continuous-Time Linear Multi-Agent Systems with Specified Convergence Speed. *Sensors (Basel)*, *23*(5), 2696. doi:10.339023052696 PMID:36904911

Yilmaz, G. E., & Akar, A. R. (2020). Decentralized auction-based task allocation in swarm robotic systems. *Robotics and Autonomous Systems*, *130*, 103538. doi:10.1016/j.robot.2020.103538

Yousefizadeh, S., Flores Mendez, J. D., & Bak, T. (2019). Trajectory adaptation for an impedance controlled cooperative robot according to an operator's force. *Automation in Construction*, *103*, 213–220. doi:10.1016/j.autcon.2019.01.006

Zhang, S., Wang, J., Luo, L., & Qiu, C. (2021). Aerial-robot-assisted multi-robot assembly system with real-time task allocation. *International Journal of Advanced Robotic Systems*, *18*, 1–12. doi:10.1177/1729881421993856

Zhou, C., Luo, H., Fang, W., Wei, R., & Ding, L. (2019). Cyber-physical-system-based safety monitoring for blind hoisting with the internet of things: A case study. *Automation in Construction*, *97*, 138–150. doi:10.1016/j.autcon.2018.10.017

APPENDIX

The server was developed in Python 3.7 under the code provided below:

```python
import socket
def open_socket(port):
    server_socket = socket.socket(socket.AF_INET, socket.SOCK_STREAM)
    server_socket.bind(('cooperativo', port))
    server_socket.listen()
    conn, addr = server_socket.accept()
    return conn
conn_input  = open_socket(3001)
conn_output = open_socket(3002)
print ('Connection with Robots established')
buffer_length = 512
while True:
    data, ret = conn_input.recvfrom(buffer_length)
    data_length = len(data)
    if data_length > 0:
        print('Leader Position: ' + data.decode() + ' received through port
3001')
        while len(data) < buffer_length:
            data = data + b' '
        conn_output.sendall(data)
        print('\nInformation transmitted to follower through port 3002')
server_socket_input.close()
server_socket_output.close()
```

Chapter 13
Impact of UAVs in Agriculture

Megha Bhushan
DIT University, India

Arun Negi
Deliotte USI, India

ABSTRACT

Agriculture is mostly practiced in rural areas where there is less population and no proper scouting. Unmanned aerial vehicle (UAVs) can reduce human involvement in agriculture and solve many issues such as monitoring water levels, detecting crop disease, controlling the consumption of water and many more. UAVs application has contributed to many areas of agriculture such as insecticide as well as fertilizer prospecting and spraying, seed planting, weed recognition, soil mapping using aerial imaging, crop forecasting and so forth. Through these methods, crops can be cultivated without making excess use of water and chemicals which keep them safe and strong. Further, UAVs are replacing the man-made aircrafts because of their peculiar feature of capturing high resolution imagery below cloud level and its flexibility to work on different geographical locations. The multifunctioning UAVs reduce time and increase productivity. Therefore, this chapter provides a review on a smarter agricultural system using UAVs in order to enhance food productivity.

INTRODUCTION

As stated by 'Agriculture in 2050 Project,' it has been estimated that the population will increase to 10 billion by the year 2050. In light of this growth, the food production will be boosted by ~70% and proper water management will be in-disposable (Jha et al., 2019; Kim et al., 2019; Niu et al., 2019). Agriculture is the backbone as it serves as the primary source of food production. However, there are many challenges in this area such as predicting the type of soil, weather forecasting, crop outcome, monitoring water levels, scheming a proper irrigation plan, detecting crop diseases (Daponte et al., 2019; Pawar et al., 2020), type of soil and removal of pests and weeds (Ju & Son, 2018; Kulbacki et al., 2018; Lakshmi & Naresh, 2018). Many techniques are being currently used for enhancing agricultural productivity. Precision agriculture is one of these modern farming practices that can make production more efficient

DOI: 10.4018/978-1-6684-8785-3.ch013

and effective by perceiving, measuring and reacting to intra as well as inter-field crop variability. The incorporation of UAV-based technology will reduce farmers' expenditure along with the benefit of being more user-friendly and viability in the long run.

Initially, precision agriculture was implemented through satellites; however, obtaining satellite data is not conventionally simple or easily economically viable. Moreover, the data so obtained from these recourses is incomprehensible to farmers for direct application due to the fact that the majority of the rural places possess deficient telecommunications infrastructure. Therefore, the need for an easily accessible technique that can work in any location, regardless of internet availability, is required. Thus, the UAVs are being considered as an alternative to replace the manmade aircrafts (Shruthi et al., 2019) to extend the potential of the smart and precise data accumulation and analysis with economic viability. The UAVs have significantly decreased the amount of time for the same which results in an increase in stability, accuracy, and productivity (Kim et al., 2019). The main advantage of UAVs is that can navigate through the 3D space in tailor made trajectories according to the specific data requirements. Depth sensors (Bhushan et al., n.d.; R. Goel et al., 2020; Kumar et al., 2023; Samant et al., 2021) are incorporated so as to facilitate the generation of maps and 3D templates. They are relatively easy to operate and provide temporal as well as high spatial resolutions with a wide spatial coverage.

UAVs and their control systems together constitute the Unmanned Aerial System (UAS) which primarily relies on the utilization of Machine Learning (ML) techniques (Bhushan et al., 2023; Kedia & Bhushan, 2022; Nalavade et al., 2020; Rana & Bhushan, 2022; Singh & Bhushan, 2022; Suri et al., 2022; Verma et al., 2022) through which high resolution image data is obtained at more frequent periodic intervals when compared to images taken by the satellite. It is important to check the quality of data (Bhushan et al., 2020; Bhushan, Ángel Galindo Duarte, et al., 2021; Bhushan, Kumar, et al., 2021; S. Goel, 2012; Negi & Kaur, 2017) which is being used for the application of ML techniques as this aspect directly affects the results derived from the data so collected. Due to its peculiar feature or ability of capturing high resolution imagery data (Norasma et al., 2019; Rao & Rao, 2019) below cloud levels, UAVs are used to collect Very High Resolution (VHR) images thereby providing an economically effective data analysis techniques along with sensing capabilities (Feng & Li, 2019; Lu, 2019). There are different aspects where UAVs especially in precision agriculture as it comprises of various processes such as soil mapping, water management, weather monitoring, etc.

UAVs such as crop monitoring drones (Shruthi et al., 2019) are utilized for planting seeds by shooting them, planting nutrients onto the field, and determining the strength of the nutrient uptake, which can optimize expenses to by a factor of ~70 percent (Huuskonen & Oksanen, 2018). Similarly, irrigation and application of pesticides are an essential part of agriculture which also utilize extensive utilization of drones for farming processes involving fluids like evapotranspiration (combination of evaporation and transpiration) (Gago et al., 2015; Niu et al., 2019). This helps to reduce adverse health problems faced by farmers due to excessive contact with harmful chemicals especially in pesticides, and weedicides while being manually sprayed in the fields. This application further extends for its utilization for semi-autonomous fertilizer prospecting and spraying of chemicals (Rao & Rao, 2019). The UAVs are, thus, a better choice for smart farming techniques since they can be used easily while providing high spatial coverage (Hassler & Baysal-Gurel, 2019; Vu et al., 2019; Xu et al., 2019). However, even though they provide a lot of capabilities, they are still bound by several limitations including but not limited to low battery efficiency, low flight time, less communication distance, low payload, etc. One of the solutions to these problems is the utilization of multiple UAVs systems for agriculture which increase the overall work efficiency and productivity (Barbedo, 2019; Mogili & Deepak, 2018; Raparelli & Bajocco, 2019).

LITERATURE REVIEW

In (Gago et al., 2015), various applications of UAVs using several remote sensors have been discussed. They considered the most recent remote sensing occurrences, and used them in agriculture with different kinds of UAVs. Plant water condition was examined with crop scale. They stated that UAVs were important tools for water management and precision agriculture. The application of drones in various aspects of agriculture and the implementation of UAVs for advancing Indian agriculture has been briefed in (Lakshmi & Naresh, 2018). Prediction, weather forecasting, crop outcome, monitoring water levels, detecting crop diseases, type of soil and removal of weeds are some of the major areas where UAVs are used in the farming sector. UAV generated maps and images are used to process the computations.

The authors of (Ju & Son, 2018) have discussed how Multi UAVs are better than Single UAVs in the agricultural sector. The Multi-UAV system was developed based on the dispensed swarm control algorithm and examined through 4 experimental cases and 6 performance metrics by keeping remote sensing as the benchmark test. The issues on technologies like remote sensing and UAS platforms which are used to develop the agricultural sector have been addressed in (Kulbacki et al., 2018). While emphasizing on the applications of remote sensing technology, drones are used in smart farming. The global reports indicated that to improve farm productivity and increase yields, there is a necessity of technology like remote sensing using drones.

The soil field plot to form a color map of soil has been discussed in (Huuskonen & Oksanen, 2018). This soil sampling works as a means to collect the information regarding fertilization of fields. A complete cycle of precision agriculture from plowing the field to soil analysis was to form a topsoil map with management zones for farming purpose in precision agriculture. A review on the use of UAVs for monitoring crops and spraying pesticides in precision agriculture has been done in (Mogili & Deepak, 2018). The sprinkling system in UAVs which is navigated with Global Positioning System (GPS) tracks the proper places to spray pesticides which reduce the wastage of water and chemicals. Therefore, an automated drone system is developed in UAVs for crop monitoring and pesticide spraying. The most recent trends as well as applications of the major technologies associated with agricultural UAVs have been discussed in (Kim et al., 2019). Application of UAVs has proven to contribute in many areas in the agricultural sector such as in planting seeds, mapping and many more.

A review of various UAVs based approaches of ET estimations and the implementation of Remote Sensing Energy Balance (RSEB) algorithms have been done in (Niu et al., 2019). They observed that when compared with other satellites based remote sensing methods, UAV platform and light weight sensors can ensure better quality, higher spatial and temporal resolution images. A brief outline of the current enactment of automation in agriculture is discussed in (Jha et al., 2019). The proposed system can be implemented in botanical farms for flower and leaf identification and watering using IoT is discussed. It improves the complete progress of the country. The authors in (Daponte et al., 2019) have reviewed the area of precision agriculture using drones. The drones were equipped with tools like thermal, visible and multispectral cameras for precision agriculture monitoring. The design was done on the basis of time, measurement, and usage of power with information on the control system architecture and the parameters required before a flight were also discussed.

In (Norasma et al., 2019), authors have discussed UAV as an alternative to increase yield. They have investigated UAVs to ensure data quality and analysis so that it will be easy for us to acquire précised analysis and accurate data if a suitable sensor is used. The mission planner software was used for spraying pesticide manually or semi-autonomously with the help of low-cost UAVs in (Rao & Rao, 2019).

It diminishes the adverse effects of the pesticides on humans, can spray the large area in a short duration and acts both as a quadcopter and a spraying device. A mobile or a radio controller can be used to operate i.e. method of species mapping from images provided by UAVs over metropolitan areas was discussed in (Feng & Li, 2019). These are proposed by method of evaluating different plant species using ortho-images and Digital Surface Model (DSM) in mapping. It provides information about spatial allotment of tree species. A four-rotor UAV was designed and the purpose of capturing aerial images of the agricultural land was discussed in (Lu, 2019). The purpose of the four rotor UAV aerial acquisition element is to rectify crop growth status, physiological indicators or to estimate crop yield. For aerial photography and wireless communication many processors are installed.

In (Vu et al., 2019), authors have discussed developing a drone, but it used to be quite expensive, so a low cost-effective quadcopter was proposed. This drone mobility helps to monitor large areas, solve problems and minimize crop care costs. A brief explanation on the advantages and disadvantages of the various technologies in UAVs like field mapping, inventory counting, biomass estimation, chemical spraying, weed management, plant stress detection and looks into the future development of research on UAS in agriculture was given in (Hassler & Baysal-Gurel, 2019). The major goal is to cover some areas in use, physical components of systems, processing and gathering of data and some information about the legal, economic and integration factors.

The separation of cultivated land from other terrains was discussed in (Xu et al., 2019). For distinguishing vegetation and non-vegetation indexes they have used Visible-Band Difference Vegetation Index (VDVI) and ExG index. The use of UAVs in agriculture and forestry was discussed in (Raparelli & Bajocco, 2019). They concluded that it is mainly focused on precision farming and crop monitoring and have reviewed the radar and laser data for canopy structural analysis and vegetation mapping. In (Barbedo, 2019), authors have discussed different solutions that have extracted data from the images taken along UAVs flights. An insight on the image sensors was provided that concerned the challenges related to UAVs, sensors, and their calibration that in (Shruthi et al., 2019), authors have suggested using UAVs for plant disease identification using machine learning. It was observed that Convolutional Neural Network (CNN) will give high accuracy and it has the capability to detect more diseases in multiple crops. This use of latest technology, Precision Agriculture will be a great help to the farmers to take better decisions

Table 1. Summary of the existing works

Article	Technique(s)	Merits	Demerits
Kim et al. (2019)	Agricultural UAV, navigation control, imaging sensors	Working hours are reduced, Stability is increased, Measuring accuracy is increased, and Production is increased, Less expensive than most other agriculture machines, Easily operated.	Low battery life, Less flight duration, Less interaction distances, Less loading capacity
Niu et al. (2019)	Different models of UAVs for high resolution pictures, Different sensors for mapping of evapotranspiration.	Can be operated at any time, Cost efficient, Higher resolution pictures	Each model of UAV drones has their own advantages and disadvantages, METRIC which is based-on satellite platforms, so there are limitations regarding time and frequency, Two source energy balance (TSEB) which is less widely known even though it's the most efficient, UAVs depend on weather conditions, and They rely on field scale and flight time.
Jha et al. (2019)	UAV's, Quadcopters	Reducing crop disease, Storage management, Pesticide control, water management, Getting better crops, Better field management, Removal of weeds.	Excessive cost, Uncertain Technical issues, Reduces the employment count.
Daponte et al. (2019)	UAV, Drones	Helps the farmers to inspect smart and targeted irrigation and civilization, can detect the areas where sufficient irrigation is required and where disease is scattering, save water resources, Reduce agrochemical products, advanced techniques to improve crop productivity and quality, can fix any problem related to resources.	They can do only one task at a time, Additional load on humans, Does not recognize proper pesticides and fertilizers for respective farms, Low payload.
Lakshmi & Naresh (2018)	UAV	Collect images for training data set in order to train the model to know the changes in Growth of crop, visualize diseases, Control weeds, Generating maps from smaller UAVs, Monitoring water levels, Detecting the type of soil.	Cannot be used in extreme weather conditions, UAVs with more features are expensive.
Ju & Son (2018)	A Multi-UAV system for agriculture using the distributed Swarm Control Algorithm	Working hours and labor requirements have been significantly reduced, Efficiency of agricultural works has improved significantly, and Agricultural tasks can be completed easily, Consuming less time to complete the task.	Less payload, Less flight time
Kulbacki et al. (2018)	Agricultural UAV, navigation control, imaging sensors	Collected data, mapping of field, 3D models, supported early soil analysis, acquired data in the form of images are useful in irrigation, determining the concentration of nutrient uptake in various fields, estimating changes for crops, predicting the right moment in harvesting, Crops monitoring drones were used for shooting seeds and plant nutrients into the soil	Limited load capacity, does not support in extreme weather conditions, UAVs with extra features are costly, Low flight time, Less communication distance

Continued on following page

Table 1. Continued

Article	Technique(s)	Merits	Demerits
Shruthi et al. (2019)	Plant disease detection, classification, ML	SVM classifiers are used by many authors compared with other classifiers, CNN detects more diseases with high accuracy.	Cannot perform the techniques in ML like Decision Trees (DT) and Naive Bayes classifier.
Norasma et al. (2019)	UAVs with standard assigned sensors applicable in vast agricultural lands	UAVs will solve specific issues in agriculture, The integrated technology will help farmers in field operations.	UAVs cannot solve all problems, Tasks like image processing should be improved, and Data should be accurate.
Rao & Rao (2019)	UAVs Used for Spraying Pesticides	Controlling the deaths of farmers, Less Consumption of time, Fastness and Accuracy of Spraying, Pesticide Control, Crop Health and Yield can be estimated.	The UAVs can estimate but they cannot tell the accurate number, Excess use of sprayers can cause damage to the crop.
Feng & Li (2019)	Simple Thresholding Method	The result in the method of all 4 tree species, together with an improvement of 0.61-5.81% in overall accuracy and an accuracy assessment.	Study area with limited coverage
Lu (2019)	Four-Rotor UAV, Farmland monitoring, Aerial photography, STM32	UC/OC-II enhances the stability and it is efficient; The design of the controller is not complex, aerial photography is simple and it does not occupy a large amount of time in the Central Processing Unit (CPU); Module of aerial photography wireless transmission is omitted which costs less.	The work done in the study is technical and may require basic of technical skills for the adequate throughput of the provided model.
Huuskonen & Oksanen (2018)	Augmented reality as wearable technology in agriculture.	During application only, we can collect the sample points and it requires some inputs from the user, Light rain preceded plowing and therefore, the moisture of the soil had stabilized before the operation.	It requires some input from the user all the time, Heavy rain cannot be stabilized before the operation, It requires some input from the user all the time, Heavy rain cannot be stabilized before the operation.
Gago et al. (2015)	UAV and Remote sensing	Great stability, No take-off or landing requirements, No runway requirement, More user-friendly, Low price, High amperes can be handled by the motors, Has a capacity of 1kg payload.	Less flight times, Only stay in air for 4-5mins, Doesn't cover all types of interests.
Vu et al. (2019)	UAVs for monitoring	These drones can capture large areas, Cost effective Quad-copter, UAV photogrammetry for mapping and modeling application.	Quadcopter cannot afford a complex survey mission.
Hassler & Baysal-Gurel (2019)	UAV, imaging devices, gripping tools, spraying equipment	UAVs flies at different trajectories, Depth sensors are incorporated to create maps and 3D models for information, Existing Wireless Sensor Networks (WSN) can be connected to the system, which is an integrated approach for precision farming, Potential in UAS agriculture is very high.	Short flight time, Large payloads, Dependent on weather conditions, costly methods, Time consumed to produce high resolution pictures.

Continued on following page

Table 1. Continued

Article	Technique(s)	Merits	Demerits
Xu et al. (2019)	High Resolution Imaging	It has high spatial resolution, flexibility and high timeliness, Suitable for any type of runways for landing and take-off, Cost and difficulty of collecting remote sensing images reduced.	Visible remote sensing data lacks spectral information in the infrared band.
Raparelli & Bajocco (2019)	Different model of UAVs for bibliometric analysis	Overall structural analysis and vegetation mapping.	In some way the accuracy will be missed.
Barbedo (2019)	UAV, imaging sensors	Improved balance between usability and safety, capable of covering more area and faster than people, Faster detection of problems, much easier to handle.	Hindered by many technical difficulties, Difficult to address all the problems, High diversity of targeted applications, Each problem has a specific characteristic that has to be taken into account.
Mogili & Deepak (2018)	UAV, Drones	More machines in places of less labor, The camera helps to find the areas where pesticides are needed to be sprayed through geographic indicators, Reduction of water wastage and chemicals by auto navigation of UAV sprinkling systems, UAVs are used to spray pesticides to the crops in the correct amount through controllers to protect the crops as well as humans from disease or allergic reactions.	They can estimate but cannot tell the accurate number, Excess use of sprayers can cause damage to the crop, Drones must be standard i.e., if anything gets damaged there will be a halt in the agricultural process and crops may get damaged, Covers only short distances
Lambertini et al. (2022)	UAV, Thermal Imaging	The quality of the results that can be achieved with commercial off-the-shelf sensors and UAVs coupled with their extensive possible future developments to generate soil maps with increased autonomy of these systems.	The end user or the farmers needs some basic degree of computational and technical knowledge in order to maximize the throughput of the said instruments.
Vélez et al. (2022)	UAV, Canopy and structure from motion analysis	The study can be utilized for segmenting the image and studying the effect of ground shadows on the image and their relationship with its agronomic and biophysical parameters, testing algorithms or workflows for real-time applications, Data fusion or combination with freely available open-access satellite images, such as Sentinel or Landsat imagery.	The canopy/crown monitoring takes a lot of time and labor because technicians usually measure tree dimensions in the field for cross verification with the data gathered from UAV in poor environmental conditions.
Mukhamediev et al. (2023)	UAV, Coverage based path planning, aerial monitoring	UAVs can collect data for decision-making and carry out individual agricultural production activities, the use of back-and-forth method based algorithm is well suited for flying over convex fields without obstacles and smaller symmetrical fields.	The effective use of UAV requires to solve the tasks of flight planning, taking into account the heterogeneity of the available attachments and the problem solved in the process of the overflight requires technical expertise.

Continued on following page

CONCLUSION AND FUTURE WORK

A drastic change is observed in the field of agriculture through the usage of UAVs. These are used for implementing methods like soil mapping, pesticide spraying, crop monitoring, and water management. However, these methods have their own advantages and disadvantages. This paper provides a review on the latest research works including various applications of UAVs in the field of agriculture in order to enhance food productivity along with advantages in the usage of UAVs over other man made aerial vehicles (satellites). The future work may include developing a robust technology in UAVs for good images, efficient battery usage and spraying. Further, automating the process of farming and using multiple UAVs for collecting crops and designing automated UAVs for spraying pesticide. This is, thus a very interesting and rapidly expanding sub-domain of IoT based precision agriculture.

REFERENCES

Barbedo, J. G. A. (2019). A review on the use of unmanned aerial vehicles and imaging sensors for monitoring and assessing plant stresses. *Drones (Basel)*, *3*(2), 40. doi:10.3390/drones3020040

Bhushan, M., Ángel Galindo Duarte, J., Samant, P., Kumar, A., & Negi, A. (2021). Classifying and resolving software product line redundancies using an ontological first-order logic rule based method. *Expert Systems with Applications*, *168*, 114167. doi:10.1016/j.eswa.2020.114167

Bhushan, M., Iyer, S., Kumar, A., Choudhury, T., & Negi, A. (n.d.). *Artificial Intelligence for Smart Cities and Villages: Advanced Technologies*.

Bhushan, M., Kumar, A., Samant, P., Bansal, S., Tiwari, S., & Negi, A. (2021). Identifying Quality Attributes of FODA and DSSA Methods in Domain Analysis using a Case Study. *2021 10th International Conference on System Modeling & Advancement in Research Trends (SMART)*, 562–567. 10.1109/SMART52563.2021.9676289

Bhushan, M., Negi, A., Samant, P., Goel, S., & Kumar, A. (2020). A classification and systematic review of product line feature model defects. *Software Quality Journal*, *28*(4), 1507–1550. doi:10.100711219-020-09522-1

Bhushan, M., Pandit, A., & Garg, A. (2023). Machine learning and deep learning techniques for the analysis of heart disease: A systematic literature review, open challenges and future directions. *Artificial Intelligence Review*, 1–52. doi:10.100710462-023-10493-5

Daponte, P., De Vito, L., Glielmo, L., Iannelli, L., Liuzza, D., Picariello, F., & Silano, G. (2019). A review on the use of drones for precision agriculture. *IOP Conference Series. Earth and Environmental Science*, *275*(1), 12022. doi:10.1088/1755-1315/275/1/012022

Feng, X., & Li, P. (2019). A tree species mapping method from UAV images over urban area using similarity in tree-crown object histograms. *Remote Sensing (Basel)*, *11*(17), 1982. doi:10.3390/rs11171982

Gago, J., Douthe, C., Coopman, R. E., Gallego, P. P., Ribas-Carbo, M., Flexas, J., Escalona, J., & Medrano, H. (2015). UAVs challenge to assess water stress for sustainable agriculture. *Agricultural Water Management*, *153*, 9–19. doi:10.1016/j.agwat.2015.01.020

Goel, R., Jain, A., Verma, K., Bhushan, M., Kumar, A., & Negi, A. (2020). Mushrooming trends and technologies to aid visually impaired people. *2020 International Conference on Emerging Trends in Information Technology and Engineering (Ic-ETITE)*, (pp. 1–5). IEEE. 10.1109/ic-ETITE47903.2020.437

Goel, S. (2012). Transformation from LEL to UML. *International Journal of Computer Applications*, *48*(12), 888–975.

Hassler, S. C., & Baysal-Gurel, F. (2019). Unmanned aircraft system (UAS) technology and applications in agriculture. *Agronomy (Basel)*, *9*(10), 618. doi:10.3390/agronomy9100618

Huuskonen, J., & Oksanen, T. (2018). Soil sampling with drones and augmented reality in precision agriculture. *Computers and Electronics in Agriculture*, *154*, 25–35. doi:10.1016/j.compag.2018.08.039

Jha, K., Doshi, A., Patel, P., & Shah, M. (2019). A comprehensive review on automation in agriculture using artificial intelligence. *Artificial Intelligence in Agriculture*, *2*, 1–12. doi:10.1016/j.aiia.2019.05.004

Ju, C., & Son, H. Il. (2018). Performance evaluation of multiple UAV systems for remote sensing in agriculture. *Proceedings of the Workshop on Robotic Vision and Action in Agriculture at the IEEE International Conference on Robotics and Automation (ICRA)*, Brisbane, Australia.

Kedia, S., & Bhushan, M. (2022). Prediction of mortality from heart failure using machine learning. *2022 2nd International Conference on Emerging Frontiers in Electrical and Electronic Technologies (ICEFEET)*, 1–6.

Kim, J., Kim, S., Ju, C., & Son, H. (2019). Unmanned aerial vehicles in agriculture: A review of perspective of platform, control, and applications. *IEEE Access : Practical Innovations, Open Solutions*, *7*, 105100–105115. doi:10.1109/ACCESS.2019.2932119

Kulbacki, M., Segen, J., Knieć, W., Klempous, R., Kluwak, K., Nikodem, J., Kulbacka, J., & Serester, A. (2018). Survey of drones for agriculture automation from planting to harvest. *2018 IEEE 22nd International Conference on Intelligent Engineering Systems (INES)*, 353–358.

Kumar, A., Bhushan, M., Galindo, J. A., Garg, L., & Hu, Y.-C. (2023). *Machine Intelligence, Big Data Analytics, and IoT in Image Processing: Practical Applications*. John Wiley & Sons. doi:10.1002/9781119865513

Lakshmi, J. V. N., & Naresh, G. V. N. (2018). *A Review on Developing Tech-Agriculture using Deep Learning Methods by Applying UAVs*. IJSRCSEIT.

Lambertini, A., Mandanici, E., Tini, M. A., & Vittuari, L. (2022). Technical Challenges for Multi-Temporal and Multi-Sensor Image Processing Surveyed by UAV for Mapping and Monitoring in Precision Agriculture. *Remote Sensing (Basel)*, *14*(19), 4954. doi:10.3390/rs14194954

Lu, C. (2019). Design of a Simple Aerial Photo UAV for Agriculture. *Revista de la Facultad de Agronomía*, *36*(3), 781–792.

Mogili, U. M. R., & Deepak, B. (2018). Review on application of drone systems in precision agriculture. *Procedia Computer Science*, *133*, 502–509. doi:10.1016/j.procs.2018.07.063

Mukhamediev, R. I., Yakunin, K., Aubakirov, M., Assanov, I., Kuchin, Y., Symagulov, A., Levashenko, V., Zaitseva, E., Sokolov, D., & Amirgaliyev, Y. (2023). Coverage path planning optimization of heterogeneous UAVs group for precision agriculture. *IEEE Access : Practical Innovations, Open Solutions*, *11*, 5789–5803. doi:10.1109/ACCESS.2023.3235207

Nalavade, A., Bai, A., & Bhushan, M. (2020). Deep learning techniques and models for improving machine reading comprehension system. *IJAST*, *29*(04), 9692–9710.

Negi, A., & Kaur, K. (2017). *Method to resolve software product line errors*. Information, Communication and Computing Technology: Second International Conference, ICICCT 2017, New Delhi, India.

Niu, H., Zhao, T., Wang, D., & Chen, Y. (2019). Estimating evapotranspiration with UAVs in agriculture: A review. *2019 ASABE Annual International Meeting*, 1. ASABE.

Norasma, C. Y. N., Fadzilah, M. A., Roslin, N. A., Zanariah, Z. W. N., Tarmidi, Z., & Candra, F. S. (2019). Unmanned aerial vehicle applications in agriculture. *IOP Conference Series. Materials Science and Engineering*, *506*(1), 12063. doi:10.1088/1757-899X/506/1/012063

Pawar, S., Bhushan, M., & Wagh, M. (2020). The plant leaf disease diagnosis and spectral data analysis using machine learning–A review. *International Journal of Advanced Science and Technology*, *29*(9s), 3343–3359.

Rana, M., & Bhushan, M. (2022). Machine learning and deep learning approach for medical image analysis: Diagnosis to detection. *Multimedia Tools and Applications*, 1–39. doi:10.100711042-022-14305-w PMID:36588765

Rao, V. P. S., & Rao, G. S. (2019). Design and modelling of anaffordable uav based pesticide sprayer in agriculture applications. *2019 Fifth International Conference on Electrical Energy Systems (ICEES)*, (pp. 1–4). IEEE.

Raparelli, E., & Bajocco, S. (2019). A bibliometric analysis on the use of unmanned aerial vehicles in agricultural and forestry studies. *International Journal of Remote Sensing*, *40*(24), 9070–9083. doi:10.1080/01431161.2019.1569793

Samant, P., Bhushan, M., Kumar, A., Arya, R., Tiwari, S., & Bansal, S. (2021). Condition monitoring of machinery: A case study. *2021 6th International Conference on Signal Processing, Computing and Control (ISPCC)*, 501–505.

Shruthi, U., Nagaveni, V., & Raghavendra, B. K. (2019). A review on machine learning classification techniques for plant disease detection. *2019 5th International Conference on Advanced Computing & Communication Systems (ICACCS)*, (pp. 281–284). IEEE.

Singh, S. N., & Bhushan, M. (2022). Smart ECG monitoring and analysis system using machine learning. *2022 IEEE VLSI Device Circuit and System (VLSI DCS)*, (pp. 304–309). IEEE.

Suri, R. S., Dubey, V., Kapoor, N. R., Kumar, A., & Bhushan, M. (2022). Optimizing the Compressive Strength of Concrete with Altered Compositions Using Hybrid PSO-ANN. *Information Systems and Management Science: Conference Proceedings of 4th International Conference on Information Systems and Management Science (ISMS) 2021*, (pp. 163–173). IEEE.

Vélez, S., Vacas, R., Martín, H., Ruano-Rosa, D., & Álvarez, S. (2022). High-Resolution UAV RGB Imagery Dataset for Precision Agriculture and 3D Photogrammetric Reconstruction Captured over a Pistachio Orchard (Pistacia vera L.) in Spain. *Data*, *7*(11), 157. doi:10.3390/data7110157

Verma, U., Garg, C., Bhushan, M., Samant, P., Kumar, A., & Negi, A. (2022). Prediction of students' academic performance using Machine Learning Techniques. *2022 International Mobile and Embedded Technology Conference (MECON)*, (pp. 151–156). IEEE. 10.1109/MECON53876.2022.9751956

Vu, M. T., Vuong, Q. H., Nguyen, V. T., Quach, C. H., Truong, N. T., & Pham, M. T. (2019). *Design and Implement Low-cost UAV for Agriculture Monitoring*.

Xu, W., Lan, Y., Li, Y., Luo, Y., & He, Z. (2019). Classification method of cultivated land based on UAV visible light remote sensing. *International Journal of Agricultural and Biological Engineering*, *12*(3), 103–109. doi:10.25165/j.ijabe.20191203.4754

KEY TERMS AND DEFINITIONS

Aerial Monitoring: Aerial monitoring is surveillance that is often carried out by reconnaissance aircraft or unmanned aerial vehicles.

Drones: A craft that flies without a human pilot, flight crew, or passengers is called a drone.

Navigation: It is the practice of locating a ship, aero plane, or other vehicle and directing it to a certain location.

Payload: The weight of a vehicle without the required items for its operation.

Quadcopters: A kind of chopper having four rotors is known as a quadcopter.

Remote Sensing: The method of identifying and keeping track of an area's physical features by measuring it's reflected and emitted radiation from a distance (usually from an orbiting satellite or an aero-plane).

Segmentation: It means to split the subject into pieces, or segments, that can be identified based on some parameters.

Chapter 14
A Survey on Diagnosis of Hazardous Gas Emission Using AI Techniques

N. Madhuram
Puducherry Technological University, India

R. Kalpana
Puducherry Technological University, India

ABSTRACT

Artificial intelligence (AI) techniques play an important role in predicting, diagnosing, and monitoring the emission of hazardous or toxic gases emitted from industries or various places. Also, it helps in continuous monitoring of environmental pollutants based on time series. Leakage of gas or toxic gases leads to health-related issues to human beings. Deep learning is a part of AI, and several deep learning algorithms are used to detect the emission of toxic gases in industries, indoors, refineries, etc. This chapter gives a detailed survey on different sensors and different deep learning algorithms used for diagnosing emission of hazardous gases.

INTRODUCTION

In recent years Artificial Intelligence has been used in real time applications worldwide. It is a field that deals with machine's Intelligence behaviour like Problem solving and learning. With the advancement of science and technology various environmental pollution problems occurs. Hazardous gases such as Volatile Organic Compounds, flammable, explosive gases, etc are emitted during industrial accidents, product manufacturing and transportation (Trivedi P, Purohit D, Soju A, Tiwari RR, 2014). Various Machine Learning and Deep Learning algorithms has been used to diagnosis the emission of hazardous gases both in environment and in Industries. Emission of these gases over a tolerant level may cause severe damage to human health. So, diagnosis on emission of these gases in lower level is more important (Narkhede et al., 2021). There are numerous of Hazardous gases which primarily includes, Hydrogen

DOI: 10.4018/978-1-6684-8785-3.ch014

Copyright © 2023, IGI Global. Copying or distributing in print or electronic forms without written permission of IGI Global is prohibited.

Sulphide, Carbon Monoxide, Nitrogen Oxides, Ozone, Solvents (PennEHRS, 2020). These Hazardous gases are categorised into four types which includes, acutely toxic, corrosive, flammable, dangerously reactive and oxidizing gases as labelled in **Figure 1**.

Figure 1. Types of toxic gases

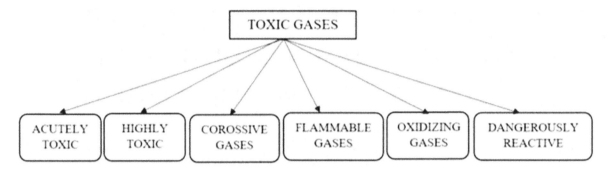

Acutely Toxic Gases

It can be any gas with median lethal concentration of LC50 of 500 ppm or less. These gases lead to combustion while reacting with oxygen as source. Acutely Toxic Gases reacts rapid and violent with combustible substances and flammable vapours. Example of this oxiding gases includes fluorine, chlorine dioxide, nitrogen oxides (NOx), and chlorine trifluoride.

Toxic Gas

Has median Lethal Concentration of LC50 in air of more than 200 ppm and less than 2,000 ppm.

Highly Toxic Gas

Toxic gas with median lethal concentration (LC50) in air of 200 ppm or less.

Corrosive Gases

Corrosive Gases has the ability to burn or destroy the organic tissues and can also corrode metals. Corrosive gases include Ammonia, Chlorine, Hydrogen Chloride and Sulphur Dioxide.

Flammable and Pyrophoric Gases

A flammable gas is gas material which is a gas at 20°C (68°F) and is ignitable at an absolute pressure of 14.7 psi when in a mixture of 13% or less by volume with air this gas has a flammable range at absolute pressure of 14.7 psi with air of at least 12%, regardless of lower limit. Some of the Flammable gases are Acetylene, Butane, Carbon monoxide, Ethylene, and Hydrogen.

Dangerously Reactive

Some of the gases are chemically unstable. If exposed to slight change in temperature or variation in pressure or a mechanical shock, they spontaneously undergo some chemical reactions such as decomposition resulting in fire explosion. Examples of these gases are Methyl acetylene, Vinyl chloride, Tetrafluoroethylene and Vinyl fluoride.

Oxidizing Gases

Includes any gases containing higher atmospheric concentrations greater than 23-25%, Nitrogen oxides, chlorine, and fluorine. Carbon-containing substances such as most flammable gases, flammable and combustible liquids, oils, greases, many plastics, fabrics, finely divided metals, other oxidizable substances such as hydrogen, hydrides, sulphur or sulphur compounds, silicon and ammonia or ammonia compounds leading to fire explosion.

The Architecture for diagnosing emission of hazardous gases is labelled in **Figure 2.**

Figure 2. System architecture

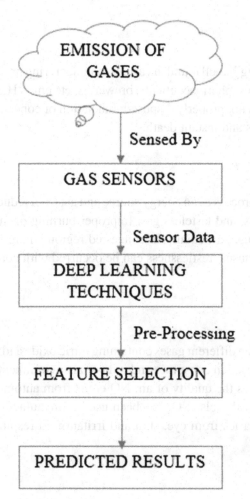

IMPACT OF HAZARDOUS GASES IN HUMANS

Based on industries, we might come in contact with various harmful gases regularly (Us & Gas, 2020). Some of the hazardous gases shown in Figure 3 and its impact in humans are:

Figure 3. Types of hazardous gases

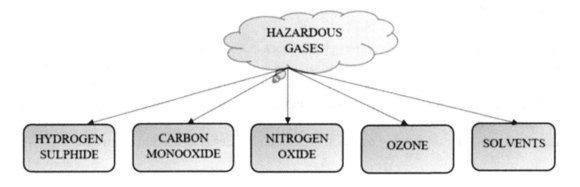

Hydrogen Sulphide

H_2S also identified as "rotten egg" smell found in various manufacturing process and chemical substances. Products like pesticides, plastics, pharmaceuticals, breweries, etc emits H_2S as its by-product. The toxic level of H_2S is very high if it is not properly disposed. Inhalation of concentrated H_2S leads to irritation, unconsciousness, memory loss and instant death.

Carbon Monoxide

Commonly used in industrial process as an energy source and acts as a reducing agent. Carbon monoxide (CO) is a colourless, odourless, and tasteless gas. Improper burning of materials leads to emission of CO which are highly poisonous, particularly in congested region where exposure of humans can't be observed. Health issues like nausea, restlessness can be occurred with continuous exposure of this CO can leads to death.

Nitrogen Oxides

Nitrogen oxides comprises seven different gases containing nitric oxide and nitrogen dioxide as the major forms. These compounds are used in consumer and industrial environments which is a major contribution of air pollution and reduces the quality of air. NO emits from automobiles, cultivating process or as a by-product of ignition fossil fuels. NO_2 has been used in manufacturing of fuels in missiles and in explosives. Health issues varies from eye, skin and irritation in respiratory system around acutely harmful resources.

Ozone

Ozone is mainly harmful to plant life and humans. Ozone is a toxic gas made up of trioxide i.e., three oxygen atoms (O_3) and commonly seen in the form of smog across skyline. It appears in the upper layer of the atmosphere stratosphere, chemical actions emitted from vehicles or gas fuels leads to major contribution of ozone gas at ground level. Health defects such as lowering function of lungs, respiratory infection, and high exposure to UV (Ultra-Violet radiation) leads to skin cancer in humans and makes ozone as a toxic substance which needs continuous monitoring.

Solvents

In summation to highly combustible, the effects of organic solvents are more poisonous in nature. These are carbon-based materials which has the ability to disintegrate or dissipate with one or more other substances and generally exists in gasoline, fossil fuels, paint and grease removers. Excessively combustible at high absorption can infect our central nervous system additional side effects will be giddiness, doziness, absence of attentiveness, confusion, pain in head, unconsciousness and finally leads to death when exposed to high duration of these solvents.

RELATED WORKS

The hazardous gases emitted from industries can cause Synthetic or irradiated damage to the surrounding nature. In addition of rapid advancements in industries and automated systems in chemical plants leakage of gas is a general problem (Zhou et al., 2016). Blasts, fires, leakages, and discharge of waste are few effects of accidents occurred in industries. Generally, number of chemical accidents occurs in a year is unrevealed as many of the accidents are not reported. Information from National Disaster Management Authority (NDMA) says in last ten years 130 chemical incidents has been reported leads to 259 deaths and 563 persons with major injuries(Made, n.d.). Intervention of humans is always impossible in leakage circumstances, because of toxic nature of the gases. Therefore, detecting leakage of gases and explosions within less time period is most important. Timely detection of gas leakage with accurate detection and high reliability using modern techniques is necessarily required with supporting technological solution. Detection of leakage of gas or mixture of gases is also a challenging task and requires technological assistance.

There are standard systematic methods for example optical spectroscopies, gas chromatography and mass spectrometry, which can be utilized to distinguish these gases are complicated, costly and can't be utilized in real-time. In Colorimetric tape a dry material of the tape responds with the gases being released and makes a different stain for different gases. If the concentration of the gas is more there will be more stain on the tape and can be measured in ppm(parts per billion) at low levels (Technology & Guide, n.d.).

Gas Chromatography is another technique that splits mixture of gases and differentiates them using parameters like boiling points, polarity, and vapor pressure (Directions & Reading, n.d.). The mixture of gases is allowed to pass through a stationary liquid or solid phase or gaseous liquid phase. This method is used to separate gaseous mixture like Hydrogen Sulphide (H_2S), Carbon monoxide (CO)and Carbon dioxide (CO_2) and other rare gases but requires huge apparatus and work force to operate (Wang et al., 2018) (Harvey, 2019).

Laser Absorption Spectroscopy is a methodology where different gas molecules absorb specific light spectrum and the amount of energy absorbed by the gas indicates the characteristics of the gaseous mixture (Bogue, 2015). It is used to measure gases that has low concentration levels such as methane, Ammonia, Carbon dioxide and water vapor. This method is expensive, large and requires huge maintenance.

Apart from traditional methods and advancement in technologies, AI based techniques can also be used to detect the emission of gases. Isha Jagtap et al. (Jagtap & Babbar, 2021) Data Mining techniques such as Decision trees, Random Forest, Machine Learning Techniques like Neural Network and Support Vector Machines for the prediction of pollutants in air. The limitations are, it is used to predict air pollutants only and requires more processing time.

Jifeng Chu et al. (Chu et al., 2021) provides a methodology that uses sensor array comprises of four different gas sensors that is used to detect the gaseous mixture of Nitrogen dioxide (NO_2) and Carbon monoxide (CO). BPNN is used to identify the gaseous mixture and Genetic Algorithm (GA) is used for optimization. Their performance measures are insufficient as it fails to detect other gaseous mixture.

Lu Han et.al. (L. Han et al., 2019) designs a method to classify the gaseous mixture by using existing CNN (convolution Neural Network) model. This model is used to classify the mixed gases using CNN for time series classification. However, results in high error rates and decrease in accuracy due to insufficient number of samples.

Yujiao Wu et.al. (Wu et al., 2019) designs a Multiple Hazard Gas Detector for air quality monitoring in residential environments using artificial intelligent methods. The authors compared with machine-learning algorithms such as Support Vector Machines (SVM), k-Nearest Neighbour algorithm (kNN) and SoftMax Regression. Yet, the sensing is less in sophisticated environments and different combination of features.

Khalaf et.al. (Khalaf, 2015) proposes an electronic nose system for classification and concentration of gaseous mixtures in refineries. Least Square Regression method was used for estimation which provides better accuracy and

Pan et al. (Pan et al., 2019) implements a deep learning framework that includes Convolutional Neural Networks (CNN) and Long Short-Term Memory (LSTM) which extracts sequential information from transient response curves. The author proposes an algorithm Fast Gas Recognition that hybrids CNN and Recurrent Neural Networks (RNN). Their performance is compared across machine learning techniques such as SVM, Random Forest, k-nearest neighbours. There are various issues in gas sensor based diagnosing system such as, gas proportion in air is low and the standard gas sensors fails to detect the gases leading to less accuracy.

Another technique used for gas detection is thermal imaging (Hamilton & Charalambous, 2013)(Noacket, 2005). When leakage of gas is detected, the surrounding temperature will be gradually increases than the normal temperature. This temperature increases can be identified and examined by thermal imaging cameras. Shreeram Marathe et al. (Marathe, 2019) proposed a system for leakage of Methane/Ethane gas leakage in pipelines using drones.

The following are the primary limitations of traditional methods, hybrid and unimodal methods for the prediction and diagnosing of gas leakages, as determined by the preceding review of current literature:

- Most of the traditional methods requires large apparatus and complex mathematical calculations for the diagnosing of gaseous mixtures.
- In Gas sensor-based identification system, sometimes the gases leaked can't be sensed by the sensors which leads to low accuracy of the system.

- Hybrid methods acquires higher accuracy but, fails to detect the gaseous mixtures of other compounds since there are less samples.

Table 1. Performance of different methodologies

ANALYSIS TYPE	SENSORS USED	METHODOLOGY	PERFORMANCE	DEMERITS
Predicting Pollutants in air	PM2.5 Sensor	Decision Trees, Neural Network, Support Vector Machines, Random Forest	ACC 91.06%	It is used to predict air pollutants only and requires more processing time.
Identification of gas mixtures of CO and NO_2	MOS Sensors	Principal component analysis (PCA), C-means clustering and Back Propagation Neural Network (BPNN)	ACC 100%	Used to predict only mixtures of CO and NO_2.
Identification of Mixed Gas with time series classification	MOS Sensors	CNN, ResNet and VGG-NET	ACC 96.67%	High error rates and decrease in accuracy due to insufficient number of samples.
Indoor Air Quality Monitoring	IR Sensors	Support Vector Machines (SVM), k-Nearest Neighbour algorithm (kNN) and SoftMax Regression	ACC 99.33%	Sensing is less in sophisticated environments.
Safety Monitoring at Refineries	Gas Sensors	Least Square Regression	ACC 95.63%	Failed to predict for other gaseous concentration.
Toxic Gas Recognition	MOS Sensors	Hybrid Convolutional and Recurrent Neural Network	ACC 93.2%	The standard gas sensors fail to detect various gases and its proportion in air.

SENSORS USED

There are several gas sensors that can be used to predict and diagnose the emission of gases.

Sensing Element (Gas Sensor)

An electronic device used to detect and identify different types of gas mainly, to measure the concentration of the toxic gas emitted and detect the leakage of those gases (Images, 2019). Many technologies have been used for production of gas sensors based on these different sensor detecting measures can be used. Each gas sensors have some specific characteristics which can be used for particular environment and gases (Torres-Martínez et al., 2019). Most commonly five sensing principles are used for the emission:

Catalytic Sensors

This kind of gas sensor measures the difference in temperature between two beads (may be a catalytic substance) in different heat responses (Padvi et al., 2021). Catalytic Bead Method as in Fig.4 is mainly used for detecting combustible gases like methane, propane, butane, and hydrogen. Particularly this method is used for detecting hydrocarbons of Low Explosive Level (LEL). The advantages are Less expensive and detects most of the combustible gases (*Understanding How A Catalyst Sensor Works | GDS Corp*, n.d.). The limitations of these sensors are consuming some electric power for heating, loses its sensitivity when exposed to high concentration of gases and needs to be replaced periodically. Recent advancements are Micro Electromechanical Systems based catalytic sensor shows better specificity with high responsiveness and consumption power is low. Ning Lu et. al (N. Lu et al., 2021) designed a methane gas sensor with printed catalytic films which acts as an active filters for monitoring of methane gas.

Figure 4. Catalytic sensor

Electrochemical Sensors

Also known as EC Sensors as in Figure 5 uses oxidization process to produce current in electrodes. The produced current will be amplified and computed by an external circuit (Property et al., 2001). The principle behind working of EC sensors is based on gas diffusion. When the gases pass over the gas chamber, reaches the active electrode and electrochemical reaction takes place depending on the gas type (Baranwal et al., 2022). Some advantages are low power consumption, low cost, and High sensitivity. The drawbacks of these sensors are rise in temperature and if the concentration of gas is high makes electrode toxic. The latest advancement is Black Phosphorous based electrochemical sensors where synthesis, enhancement strategy for stability were discussed (Li et al., 2021).

Figure 5. Electrochemical sensor

Photoionization Sensors

This type of sensors as in Figure 6 uses light ionization properties of gases to produce electric current coming from positive and negative ions (Szulczyński & Gębicki, 2017). Mainly used for detecting harmful Volatile Organic Compounds present in the environment. Advanced Proportional Integral and derivative (PID) solutions are efficient to compute organic compounds concentration ranges from 1-10ppb and have less response time (Spinelle et al., 2017). Researchers has used Photoionization sensors for monitoring Volatile Organic pollutants in Ambient air (Bílek et al., 2022).

Figure 6. Photoionization sensors

Infrared Sensors

The concept of this sensor as in Figure 7 is based on the relation between actively absorbed wavelength with the stored wavelength that are not absorbed by the gases (Aldhafeeri et al., 2020). The Non-Dispersive InfraRed (NDIR) detectors is the methodology for calculating the concentrations of carbon oxides such as CO and CO_2 (Jia et al., 2019). This method can be used for the gases that absorbs infrared radiations of familiar wavelengths. Recent advancement includes a gas detection system that analysis the gases emitted during automobile exhausts (Xu et al., 2022).

Figure 7. Infrared sensor

MOX Semiconductor Sensors or Chemical Sensors

Metal Oxide sensors in Fig.8 works on resistive principle. The heated surface of metal oxide changes its electrical resistance based on the oxygen content on its surface (Dey, 2018). These sensors can be useful in gas detection like monoxide, ammonia, nitrogen oxides, and carbon dioxide. MOX sensors are highly sensitivity and can be used for the determination of small concentrations (Lin et al., 2019). These sensors are easy to deploy, low cost and can detect the gas leaks at fewer level over a large area (Pashami et al., 2012). Recent Methods uses Calibration Time Reduction for Metal Oxide Semiconductor Gas Sensors using Deep learning and Transfer Learning (Robin et al., 2022).

Figure 8. Metal oxide sensor

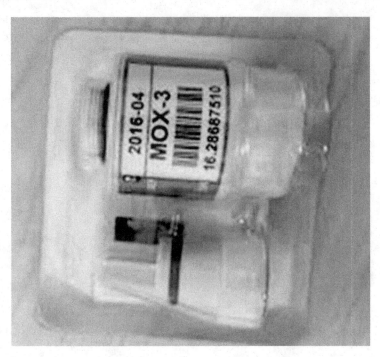

Carbon Nanotubes Gas Sensors

Carbon Nanotubes as in Figure 9 are attractive sensors possess highly sensitive and has unique characteristics. These sensors are highly responsive to small quantity of gases such as ammonia, carbon dioxide, alcohol, and nitrogen oxide at room temperature (Rana et al., 2017). The principle behind CNTs depends on nanotechnology where the properties of nanotubes can be changed or adjusted by altering the size, shape, Synthesis, and reaction conditions (T. Han et al., 2019).

Figure 9. Single walled and multi-walled carbon nanotubes

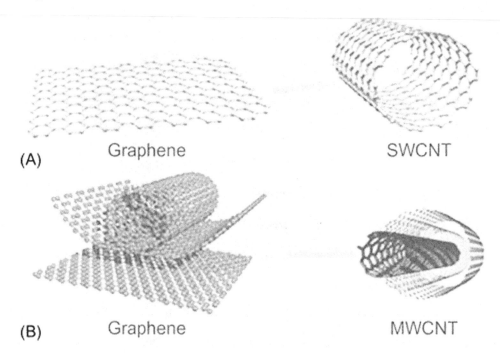

The following table shows the comparison of different sensors, the types of gases used to detect, its merits and applications.

Table 2. Comparison of different sensors and its applications

Sensor Types	Types of gas	Merits	Applications
Catalytic	Combustible Gases	Long life, Less sensitivity to Temperature, Pressure and Humidity changes, Fast Response.	Portable Gas monitors,
Electrochemical	Electrochemically active gases	High Sensitivity and easy to handle.	Agriculture, Food, Oil Industries, Environmental and Biomedical applications.
Photoionization	Volatile Organic Compounds	Very low detection level and can detect large number of substances.	Monitor potential VOCs exposure in industries and detects leakage of hazardous materials, oil, or gasoline.
Infrared Sensors	CO_2 and Hydrocarbons	High Accuracy and Selectivity, Low maintenance, highly resistant to chemical poison.	Gas warning devices, Gas analysers, medical gas measurement technology, Flame detectors
MOX Semiconductor	Combustible and Oxidizing gases	Low cost, Short Response Time, Wide Range, Long life time.	Industrial Applications.
Carbon Nanotubes	NO_2, ammonia, Volatile Organic Compounds	Quick Response Time, Low weight, Highly Sensitive and absorptive capacity	Detection of partial Discharge

DIFFERENT TYPES OF ALGORITHMS

The different type of algorithms, their working and their applications in chemical industry field are as follows.

Decision Trees

A supervised Machine learning algorithm used for classification problems. The decision tree is organized in tree structure shown in Figure 10 which contains leaf nodes, a root node and branches (decision nodes). This algorithm is used for making decisions and their consequences (Somvanshi, 2016). Mostly utilized in making decisions like which strategy is best to achieve the targeted goal. Researchers has successfully used hybrid decision trees (H. Lu & Ma, 2020) to predict the quality of water.

Figure 10. Decision trees

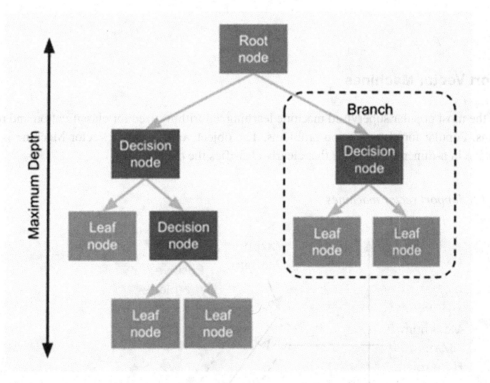

Random Forest

A supervised machine learning algorithm for both classification and regression problems. The algorithm starts with selecting of random samples from the given dataset and construct decision trees accordingly as in Figure 11. Then the prediction results are obtained from every decision tree and voting will be performed for the obtained predicted results. Finally, maximum voted predicted result will be selected as final predictive result (Tyralis et al., 2019). G. Uganya et al. (Uganya et al., 2022) uses machine learning algorithms Random Forests for predicting waste with IoT based Intelligent waste Management System.

Figure 11. Random forests

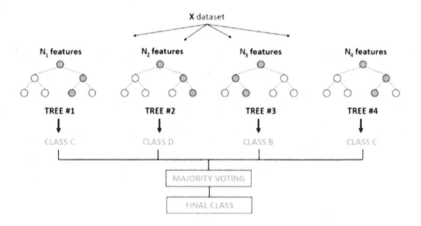

Support Vector Machines

One of the most popular supervised machine learning algorithms used for classification and regression problems. Popular for classification problems. The objective of Support Vector Machine is to find a hyperplane in n-dimensional space that clearly classifies the data point.

Figure 12. Support vector machines

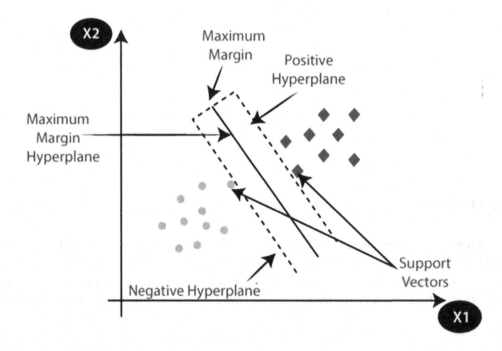

Hyperplanes are the decision boundaries that classifies the data points as in Figure 12. Support vectors are nothing but the data points that are closest to the hyperplanes which affects the position of hyperplane. The best hyperplane is the one which represents largest separation or margin between two classes (Vectors, n.d.). Shisheng Chen et al. (Chen et al., 2018) proposes a time series prediction model for CO2, TVOC and HCHO using different machine learning algorithms at different sampling points.

Artificial Neural Networks

Artificial Neural Networks inspired from Biological neural networks that builds the structure of human brain, which contains inputs, nodes, weights, and outputs. Any Artificial neural network consists of three layers an input layer, the hidden layer, and an output layer as in Figure 13. Input layer accepts the input in different formats given by the programmer.

Figure 13. Artificial neural networks

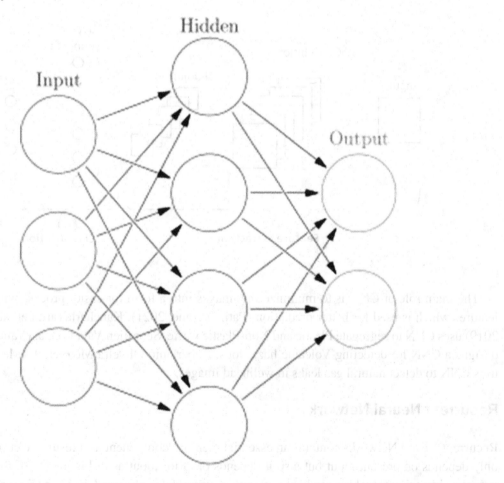

Hidden Layer performs all calculations and identifies hidden patterns and features and presents between input and output layer. The output layer where the desired predictions and classifications are obtained (Park & Lek, 2016). Rongxiao Wang et al. (Qiu et al., 2018) uses ANN model to predict the atmospheric dispersion of hazardous gases and hybrid algorithm is used to improve estimation accuracy.

Convolution Neural Network

Each image captured by Thermal camera has some non-linear features that are digitally stored in RGB format (Havens & Sharp, 2015). Simple Neural network can't have the capability to generalize the complex pattern in images. Convolution Neural Networks adept to precise differences in images and their patterns. CNN layer includes, Convolution, Max-Pooling, Flattening, and ANN layers as in Fig.14. CNN finds or extracts features in images using feature extractor and constructs a feature map (O'Shea & Nash, 2015).

Figure 14. Convolutional neural networks

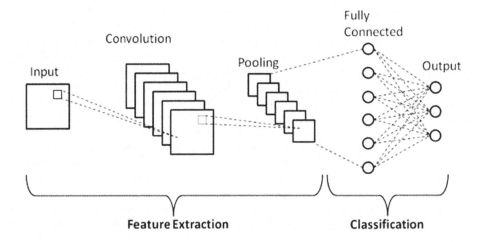

The main role of CNN is to minimize the images into a form for easier process without losing its features which is used for better prediction (Patil & Rane, 2021). KancharlaTarun et. al. (Tarun et al., 2019) uses CNN to segregate Plastic and Non-Plastic waste, Sébastien Valade et. al (Valade et al., 2019) designs a CNN for detecting Volcanic hazardous. Roberlânio Oliveira Melo et al. (Melo et al., 2020) uses CNN to detect natural gas leaks in wellhead images.

Recurrent Neural Network

Recurrent Neural Networks contains an essential memory component as a result the current output not only depends on present input but also it depends on prior input as in Figure 15. If, the size of input sequence increases vanishing gradient problem occurs and it is observed during backpropagation (Jafari et al., 2020).

Figure 15. Recurrent neural network

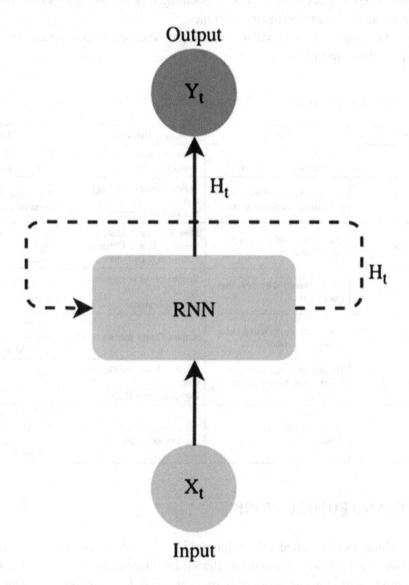

Dongseok Kwon et al. (Kwon et al., 2021) proposed an reliable low power gas sensing model using RNN for sensing Nitrous oxide and Hydrogen Sulphide gas emissions. To overcome these problems, variations of Recurrent Neural Networks known as LSTM were introduced.

LSTM (Long-Short Term Memory) consists of gates and memory components. The gate in LSTM helps to regulate and extract information from the input and allows gradients to succeeding nodes thus allowing new sequence to be trained as equal as the prior sequence and categorise learning (Van Houdt et al., 2020).

LSTM are more efficient than Conventional Recurrent Neural Networks. Surbhi Kumari et al. (Kumari & Singh, 2022) proposes a time series-based machine learning model for the prediction of effective CO_2

emissions in India. Doga Cagdas Demirkan et al. (Demirkan et al., 2022) uses modified LSTM model for predicting methane gas in underground coal mines.

The Following table shows the comparison of different Machine Learning and Deep Learning techniques and its applications in Industries.

Table 3. Comparison of different machine learning and deep learning techniques

Techniques	Merits	Demerits	Applications
Decision Trees	Easy to forecast. No effect on order of training.	Errors in training set leads to wrong final decision.	Can be used for Predictions like, Quality of water, etc.,
Random Forest	Provides Accurate Predictions. Parameters can be set easily.	Training complexity can be high. Not very interpretable.	Predicting Waste, Weather Forecast, etc.,
Support Vector Machines	Durable to noise. Avoids overfitting.	Slow Training Process. Determination of Optimal Parameter is Difficult.	Used as Time series Prediction Model for Detecting Hazardous Gases.
Artificial Neural Networks	Can be classified by more than one output.	Requires more computation power. Time consuming	Prediction of atmospheric dispersion of hazardous gases
Convolution Neural Network	Filters help in extracting exact from the input data	Requires Large data for training.	Detecting Volcanic hazardous and natural gas leaks in wellhead images.
Recurrent Neural Network	Handles any length of Input and the model size remains Constant.	Leads to Vanishing Gradient problem. Computation is High.	Can be used for sensing gas emissions in Industries.
LSTM	Used for Time-Series Predictions.	Takes longer to train and requires more memory.	Used in Time Series Classification of data. Prediction of Environmental Pollutants, etc.,

CONCLUSION AND FUTURE WORK

Deep Learning techniques has ignited a lot of interest in each field where traditional machine learning methodologies were used previously. Finally Artificial Intelligence and deep learning techniques have vast potential to produce more accurate results across various fields. Another Technique used for detecting the emission of gases is Transfer Learning can be defined as the capability of a model to learn or transfer knowledge from one task to another Several Transfer Learning techniques used across various applications like Medical, Environmental, Industrial, etc. In transfer learning the model is trained using mass volume of data and learns the bias and weights during the training process. Again, these weights are transferred to different networks for training a model rather than training from the base (R. & R., 2022). Several Pre-trained models are used to identify different tasks and does not need training from the base. These models are more useful when there are less training samples. The pre-trained models can help with generalization of neural networks and step-up convergence. J. Martinka et al. (Martinka et al., 2022) uses different CNN architectures to identify a burning substance and its accuracy under laboratory conditions. Maqsood Ahmed et al. (Ahmed et al., 2022) proposes a model P-CNN for estimating PM2.5 Concentrations. Applications of these techniques will help in improving manufacturing process in chemical industries, diagnosis the emission of gases at earlier stage, quality control, increases production speed, public safety, etc. Artificial Intelligence and Machine Learning changes the way of thinking and working gradually. Deep Learning Techniques will continue to be in trend for current and future research and development process, with its rapid development process.

REFERENCES

Ahmed, M., Xiao, Z., & Shen, Y. (2022). Estimation of Ground PM2.5 Concentrations in Pakistan Using Convolutional Neural Network and Multi-Pollutant Satellite Images. *Remote Sensing (Basel)*, *14*(7), 1735. doi:10.3390/rs14071735

Aldhafeeri, T., Tran, M. K., Vrolyk, R., Pope, M., & Fowler, M. (2020). A review of methane gas detection sensors: Recent developments and future perspectives. *Inventions (Basel, Switzerland)*, *5*(3), 1–18. doi:10.3390/inventions5030028

Baranwal, J., Barse, B., Gatto, G., Broncova, G., & Kumar, A. (2022). Electrochemical Sensors and Their Applications: A Review. *Chemosensors (Basel, Switzerland)*, *10*(9), 363. doi:10.3390/chemosensors10090363

Bílek, J., Maršolek, P., Bílek, O., & Buček, P. (2022). Field Test of Mini Photoionization Detector-Based Sensors—Monitoring of Volatile Organic Pollutants in Ambient Air. *Environments - MDPI, 9*(4). doi:10.3390/environments9040049

Bogue, R. (2015). Detecting gases with light: A review of optical gas sensor technologies. *Sensor Review*, *35*(2), 133–140. doi:10.1108/SR-09-2014-696

Chen, S., Mihara, K., & Wen, J. (2018). Time series prediction of CO2, TVOC and HCHO based on machine learning at different sampling points. *Building and Environment*, *146*(July), 238–246. doi:10.1016/j.buildenv.2018.09.054

Chu, J., Li, W., Yang, X., Wu, Y., Wang, D., Yang, A., Yuan, H., Wang, X., Li, Y., & Rong, M. (2021). Identification of gas mixtures via sensor array combining with neural networks. *Sensors and Actuators. B, Chemical, 329*, 129090. doi:10.1016/j.snb.2020.129090

Demirkan, D. C., Duzgun, H. S., Juganda, A., Brune, J., & Bogin, G. (2022). Real-Time Methane Prediction in Underground Longwall Coal Mining Using AI. *Energies, 15*(17), 1–12. doi:10.3390/en15176486

Dey, A. (2018). Semiconductor metal oxide gas sensors: A review. *Materials Science and Engineering B: Solid-State Materials for Advanced Technology, 229*(December 2017), 206–217. doi:10.1016/j.mseb.2017.12.036

Directions, F., & Reading, F. (n.d.). Gas Chromatography and Gas Chromatography } Mass Overview of Derivatization of Sugars for GC, GC-MS or GC-MS-MS Analysis. *Journal of Capillary Electrophoresis*, 2211–2223.

Hamilton & Charalambous. (2013). *Leak Detection Technology and Implementation*. IWA Publishing.

Han, L., Yu, C., Xiao, K., & Zhao, X. (2019). A new method of mixed gas identification based on a convolutional neural network for time series classification. *Sensors (Basel)*, *19*(9), 1960. doi:10.339019091960 PMID:31027348

Han, T., Nag, A., Chandra Mukhopadhyay, S., & Xu, Y. (2019). Carbon nanotubes and its gas-sensing applications: A review. *Sensors and Actuators. A, Physical*, *291*, 107–143. doi:10.1016/j.sna.2019.03.053

Harvey, D. J. (2019). Gas chromatography | Gas chromatography/mass spectrometry. Encyclopedia of Analytical Science, 169–179. Science Direct. doi:10.1016/B978-0-12-409547-2.14103-4

Havens, K. J., & Sharp, E. J. (2015). Thermal Imaging Techniques to Survey and Monitor Animals in the Wild: A Methodology. In Thermal Imaging Techniques to Survey and Monitor Animals in the Wild: A Methodology.

Getty Images. (2019). *What is a Gas Sensor ? Micron ships 232-layer NAND SSD for PCs, laptops.* Getty Images.

Jafari, R., Razvarz, S., Gegov, A., & Vatchova, B. (2020). Deep Learning for Pipeline Damage Detection: An Overview of the Concepts and a Survey of the State-of-the-Art. *2020 IEEE 10th International Conference on Intelligent Systems, IS 2020 - Proceedings*, (pp. 178–182). IEEE. 10.1109/IS48319.2020.9200137

Jagtap, I., & Babbar, N. (2021). Predicting Air Pollutant using Data Mining and Machine Learning Algorithms. *Journal of Science and Technology*, *06*, 25–30. doi:10.46243/jst.2021.v6.i04.pp25-30

Jia, X., Roels, J., Baets, R., & Roelkens, G. (2019). On-chip non-dispersive infrared CO2 sensor based on an integrating cylinder†. *Sensors (Basel)*, *19*(19), 1–14. doi:10.339019194260 PMID:31575053

Kumari, S., & Singh, S. K. (2022). Machine learning-based time series models for effective CO2 emission prediction in India. *Environmental Science and Pollution Research International*, *0123456789*. Advance online publication. doi:10.100711356-022-21723-8 PMID:35780266

Kwon, D., Jung, G., Shin, W., Jeong, Y., Hong, S., Oh, S., Bae, J. H., Park, B. G., & Lee, J. H. (2021). Low-power and reliable gas sensing system based on recurrent neural networks. *Sensors and Actuators, B: Chemical, 340*(September 2020), 129258. doi:10.1016/j.snb.2020.129258

Li, Q., Wu, J. T., Liu, Y., Qi, X. M., Jin, H. G., Yang, C., Liu, J., Li, G. L., & He, Q. G. (2021). Recent advances in black phosphorus-based electrochemical sensors: A review. *Analytica Chimica Acta, 1170*(xxxx), 338480. doi:10.1016/j.aca.2021.338480

Lin, T., Lv, X., Hu, Z., Xu, A., & Feng, C. (2019). Semiconductor metal oxides as chemoresistive sensors for detecting volatile organic compounds. In Sensors (Switzerland), 19(2). doi:10.339019020233

Lu, H., & Ma, X. (2020). Hybrid decision tree-based machine learning models for short-term water quality prediction. *Chemosphere*, *249*, 126169. doi:10.1016/j.chemosphere.2020.126169 PMID:32078849

Lu, N., Fan, S., Zhao, Y., Yang, B., Hua, Z., & Wu, Y. (2021). A selective methane gas sensor with printed catalytic films as active filters. *Sensors and Actuators. B, Chemical*, *347*(August), 130603. doi:10.1016/j.snb.2021.130603

Marathe, S. (2019). Leveraging drone based imaging technology for pipeline and RoU monitoring survey. *Society of Petroleum Engineers - SPE Symposium: Asia Pacific Health, Safety, Security, Environment and Social Responsibility 2019*. One Petro. 10.2118/195427-MS

Martinka, J., Nečas, A., & Rantuch, P. (2022). The recognition of selected burning liquids by convolutional neural networks under laboratory conditions. *Journal of Thermal Analysis and Calorimetry*, *147*(10), 5787–5799. doi:10.100710973-021-10903-2 PMID:34177362

Melo, R. O., Costa, M. G. F., & Costa Filho, C. F. F. (2020). Applying convolutional neural networks to detect natural gas leaks in wellhead images. *IEEE Access : Practical Innovations, Open Solutions*, *8*, 191775–191784. doi:10.1109/ACCESS.2020.3031683

Narkhede, P., Walambe, R., Mandaokar, S., Chandel, P., Kotecha, K., & Ghinea, G. (2021). Gas detection and identification using multimodal artificial intelligence based sensor fusion. *Applied System Innovation*, *4*(1), 1–14. doi:10.3390/asi4010003

O'SheaK.NashR. (2015). *An Introduction to Convolutional Neural Networks*. 1–11. https://arxiv.org/abs/1511.08458

Padvi, M. N., Moholkar, A. V., Prasad, S. R., & Prasad, N. R. (2021). A critical review on design and development of gas sensing materials. *Engineered Science*, *15*, 20–37. doi:10.30919/es8d431

Pan, X., Zhang, H., Ye, W., Bermak, A., & Zhao, X. (2019). A Fast and Robust Gas Recognition Algorithm Based on Hybrid Convolutional and Recurrent Neural Network. *IEEE Access : Practical Innovations, Open Solutions*, *7*, 100954–100963. doi:10.1109/ACCESS.2019.2930804

Park, Y. S., & Lek, S. (2016). Artificial Neural Networks: Multilayer Perceptron for Ecological Modeling. In Developments in Environmental Modelling (Vol. 28). Elsevier. doi:10.1016/B978-0-444-63623-2.00007-4

Pashami, S., Lilienthal, A. J., & Trincavelli, M. (2012). Detecting changes of a distant gas source with an array of MOX gas sensors. *Sensors (Basel)*, *12*(12), 16404–16419. doi:10.3390121216404 PMID:23443385

Patil, A., & Rane, M. (2021). Convolutional Neural Networks: An Overview and Its Applications in Pattern Recognition. *Smart Innovation. Systems and Technologies*, *195*, 21–30. doi:10.1007/978-981-15-7078-0_3

PennEHRS. (2020). *SOP: Hazardous and Highly Toxic Gases*. PennEHRS. https://ehrs.upenn.edu/health-safety/lab-safety/chemical-hygiene-plan/standard-operating-procedures/sop-hazardous-and

Property, P., Sensors, G., Flow, S. G., Conductivity, T., Sensors, S., Sensors, G. C., Micro, M., Analyzers, G., Absorption, O. S., Spectrometry, M., Chromatography, G., Analysis, M., Positives, F., & Reading, F. (2001). Gas sensors. In Semiconductor International, 24(6). doi:10.1016/B978-0-12-803581-8.00548-8

Qiu, S., Chen, B., Wang, R., Zhu, Z., Wang, Y., & Qiu, X. (2018). Atmospheric dispersion prediction and source estimation of hazardous gas using artificial neural network, particle swarm optimization and expectation maximization. *Atmospheric Environment*, *178*, 158–163. doi:10.1016/j.atmosenv.2018.01.056

R., S., & R., K. (2022). *Survey or Review on the Deep Learning Techniques for Retinal Image Segmentation in Predicting/Diagnosing Diabetic Retinopathy*. IGI Global. doi:10.4018/978-1-6684-4405-4.ch010

Rana, M. M., Ibrahim, D. S., Asyraf, M. R. M., Jarin, S., & Tomal, A. (2017). A review on recent advances of CNTs as gas sensors. *Sensor Review, 37*(2), 127–136. doi:10.1108/SR-10-2016-0230

Robin, Y., Amann, J., Goodarzi, P., Schneider, T., Schütze, A., & Bur, C. (2022). Deep Learning Based Calibration Time Reduction for MOS Gas Sensors with Transfer Learning. *Atmosphere (Basel), 13*(10), 1614. doi:10.3390/atmos13101614

Somvanshi, M. (2016)... *Iccubea., 2016*, 7860040.

Spinelle, L., Gerboles, M., Kok, G., Persijn, S., & Sauerwald, T. (2017). Review of portable and low-cost sensors for the ambient air monitoring of benzene and other volatile organic compounds. *Sensors (Basel), 17*(7), 1520. doi:10.339017071520 PMID:28657595

Szulczyński, B., & Gębicki, J. (2017). Currently commercially available chemical sensors employed for detection of volatile organic compounds in outdoor and indoor air. *Environments - MDPI, 4*(1), 1–15. doi:10.3390/environments4010021

Tarun, K., Sreelakshmi, K., & Peeyush, K. P. (2019). Segregation of Plastic and Non-plastic Waste using Convolutional Neural Network. *IOP Conference Series. Materials Science and Engineering, 561*(1), 1–7. doi:10.1088/1757-899X/561/1/012113

Torres-Martínez, L. M., Kharissova, O. V., & Kharisov, B. I. (2019). Handbook of ecomaterials. Handbook of Ecomaterials, 1(December 2020), 1–3773. doi:10.1007/978-3-319-68255-6

Trivedi, P., Purohit, D., Soju, A., & Tiwari, R. R. (2014). *Major Industrial Disasters in India., 9*(4), 7.

Tyralis, H., Papacharalampous, G., & Langousis, A. (2019). A brief review of random forests for water scientists and practitioners and their recent history in water resources. *Water (Basel), 11*(5), 910. doi:10.3390/w11050910

Uganya, G., Rajalakshmi, D., Teekaraman, Y., Kuppusamy, R., & Radhakrishnan, A. (2022). A Novel Strategy for Waste Prediction Using Machine Learning Algorithm with IoT Based Intelligent Waste Management System. *Wireless Communications and Mobile Computing, 2022*, 1–15. doi:10.1155/2022/2063372

GDS Corp. (n.d.). *Understanding How A Catalyst Sensor Works.* GDS Corp.

Us, A., & Gas, T. (2020). *5 Types of Toxic Gas & Their Health Effects.* GDS Corp. https://www.gdscorp.com/blog/gas-emission/5-types-of-toxic-gas-their-health-effects/

Valade, S., Ley, A., Massimetti, F., D'Hondt, O., Laiolo, M., Coppola, D., Loibl, D., Hellwich, O., & Walter, T. R. (2019). Towards global volcano monitoring using multisensor sentinel missions and artificial intelligence: The MOUNTS monitoring system. *Remote Sensing (Basel), 11*(13), 1–31. doi:10.3390/rs11131528

Van Houdt, G., Mosquera, C., & Nápoles, G. (2020). A review on the long short-term memory model. *Artificial Intelligence Review, 53*(8), 5929–5955. doi:10.100710462-020-09838-1

Vectors, S. (n.d.). Support Vector Machine — Introduction to Machine Learning Algorithms SVM model from scratch Introduction What is Support Vector Machine ? *Hyperplanes and Support Vectors Cost Function and Gradient Updates.*

Wang, T., Wang, X., & Hong, M. (2018). Gas leak location detection based on data fusion with time difference of arrival and energy decay using an ultrasonic sensor array. *Sensors (Basel)*, *18*(9), 2985. doi:10.339018092985 PMID:30205433

Wu, Y., Liu, T., Ling, S. H., Szymanski, J., Zhang, W., & Su, S. W. (2019). Air quality monitoring for vulnerable groups in residential environments using a multiple hazard gas detector. *Sensors (Basel)*, *19*(2), 362. Advance online publication. doi:10.339019020362 PMID:30658412

Xu, M., Peng, B., Zhu, X., & Guo, Y. (2022). Multi-Gas Detection System Based on Non-Dispersive Infrared (NDIR) Spectral Technology. *Sensors (Basel)*, *22*(3), 1–13. doi:10.339022030836 PMID:35161584

Zhou, Y., Zhao, X., Zhao, J., & Chen, D. (2016). Research on fire and explosion accidents of oil depots. *Chemical Engineering Transactions*, *51*, 163–168. doi:10.3303/CET1651028

Chapter 15
IoVST:
Internet of Vehicles and Smart Traffic – Architecture, Applications, and Challenges

Kavitha Rajamohan
https://orcid.org/0000-0002-7803-8901
CHRIST University (Deemed), India

Sangeetha Rangasamy
https://orcid.org/0000-0002-1850-232X
CHRIST University (Deemed), India

Nikhil Anthony Pinto
CHRIST University (Deemed), India

B. E. Manoj
CHRIST University (Deemed), India

Debanjana Mukherjee
CHRIST University (Deemed), India

Jimit Shukla
CHRIST University (Deemed), India

ABSTRACT

The internet of things (IoT) is the network of sensors, devices, processors, and software, enabling connection, communication, and data transfer between devices. IoT is able to collect and analyze large amounts of data which can then be used to automate daily tasks in various fields. IoT holds the potential to revolutionise and create many opportunities in multiple industries like smart cities, smart transport, etc. Autonomous vehicles are smart vehicles that are able to navigate and move around on their own on a well-planned road.

DOI: 10.4018/978-1-6684-8785-3.ch015

1. INTRODUCTION

The Internet of things is a collaborative network of sensors, devices, processors and required software for connection and data transfer between devices. IoT enables us to collect and analyse large amounts of data which can be used to automate daily tasks in various fields. IoT holds the potential to revolutionise and create many opportunities in multiple industries. IoT plays a crucial role in the vehicles and traffic management domain, mainly known as the Internet of Vehicles and Smart Traffic(IoVST).

Figure 1. Smart traffic management using IoV

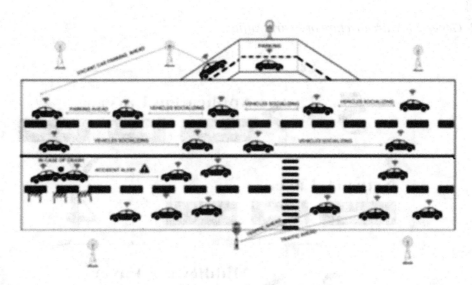

The Internet of vehicles enables autonomous driving in the collective inter-vehicle network using cloud technologies. Data can be shared between vehicles and even the central processors, and using machine learning to refine and update the onboard algorithms. Further, this network can help the vehicles communicate, thereby streamlining traffic flow by broadcasting the proposed movement route. IoV creates a seamless and interconnected transportation system that harnesses the power of data and technology. By linking devices, and vehicles, the IoV thrives to enhance transportation safety such as sensors, communication protocols and data analytics. One of the most prominent applications of the IoV is in developing smart traffic systems. Smart traffic systems utilise data analytics and other technologies to optimise traffic flow and reduce congestion. Through real-time data collection from sources such as sensors, cameras, and GPS services, smart traffic systems monitor traffic patterns and adjust traffic flows, as a result, travel times are reduced, and transportation safety is improved. Combining the IoT and smart traffic systems offers an impactful solution to worldwide transportation challenges. By leveraging data and technology, these systems enhance safety, reduce congestion, and optimise transportation systems to meet the growing demands of an evolving world. Continued developments in IoV and smart traffic systems lead to more efficient, safer, and sustainable transportation systems that benefit everyone. Using measures such as dynamic routing, intelligent traffic signals, smart vehicles, smart car parking and real-time traffic monitoring, implementing smart traffic solutions can lead to reduced emissions, improved air quality, and decreased travel times, contributing to a more sustainable and liveable city. Data analysis helps increase the effectiveness, safety, and sustainability of transportation networks and is called smart traffic management. Figure 1 shows an example of a real-time deployment of Smart Traffic Management using IoV. As shown in the figure, cars will be able to communicate easily with each other, over a wireless network, to increase the efficiency of travel, decrease travel time, and avoid congestion on the road.

2. IOV ARCHITECTURE FOR SMART TRAFFIC

Figure 2. Generic IoV architecture of smart traffic

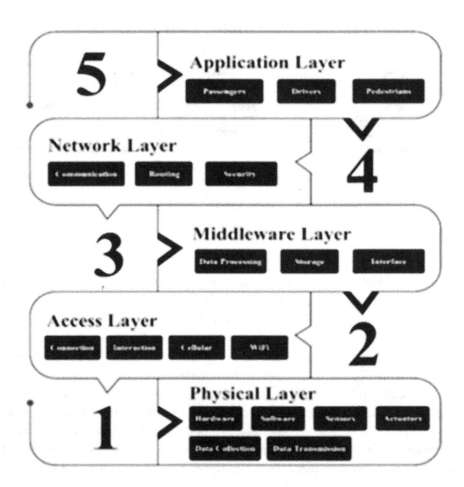

The contribution of IoT in smart traffic is increasing gradually day by day. According to (Elsagheer Mohamed & AlShalfan, 2021), traditional traffic lights continue to face issues despite significant efforts to improve traffic flow in many countries, including industrialised ones. These issues include ineffective traffic management at intersections, lack of intelligence and adaptability, need for signals that respond to the presence or absence of vehicles, vulnerability to environmental factors, absence of systems to prioritise emergency vehicles, and inability to provide dynamic and adaptive services. Knowing the disadvantages of the conventional traffic management system that have already been stated, it is evident that adding intelligence to the system will significantly enhance its functionality by allowing for dynamism and adaptivity. According to (Elsagheer Mohamed & AlShalfan, 2021), the required intelligence can be added as an IoV architecture to manage traffic smartly. According to (Abbasi et al., 2021), architecture is a framework that includes protocols and structure identification for connecting and communicating between various components in a given environment. (Karim et al., 2022) states, the architecture of the Internet of Vehicles (IoV) must address several issues, including security and privacy concerns regarding data exchange, data quality for real-time applications, management of massive amounts of data, interoperability between devices using different technologies and protocols, and decentralisation to balance factors such as latency, resource constraints, and load balancing in planned networks. An IoV architecture typically consists of five layers, including the application, network, middleware, access, and physical layer, as depicted in Figure 2. The application layer provides various services and applications to drivers, passengers, pedestrians, and other stakeholders. This layer processes and analyses the data collected from the lower layers and uses it to offer services such as navigation, traffic management, and remote diagnostics. The network layer manages the communication protocols and routing algorithms for different vehicular networks. It ensures reliable and secure data transmission between vehicles and other devices or infrastructures. The middleware layer provides common functionalities and interfaces for data processing, storage, security, and other services. It allows the different layers to communicate with each other seamlessly. The access layer enables the connection and interaction between vehicles and other devices or infrastructures using various technologies such as cellular networks, Wi-Fi, and dedicated short-range communication (DSRC). The physical layer defines the hardware and software specifications for vehicles and devices. It includes sensors, actuators, communication modules, and other components that enable data collection, processing, and transmission. In addition to the five layers, an IoV architecture also includes a protocol stack that specifies the management, operational, and security plans for each layer. The protocol stack ensures that the different layers work together seamlessly and that data is transmitted reliably and securely. Finally, an IoV architecture can be modelled based on three network elements: cloud, connection, and client.

Figure 3. Elements of IoV architecture for smart traffic

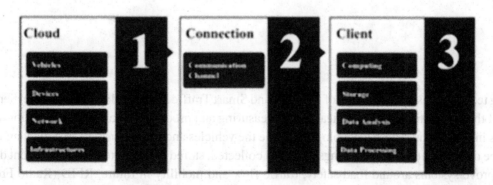

As seen in Figure 3, the cloud element provides computing and storage resources for data processing and analysis. The connection element provides the communication channels for data transmission, and the client element includes the different vehicles, devices, and infrastructures that exchange data over the network.

3. BASIC TECHNOLOGIES

IoT involves connecting physical objects to the Internet, allowing them to exchange data and communicate with each other. Some essential IoT technologies include sensors, connectivity options like Wi-Fi, Bluetooth, cloud computing, data analytics, security measures and machine learning. These technologies work together to enable IoT devices to collect, process and analyse data allowing organisations to make better decisions and optimise performance. The classification of sensors can be based on their specifications, conversion method, type of material used, sensing of physical phenomenon, and properties.

3.1 Sensing technologies

Figure 4. Sensing technologies in smart vehicle

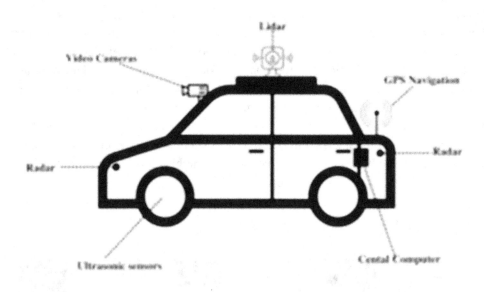

Sensing technologies in the Internet of Vehicles and Smart Traffic systems play a crucial role. Sensors are a part of the automation of any application by measuring and processing the collected data for detecting changes in conditions. These technologies enable the vehicles and traffic systems to collect and process real-time data about their surroundings, which is collected, stored and used to make intelligent decisions and improve systems around road safety, traffic flow and mobility in future. RFID (Radio Frequency Identification sensor) sensors can be attached to vehicles to track their location and movements in real-time, allowing traffic management systems to monitor and optimise traffic flow, reduce congestion and improve safety. Radar (Radio Detection and Ranging sensor) detect objects' distance, speed, and direction, which can be used to improve road safety and traffic flow. They are classified based on their range and accuracy. Radar consists of a transmitter and receiver. The transmitter produces electromagnetic waves which determine the direction and distance and other information by the reflected radio waves which were reflected back by the objects. The figure 4 displays the placement and application of sensing technologies in smart vehicles. An ultrasonic sensor is a piece of technology that uses ultrasonic sound waves to measure a target object's distance and then turns the sound that is reflected back into an electrical signal. Most often, proximity sensors are combined with ultrasonic sensors. They are present in anti-collision safety systems and self-parking automotive technologies. Robotic obstacle detection systems and manufacturing technology both use ultrasonic sensors. Ultrasonic sensors in proximity sensing applications are less prone to interference from smoke, gas, and other airborne particles than IR sensors though the physical components are still affected by variables such as heat. LiDARs are very accurate in determining an object's position but significantly less accurate than radars in measuring their velocities. LiDAR has good directivity and high measuring precision and is unaffected by road clutter. It is also used to gather information about the environment around it. LiDAR can gather information about road traffic and conditions. This information can be used to optimise traffic flow, reduce traffic congestion, and improve safety on the roads.GPS(Global Positioning System) plays a crucial role in smart traffic management and autonomous driving by providing accurate and reliable information about the location and movement of the vehicles on the roads, which is all collected and analysed to make better decisions and improve the safety on the roads. The quality of shipping conditions can be improved by monitoring the number of vibrations and strokes. Items can be traced easily using GPS-like systems. Environmental sensors gather information about weather conditions, road surface conditions and other factors like wind, rain, snow, and fog. Environmental sensors provide information about the road environment and collect and analyse it to make informed decisions and improve road safety and efficiency.

3.2 Communication Technologies

Communication technologies enable Vehicles, Traffic management systems, and other connected devices to connect one another and facilitate real-time, allowing for more efficient and safer operation on roads. Key Communication technologies used in IoVST includes Vehicle-to-Vehicle (V2V) Communication, Vehicle to Infrastructure(V2I) Communication and 5G. These technologies enables vehicles to communicate with each other directly with the feature to exchange information about road conditions, traffic congestion and other factors. All other vehicle on-board systems can use this information to make decisions based on the information received to control the vehicle's motion and improve road safety and efficiency. The representation of this communication can be observed in the Figure 5. These technologies also provides low latency and high-speed communication between traffic management systems, vehicles and other connected devices.

Figure 5. Communication technologies in IoVST

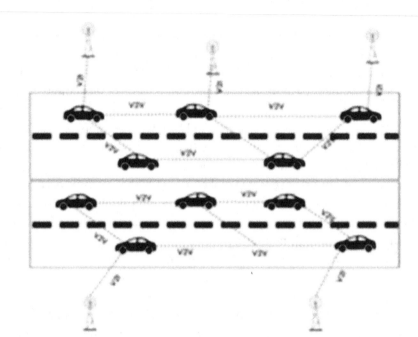

3.3 Learning for the Future

A rapid development offers a considerable potential for learning and professional development in the IoVST domain. Navigation, perception, decision-making, and communication technologies are among the crucial areas of concentration and development. The need for real-time traffic updates, improved road safety and efficiency, and congestion reduction are all made possible by optimising traffic management systems to improve road travel efficiency, it is also increasingly important to comprehend how autonomous vehicles communicate with one another and with traffic control systems.

We must be aware of the security risks and difficulties that these systems encounter to rely on linked technologies in the Internet of Things and Smart Traffic. We must then design strategies to address these issues and guarantee the security and privacy of connected devices. Massive amounts of data generated by IoV's and traffic management systems must be examined in order to get precise insights and make wise decisions to increase road efficiency and safety. IoVST is a fascinating area for future research and development thanks to developments in Big Data, AI, and IoT that have greatly increased the potential for minimising problems with road traffic management.

4. ARTIFICIAL INTELLIGENCE FOR SMART TRAFFIC

Artificial Intelligence (AI) is a rapidly growing field that has the potential to revolutionise many

Figure 6. Artificial intelligence algorithms for IoVST

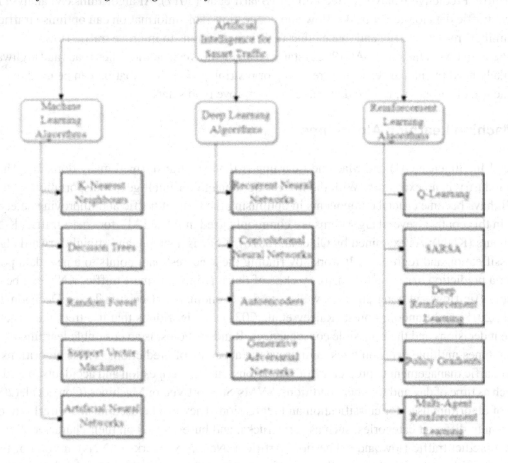

industries, including smart traffic. AI can be used in smart traffic to improve safety, reduce congestion, optimise traffic flow, and enhance the driving experience. Figure 6 gives a bird's eye view of various available artificial intelligence algorithms for IoV and smart traffic control. Here are some ways that AI is being used in transport domain.

1. Autonomous Vehicles: According to Ranka et al. (2020), AI algorithms are the backbone of autonomous vehicles. These algorithms are used to perceive the environment, make decisions, and control the vehicle's movements. With the development of autonomous vehicles, it is expected that the smart traffic industry will become safer, more efficient, and more convenient.
2. Object Detection and Recognition: According to Amitha & Narayanan (2021), AI algorithms can detect and recognize objects, such as vehicles and road signs, in real-time video footage captured by cameras mounted on roads and highways. This information can improve road safety, optimise traffic flow, and enhance the driving experience.
3. Intelligent Transportation Systems: According to Kaluvan et al. (2021), AI algorithms can be used to power intelligent transportation systems (ITS) that help to manage and optimise traffic flow. For example, AI algorithms can predict and prevent congestion, coordinate the behaviour of autonomous vehicles, and optimise traffic signal timings.

4. Traffic Predictive Analytics: According to Sharif et al. (2017), AI algorithms can analyse historical traffic data to predict traffic flow and congestion. This information can optimise traffic signal timings, reroute traffic, and make other real-time decisions to improve traffic flow.
5. Predictive Maintenance: DAC (2022) states AI algorithms can predict when roads and highways will likely need maintenance, such as repaving or restriping. This information can be used to optimise the use of resources, reduce downtime, and improve road safety.

4.1 Machine Learning Algorithms

Artificial Intelligence (AI) and Machine Learning (ML) have transformed many domains. The smart traffic industry is no exception. With the increasing demand for intelligent transportation systems, AI and ML have become crucial components in optimising the flow of traffic and improving safety on the roads. In this context, several algorithms are commonly used in AI and ML for smart traffic. K-Nearest Neighbours (KNN): As examined by Oh et al. (2016), KNN is a simple and straightforward algorithm for classification and regression. It works by finding the k-nearest data points to a new data point and making a prediction based on the majority class of those points. In smart traffic, KNN can be used to predict traffic flow and congestion, as well as to detect anomalies. Decision Trees: Decision Trees(R N et al., 2018) and Random Forest (Rajeev et al., 2021) are algorithms that use tree-like structures to represent decisions and their possible consequences. Random Forest is an ensemble learning algorithm that combines multiple decision trees to improve the accuracy of predictions. These algorithms can be used in traffic management to predict traffic conditions, such as congestion and accidents, based on factors such as time of day and weather conditions. SVMs-Support Vector Machines (Cong et al. (2016) are a type of algorithm used for classification and regression. They can be used in smart traffic to classify vehicles into different categories, such as cars, trucks, and buses, based on their characteristics. SVMs can also predict traffic flow and congestion. Artificial Neural Networks (ANNs) are one of the most popular algorithms used in AI and ML for traffic management (Ashifuddin Mondal & Rehena, 2019). They are modelled after the human brain and are designed to learn and make decisions based on input data. ANNs can predict traffic flow, detect anomalies, and classify traffic patterns.

4.2 Deep Learning Algorithms

Deep Learning (DL) algorithms are a subset of AI &ML that are specifically designed to handle large and complex data sets. These algorithms have proven highly effective in various applications, including the smart traffic industry. Here are some of the most commonly used DL algorithms in AI for smart traffic are Recurrent Neural Networks (RNNs), Convolutional Neural Networks (ConvNets or CNNs), Autoencoders, Generative Adversarial Networks (GANs). RNNs are a type of neural network that is designed to process sequences of data such as time-series data(Tian & Pan, 2015). In smart traffic, RNNs can predict traffic flow, congestion, and travel time, considering historical data and patterns. CNNs are deep neural networks designed to analyse and classify images(Kurniawan et al., (2018). They can be used in smart traffic to detect and recognize objects, such as vehicles and road signs, in real-time video footage captured by cameras mounted on roads and highways. Autoencoders are neural networks for unsupervised learning(Wei et al., 2019),. They can be used in smart traffic to identify patterns in traffic data, such as detecting congestion, estimating travel time, and identifying anomalies. GANs are a type of deep learning algorithm that involves two neural networks competing against each other to generate realistic data(Chen et al. (2021). In smart traffic, GANs can generate synthetic traffic data for training and testing machine learning models.

4.3 Reinforcement Learning Algorithms

Reinforcement Learning (RL) is a type of AI &ML that involves training agents to make decisions and take actions in an environment to maximise a reward signal. In the smart traffic industry, RL algorithms can optimise traffic flow and reduce congestion, making the roads safer and more efficient. Here are some of the most commonly used RL algorithms in AI for smart traffic:

- Q-Learning: It is a type of RL algorithm that involves learning the best action to take in a given state to maximise a reward signal. Q-Learning can be used in smart traffic to regulate the timing of traffic signals, regulate the speed of autonomous vehicles, and make other real-time decisions to enhance traffic flow(Agafonov & Myasnikov, 2021).
- SARSA: A type of RL algorithm similar to Q-Learning, but it uses the expected value of the following state and action to update the current state-action value(Kekuda et al., 2021). In smart traffic, SARSA can optimise traffic signal timings, control the speed of autonomous vehicles, and make other real-time decisions to improve traffic flow.
- Deep Reinforcement Learning (DRL): Jang et al. (2018) state DRL is a type of RL algorithm that involves training deep neural networks to make decisions and take actions in an environment. DRL is a method that enables smart traffic management by fine-tuning the duration of traffic signals, modulating the velocity of autonomous vehicles, and making other instantaneous decisions to enhance traffic flow.
- Policy Gradients: It is a type of RL algorithm that directly optimises the policy that maps states to actions rather than estimating the value of each state-action pair (Mousavi et al., 2017). They can be applied to smart traffic problems, such as adjusting traffic signal timings, regulating the speed of autonomous vehicles, and making other real-time decisions that enhance traffic flow.
- Multi-Agent Reinforcement Learning (MARL): According to Vidhate & Kulkarni (2017), MARL is a type of RL algorithm that involves training multiple agents to cooperate and compete in an environment. In smart traffic, MARL can be used to coordinate the behaviour of autonomous vehicles and optimise traffic flow in real time.

4.4 Case Study

One example of AI being used in smart traffic is the application of AI in managing traffic in the city of Moscow, Russia as found by Yuloskov et al. (2021). In 2018, the Moscow government launched a program to reduce congestion and optimise traffic flow in the city. The program involved the deployment of AI algorithms and cameras at intersections to monitor traffic flow in real-time. The AI algorithms analyse the video footage to predict traffic flow and congestion. Based on this information, the algorithms make real-time decisions to optimise traffic signal timings, reroute traffic, and control the speed of vehicles. The program has been highly successful in reducing congestion, improving traffic flow, and enhancing road safety in the city.

Another example of AI being used in smart traffic is the application of AI in the management of autonomous vehicles. In recent years, several companies have developed autonomous vehicles that use

AI algorithms to perceive the environment, make decisions, and control the vehicle's movements. These vehicles can potentially revolutionise the smart traffic industry by improving safety, reducing congestion, and enhancing the overall driving experience. For example, Waymo as described by Muiruri (2021), a subsidiary of Alphabet Inc., has developed a self-driving taxi service that uses AI algorithms to power its vehicles. The service has been tested in several cities and has received positive feedback from riders for its safety and convenience.

The case study report (AI Integration for Enhancing Road Safety in India: Three Case Studies, 2020) explains about Driveri and PathPartner Technology. Driveri is an AI-powered fleet safety platform that provides complete hardware and software solutions for enhancing fleet safety. It uses AI and edge computing to analyse data in real time, providing deep insight into the driving environment and identifying areas for improvement and driver recognition opportunities. The platform alerts managers in real time of any driving violations and provides in-cab audio alerts to drivers for distracted driving. It measures and alerts speed; and provides real-time scores and tips on how to improve driving performance. The platform also has a virtual coach that gives an automated coaching workflow for drivers.

The PathPartner Technology is an embedded systems based company in Bangalore that specialises in developing Driver Monitoring Systems (DMS). The DMS uses advanced computer vision algorithms to provide reliable driver monitoring on edge. It is capable of precise monitoring, such as eye tracking and facial analysis, even under challenging operating conditions. The system uses a single low-power camera inside the vehicle to provide analysis of driver identification, drowsiness detection and prediction, driver distraction alert, driver action classification, driver emotion estimation, and in-cabin occupancy. The system gives reliable performance at an affordable cost and triggers audio and visual alerts if the driver is found to be drowsy or distracted. It is suitable for integration into a wide range of Advanced Driver Assistance Systems (ADAS) and Autonomous Driving (AD) applications. It is platform agnostic, working on hybrid architectures, including ARM, Digital Signal Processor (DSP), Draphic Processing Unit (GPU), Field-Programmable Gate Array (FPGA), and vision accelerators.

These are just a few examples of how AI is being used to manage traffic. With its ability to analyse real-time data and make predictions, AI has the potential to improve the efficiency and safety of the road networks significantly.

5. APPLICATION OF SMART TRAFFIC

Traffic management is one of the biggest challenges for a city, although there are numerous rules and regulations regarding Traffic Management still there are multiple fatalities. Smart Traffic applications aim to reduce fatalities, efficient resource allocations and environmental benefits. There are various ways to achieve smart traffic management. A few of them are discussed in detail in the following sections.

5.1 Traffic Congestion Management

Traffic Congestion refers to the long queueing and slower traffic conditions contributing to longer trip times. With the increase of population in cities traffic is also ever increasing. The most recurring problem in the cities is traffic congestion. It is a disadvantageous situation, and there need to be smart ways to handle this problem. Over the past few decades, numerous research projects in traffic congestion control have yielded several solutions such as using the integrating connected devices, cameras, sensors, and cloud-based analytics to improve traffic flow and reduce congestion. The aspects of IoT congestion managements by Sadhukhan and Gazi (2018) are real-time traffic monitoring, smart traffic lights, V2V communication, predictive Analysis of traffic, and dynamic route planning. These aspects are used to increase/decrease waiting time of traffic light, optimise safety and decrease traffic, provide an alternative path to avoid congestion.

Table 1. Machine learning algorithms for IoVST applications

No.	Application	Reference	Algorithm Type	Learning Type
1	Traffic Signal Optimization	Yuloskov et al. (2021)	Deep learning	Supervised learning
2	Driver Monitoring Systems	(Saleh & Fathy, 2023)	Deep Learning	Supervised learning
3	Autonomous Vehicle Management	Muiruri (2021)	Population-Based Training	Reinforcement learning
4	Resource Allocation in fog computing	Lee & Lee(2020)	Deep learning	Reinforcement learning
5	Sustainable and Secure Resource Management	Badidi (2022)	CNN at edge nodes	Supervised learning
6	Advanced Driver Assistance System	Zantalis et al. (2019)	Ensemble Adaptive Boost	Supervised learning
7	Traffic Flow Detection	Zantalis et al. (2019)	Random forest	Supervised learning
8	Short Term Traffic Flow	Zantalis et al. (2019)	Bayesian Network	Supervised learning
9	Pavement Distress Detection	Gopalakrishnan (2018)	CNN and DCNN	Reinforcement learning
10	Smart Parking	Rizvi et al. (2019)	CNN	Reinforcement learning
11	Traffic Congestion prediction	Zantalis et al. (2019)	Regression Analysis and SVM	Reinforcement learning
12	Accident Detection/Prevention	Liu et al. (2014)	CHMM and SVM and ANN	Supervised learning

5.2 Intelligent Transport Systems

Intelligent transport system(ITS) refers to the advanced technologies for the transportation system that aims at improving safety of the people, efficiency of the vehicle, and sustainability. ITS attempts to deliver cutting-edge services connected to various modes of transportation and traffic management. It empowers users to use transportation networks in a safer, organised, and "smarter" manner. ITS encompasses technologies related to connected vehicles such as ADAS and V2V communications, which helps collect data that results in better transportation management. ITS enables smart traffic technologies. ADAS is one of the most used systems enabling smart vehicle production. Any of a range of electronic technologies that help drivers with driving and parking functions is known as an ADAS. It improves auto and road safety by providing a secure human-machine interface. ADAS employs automated technology, like cameras and sensors, to recognise surrounding impediments or driving errors and take appropriate action. Depending on the elements installed in the car, ADAS can enable varying degrees of autonomous driving (Advanced Driver Assistance Systems) – Overview of ADAS Applications | Synopsys, 2023). The applications of ADAS by companies like Tesla, Nissan, and Mercedes include: Adaptive Cruise Control, Glare-Free High Beam and Pixel Light, Adaptive Light Control, Automatic Parking, Autonomous Valet

Parking, Navigation System, Night Vision, Unseen Area Monitoring, Automatic Emergency Braking, Crosswind Stabilization, Driver Drowsiness Detection, Driver Monitoring System, 5G and V2X.

5.3 Smart Parking Solutions

Smart parking aims at solving parking spot problems. Since traffic is ever-increasing, the search for parking spots is a time taking and tedious task. Smart parking solutions are efficient ways of handling parking in smart cities. Smart parking solutions aim to increase parking's effectiveness and convenience while lowering traffic and pollution. Smart traffic systems include the variety of smart parking solutions. Monitoring parking lots using cameras and sensors and dynamically provide the parking space information to the drivers. Collecting parking patterns and integrating them for better and more effective parking. Optimising Parking spaces with the help of predictive analysis. Smart parking lots may include dynamic lighting to reduce emissions. Electronic parking ticket collection. E-payments to optimise time. Smart Monitoring of NO parking areas to regulate traffic rules. Hong Kong operator HKT and parking and ticketing systems(Waring et al., 2019) provider Flowbird on January 20, 2021, launched "HKe", a smart parking metre system. The system is available as a mobile application. Vehicle sensors will detect the availability of parking spaces and communicate the information in real time to motorists via the app. They can pay parking fees with multiple payment options and top-up payments remotely through the app.

5.4. Real-Time Traffic Monitoring and Control

Real-time traffic monitoring refers to data collected on traffic flow, traffic speed, and congestion levels in real time using various technologies such as sensors, cameras, GPS, and linked vehicles. Real-Time traffic monitoring enables all the applications in Smart Traffic Management. When the traffic stages are monitored, and patterns are analysed through the processing unit, the systems can get "smarter" and make effective decisions. Real-time traffic monitoring can assist in preventing accidents and enhancing road safety with the help of information collected on traffic flow and congestion levels. Real-time traffic monitoring can help improve traffic management systems and can ease congestion by collecting and analysing data on traffic flow and speed. Real-time traffic information can be utilised to plan routes more efficiently, and can be utilised to shorten commute times and enhance the overall driving experience. The operation of public transportation networks can be optimised,by lowering costs, and by increasing efficiency, with the help of real-time traffic monitoring. Real-time traffic monitoring can detect accidents in real-time and respond to it. The National Institute of Urban Affairs under the Ministry of Urban and Housing Affairs (Niua et al., 2019) has proposed a Real-time Traffic monitoring system in Pune, India. The system was just proposed in 2023 with smart cities as an aim. Real-time traffic monitoring is a crucial system to optimise traffic management. To increase efficiency and reduce congestion of traffic.

5.5 Predictive Traffic Analysis

Predictive analysis refers to the analysis done on the collected data to predict the traffic flow of a particular region. There are various methods to carry out predictive analysis using various machine learning algorithms. AI and ML made possible for the traffic behaviour prediction where the data is used by the traffic management applications to optimise traffic behaviours and efficiently manage traffic in cities. The predictive analysis can be classified as Traffic Flow Prediction, Traffic Speed Prediction, Traffic

Pattern Prediction, Traffic state and parameter prediction, Traffic Density Estimation, Traffic congestion, Travel time prediction, Accident Prone Area Prediction, and User behaviour and awareness prediction. Predictive analysis helps improve traffic route predictions, Increase the efficiency of vehicles, increase incident response, avoid accidents, avoid congestion and overall improve travel time effectively. Microsoft Bing Maps: suggesting routes based on traffic, JamBayes: forecast future traffic, ClearFlow project of Microsoft: predict the traffic flows on all street segments of a greater city area are the real-time implementation with predictive traffic analysis (Predictive Analytics for Traffic - Microsoft Research, 2019).

5.6 Automated-Road Incident Detection and Response

IoT-enabled automated road incident detection systems can detect incidents, such as accidents and collisions in roads, in real-time. It can respond to the incident in real-time and it reduces the impact of incidents on traffic flow. Road Accident Detection can be accomplished by positioning sensors and cameras over accident-prone locations and enabling systems to send alerts to hospitals and road control units for assistance. When a traffic collision occurs, there are infrastructures that will broadcast data through the routing systems to minimise congestion. Vehicles can detect the signals from over a kilometre away, allowing them to reroute if possible. Predictive traffic analysis can aid in detecting potential incidents such as accidents and alert drivers for precautionary actions. Real-time traffic monitoring can detect incidents, such as accidents in real-time and respond to it. These systems improve safety and lessen the impact of accidents on traffic flow by using real-time data from connected vehicles and other sources to detect and respond to issues and crises. Systems using Arduino, GSM, Vibration sensors and GPS can be made to predict accidents on road by Baballe (2022). However, these systems are still in their infancy and will be implemented in real-time.

5.7 Electronic Toll Collection Systems

An e-toll collection involves collecting tolls on roads and highways using electronic devices like RFID (Radio-Frequency Identification) and GPS. Traditional manual toll collection methods are replaced by e-toll collection, which has various advantages. One of the most implemented IoT technologies, e-toll collection is convenient and has made toll management much more efficient in terms of revenue collection and utilisation of time. Toll booths have sensors to scan and generate the identification of the vehicle with the RFID chip. Using GPS, it tracks the vehicle. There are various applications with which there can be pre–payment for the particular toll booth if the vehicle is taking the same highway. When a route is set from source to destination, a vehicle can easily pay the toll taxes beforehand, and with the RFID scanner, the data is recognised at the e-toll booth. The e-toll collection speeds up traffic because cars don't have to stop at toll booths to pay. In order to reliably track toll payments and lower the risk of fraud and errors. Data on traffic flow, toll payments, and vehicle usage can be collected by e-toll collecting systems and used to enhance traffic management and road design which can in turn, improve data analysis. By lowering the cost of toll collecting and by increasing the accuracy of toll payments. FASTag is the real-time implementation of e-toll collection in India as stated by (Arya Omnitalk Wireless Solutions Pvt Ltd, 2021). FASTag is a device that employs RFID technology for making toll payments directly while the vehicle is in motion. FASTag is affixed on the windscreen of the vehicle and enables a customer to make toll payments directly from the account which is linked to FASTag. The first interoperable ETC Pilot Project between Mumbai and Vadodara was successfully rolled out in the year 2013. NHAI(National Highway Authority of India) is the nodal agency to implement e-toll on a Pan-India basis.

5.8 Connected and Autonomous Vehicles

Automation is the future for vehicles. Connected Autonomous Vehicles(CAV)s are automobiles that can operate entirely or partially without a driver. CAVs are outfitted with cutting-edge technologies that allow them to operate independently and communicate with other vehicles and infrastructure, such as sensors, cameras, and communication systems. CAVs are designed to increase driving comfort, lessen congestion, and improve road safety. CAVs are deployed with cutting-edge safety technology sensors and cameras that can recognise hazards. This connects with the infrastructure of other cars to improve traffic flow, decrease congestion, and increase the effectiveness of transportation systems. Autonomous Vehicles provide access to transportation to those who cannot access it such as the elderly and the disabled. It can improve traffic flow, can shorten travel times, and it also can enhance the effectiveness of transportation networks. It can decrease traffic accidents significantly as 90% of road accidents are caused by human errors. CAVs will increase automation and efficiency of transportation. Partially automated vehicles are available in the European Union, and by 2030 there will be a significant increase in the number of CAVs (Connected Autonomous Vehicles - Ferrovial, 2020).

5.9 Smart-Road Lighting Systems

Smart road lighting systems are introduced to be more energy efficient. Lights on the roads are equipped with sensors to detect motion and switch on the lights in the roads that are usually inactive. Street lights are only powered during a fixed time to reduce the overconsumption of electricity. Lights on Highways are adjusted according to traffic congestion and environmental factors. Smart Road lighting systems will provide adequate lighting on the roads to help vision for the travellers. Real-Time Implementation Tata Communications has deployed smart streetlights in several cities, including Noida, NCR, Ahmedabad, Nashik, and Kolkata. Delhi already has around 80,000 smart streetlights, which have been live since August 2019. The city of Jamshedpur is deployed with 300 smart streetlights already installed for the Jamshedpur Utilities and Services Co. Ltd (JUSCO), enabling a reduction in their carbon footprint (Mishra et al., 2021).

5.10. Green and Sustainable Transportation Planning and Management

The broader aim of smart cars and IoT-based smart traffic management is sustainability. Traffic management and planning will play a significant contributing factor to emissions and electricity management. Green and sustainable planning will lead to safer cities, healthy citizens and better life quality in cities Shah et al. (2021). The role of IoT broadly is to maintain resources sustainably. Vehicle emissions contribute majorly to air pollution. This is harmful and hazardous to human beings. Smart ways to manage vehicular emissions are the only way to a green and pollution-free earth.

6. RECENT TECHNOLOGIES IN SMART TRAFFIC

There are many applications of IoT in smart traffic that already exist, including license plate recognition, traffic monitoring cameras, GPS, and mapping technologies. Still, more is needed to realize the complete potential of smart traffic systems. For these applications to expand, it is necessary for the various components, such as traffic lights, cars, and cameras, to communicate with each other. So, the latest research focuses on technologies enabling V2I and Vehicle V2V. In current time, the frontrunners in smart transport system technologies are 5G, cloud/fog computing, and blockchain.

6.1 Cloud/Fog Computing

Cloud computing is a method where data is sent via a network to a centralized system for storage and processing. On the other hand, Fog computing sends edge data to a decentralized network of processing devices geographically closer to each other. It is evident that individual fog nodes may not be as powerful as the cloud but benefit from reduced latency and faster responses. A comprehensive smart traffic system would have millions of sensors constantly collecting huge amounts of data. So, Big Data analysis becomes important, but the storage and processing needed cannot be feasibly implemented within vehicles and edge infrastructure. So, it is seen that most smart traffic solutions utilize remote processing via roadside units, which take data from the various edge devices, and store and process it. Such cloud systems can focus solely on analyzing the data with powerful processors and large amounts of storage. It becomes possible to run ML and AI algorithms in these systems, which help in traffic prediction and management, and aid driverless vehicle decisions (Ali et al., 2022). But cloud computing holds the drawback of having higher latency, which is a disadvantage in traffic systems that require immediate real-time responses. This is due to the distance between the edge and the cloud traversed by multiple signals. Fog computing may thus be a better alternative. Today's vehicles already possess many sensors for data collection, and upgrading these, along with other small infrastructure upgrades at minimal cost, can create a Vehicular Fog Network (VFN).And additional Fog computing units may be used along the traffic ways to handle higher loads, thereby distributing processing across a larger area and reducing latency by reducing node-node distances (Lee & Lee, 2020). While fog reduces the dependency on a single system, the cloud can be used in parallel to enhance lower-priority computing.

6.2 5G and its Potential

Communication capabilities are imperative to realise the above Cloud/Fog systems. Current networks do not have the capacities and speeds needed for efficient networks, so 5G has been proposed for use. 5G promises higher speeds, lower latency, and greater carrying capacity. Using higher capacity millimetre waves and a greater number of smaller cells helps to accomplish this.

The current cloud-dependent traffic systems suffer from latency and low connectivity. This is inefficient and unsafe in a fully smart, autonomous traffic system. To this end, 5G can play a huge role. As mentioned by Yuan et al. (2020) and Pang et al. (2021), the ability of 5G to connect with a huge number of devices allows for the extension and creation of strong V2X (vehicle to everything) systems. In particular, using 5G to create distributed and decentralised V2V relay networks to improve signal delivery shows great promise. As mobile hardware improves, it can allow distributed processing to make timely decisions and exchange driving information between nodes. Ray et al. (2022) demonstrate experimentally that realizing such networks is feasible in real life, even in areas with poor connectivity via small-scale vehicle-to-vehicle communication, which works even without line-of-sight.

Figure 7. Network of 5G enabled vehicles, infrastructure, and cloud/fog units

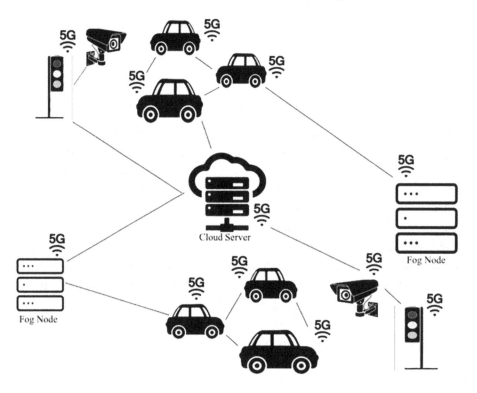

Thus, 5G can enable a capable, high-capacity, fast network for V2X communications and inter-infrastructure communications required to realize a smart traffic system. Figure 7 shows an overview of connections of 5G-enabled vehicles and other sensors, along with cloud/fog units. However, there are still a few points to keep in mind. As explained, a 5G network alone is not good enough to fully process the data (Yuan et al., 2020), (Al Ridhawi, Alogaily, Boukerche & Jararweh, 2021). As an emerging technology, little is known about the possible cyber threats and security measures pertaining to 5G. In the next section, blockchain technology is discussed to secure 5G traffic networks.

6.3 Blockchain

Blockchain is a decentralized digital ledger that enables secure, transparent records keeping and trans-actions. Blockchain provides the benefits of decentralisation, cryptography and data protection, and immutability over other data verification methods. It works via a consensus of credibility achieved via algorithms spread across multiple nodes. If the data transaction is valid and credible, the records and the data are grouped into a block and linked onto a chain of similar blocks. Here, all nodes manage copies of valid transactions, ensuring transparent and robust security. It is essential to secure the huge amounts of data transmitted in a smart traffic systems and the use of blockchain for data verification and security has been proven to work in V2X networks, ensuring reliable, consistent, and secure data transmission between vehicles and infrastructure (Wazid et al., 2022). It would also be able to identify and deal with any compromised nodes and malicious agents (Badidi, 2022).

Another possible blockchain application relates to smart traffic morals, especially in autonomous vehicles. Adding an incentive system, vehicles, and other components could be encouraged to make and learn better choices (Khalid et al., 2021). Blockchain can provide a secure and transparent record of "reputation", allowing nodes to judge the reliability of information and make better decisions. Hence, using blockchain can greatly increase the security of data transmission in 5G-enabled Cloud/Fog networks, allowing for safer and more optimal real-time responses.

7. CHALLENGES AND THE WAY FORWARD

Smart traffic systems, even in their current state, hold much promise, and already plenty of research is being done into realizing and creating feasible and robust working systems. However, there are still many problems which would need to be solved as shown in Figure 8. In the field of smart transport systems, most research assumes fully automated systems without human control and ideal conditions for data sharing, including complete and secure data transmission. However, these assumptions represent experimental ideal conditions rather than the realities of the real world. Adapting experimental results to real-world complexities is the primary challenge in the field, necessitating further technological advancements. Overcoming this challenge requires the development of advanced technologies that can effectively translate experimental findings into practical applications (Wazid et al., 2022; Badidi, 2022).

Figure 8. Challenges of IoVST

During the transitional period of adopting Internet of Vehicle and Smart Transportation (IoVST) systems, it is essential to carefully consider the coexistence of "smart" and non-smart segments within the transport network. This phase is expected to be chaotic, characterized by a high level of interactions between human-controlled vehicles and smart vehicles. Consequently, a thorough examination of potential challenges arising from various domains, both technical and non-technical, becomes imperative. Deep thought must be given to anticipate and address the complexities that may emerge, ensuring a smooth transition into a fully integrated and functional smart transport system.

Ethics and morals are significant considerations in the development and implementation of smart transport systems. In the event of an accident or system fault, determining who should be held responsible becomes a crucial question. Additionally, the decision-making process of the vehicles themselves requires careful consideration. To address these ethical and moral complexities, artificial intelligence (AI) and machine learning (ML) algorithms need to be trained and implemented effectively. By incorporating ethical principles into the algorithms, the aim is to reduce potential moral dilemmas and ensure responsible decision-making within the smart transport system.

The technology itself presents another crucial aspect to consider in the development of smart transport systems. In addition to necessary technological advancements, it is essential to address the challenges associated with legacy systems and interoperability. Ensuring compatibility between newer and older systems is vital to facilitate seamless integration. Furthermore, algorithms must possess the capability to adapt effectively to the data collected by the hardware. By tackling these technology-related issues, smart transport systems can achieve harmonious coexistence with existing infrastructure and enable successful adaptation to various data sources.

The question now is how to fully realize these smart transport systems. As with many things in life, the path forward is difficult, but not impossible. In order to realise these smart traffic systems, many around the world have begun researching and experimenting. Various research have already proposed solutions in the form of experimental results and theoretical algorithms. More research has also been called for and definitely will be done. The way forward demands research, experiments and the continuous development of hardware and software technology.

More advanced and efficient processors and storage devices need to be developed on the hardware side. Sensors and cameras also need to be refined and precise to collect more accurate and usable data. At the same time, it would be ideal for all these to improve while getting smaller and lighter.

On the software side, ML and other algorithms need to be trained to be faster, more accurate and with simulated moral capabilities. Given the potential amounts of data, efficient algorithms will be required. It is also known that mobile networks are also continuously being upgraded, and 5G cellular networks are already being implemented worldwide for cell phones and smart home networks. In this regard, the required steps for a smart traffic system would be to solve the issues of scale and power so that they can support and continuously power the large networks required. Perhaps one final step towards realising the smart traffic systems is that of education. Humans may become less active participants in traffic in near future, but there will still be some unavoidable actions people would have to take, and not to mention that pedestrians are also a part of traffic. Thus, it becomes important to ensure that all people who are not part of the connected system, but moving around independently, are educated and informed of the behaviours and decision-making probabilities of the smart traffic system.

CONCLUSION

IoVST refers to the application of sophisticated technology, such as connected vehicles, sensors, and communication systems, to enhance the effectiveness, safety, and sustainability of transportation systems. IoVST can be implemented using cutting-edge algorithms and technologies with the help of Machine Learning algorithms. Predictive analysis can be done using several paradigms. There are numerous applications of IoVST which improves transportation management. These applications have the power to completely change the way we travel by enhancing traffic flow, reducing congestion, and giving drivers and traffic authorities real-time information. It is utilised to improve road design and traffic management systems, further boosting the effectiveness and safety of transportation networks, by gathering data on traffic patterns, road conditions, and vehicle usage. The broader aim of IoVST is sustainability. Vehicle emissions contribute majorly to air pollution. This is harmful and hazardous to human beings. Smart ways to manage vehicular emissions are the only way to a green and pollution-free earth.

REFERENCES

Abbasi, S., Rahmani, A. M., Balador, A., & Sahafi, A. (2021). Internet of Vehicles: Architecture, services, and applications. *International Journal of Communication Systems*, *34*(10). doi:10.1002/dac.4793

Advanced Driver Assistance Systems – Overview of ADAS Applications. (2023). Synopsys. https://www.synopsys.com/automotive/what-is-adas.html

Agafonov, A., & Myasnikov, V. (2021). *Traffic Signal Control: a Double Q-learning Approach. Annals of Computer Science and Information Systems*. Fed CSIS. doi:10.15439/2021F109

Ali, M. H., Jaber, M. M., Abd, S. K., Alkhayyat, A., & Albaghdadi, M. F. (2022). Big data analysis and cloud computing for smart transportation system integration. *Multimedia Tools and Applications*, 1–18. doi:10.100711042-022-13700-7

Amitha, I. C., & Narayanan, N. K. (2021). Object Detection Using YOLO Framework for Intelligent Traffic Monitoring. *Lecture Notes in Electrical Engineering*, (pp. 405–412). IEEE. doi:10.1007/978-981-16-5078-9_34

Arya Omnitalk Wireless Solutions. (2021, January 30). *Electronic Toll Collection System*. Fastag. Arya Omnitalk. https://aryaomnitalk.com/electronic-toll-collection-system/

Ashifuddin Mondal, M., & Rehena, Z. (2019). Intelligent Traffic Congestion Classification System using Artificial Neural Network. *Companion Proceedings of the 2019 World Wide Web Conference*. ACM. 10.1145/3308560.3317053

Baballe, M. A. (2022b, July 26). *Accident Detection and Alerting Systems: A Study*. Research Gate. https://www.researchgate.net/publication/362264272_Accident_Detection_and_Alerting_Systems_A_Study

Badidi, E. (2022). Edge AI and Blockchain for Smart Sustainable Cities: Promise and Potential. *Sustainability (Basel)*, *14*(13), 7609. doi:10.3390u14137609

Chen, Q., Wang, W., Huang, K., De, S., & Coenen, F. (2021). Multi-modal generative adversarial networks for traffic event detection in smart cities. *Expert Systems with Applications, 177*, 114939. doi:10.1016/j. eswa.2021.114939

Cong, Y., Wang, J., & Li, X. (2016). Traffic Flow Forecasting by a Least Squares Support Vector Machine with a Fruit Fly Optimization Algorithm. *Procedia Engineering, 137*, 59–68. doi:10.1016/j. proeng.2016.01.234

Connected Autonomous Vehicles. (2020, April 21). Ferrovial. https://www.ferrovial.com/en/innovation/ technologies/connected-autonomous-vehicles/

DAC. (2022, September 27). *Predictive and preventive maintenance in smart cities.* DAC .Digital. https:// dac.digital/predictive-and-preventive-maintenance-in-smart-cities/

Elsagheer Mohamed, S. A., & AlShalfan, K. A. (2021). Intelligent Traffic Management System Based on the Internet of Vehicles (IoV). *Journal of Advanced Transportation, 2021*, 1–23. doi:10.1155/2021/4037533

Gopalakrishnan, K. (2018). Deep Learning in Data-Driven Pavement Image Analysis and Automated Distress Detection: A Review. *Data, 3*(3), 28. doi:10.3390/data3030028

I., P. T. (n.d.). *India's first electronic toll collection system launched on Ahmedabad-Mumbai Highway.* Indian Tollways - an e-News Magazine on BOT Road Projects. https://www.indiantollways.com/indias-first-electronic-toll-collection-system-launched-on-ahmedabad-mumbai-highway/

Jang, I., Kim, D., Lee, D., & Son, Y. (2018). An Agent-Based Simulation Modeling with Deep Reinforcement Learning for Smart Traffic Signal Control. *2018 International Conference on Information and Communication Technology Convergence (ICTC).* IEEE. 10.1109/ICTC.2018.8539377

Kaiwartya, O., Abdullah, A. H., Cao, Y., Altameem, A., Prasad, M., Lin, C. T., & Liu, X. (2016). Internet of Vehicles: Motivation, Layered Architecture, Network Model, Challenges, and Future Aspects. *IEEE Access : Practical Innovations, Open Solutions, 4*, 5356–5373. doi:10.1109/ACCESS.2016.2603219

Kaluvan, H., Baskar, P. K., Ramanathan, S., Sundar, S., & Jeyaram, S. (2021). Intelligent transportation system and smart traffic flow with IOT. *Indian Journal of Radio & Space Physics, 50*. http://op.niscpr. res.in/index.php/IJRSP/article/view/62096/465480553

Karim, S. M., Habbal, A., Chaudhry, S. A., & Irshad, A. (2022). Architecture, Protocols, and Security in IoV: Taxonomy, Analysis, Challenges, and Solutions. *Security and Communication Networks, 2022*, 1–19. doi:10.1155/2022/1131479

Kekuda, A., Anirudh, R., & Krishnan, M. (2021). Reinforcement Learning based Intelligent Traffic Signal Control using n-step SARSA. *2021 International Conference on Artificial Intelligence and Smart Systems (ICAIS).* IEEE. 10.1109/ICAIS50930.2021.9395942

Khalid, A., Iftikhar, M. A., Almogren, A., Khalid, R., Afzal, M., & Javaid, N. (2021). A blockchain based incentive provisioning scheme for traffic event validation and information storage in VANETs. *Information Processing & Management, 58*(2), 102464. doi:10.1016/j.ipm.2020.102464

Kurniawan, J., Syahra, S. G., Dewa, C. K., & Afiahayati. (2018). Traffic Congestion Detection: Learning from CCTV Monitoring Images using Convolutional Neural Network. *Procedia Computer Science*, *144*, 291–297. doi:10.1016/j.procs.2018.10.530

Lee, S., & Lee, S. (2020). Resource Allocation for Vehicular Fog Computing Using Reinforcement Learning Combined With Heuristic Information. *IEEE Internet of Things Journal*, 7(10), 10450–10464. doi:10.1109/JIOT.2020.2996213

Liu, W., Kim, S., Marczuk, K. A., & Ang, M. H. (2014). *Vehicle motion intention reasoning using cooperative perception on urban road*. IEEE. doi:10.1109/ITSC.2014.6957727

Mehra, S. (2020, December 20). AI integration for enhancing road safety in India: Three case studies. *Indiaai.Gov.In*. https://indiaai.gov.in/article/ai-integration-for-enhancing-road-safety-in-india-three-case-studies

Microsoft. (2019, August 20). *Predictive Analytics for Traffic*. Microsoft Research. https://www.microsoft.com/en-us/research/project/predictive-analytics-for-traffic/

Mishra, L. (2021, September 6). *Smart lighting the road to sustainability*. Tata Communications New World. https://www.tatacommunications.com/blog/2020/08/smart-lighting-the-road-to-sustainability/

Mousavi, S. S., Schukat, M., & Howley, E. (2017). Traffic light control using deep policy-gradient and value-function-based reinforcement learning. *IET Intelligent Transport Systems*, *11*(7), 417–423. doi:10.1049/iet-its.2017.0153

Muiruri, L. (2021, November 23). *What You Need to Know About Waymo's New Self-Driving Taxi Service*. Makeuseof. https://www.makeuseof.com/waymos-new-self-driving-taxi-service/

Nesa, N., & Banerjee, I. (2017). IoT-based sensor data fusion for occupancy sensing using Dempster–Shafer evidence theory for smart buildings. *IEEE Internet of Things Journal*, *4*(5), 1563–1570. doi:10.1109/JIOT.2017.2723424

Niua. (2019). ENABLE : REAL-TIME TRAFFIC MONITORING TOOL FOR EFFECTIVE AND EFFICIENT DECISION MAKING. *Niua*. https://www.niua.org/iscfip/compendium/project/enable-real-time-traffic-monitoring-tool-effective-and-efficient-decision-making

Oh, S., Byon, Y. J., & Yeo, H. (2016). Improvement of Search Strategy With K-Nearest Neighbors Approach for Traffic State Prediction. *IEEE Transactions on Intelligent Transportation Systems*, *17*(4), 1146–1156. doi:10.1109/TITS.2015.2498408

Pang, S., Wang, N., Wang, M., Qiao, S., Zhai, X., & Xiong, N. (2021). A Smart Network Resource Management System for High Mobility Edge Computing in 5G Internet of Vehicles. *IEEE Transactions on Network Science and Engineering*, *8*(4), 3179–3191. doi:10.1109/TNSE.2021.3106955

Predictive Analytics for Traffic. (2019b, August 19). Microsoft Research. https://www.microsoft.com/en-us/research/project/predictive-analytics-for-traffic/

(1522). R N, R., R., V., & M. R., A. (2018). Autonomous Traffic Signal Control using Decision Tree. [IJECE]. *Iranian Journal of Electrical and Computer Engineering, 8*(3). doi:10.11591/ijece.v8i3. pp1522-1529

Rajeev, G. L., & Nancy, R. S, M., John, J., & John, N. M. J. N. M. (2021). Traffic Flow Prediction using Random Forest and Bellman Ford for Best Route Detection. *International Journal of Engineering Research and Technology, 9*(13). https://www.ijert.org/research/traffic-flow-prediction-using -random-forest-and-bellman-ford-for-best-route-detection-IJE RTCONV9IS13021.pdf

Ranka, S., Rangarajan, A., Elefteriadou, L., Srinivasan, S., Poasadas, E., Hoffman, D., Ponnulari, R., Dilmore, J., & Byron, T. (2020). A Vision of Smart Traffic Infrastructure for Traditional, Connected, and Autonomous Vehicles. *2020 International Conference on Connected and Autonomous Driving (MetroCAD).* IEEE. 10.1109/MetroCAD48866.2020.00008

Ray, J. K., Biswas, A. S., Sil, S., Bera, R., Shome, S., Biswas, P., & Mitra, M. (2022). Realization of 5G V2V communication system at 28 GHz for smart vehicle. *Innovations in Systems and Software Engineering.* doi:10.100711334-022-00435-9

Ridhawi, I. A., Aloqaily, M., Boukerche, A., & Jararweh, Y. (2021). Enabling Intelligent IoCV Services at the Edge for 5G Networks and Beyond. *IEEE Transactions on Intelligent Transportation Systems, 22*(8), 5190–5200. doi:10.1109/TITS.2021.3053095

Sadhukhan, P., & Gazi, F. (2018b). *An IoT based Intelligent Traffic Congestion Control System for Road Crossings.* IEEE. doi:10.1109/IC3IoT.2018.8668131

Saleh, S. N., & Fathy, C. (2023). A Novel Deep-Learning Model for Remote Driver Monitoring in SDN-Based Internet of Autonomous Vehicles Using 5G Technologies. *Applied Sciences (Basel, Switzerland), 13*(2), 875. doi:10.3390/app13020875

Shah, K. J., Pan, S., Lee, I., Kim, H., You, Z., Zheng, J., & Chiang, P. (2021). Green transportation for sustainability: Review of current barriers, strategies, and innovative technologies. *Journal of Cleaner Production, 326,* 129392. doi:10.1016/j.jclepro.2021.129392

Sharif, A., Li, J., Khalil, M., Kumar, R., Sharif, M., & Sharif, A. (2017). *Internet of things — smart traffic management system for smart cities using big data analytics.* Active Media Technology. doi:10.1109/ ICCWAMTIP.2017.8301496

Tian, Y., & Pan, L. (2015). Predicting Short-Term Traffic Flow by Long Short-Term Memory Recurrent Neural Network. *2015 IEEE International Conference on Smart City/SocialCom/SustainCom (SmartCity).* IEEE. 10.1109/SmartCity.2015.63

Vidhate, D. A., & Kulkarni, P. (2017). Cooperative multi-agent reinforcement learning models (CMRLM) for intelligent traffic control. *2017 1st International Conference on Intelligent Systems and Information Management (ICISIM).* IEEE. 10.1109/ICISIM.2017.8122193

Waring, J. (2019, May 21). HKT selected for smart parking system. *Mobile World Live.* https://www. mobileworldlive.com/apps/news-apps/hkt-selected-for-smart-parking-system/

Wazid, M., Bera, B., Das, A., Mohanty, S. P., & Jo, M. (2022). Fortifying Smart Transportation Security Through Public Blockchain. *IEEE Internet of Things Journal*, 9(17), 16532–16545. doi:10.1109/JIOT.2022.3150842

Wei, W., Wu, H., & Ma, H. (2019). An AutoEncoder and LSTM-Based Traffic Flow Prediction Method. *Sensors (Basel)*, 19(13), 2946. doi:10.339019132946 PMID:31277390

Yuan, W., Li, S., Xiang, L., & Ng, D. W. K. (2020). Distributed Estimation Framework for Beyond 5G Intelligent Vehicular Networks. *IEEE Open Journal of Vehicular Technology*, 1, 190–214. doi:10.1109/OJVT.2020.2989534

Yuloskov, A., Bahrami, M. R., Mazzara, M., & Kotorov, I. (2021). Smart Cities in Russia: Current Situation and Insights for Future Development. *Future Internet*, 13(10), 252. doi:10.3390/fi13100252

Zantalis, F., Koulouras, G., Karabetsos, S., & Kandris, D. (2019). A Review of Machine Learning and IoT in Smart Transportation. *Future Internet*, 11(4), 94. doi:10.3390/fi11040094

Chapter 16
Smart Cities:
Redefining Urban Life Through IoT

Kavitha Rajamohan
CHRIST University (Deemed), India

Sangeetha Rangasamy
CHRIST University (Deemed), India

Alvis Abreo
CHRIST University (Deemed), India

Rohit Upadhyay
CHRIST University (Deemed), India

Raison Sabu
CHRIST University (Deemed), India

ABSTRACT

A smart city is designed using acceptable internet of things (IoT) technologies that solve urban life problems and provide quality of life to the residents. IoT refers to a network of physical devices that are capable of gathering and sharing data and expediting numerous functions without human assistance. IoT supports smart home builders and managers by providing an efficient ecosystem in terms of less operating cost and improvising residence services. In recent days, the initiative of smart homes/buildings/ cities is increasing gradually around the globe. The inclined population in an urban area also expects well-managed automated services in their everyday life.

DOI: 10.4018/978-1-6684-8785-3.ch016

1. INTRODUCTION

1.1 Smart City

A smart city is a city that utilizes technology and data analysis to improve the quality of life for its citizens, enhance sustainability, and streamline urban services. The concept of a smart city involves the integration of information and communication technology (ICT) and the Internet of Things (IoT) to manage a city's assets and infrastructure, including transportation, energy, healthcare, public safety, and waste management. The goal of a smart city is to create efficient systems that work together to solve urban problems and improve the lives of citizens. Smart city initiatives can include implementing smart traffic management systems, using renewable energy sources, implementing e-governance solutions, and providing citizens with access to high-speed internet. The data generated from these systems can be analysed to identify patterns and improve decision-making, leading to better resource allocation and overall city planning.

1.2 Internet of Things

The Internet of Things (IoT) is a network of physical objects, including furniture, cars, home appliances, and other things, that can connect to one another and share data. These objects are implanted with electronics, software, sensors, and connectivity. The IoT allows for the seamless and automated exchange of information between devices, enabling them to work together to achieve a common goal without the need for human intervention. The IoT has the potential to revolutionize the way we live, work, and communicate by creating a connected world where devices can communicate with each other to provide a seamless and efficient experience for users. IoT devices can range from everyday household items such as smart thermostats and lighting systems, to industrial equipment, such as machines on a factory floor, to healthcare devices such as wearable fitness trackers.

1.3 IoT in Smart Cities

The integration of IoT into smart city initiatives is a key aspect of creating a more livable, efficient, and sustainable urban environment. Cities may gather and analyse vast volumes of data from a variety of sources, such as traffic patterns, energy use, and waste management systems, by implementing IoT technology. This data can then be used to inform decision-making, leading to improved resource allocation and city planning. IoT devices can range from smart traffic lights that adjust their timing based on real-time traffic data, to sensors that monitor air quality, to smart waste management systems that optimize pick-up routes based on the level of garbage in each bin. By connecting these devices and systems, cities can create a more integrated and efficient urban infrastructure. BECKETT and CAMARATA(2022) has proposed a study of two cities in Pennsylvania named Harrisburg and Pittsburgh which are the emerging leaders in the field of smart cities. This study describes the unique approaches they have taken in terms of smart and connected technology design around mobility, public safety and sustainability. The use of these technologies was found to raise the standard of living for the local populace.

2. IoT ARCHITECTURE IN SMART CITIES

Smart cities typically use the Internet of Things (IoT) architecture to collect, process, and analyse data from various sources to improve the quality of life for their residents. The IoT architecture for smart cities typically consists of five layers namely Sensing layer, Network layer, Middle layer, Application layer and Business layer. The generic IoT architecture for smart city is depicted in Figure 1. This layered architecture components are explained as follows:

- Sensors and Devices: This component consists of a large number of sensors and devices that collect data from various sources such as traffic, weather, energy usage, and air quality. These devices can be connected to the Internet via Wi-Fi, cellular, or other communication technologies.
- Edge Computing: Edge computing is a sort of distributed computing that processes data locally, as opposed to centrally, at the source of the data. To reduce latency and bandwidth needs for transferring data to a central location, edge computing is utilised in a smart city to process data from IoT devices in real-time.
- Gateway: The gateway is a device that acts as a bridge between the IoT devices and the central computing infrastructure. It is responsible for filtering and pre-processing data, as well as managing the communication between the devices and the central infrastructure.
- Cloud or Data Center: This component is responsible for storing and processing large amounts of data collected from IoT devices. The cloud or data center can use big data analytics and machine learning algorithms to analyze the data and provide insights that can be used to improve city services.
- Application Layer: This component consists of applications and services that use the data from the IoT devices to provide value to residents and city officials. For example, traffic management applications can use data from traffic sensors to optimize traffic flow, while energy management applications can use data from energy meters to improve energy efficiency.
- User Interface: The user interface provides a way for city officials and residents to access the data and insights generated by the IoT architecture. This can be done through a web portal, mobile app, or other user-friendly interface.

Figure 1. Generic IoT architecture for smart city

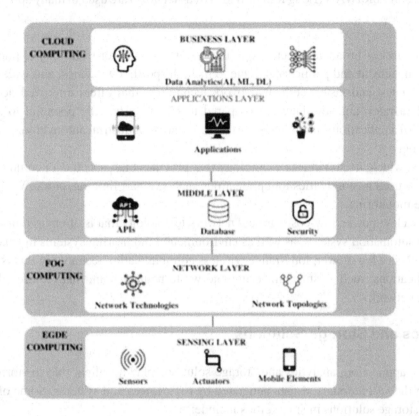

2.1 IoT Devices

In a smart city architecture, a variety of IoT devices are used to improve the quality of life for citizens, increase efficiency and sustainability, and enhance the overall management of the city. Sensor deployed devices are used in various applications. IoT-enabled streetlights that can be controlled and monitored remotely to conserve energy and improve public safety. IoT sensors and smart bins that help to optimize waste collection and reduce litter. IoT devices such as traffic sensors, GPS, and cameras helps to reduce congestion, improve road safety, and monitor emissions. Another set of IoT sensors that measure air quality, temperature, and other environmental factors to help city authorities understand and mitigate environmental challenges. IoT-powered cameras, sensors, and other devices that help to improve public safety by providing real-time information to law enforcement agencies. Wearable medical devices, such as smartwatches and fitness trackers, can monitor a patient's vital signs, activity levels, and sleep patterns, providing important information to healthcare providers for early detection of potential health problems.

2.2 Network Communication

In a smart city architecture, network communication is the backbone that connects different IoT devices and systems to enable data exchange and interoperation. In past decade various network technologies are introduced for smart city applications. Out of that LTE (Long-Term Evolution), 5G (Fifth Generation),

Wi-Fi, Zigbee, and LoRaWAN (Long Range Wide Area Network) are used in many applications(Sharma et.al,. 2021).

- LTE is a wireless broadband technology that is widely used for smart city applications, such as traffic management and public safety, due to its high speed, low latency, and wide coverage.
- 5G is the latest and fastest cellular network technology that offers improved network capacity, higher data rates, and low latency compared to LTE. 5G has the potential to support a wide range of IoT applications, including smart transportation, industrial automation, and remote health monitoring.
- Wi-Fi is a widely used wireless technology that provides high-speed data communication and is commonly used in smart cities for applications such as smart lighting, public Wi-Fi hotspots, and real-time monitoring.
- Zigbee is a low-power, low-cost wireless network technology that is often used in smart home and building automation systems, as well as environmental monitoring systems in smart cities.
- LoRaWAN is a low-power, long-range wireless communication technology that is well-suited for IoT applications, such as smart metering and waste management, that require a long range and low-cost network.

2.3 Analytics and Storage Solutions

In a smart city application, analytics and storage solutions play a critical role in storing, processing, and managing the vast amounts of data generated by IoT devices and systems. Some of the most used analytics and storage solutions in smart cities include:

Big Data platforms: Big Data platforms such as Hadoop and Spark are used to process and analyse the large amounts of data generated by IoT devices in real-time, enabling city authorities to make data-driven decisions.

Cloud storage: For the enormous amounts of data produced by smart city technologies, scalable and secure storage is provided via cloud storage solutions like Amazon Web Services (AWS) and Microsoft Azure.

Predictive analytics: Predictive analytics tools, such as IBM Watson and Google Cloud AI, are used to analyse data from IoT devices and make predictions about future trends, enabling city authorities to proactively address potential issues.

Real-time analytics: Real-time analytics solutions, such as Apache Flink and Apache Storm, are used to process and analyse data from IoT devices in real-time, allowing city authorities to respond to events and incidents in real-time.

Data visualization: Data visualization tools, such as Tableau and PowerBI, are used to display and interact with data from IoT devices and systems in a visually appealing and intuitive manner, enabling city authorities to make informed decisions.

NoSQL databases: NoSQL databases, such as MongoDB and Cassandra, are used to store and manage unstructured and semi-structured data generated by IoT devices, as well as provide scalability and performance for real-time data processing.

3. APPLICATIONS OF IOT IN SMART CITIES

IoT has a significant impact on smart cities by enabling the development of advanced, interconnected systems that improve the quality of life for citizens. IoT applications in smart cities range from environmental monitoring to public safety. For example, IoT-enabled sensors and devices can be used to monitor air quality, optimize waste collection, improve traffic flow, and enhance public safety by providing real-time information to authorities. With the help of IoT, smart cities can become more efficient, sustainable, and responsive to the needs of their citizens, while also improving the overall management and decision-making of city authorities. As depicted in Figure 2 IoT application in smart city includes smart healthcare, smart energy, smart living, smart infrastructure, smart economy, smart environment, smart people, and smart mobility.

3.1 Smart Healthcare

Smart healthcare is an important application of the Internet of Things (IoT) in smart cities. IoT-enabled devices and systems can improve the delivery of healthcare services by providing real-time patient monitoring, remote consultation, and personalized treatment. The research by Trovato et al. (2022) shows how sensors have also been deployed into textile fabrics and are used to monitor biometric parameters. Some examples of IoT technologies in smart healthcare are telemedicine, smart hospital system and mHealth.

Figure 2. Applications of IoT in smart cities

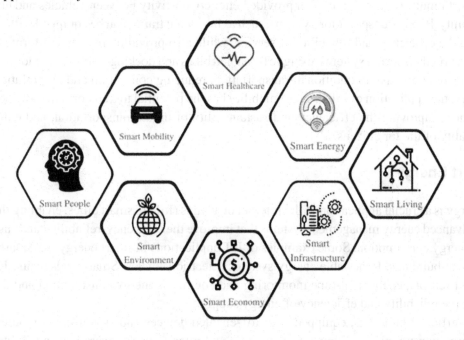

Telemedicine technologies, such as video conferencing and remote monitoring, allow patients to receive healthcare services from their homes, reducing the need for in-person visits and improving access to care for those who live in remote areas. Smart hospital systems, such as IoT-enabled medical equipment and electronic health records, improve the efficiency and quality of care by providing real-time patient data and enabling healthcare providers to quickly and easily access and share important information. According to Tavakoli, Carriere and Torabi (2020) since the Covid-19 pandemic, the use of robotics along with smart wearables can help in providing the patients better treatment along with reducing the risk of infection as well as transmission. mHealth, or mobile health, leverages mobile technology and IoT devices to provide health services and information to patients on-the-go, improving access to care and enabling patients to better manage their health. In summary, the application of IoT in smart healthcare has the potential to transform the delivery of healthcare services, improving access to care, reducing costs, and enhancing the overall quality of life for patients.

3.2 Smart Mobility

Smart mobility is a key application of the Internet of Things (IoT) in smart cities, enabling the development of advanced transportation systems that improve the efficiency, safety, and sustainability of urban mobility. Some examples of IoT applications in smart mobility are Intelligent transport system, Connected vehicles, Public transportation, Smart bike-sharing. Intelligent transportation systems, such as real-time traffic management and smart parking systems, use IoT-enabled sensors and tools to improve city transportation efficiency by streamlining traffic and easing congestion. Connected vehicles, equipped with IoT devices and sensors, can communicate with each other and with smart infrastructure, improving road safety, reducing fuel consumption, and enabling new mobility services such as car-sharing and autonomous driving. According to K. Guan et al. (2020) wireless connectivity between smart objects equipped with multi-sensor systems can provide better connectivity between vehicles and enable intelligent mobility. Public transportation systems, such as buses and trains, can be equipped with IoT devices to improve the efficiency and reliability of service, while also providing real-time information to passengers. Smart bike-sharing systems, using IoT-enabled bikes and docking stations, provide a sustainable and convenient alternative to traditional transportation, reducing congestion and air pollution in cities. In summary, the application of IoT in smart mobility has the potential to transform the way people move around cities, improving the efficiency and sustainability of urban transportation, and enhancing the overall quality of life for citizens.

3.3 Smart Energy

Smart energy is a crucial application of the Internet of Things (IoT) in smart cities, enabling the development of advanced energy management systems that improve the efficiency, reliability, and sustainability of urban energy consumption. Some examples of IoT applications in smart energy are Smart grids, Energy efficient buildings, Renewable energy systems, Electric vehicles. Smart grids, using IoT-enabled devices and sensors, enable real-time monitoring and control of energy consumption and distribution, improving the reliability and efficiency of energy supply.

Energy efficient buildings, equipped with IoT-enabled devices and systems, can monitor, and optimize energy consumption, reducing energy waste and improving the overall sustainability of a city. An algorithm that aids in decreasing energy usage and costs while taking into account the availability of renewable energy sources and user activity has been proposed by Gutierrez-Martinez et al. (2019). Renewable energy systems, such as solar and wind, can be equipped with IoT devices to optimize energy generation, storage, and distribution, reducing the dependence on fossil fuels, and improving the overall sustainability of a city. Electric vehicles, equipped with IoT-enabled charging stations and batteries, can provide a sustainable alternative to traditional transportation, reducing air pollution and dependence on fossil fuels. In summary, the application of IoT in smart energy has the potential to transform the way energy is consumed and managed in cities, improving the efficiency and sustainability of urban energy systems, and enhancing the overall quality of life for citizens.

3.4 Smart Environment

Smart environment is a vital application of the Internet of Things (IoT) in smart cities, enabling the development of advanced environmental monitoring and management systems that improve the sustainability and quality of life in urban areas. Some examples of IoT applications in smart environment are environmental monitoring, smart waste management, sustainable agriculture, and climate resilience. In environmental monitoring, IoT-enabled sensors and devices can provide real-time information on air and water quality, temperature, and other environmental factors, enabling early detection and response to potential environmental issues. Smart waste management using IoT-enabled waste bins and trucks, can optimize waste collection, reducing waste and improving the overall sustainability of a city. Sustainable agriculture is made using IoT-enabled devices and systems, that can monitor soil and weather conditions, improving crop yields and reducing waste, while also improving the overall sustainability of urban agriculture. In climate resilience, the IoT-enabled sensors and systems, can provide real-time information and early warning systems, improving the ability of cities to respond to and recover from natural disasters and climate change. In summary, the application of IoT in smart environment has the potential to transform the way cities manage and monitor their environment, improving the sustainability and quality of life for citizens, and enhancing the overall resilience of urban areas to environmental and climate challenges.

3.5 Smart People

Smart people is a key application of the Internet of Things (IoT) in smart cities, enabling the development of advanced systems and services that improve the safety, health, and well-being of urban residents. public safety, healthcare, smart homes, and citizen engagement are Some examples of IoT applications in smart people. Public safety is taken care by using IoT-enabled cameras and sensors. This system improves the response time and efficiency of emergency services, and also provides real-time information to the citizens in the case of an emergency. IoT-enabled devices and systems in smart home can improve the comfort, safety, and energy efficiency of residential buildings, while also enabling new services such as home automation and smart home security. In citizen engagement process the IoT-enabled platforms and devices, can improve the ability of cities to engage and communicate with citizens, enabling new forms of participation and collaboration in city decision-making processes. In summary, the application of IoT in smart people has the potential to transform the way cities provide services and support to their citizens, improving the safety, health, and well-being of urban residents, and enhancing the overall quality of life in smart cities.

3.6 Smart Economy

Smart economy is a crucial application of the Internet of Things (IoT) in smart cities, enabling the development of advanced systems and services that drive economic growth and competitiveness in urban areas. Some examples of IoT applications in the smart economy are Smart logistics, Smart manufacturing, Smart tourism, and Smart financial services. Smart logistics, using IoT-enabled devices and sensors, can optimize supply chain management and delivery systems, reducing costs and improving the efficiency of goods and services. IoT-enabled devices and systems, can optimize production processes, reducing waste and improving the efficiency of production facilities in smart manufacturing. IoT-enabled services also improves the quality and accessibility of tourism services, while also providing real-time information to visitors and tourists as a smart tourism. Smart financial services, using IoT-enabled devices and platforms, can improve the accessibility and security of financial services, while also enabling new forms of financial transactions and services. The application of IoT in smart economy has the potential to transform the way cities drive economic growth and competitiveness, improving the efficiency and sustainability of urban economic systems and enhancing the overall quality of life for citizens.

3.7 Smart Living

Smart living is a the most interesting application of the Internet of Things (IoT) in smart cities, This system enables the development of advanced systems and services that improve the comfort, quality, and sustainability of urban life in a smart city. Various applications are proposed related to smart living, out of that smart building, smart home, quality of life are some examples of IoT applications in smart living. IoT embedded smart buildings are developed to improve the comfort, safety, and energy efficiency of buildings. This also enables the traditional building to the new forms of building with automation and management. A research by Aryal et al. (2019) provides an example of how IoT-enabled devices can improve comfort and promote the health of its users through the use of Smart Desks that can be personalized according to the user's preference. Inorder to improve the efficiency, reliability, and sustainability of urban infrastructure an IoT-enabled Smart city infrastructure is introduced. This system monitor the structural damage of the building and report to the authority when attention need. As more systems are being developed to transform homes into smart homes systems which are low-cost, expandable along with being energy efficient are also being developed. This strategy was used by Mihalache (2017) in their research to create an inexpensive and energy-efficient home automation system using an Arduino Uno integrated with pertinent modules.By delivering real-time information and services that improve the comfort and sustainability of urban living, quality of life can be improved by deploying IoT-enabled devices and services. In summary, the application of IoT in smart living has the potential to transform the way cities and citizens live, work, and interact, improving the overall quality of life in smart cities and enhancing the sustainability and resilience of urban areas.

3.8 Smart infrastructure

Smart infrastructure is a crucial application of the Internet of Things (IoT) in smart cities, enabling the development of advanced systems and services that improve the efficiency, reliability, and sustainability of urban infrastructure. Some examples of IoT applications in smart infrastructure are safety and security monitoring of the building, auto ON/OFF street lights, water and energy management of the entire city.

Smart energy, using IoT-enabled devices and systems, can improve the efficiency, reliability, and sustainability of energy systems, such as power grids, renewable energy systems, and energy storage systems. Smart water management system is designed for sustainable water systems, such as water treatment and distribution of the water resource.

4. EMERGING IoT TECHNOLOGIES IN SMART CITIES

The field of the Internet of Things (IoT) is expanding quickly and has the potential to completely change how we live and work. In the context of smart cities, IoT technologies enable cities to become more connected, efficient, and sustainable. This includes the deployment of connected devices and sensors to gather data, real-time monitoring and control of critical infrastructure, and the use of smart systems to manage traffic, energy, and waste. Some examples of emerging IoT technologies in smart cities include smart lighting, air quality monitoring, smart waste management, and autonomous vehicles. These innovations could raise living standards, lessen their negative effects on the environment, and increase public safety.

4.1 IoT Devices

There are several IoT devices that are emerging as key technologies in smart cities. Some of the most significant include:

Smart sensors:These are devices that can gather data about the environment, such as temperature, humidity, and air quality, and sending that data to a central management system.
Smart meters:These are devices that are used to measure and monitor energy usage in real-time, allowing cities to manage energy more efficiently and reduce waste.
Smart lighting:This technology uses connected lighting systems to manage streetlights and other public lighting, reducing energy consumption and improving safety.

These are just a few examples of the many IoT devices that are emerging as key technologies in smart cities.

4.2 Edge/Cloud Computing

Cloud computing and edge computing are two emerging technologies that are having a significant impact on the development of smart cities.

Cloud computing: This technology enables cities to store, process, and manage vast amounts of data in a centralized, cloud-based environment. This data can then be analysed and used to improve city planning and management.
Edge computing: Edge computing technology enables data processing to occur at the edge of a network, closer to the source of the data. This reduces the amount of data that needs to be transmitted to the cloud, improving the speed and reliability of data processing. These technologies are being used in a variety of ways to improve city operations and services. For example, edge computing can be used to support real-time monitoring and control of critical infrastructure, such as traffic manage-

ment systems, while cloud computing can be used to store and analyze data from many connected devices and sensors. Together, cloud computing and edge computing are helping to create a more connected, efficient, and sustainable future for cities around the world. The accepted use of edge computing-based IoT devices(Renata Walczak, 2023) are deployed to gather information in smart cities. This study held at Warsaw University of Technology also this study prevents city dwellers from embracing data collection technology.

4.3 Internet Protocols

The Internet Protocol (IP) is the foundation of the modern internet, and there are several emerging IP technologies that are actively contributingto the development of smart cities. These include:

IPv6: This is the latest version of the Internet Protocol, and it provides a much larger address space than its predecessor, IPv4. This is important in smart cities because it enables the deployment of many connected devices and sensors.

Low-Power Wide-Area Networks (LPWAN): This is a type of wireless networking technology that is designed to support many connected devices over a long-range using very little power. This makes it ideal for use in smart cities, where the deployment of large numbers of sensors is crucial.

5G: The fifth generation of mobile networks, 5G provides faster speeds and lower latency than previous generations of mobile networks. This is important for smart cities because it enables the deployment of real-time applications and services, such as autonomous vehicles and traffic management systems.

MQTT: This is a lightweight messaging protocol that is designed for use in IoT systems. It is ideal for use in smart cities because it enables the efficient transmission of data between devices and systems, even over low-bandwidth networks.

These IP technologies are helping to create a more connected, efficient, and sustainable future for cities around the world.

4.4 Artificial Intelligence

The field of artificial intelligence (AI), which is advancing quickly, has the potential to completely change how cities are managed and operated. Some of the most significant emerging AI technologies in smart cities include:

Machine learning: This technology enables cities to process and analyse large amounts of data in real-time, providing insights that can be used to improve city planning and management.

Natural language processing (NLP): NLP technologies are used to process and analyse human language, enabling cities to interact with citizens in more natural and intuitive ways.

Computer vision: Computer vision technologies enable cities to gather and analyse visual data, such as video feeds from cameras, to improve public safety and monitor critical infrastructure.

Predictive maintenance: Predictive maintenance technologies use machine learning algorithms to analyse data from connected devices and sensors, helping cities to predict and prevent equipment failures and reduce downtime.

Autonomous systems: Autonomous systems, such as self-driving vehicles, employ artificial intelligence (AI) to decide and act, improving efficiency and reducing the need for human intervention.

These AI technologies are helping to create a more efficient, sustainable, and intelligent future for cities around the world. By enabling cities to process and analyse large amounts of data in real-time, AI is transforming the way cities are managed and operated and improving the quality of life for citizens.

4.5 Big Data Analytics

Big data and analytics are emerging technologies that are playing a significant role in the development of smart cities. They enable cities to collect, process, and analyse vast amounts of data from numerous sources, providing valuable insights that can be used to improve city planning and management.

Hadoop: This is an open-source software framework that enables the storage and processing of large amounts of data. It is used in smart cities to store and process data from a variety of sources, including sensors and connected devices.

Spark: This is a fast, in-memory data processing engine that can be used to handle enormous amounts of data in real-time. It is used in smart cities to analyse and process data from a variety of sources, including sensors and connected devices.

NoSQL databases: These are databases that are designed to store and process large amounts of unstructured data. They are used in smart cities to store and process data from a variety of sources, including social media, sensors, and connected devices.

Data visualization tools: These tools enable cities to visualize and analyse large amounts of data, providing valuable insights that can be used to improve city planning and management.

These big data and analytics technologies are helping to create a more connected, efficient, and sustainable future for cities around the world. By enabling cities to process and analyse large amounts of data in real-time, these technologies are transforming the way cities are managed and operated and improving the quality of life for citizens.

4.6 Blockchain

Blockchain technology is an emerging technology that has the potential to revolutionize the way cities are managed and operated. Some of the most significant emerging blockchain technologies in smart cities include:

Decentralized Identity Management: This technology enables cities to manage and secure digital identities, providing a secure and efficient way to store and access sensitive information.

Smart Contracts: Smart contracts are self-executing contracts in which the conditions of the agreement between buyer and seller are directly incorporated into lines of computer code. They are used in smart cities to automate and streamline various processes, including property transactions and supply chain management.

Tokenization: This technology enables cities to tokenize assets, such as real estate and energy, making it easier to buy, sell, and trade these assets on a decentralized platform.

Public Record Keeping: Blockchain technology can be used to securely store and manage public records, such as property and voting records, improving transparency and security.

Supply Chain Management: Blockchain technology can be used to track goods and materials as they move through the supply chain, enabling cities to monitor and improve the sustainability of their supply chains.

These blockchain technologies are helping to create a more secure, transparent, and efficient future for cities around the world. By enabling cities to automate and streamline processes and securely store and access sensitive information, blockchain technology is transforming the way cities are managed and operated and improving the quality of life for citizens.

4.7 Security and Privacy

As cities become more connected and reliant on technology, the importance of security and privacy technologies is growing. Some of the most significant emerging security and privacy technologies in smart cities include:

Encryption:Encryption technologies are used to secure data, preventing unauthorized access, and protecting sensitive information.

Firewalls: Firewalls are network security systems that protect against unauthorized access to computer networks. They are used in smart cities to secure networks and protect against cyberattacks.

Access control: Access control technologies are used to manage and restrict access to systems and data, ensuring that only authorized individuals can access sensitive information.

Identity and Access Management (IAM): IAM technologies are used to manage and secure digital identities, ensuring that only authorized individuals can access systems and data.

Multi-Factor Authentication (MFA): MFA technologies require users to provide multiple forms of authentication, such as a password and a fingerprint, to access systems and data, providing an additional layer of security.

These security and privacy technologies are essential to ensuring the security and privacy of sensitive information in smart cities. By protecting against unauthorized access and cyberattacks, these technologies are helping to create a more secure and trustworthy future for cities around the world.

4.8 Digital Twin

Digital twin technology is an emerging technology that is transforming the way cities are managed and operated. A digital twin is a virtual replica of a physical object, system, or city, that can be used to simulate, analyze, and optimize real-world processes and systems. Some of the most significant emerging digital twin technologies in smart cities include:

Smart City Digital Twins: These digital twins are virtual replicas of entire cities, enabling city planners and managers to simulate and analyze city-wide systems, including transportation, energy, and water networks.

Building Digital Twins: These digital twins are virtual replicas of individual buildings, enabling building owners and managers to simulate and optimize building systems, including heating, cooling, and lighting systems.

Infrastructure Digital Twins: These digital twins are virtual replicas of infrastructure systems, such as water and energy networks, enabling city planners and managers to simulate and analyse these systems to improve efficiency and sustainability.

Traffic Digital Twins: These digital twins are virtual replicas of city-wide transportation systems, enabling city planners and managers to simulate and analyze traffic patterns and optimize transportation networks.

These digital twin technologies are helping to create a more efficient, sustainable, and resilient future for cities around the world. By enabling city planners and managers to simulate and analyse city-wide systems, digital twin technology is transforming the way cities are managed and operated and improving the quality of life for citizens.

5. MACHINE LEARNING ENABLED IoT IN SMART CITIES

The Internet of Things (IoT) and Machine Learning (ML) are two technologies that are often used together to create powerful and intelligent systems. In the context of smart cities, IoT and ML can be used to improve various aspects of city life, such as traffic management, energy efficiency, and public safety. IoT devices generate vast amounts of data, which can be analysed and processed using ML algorithms. For example, data from traffic sensors can be used to train ML algorithms to predict traffic patterns and optimize traffic flow. Similarly, data from smart energy meters can be used to train ML algorithms to predict energy usage patterns and optimize energy efficiency. ML can also be used to enhance the security and reliability of IoT systems. For example, ML algorithms can be used to detect and prevent cyberattacks on IoT devices, and to identify and fix problems in the network before they cause widespread disruption. Predicting rainfall in real-time is the critical role of IoT in weather forecasting. A rainfall prediction system using machine learning algorithms is proposed to get result in real-time environment(Rahman et al., 2022). This model is tested using weather data from 2015 to 2017 and this fusion based model performs well than the other models. The survey study (Band et al., 2022) mentioned that the popular machine learning algorithms are used in smart city for various applications like energy management, waste collection, intruder detection, vehicular traffic and etc. Overall, the combination of IoT and ML has the potential to create a more efficient, sustainable, and secure future for smart cities. By using IoT devices to generate data and ML algorithms to process and analyse this data, cities can gain valuable insights into the functioning of their systems, enabling them to make informed decisions and improve the quality of life for citizens.

6. SMART CITES IN INDIA

The goal of building a smart city is to improve the quality of life for its citizens, increase sustainability, and reduce costs and resource consumption. In India, the concept of smart cities has gained significant momentum in recent years, and several cities across the country have taken initiatives to adopt smart city technologies.

One of the flagship smart city projects in India is the Smart Cities Mission(Ministry of Urban Development Government of India, 2015) launched by the government of India in 2015. Under this mission, 100 cities across the country were selected to be developed as smart cities, with each city receiving central financial assistance of Rs. 500 crores to implement its smart city plan. The mission aims to promote economic growth and improve the quality of life of citizens by encouraging local area development and utilizing technology that leads to smart results.

Some of the key elements of smart cities in India include:

1. Intelligent transportation systems: The integration of technology in the transportation sector aims to improve traffic flow, reduce congestion, and provide citizens with real-time information about transportation options.
2. Smart energy management: This involves the use of smart grids, renewable energy sources, and energy-efficient buildings to reduce energy consumption as well as the city's carbon footprint.
3. Smart healthcare: The use of technology in the healthcare sector aims to improve access to healthcare services, reduce costs, and enhance the quality of care.
4. Smart governance: This involves the use of technology to improve governance and civic engagement, enhance transparency and accountability, and increase citizen participation in government processes.
5. Smart building and housing: The use of technology in the construction sector aims to improve the sustainability and energy efficiency of buildings and reduce the overall carbon footprint of the city.

In recent years, several Indian cities have taken steps to implement Internet of Things (IoT) technology as part of their smart city initiatives. Some of the cities that have already implemented IoT solutions include Bengaluru, Pune, Jaipur, Surat and Bhubaneshwar. All of the above-mentioned cities have incorporated smart lighting system and intelligent transport management in some of the areas of the city. According to (Manasi S, 2015) Green buildings are getting popular in Bengaluru with several projects initiated in the city, these buildings focus on water conservation, energy efficiency and being environmentally responsible. Another initiative this city has taken is the installation of smart e-toilets which flushes automatically and also has smart fan and lighting. The city of Surat has taken steps to provide the public with WiFi hotspots across various locations in the city to provide the citizens with better connectivity and networking. The city of Bhubaneshwar has created a citizen engagement platform to encourage the citizens to participate in the city's development and decision-making processes to improve the quality of life for its citizens.

These are just a few examples of the cities in India that have already implemented IoT solutions. Overall, the use of IoT technology in Indian cities holds great promise to improve the quality of life for citizens, increase sustainability, and drive economic growth. However, it is important to ensure that the technology is implemented in a way that is inclusive and accessible to all citizens, regardless of socio-economic background.

7. EVIDENCE FOR SMART CITIES FROM VARIOUS COUNTRIES

There are numerous examples of successful implementation of smart city initiatives in various countries around the world. Some of the popular examples include Singapore, Helsinki, Zurich, Amsterdam, Oslo, Seoul, Copenhagen, San Francisco, Berlin, London and Sydney. According to the Smart City Index 2021 report Singapore was elected as the smartest city in the world followed by Zurich in the second place and Oslo in the third place. All of the above-mentioned cities have implemented smart transportation system that includes real-time traffic monitoring, predictive monitoring and intelligent traffic management along with smart energy system that incorporates renewable energy sources, energy efficient buildings and smart grid technology. Another example of a smart city is Dubai (Kadhim, 2019) which has transformed itself from a desert legend to the first smart city in the Middle East, this city has focused most on reducing energy consumption and has taken sustainable approaches to be environmentally responsible.

These are a very few examples of the cities around the world that have implemented IoT solutions as part of their smart city initiatives. The use of IoT technology in cities holds great promise to improve the quality of life for citizens, increase sustainability, and drive economic growth. However, it is important to ensure that the technology is implemented in a way that is inclusive and accessible to all citizens, regardless of socio-economic background, and that privacy and security concerns are properly addressed.

8. CHALLENGES IN IMPLEMENTING IOT IN SMART CITIES

The implementation of IoT in smart cities can face several challenges, some of which include:

1. Privacy and security concerns: One of the biggest challenges of implementing IoT in smart cities is ensuring the privacy and security of citizens' personal and sensitive information. With the large amounts of data generated by IoT devices, it is important to ensure that this data is protected from cyber-attacks, hacking, and unauthorized access.
2. Cost: Implementing IoT technology in cities can be expensive, and many cities may struggle to find the resources to finance the development and deployment of IoT systems. This can be particularly challenging for smaller cities or cities with limited budgets.
3. Interoperability: Another challenge of implementing IoT in smart cities is ensuring that different IoT devices and systems can work together seamlessly. With many different types of devices and systems being used, it can be difficult to ensure that they all communicate effectively with one another.
4. Complexity: Implementing IoT in smart cities can be complex, with many different types of systems, devices, and technologies involved. This can make it difficult for cities to effectively manage and maintain their IoT systems over time.
5. Regulation: There are also challenges related to regulation and standardization in the implementation of IoT in smart cities. For example, cities may struggle to ensure that the data generated by IoT devices is being used in a responsible and ethical way, and that citizens' privacy rights are being protected.

Despite these challenges, the potential benefits of IoT in smart cities are significant, and many cities are continuing to work to overcome these challenges and implement IoT systems to improve the quality of life for their citizens, increase sustainability, and drive economic growth.

9. RESEARCH ISSUES AND SOLUTIONS AND OPPORTUNITIES FOR SMART CITIES

There are several research issues and opportunities in the area of smart cities:

1. Privacy and security: One of the key research issues in smart cities is the privacy and security of citizens' personal and sensitive information generated by IoT devices and other technology systems. Researchers are working to develop solutions to protect this data from cyber-attacks, hacking, and unauthorized access.
2. Interoperability: Another research issue in smart cities is ensuring that different technology systems, such as IoT devices and transportation systems, can work together seamlessly. Researchers are working on developing standards and protocols that can be used to ensure that different systems can communicate effectively with one another.
3. Energy efficiency: Smart cities use a significant amount of energy, and researchers are working on developing new technologies that can help reduce energy consumption and improve the sustainability of these cities.
4. Human-centered design: Researchers are also working on developing smart city technologies that are more centered around human needs and take into account the unique characteristics of different cities and communities.
5. Predictive analytics: Another area of opportunity in smart cities is the use of predictive analytics to make data-driven decisions. By analyzing large amounts of data generated by technology systems, researchers are working to develop models that can be used to make predictions about everything from traffic patterns to energy usage.
6. Smart transportation: There is also significant opportunity for research in the area of smart transportation, particularly in developing systems that can improve traffic flow, reduce emissions, and improve safety.
7. Inclusive urban planning: Another important research issue in smart cities is ensuring that smart city initiatives are inclusive and benefit all citizens, regardless of their socio-economic status or other factors.

These are just a few examples of the research issues and opportunities in the area of smart cities. By addressing these challenges and capitalizing on these opportunities, researchers can help to create smarter, more sustainable, and more livable cities for the future.

Smart cities present a range of opportunities for businesses, governments, and citizens alike, including:

1. Improved quality of life: Smart city initiatives can help to improve the quality of life for citizens by making cities more livable, efficient, and sustainable. This can include initiatives such as smart transportation systems, improved waste management, and more efficient energy usage.

2. Economic development: Smart cities can also help to drive economic development by attracting new businesses and investment, improving the efficiency of existing businesses, and creating new job opportunities.

3. Increased sustainability: By using technology to reduce waste and increase efficiency, smart cities can help to reduce their environmental footprint and become more sustainable over time.

4. Better governance: Smart city initiatives can also help to improve governance by providing governments with new tools and insights to make more informed decisions, and by making it easier for citizens to engage with their local government.

5. Improved public services: By leveraging technology, smart cities can also improve the delivery of public services, such as healthcare, education, and public safety.

6. New business opportunities: There is also significant opportunity for businesses in the smart city space, including in areas such as IoT, big data, renewable energy, and smart transportation.

These are just a few examples of the opportunities that smart cities present. By taking advantage of these opportunities, cities can become more livable, efficient, and sustainable, and can provide new opportunities for citizens, businesses, and governments alike.

10. CONCLUSION

The Internet of Things (IoT) is an essential component of smart cities, providing the technology and connectivity that is needed to create more efficient, sustainable, and livable cities. IoT devices and systems are used to gather data, automate processes, and connect different technology systems, all of which are critical to the success of smart cities.However, there are also several challenges that must be overcome in order to fully realize the potential of IoT in smart cities. These include privacy and security concerns, ensuring interoperability between different systems, and ensuring that smart city initiatives are inclusive and benefit all citizens.

Despite these challenges, there are also many opportunities that come with IoT in smart cities. These include improved quality of life, economic development, increased sustainability, better governance, and improved public services.

In conclusion, IoT is a critical component of smart cities, and has the potential to transform the way that cities function and provide significant benefits to citizens, businesses, and governments alike. However, careful planning, design, and implementation are needed to ensure that IoT is used effectively and responsibly in smart cities, and that all citizens are able to reap the benefits of this technology.

11. REFERENCES

Aryal, A., Becerik-Gerber, B., Anselmo, F., Roll, S. C., & Lucas, G. M. (2019). Smart Desks to Promote Comfort, Health, and Productivity in Offices: A Vision for Future Workplaces. *Frontiers in Built Environment*, *5*, 76. doi:10.3389/fbuil.2019.00076

Band, S. S., Ardabili, S., Sookhak, M., Chronopoulos, A. T., Elnaffar, S., Moslehpour, M., Csaba, M., Torok, B., Pai, H.-T., & Mosavi, A. (2022). When Smart Cities Get Smarter via Machine Learning: An In-Depth Literature Review. *IEEE Access : Practical Innovations, Open Solutions*, *10*, 60985–61015. doi:10.1109/ACCESS.2022.3181718

Beckett-Camarata, J. (2022). Smart and connected cities in Pennsylvania: A multi-case study. Smart Cities and Regional Development (SCRD). *Journal*, *6*(1), 67–78. doi:10.25019crd.v6i1.121

Guan, K., Rupp, M., Kurner, T., Briso, C., Matolak, D. W., Takada, J.-I., & Wang, W. (2020). IEEE Access Special Section Editorial: 5G and Beyond Mobile Wireless Communications Enabling Intelligent Mobility. *IEEE Access : Practical Innovations, Open Solutions*, *8*, 208892–208897. doi:10.1109/ACCESS.2020.3037635

Gutierrez-Martinez, V. J., Moreno-Bautista, C. A., Lozano-Garcia, J. M., Pizano-Martinez, A., Zamora-Cardenas, E. A., & Gomez-Martinez, M. A. (2019). A Heuristic Home Electric Energy Management System Considering Renewable Energy Availability. *Energies*, *12*(4), 671. doi:10.3390/en12040671

Iliev, Y., & Ilieva, G. (2022). A Framework for Smart Home System with Voice Control Using NLP Methods. *Electronics (Basel)*, *12*(1), 116. doi:10.3390/electronics12010116

Kadhim, W. (2019). Case study of Dubai as a Smart City. *International Journal of Computer Applications*, *178*(40), 35–37. doi:10.5120/ijca2019919291

Manasi, S. (2015). Smart Technologies in Smart Cities – Insights into Bengaluru City's initiatives. *Journal of Development, Management and Communication*.

Mihalache, A. (2017, June 30). Wireless Home Automation System using IoT. Informatica Economica, 21(2/2017), 17–32. doi:10.12948/issn14531305/21.2.2017.02

Ministry of Urban Development Government of India. (2015). *Smart Cities*. MUDGI. https://smartcities.gov.in/themes/habikon/files/SmartCityGuidelines.pdf

Rahman, A., Abbas, S., Gollapalli, M., Ahmed, R., Aftab, S., Ahmad, M., Khan, M. A., & Mosavi, A. (2022). Rainfall Prediction System Using Machine Learning Fusion for Smart Cities. *Sensors (Basel)*, *22*(9), 3504. doi:10.339022093504 PMID:35591194

Rodriguez, J. A., Fernandez, F. J., & Arboleya, P. (2018). Study of the Architecture of a Smart City. *The 2nd International Research Conference on Sustainable Energy, Engineering, Materials and Environment*. MDPI. 10.3390/proceedings2231485

Sharma, H., Haque, A., &Blaabjerg, F. (2021). Machine Learning in Wireless Sensor Networks for Smart Cities: A Survey. *Electronics, 10*(9), 1012. MDPI AG. doi:10.3390/electronics10091012

Tavakoli, M., Carriere, J., & Torabi, A. (2020). Robotics, Smart Wearable Technologies, and Autonomous Intelligent Systems for Healthcare During the COVID-19 Pandemic: An Analysis of the State of the Art and Future Vision. *Advanced Intelligent Systems*, *2*(7), 2000071. doi:10.1002/aisy.202000071

Trovato, V., Sfameni, S., Rando, G., Rosace, G., Libertino, S., Ferri, A., & Plutino, M. R. (2022). A Review of Stimuli-Responsive Smart Materials for Wearable Technology in Healthcare: Retrospective, Perspective, and Prospective. *Molecules (Basel, Switzerland)*, 27(17), 5709. doi:10.3390/molecules27175709 PMID:36080476

Walczak, R., Koszewski, K., Olszewski, R., Ejsmont, K., & Kálmán, A. (2023). Acceptance of IoT Edge-Computing-Based Sensors in Smart Cities for Universal Design Purposes. *Energies*, 16(3), 1024. doi:10.3390/en16031024

Chapter 17
IoT and Machine Learning on Smart Home–Based Data and a Perspective on Fog Computing Implementation

Asha Rajiv
School of Sciences, Jain University (Deemed), India

Dharmesh Dhabliya
Symbiosis Law School, Symbiosis International University, Pune, India

Abhilash Kumar Saxena
Teerthanker Mahaveer University, India

R. K. Yadav
Raj Kumar Goel Institute of Technology, India

Digvijay Singh
Dev Bhoomi Uttarakhand University, India

Ankur Gupta
Vaish College of Engineering, India

Aishwary Awasthi
Sanskriti University, India

ABSTRACT

This study emphasises the need for energy efficiency in buildings, focusing primarily on the heating, ventilation, and air conditioning (HVAC) systems, which consume 50% of building energy. A predictive system based on artificial neural networks (ANNs) was created to generate short-term forecasts of indoor temperature using data from a monitoring system in order to reduce this energy use. The technology seeks to estimate inside temperature in order to determine when to start the heating, ventilation, and air conditioning system, potentially reducing energy use dramatically. The chapter describes the system's code implementation, which includes data pre-processing, model training and evaluation, and result visualisation. In terms of evaluation metrics, the model performed well and revealed the potential for large energy savings in buildings.

DOI: 10.4018/978-1-6684-8785-3.ch017

INTRODUCTION

The article addresses the use of an Artificial Neural Network (ANN) system to estimate the indoor temperature of a room, using data from a complex monitoring system serving as input. The purpose is to save energy by anticipating the indoor temperature and deciding whether or not to turn on the Heating, Ventilation, and Air Conditioning (HVAC) system. Because building energy consumption accounts for a large fraction of overall global energy consumption, the article emphasises the necessity of energy efficiency-related methods and optimisation approaches in buildings to minimise global energy consumption. The installation of such devices could serve as a wake-up call for building owners, urging them to reassess their operations and begin investigating new technologies. In addition to the ANN system, the paper discusses fog computing, a decentralised computing infrastructure that allows data processing and storage to be closer to the data source. Fog computing has the ability to increase the ANN system's efficiency and accuracy, as well as other building automation systems.

Fog Computing to Analyse Smart Home Based Data

Fog computing is a paradigm that has evolved as a solution to the centralised cloud computing model's difficulties. It entails deploying computing resources near the network's edge, closer to the data source, in order to give a more distributed and decentralised approach to data processing and analysis (Lamiae et al., 2023). Fog computing can provide an efficient approach to process data received from numerous sensors and devices deployed within a building in the context of building energy management. Data may be processed and analysed in real-time by putting fog nodes closer to these devices, lowering latency and enhancing the overall efficiency of the building energy management system (Aqib et al., 2023). Fog computing can also offer a more scalable and adaptable approach to building energy management, allowing building owners to swiftly deploy new sensors and devices as needed and react to changing energy use trends. Fog computing is a type of distributed computing infrastructure that extends cloud computing to the network's edge (Katal et al., 2023). It allows for data processing and analysis to take place closer to the source of the data (Padhy et al., 2023). Fog computing can be used to process and analyse data from sensors and devices located in buildings in the context of building energy usage (Anoushee et al., 2023). Fog computing, for example, may analyse data from temperature sensors, occupancy sensors, lighting systems, and HVAC systems to find patterns and anomalies in energy consumption (Anoushee et al., 2023).

IoT in Smart Homes Data and Fog Computing

The application of machine learning and IoT on smart home data to reduce energy usage is demonstrated in this code implementation. The goal is to estimate indoor temperature using previous data and then utilise the projections to make more energy-efficient HVAC decisions. To provide a short-term forecast of indoor temperature, the system employs an Artificial Neural Network (ANN) model. This forecast enables the smart home to adjust to future temperature circumstances and operate the HVAC system in the most energy-efficient way possible. Furthermore, this application emphasises the significance of energy efficiency tactics and optimisation approaches in structures. HVAC's high power use, accounting for a large percentage of total consumption, mandates the employment of these strategies to reduce building energy consumption. The implementation analyses the data and generates helpful insights by utilising several numerical and data visualisation libraries. Data pretreatment techniques such as feature engineering, date encoding, and ordinal encoding are also used to prepare the data for the ANN model. Finally, the implementation assesses the ANN model's performance using several measures where the model's evaluation measures demonstrate its ability to predict indoor temperature and reduce energy consumption.

Applications of Internet of Things and Fog Computing in Smart Homes

Energy Management: IoT and fog computing can be used to monitor and manage smart homes' energy consumption. Homeowners may track their energy consumption and make changes to reduce it by using sensors and smart gadgets. The data collected by these sensors can be analysed in real-time using fog computing to optimise energy usage and prevent waste.

Home Security: Internet of Things-enabled security systems can identify and alert homeowners to potential security concerns such as burglaries or fires. The sensors can be linked to a fog computing network, which can process and analyse data in real time, alerting homeowners to any system irregularities.

Health Monitoring: Smart homes can be outfitted with health monitoring systems that use IoT sensors to track residents' health. This data can be analysed in real-time using fog computing to warn healthcare providers of any important health issues. IoT sensors can be used to monitor the quantity of natural light entering a room and change the inside lighting as needed. This data can be processed in real-time using fog computing to optimise illumination based on the time of day, weather conditions, and user preferences.

HVAC Optimisation: Data acquired from sensors and smart devices can be used to optimise IoT-enabled HVAC systems. Fog computing can be used to analyse this data and modify temperature settings based on occupancy, weather conditions, and energy efficiency.

Smart Appliances: Internet of Things-enabled appliances, such as refrigerators and washing machines, may interact with one another and with homeowners in order to optimise their utilisation. For example, when food is due to expire, a refrigerator may send a warning to a homeowner's phone, and a washing machine could alter its cycle based on the weight of the laundry load.

Water Management: IoT sensors can be used in smart homes to monitor water usage, detect leaks, and notify homeowners of any anomalies. This data can be analysed in real-time using fog computing to optimise water usage and reduce waste.

Elder Care: In smart homes, IoT and fog computing can be utilised to monitor the health and safety of older people. This can incorporate fall detection, medication reminders, and cognitive decline monitoring. This data can be analysed in real-time using fog computing to warn carers of any significant incidents.Personalised

Entertainment: Entertainment systems that use IoT sensors to personalise the user experience can be installed in smart homes. A smart TV, for example, may suggest shows based on the user's viewing history, while a smart speaker could play music based on the user's mood and preferences. This data can be analysed in real-time using fog computing to produce a more personalised entertainment experience.

Literature Review

Das et al., (2023) explored the difficulties, characteristics, limitations, and prospective applications of fog computing. They provided a fog computing based environment on current fog computing research and identified security, privacy, application, and communication problems as major contributors. They also highlighted possible fog computing applications, such as healthcare applications, creative city applications, and farm applications.

Hazra et al., (2023) conducted a survey of fog computing for next-generation internet of things (IoT) applications, encompassing IoT architecture, several of the evaluation metrics, and the application features of fog computing, as well as its advancement over the last four years. They also investigated numerous obstacles and potential solutions for establishing interoperable communication.

Hassan et al., (2023) presented an outline of various sectors related with the fog computing paradigm for efficiently executing internet of things (IoT) applications. They presented a remote pain monitoring application based on fog computing and evaluated the proposed policy.

Javanmardi et al., (2023) addressed the security problems of the IoT-Fog architecture, which includes fog devices at the fog layer, as well as fog layer vulnerabilities and assaults. In addition, the survey described the most popular IoT-Fog security defense measures and investigated how SDN may assist these mechanisms in IoT-Fog networks.

Gupta et al., (2023) did a survey on intelligent resource management using Reinforcement Learning in a Fog Computing-based IOT environment. The survey investigated the issues in fog computing resource management and compared RL and DRL approaches to classic algorithmic methods.

Gupta et al. (2020) developed a machine learning model for breast cancer prediction in the context of the Internet of Things (IoT) in I-SMAC. The scientists created a prediction model using several machine learning approaches such as support vector machine and logistic regression, which yielded promising accuracy results.

Gupta et al., (2022) proposed a wearable sensor-based smart home evaluation system based on sequential minimization optimisation and random forest. The authors collected data from multiple sensors and created a prediction model using random forest, which outperformed typical machine learning techniques in terms of accuracy.

Ortiz-Garcés et al., (2023) demonstrated an emergency response prototype leveraging Internet of Things in an environment based on Fog Computing. The prototype attempted to shorten response times and deliver only relevant data to cloud computing by interconnecting fog computing nodes close to the vehicle using a low-power area network protocol. A comparison of fog computing with cloud computing was performed.

Songhorabadi et al., (2023) offered a study for cutting-edge fog-based approaches, studying three classifications: service-based, and application-based and resource-based. They also researched the evaluation elements, employed tools, evaluation methodologies, the benefits and drawbacks of each class, and presented complete and distinct open concerns and challenges

Talukdar et al., (2022) demonstrated a detection based on suspicious activity and categorization system in an IoT context using a machine learning approach. The authors classified suspicious actions using several techniques such as k-nearest neighbor and random forest, demonstrating the effectiveness of their suggested approach.

Veeraiah et al,. (2022) addressed how incorporating IoT devices can improve metaverse capabilities. The authors developed a framework for integrating IoT devices to improve the functionality and capacities of the metaverse, giving IoT devices in smart homes a new dimension.

Sharma et al., (2023) proposed an optimal task scheduling two-staged approach for the task distribution that is performed within smart homes taking into consideration many fog nodes, which uses a naive-Bayes model for training first and then a hyperheuristic approach for optimisation in the second stage.

Problem Statements

Building energy use notably that of Heating, Ventilation, and Air Conditioning (HVAC) systems, accounts for a considerable fraction of global energy consumption. With rising global energy demand, it is critical to employ energy-efficient measures and optimisation techniques in buildings to reduce energy usage. Predictive systems based on Machine Learning algorithms, for example, can forecast indoor temperature. The usage of IoT devices and sensors in smart homes provides a massive amount of data that may be used to manage building energy. However, efficient data analysis and processing remains a hurdle. Furthermore, due to significant latency and network capacity limits, reliance on cloud computing for real-time processing of this data may not be suitable.

Visualisation of the Smart Home IoT Data

Figure 1 depicts the CO_2 concentration in a building's living and dining rooms over time. The x-axis is the observation index, which is effectively a time measure, and the y-axis is the CO_2 concentration in parts per million (ppm). The plot contains two lines, one for the CO_2 concentration in the dining room and the other for the CO_2 concentration in the living room. The graphic demonstrates how the CO_2 content varies over time in rooms. This could be an indicator of inadequate ventilation or air circulation in the building. It also demonstrates that the CO_2 concentration in the living room is sometimes higher than in the dining room, and vice versa.

Figure 1. A line plot visualising CO_2 concentration in facade

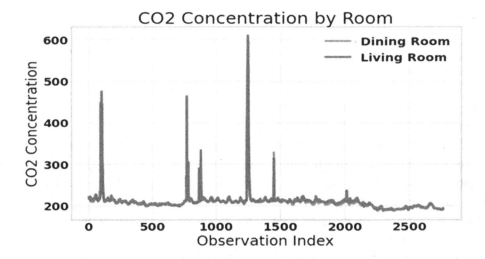

Figure 2. A scatter plot showing meteo wind data visualisation

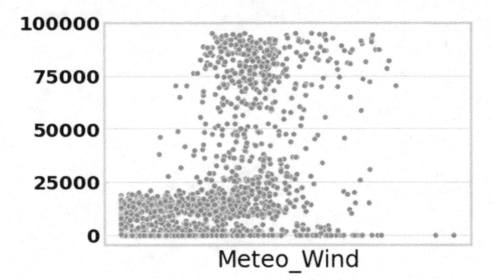

Figure 2 depicts a scatter plot of Meteo Sun Light in Smart Homes' West Facade over time. Figure 3 depicts a visualisation of the forecast components generated by the Prophet model. The magnitude of the component is shown on the y-axis of each subplot, and the time period during which the component is calculated is shown on the x-axis. The trend component is depicted as a blue line with circle marks at each observation in this graphic. The weekly and yearly components are depicted as bars with circular markers at the top. The graphic is useful for visualising the many patterns in the data and understanding how they affect the forecast. It can also aid in the detection of any anomalies or outliers in the data that may have an impact on the forecast. In Figure 4 the scatter plots depict the relationship between the indoor temperature room variable and its previous values with the lag values selected. In the scatter plot with lag=2, for example, each point represents the current value of the indoor temperature room variable displayed against its value two time steps ago. The autocorrelation of the indoor temperature room variable can be visualised using scatter plots. If there is a positive connection between the temperature room variable and its previous values, we will notice a pattern of points that are densely packed around a line in these plots. If there is no association, the points will be distributed at random.

Figure 3. Visualisation of forecast components based on days of the week

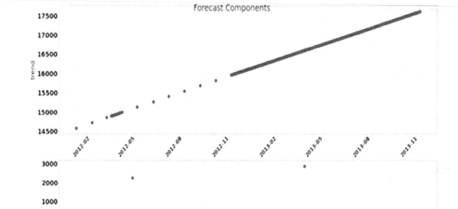

Figure 4. Figure showing lag plots of the data

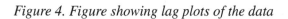

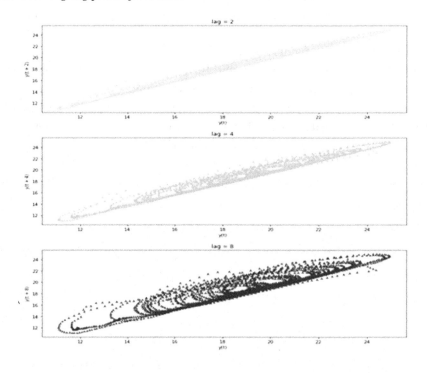

Figures 5 and 6 depict the cross-correlation of a room's interior temperature with various numerical variables in the dataset. The graphic contains a grid of subplots, each representing the cross-correlation of the interior temperature with a specific numerical feature. A cross correlation diagram depicts the relationship between two signals as a function of time delay.

Figure 5. Visualization showing different features of the dataset

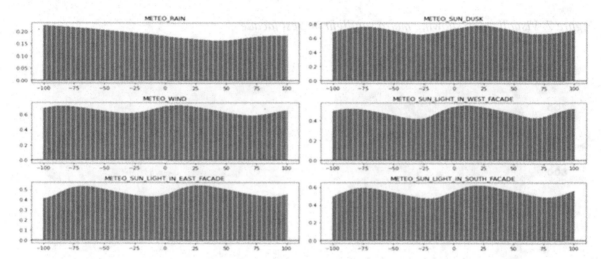

The graph allows us to see if there are any substantial lags between the indoor temperature and the other numerical variables in the dataset. For example, we can see substantial negative or positive correlations between the indoor temperature and the related numerical feature in some subplots, indicating that the two signals have a link at a specific time delay. This information can be used to select features or to create predictive models that seek to forecast the indoor temperature based on the dataset's other numerical attributes.

Figure 6. Visualization showing different features of the dataset -2

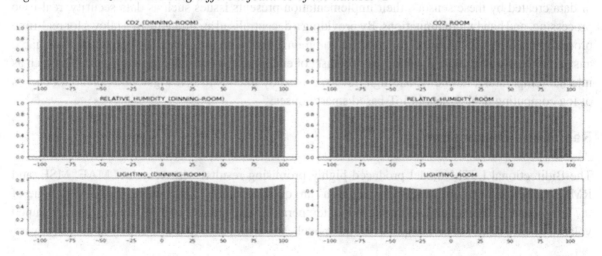

Figure 7 depicts the distribution of Meteo Sun Light in the building's West Facade. The x-axis displays Meteo Sun Light values, while the y-axis displays the frequency or number of observations at each Meteo Sun Light value. The curve in the illustration is a kernel density estimate (KDE), which is an estimate of the variable's probability density function. According to the graph, the Meteo Sun Light in the West Facade has a bimodal distribution, which implies it has two peaks. The first peak is near zero, indicating that there were several observations with low Meteo Sun Light values.

Figure 7. Visualization showing metro sun light in west facade

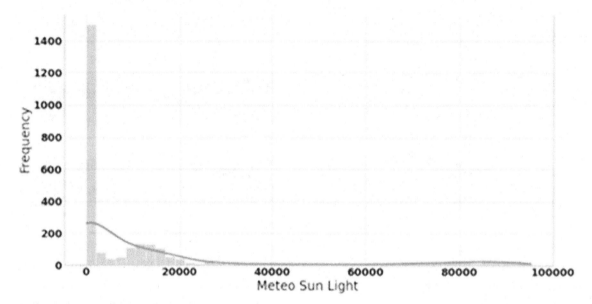

Proposed Work

The usage of IoT devices and machine learning algorithms in smart homes has provided numerous prospects for boosting energy efficiency and domestic automation. However, due to the massive amounts of data created by these sensors, their implementation presents issues such as data security, real-time processing, and bandwidth limitations. By providing a decentralised architecture that allows for real-time processing and reduces the need for constant communication with the cloud, fog computing is a viable answer to these difficulties. By analysing a dataset of energy usage and environmental variables such as internal temperature, humidity, and light levels, this study intends to investigate the potential benefits of fog computing deployment in IoT-based smart houses.

Result and Discussion

The Bidirectional LSTM model produced highly promising results, with very low MAE, MSE, and RMSE values, indicating that model predictions are quite accurate. Furthermore, the MAPE value of 0.77 suggests that the model can forecast indoor temperature with an error of less than 1%. The Explained volatility Score (EVS) of 0.98 implies that the model can explain 98% of the volatility in the target variable. Furthermore, the R2 value of 0.98 implies that the model explains 98% of the variance in the target variable. These findings show that the model worked well in predicting indoor temperature in a smart house. Figure 8 depicts the Bidirectional LSTM model's training and validation loss values during its training phase. The blue line represents the training loss curve, whereas the orange line represents the validation loss curve. The training loss of the model begins around 0.026 and rapidly reduces with each epoch, whereas the validation loss begins around 0.0037, drops until epoch 2, then slightly increases. The EarlyStopping callback is called at epoch 4, forcing the model to halt training. If the validation loss does not reduce after two consecutive epochs, this callback terminates the model's training. In epoch 4, the validation loss is 0.0056. Figure 10 represents the unidirectional LSTM.

Figure 8. Bidirectional LSTM model loss in training and validation

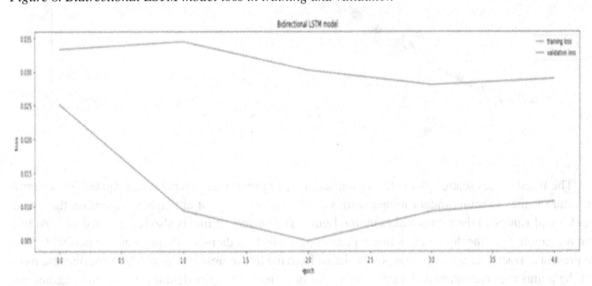

Figure 9. Actual vs predicted values of the bidirectional LSTM model

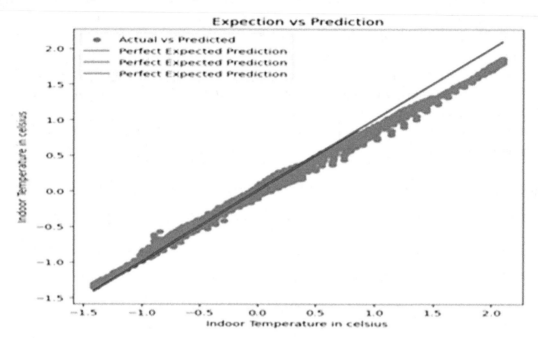

The fitted values scatter plot is the visualisation in Figure 9 that compares anticipated indoor temperature values to actual indoor temperature values. A scatter plot of anticipated values on the y-axis and actual values on the x-axis makes up the figure. The perfect fit line is also represented as a red line on the graph, with the dots on this line representing perfect predictions. Points near the perfect fit line represent correct forecasts, whereas points distant from the line represent poor predictions. Because most of the points are concentrated along the perfect fit line, the scatter plot demonstrates a good agreement between the projected and actual interior temperature values in this scenario. Similarly the scatter plot for uni-directional LSTM is shown in Figure 11.

Figure 10. Unidirectional LSTM model loss in training and validation

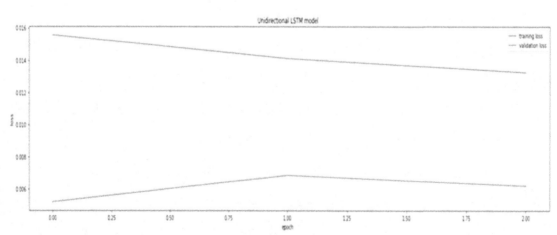

Figure 11. Actual vs predicted values of the unidirectional LSTM model

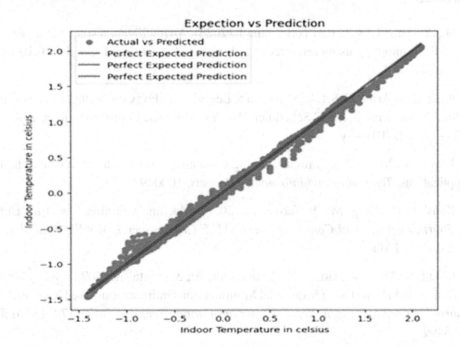

CONCLUSION

A Bidirectional LSTM model was used to estimate indoor temperature in a room based on environmental data such as outdoor temperature, humidity, and sun irradiation. We preprocessed the data by generating lag features and dividing it into training and validation sets. The model performed admirably, with an R-squared value of 0.98, suggesting that it accounts for 98% of the variance in the validation data. The model also exhibited low mean absolute error and mean squared error values, showing that the model's predictions were accurate. We can limit the quantity of data that needs to be sent to the cloud for processing by exploiting the capabilities of fog computing, which includes placing computing resources closer to the network's edge (i.e., at the device or gateway level). This can result in faster processing speeds and lower latency, which is important for real-time applications such as smart home automation.

Future Scope

This kind of project is going to have various future scopes. Exploration of more complex machine learning algorithms to increase the accuracy of temperature predictions is one prospective avenue. Another possibility is to broaden the project's scope beyond interior temperature prediction to include other aspects of energy management in buildings. Predicting energy demand, optimising the use of renewable energy sources, and controlling energy storage systems are all examples of this. Finally, the proposed system's implementation can be tested in a real-world scenario. The usage of fog computing can bring numerous advantages in terms of reducing latency and boosting overall system performance.

REFERENCES

Anoushee, M., Fartash, M., & Akbari Torkestani, J. (2023). An intelligent resource management method in SDN based fog computing using reinforcement learning. *Computing*, 1–30. doi:10.100700607-022-01141-x

Aqib, M., Kumar, D., & Tripathi, S. (2023). Machine Learning for Fog Computing: Review, Opportunities and a Fog Application Classifier and Scheduler. *Wireless Personal Communications*, *129*(2), 853–880. doi:10.100711277-022-10160-y

Das, R., & Inuwa, M. M. (2023). A review on fog computing: issues, characteristics, challenges, and potential applications. *Telematics and Informatics Reports,* 100049.

Gupta, A., Kaushik, D., Garg, M., & Verma, A. (2020). Machine Learning model for Breast Cancer Prediction. *Fourth International Conference on I-SMAC (IoT in Social, Mobile, Analytics and Cloud) (I-SMAC),* Palladam, India.

Gupta, N., Janani, S., Dilip, R., Hosur, R., Chaturvedi, A., & Gupta, A. (2022). Wearable Sensors for Evaluation Over Smart Home Using Sequential Minimization Optimization-based Random Forest. *International Journal of Communication Networks and Information Security*, *14*(2), 179–188. doi:10.17762/ijcnis.v14i2.5499

Gupta, S., & Singh, N. (2023). Toward intelligent resource management in dynamic Fog Computing-based Internet of Things environment with Deep Reinforcement Learning: A survey. *International Journal of Communication Systems*, *36*(4), e5411. doi:10.1002/dac.5411

Hassan, S. R., & Rashad, M. (2023). *Cloud Computing to Fog Computing: A Paradigm Shift*.

Hazra, A., Rana, P., Adhikari, M., & Amgoth, T. (2023). Fog computing for next-generation internet of things: Fundamental, state-of-the-art and research challenges. *Computer Science Review*, *48*, 100549. doi:10.1016/j.cosrev.2023.100549

Javanmardi, S., Shojafar, M., Mohammadi, R., Alazab, M., & Caruso, A. M. (2023). An SDN perspective IoT-Fog security: A survey. *Computer Networks*, *229*, 109732. doi:10.1016/j.comnet.2023.109732

Katal, A. (2023). Leveraging Fog Computing for Healthcare. In *Deep Learning Technologies for the Sustainable Development Goals: Issues and Solutions in the Post-COVID Era* (pp. 51–68). Springer Nature Singapore. doi:10.1007/978-981-19-5723-9_4

Lamiae, E., Hicham, G. T., Fatiha, E., & Mohammed, B. (2023, March). Smart Home and Machine Learning as a Sustainable Healthcare Solution: Review and Perspectives. In *Innovations in Smart Cities Applications Volume 6: The Proceedings of the 7th International Conference on Smart City Applications* (pp. 145-155). Cham: Springer International Publishing.

Ortiz-Garcés, I., Andrade, R. O., Sanchez-Viteri, S., & Villegas-Ch, W. (2023). Prototype of an Emergency Response System Using IoT in a Fog Computing Environment. *Computers*, *12*(4), 81. doi:10.3390/computers12040081

Ortiz-Garcés, I., Andrade, R. O., Sanchez-Viteri, S., & Villegas-Ch, W. (2023). Prototype of an Emergency Response System Using IoT in a Fog Computing Environment. *Computers*, *12*(4), 81. doi:10.3390/computers12040081

Padhy, S., Alowaidi, M., Dash, S., Alshehri, M., Malla, P. P., Routray, S., & Alhumyani, H. (2023). AgriSecure: A Fog Computing-Based Security Framework for Agriculture 4.0 via Blockchain. *Processes (Basel, Switzerland)*, *11*(3), 757. doi:10.3390/pr11030757

Padhy, S., Alowaidi, M., Dash, S., Alshehri, M., Malla, P. P., Routray, S., & Alhumyani, H. (2023). *AgriSecure: A Fog Computing-Based Security Framework for Agriculture 4*. AgriSecure.

Sharma, O., Rathee, G., Kerrache, C. A., & Herrera-Tapia, J. (2023). Two-Stage Optimal Task Scheduling for Smart Home Environment Using Fog Computing Infrastructures. *Applied Sciences (Basel, Switzerland)*, *13*(5), 2939. doi:10.3390/app13052939

Songhorabadi, M., Rahimi, M., MoghadamFarid, A. M., & Haghi Kashani, M. (2023). Fog computing approaches in IoT-enabled smart cities. *Journal of Network and Computer Applications*, *211*, 103557. doi:10.1016/j.jnca.2022.103557

Talukdar, S. B., Dhabliya, D., Ahamad, S., & Gupta, A. (2022). Machine Learning for Anomaly Detection in IoT. In Emerging Trends and Applications of Machine Learning in Data Science (pp. 335-352). IGI Global.

Talukdar, V., Dhabliya, D., Kumar, B., Talukdar, S. B., Ahamad, S., & Gupta, A. (2022, November). Suspicious Activity Detection and Classification in IoT Environment Using Machine Learning Approach. In *2022 Seventh International Conference on Parallel, Distributed and Grid Computing (PDGC)* (pp. 531-535). IEEE. 10.1109/PDGC56933.2022.10053312

Veeraiah, V., Gangavathi, P., Ahamad, S., Talukdar, S. B., Gupta, A., & Talukdar, V. (2022, April). Enhancement of meta verse capabilities by IoT integration. In *2022 2nd International Conference on Advance Computing and Innovative Technologies in Engineering* (ICACITE) (pp. 1493-1498). IEEE. 10.1109/ICACITE53722.2022.9823766

Chapter 18
Activity Recognition and IoT–Based Analysis Using Time Series and CNN

N. Beemkumar

Faculty of Engineering and Technology, Jain University (Deemed), India

Sachin Gupta

Anskriti University, India

Shambhu Bhardwaj

Teerthanker Mahaveer University, India

Dharmesh Dhabliya

Symbiosis Law School, Symbiosis International University, Pune, India

Mritunjay Rai

Shri Ramswaroop Memorial University, India

Jay Kumar Pandey

Shri Ramswaroop Memorial University, India

Ankur Gupta

Vaish College of Engineering, India

ABSTRACT

Using time series data obtained from accelerometer and gyroscope sensors on an iPhone 6s, the authors address the topic of human activity and attribute detection. The collection contains time series data from 24 subjects who completed six activities in 15 trials. The aim is to appropriately identify the six activities using machine learning techniques. The usage of a convolutional neural network (CNN) for the categorization of human activity and attribute identification data obtained from accelerometer and gyroscope sensors on an iPhone 6s is proposed in this research. The collection contains time series data from 24 subjects who completed six activities in 15 trials. The study begins by pre-processing the data by transforming the folders into class labels and plotting the time series data. The time-series data is made up of multivariate data from both the accelerometer and gyroscope sensors, totaling 12 characteristics. The accelerometer sums up two acceleration vectors, gravity, and user acceleration, which may be distinguished using core motion tracking technology.

DOI: 10.4018/978-1-6684-8785-3.ch018

INTRODUCTION

The Human Activity and Attribute Recognition: Phone Accelerometer and Gyroscope dataset, which comprises time-series data generated by accelerometer and gyroscope sensors, is one such dataset that has been widely utilised for this purpose. The dataset contains information from 24 people of various ages, genders, weights, and heights who participated in 15 trials of six distinct activities (walking downstairs, upstairs, sitting, standing, running, and walking) in the same location and conditions. For each trial, the dataset includes a multivariate time-series with 12 features such as attitude, gravity, userAcceleration, and rotationRate. The accelerometer calculates the total of two acceleration vectors: user acceleration and gravity, while the gyroscope tracks the device's orientation. In these 12 feature, the Acceleration has 3 features, X, Y and Z. As a result, the dataset records both the physical activity and the device's orientation during the activity. Human activity detection has gotten a lot of interest in recent years, both to the increasing availability of sensor data and the necessity for automated analysis of human behaviour in a variety of scenarios. The utilisation of time series data from wearable sensors, which can provide a plethora of information on human movement and behaviour, has been an important field of research. Such information has been widely used in applications, including health monitoring and rehabilitation, sports performance analysis, and security.

Human Activity Recognition Based on Sensors Data

The classification of activities conducted by individuals using sensors attached to their bodies is a component of HAR. The goal is to precisely detect individuals' actions, which can have a variety of applications in domains such as sports training, healthcare, and behaviour monitoring. The availability of low-cost sensors, their non-intrusive nature, and the potential to collect data in real-world scenarios have all contributed to the increased interest in HAR employing wearable sensors. Human activity recognition is one such application where time series data analysis is crucial. Based on data received from sensors, it recognises various human behaviours such as walking, jogging, sitting, standing, and others. Human activity recognition offers a wide range of applications, including healthcare monitoring, sports performance analysis, and geriatric care.

Human Activity Recognition (HAR) and Smartphones and Wearable Devices

HAR has significant consequences in a variety of industries, including healthcare, sports, and entertainment. In healthcare, for example, it can be used to monitor the everyday elderly people activities or patients with chronic conditions in order to spot any changes in their behaviour or prospective health problems. It can be used in sports to analyse athlete performance and give personalised training strategies. It can be used to build interactive games or immersive virtual reality experiences in the entertainment industry. The use of huge datasets that record a varied range of activities conducted by a number of persons is required for the creation of accurate and trustworthy HAR models. The use of smartphones and wearable devices in recent years has permitted the capture of large-scale sensor data in a more naturalistic setting. For example, (Gupta et al., 2022) discuss the use of artificial intelligence in medical data analysis

The analysis of sensor data supplied by accelerometer and gyroscope sensors in cell phones is the topic of this paper. We intend to investigate the feasibility of using such data to recognise human actions and infer personal qualities such as gender or personality. We anticipate that our research will add to the increasing corpus of HAR research and provide useful insights into the possibilities of IoT sensor data for human activity detection. As an example, (Talukdar ct al. 2022) and (Veeraiah et al. 2022) explore the use of machine learning and IoT for suspicious activity detection and enhancing the capabilities of the metaverse, respectively.

Applications of Human Activity Recognition (HAR) and Internet of Things Analysis

HAR offers a wide range of practical applications, from healthcare to sports, security, and environmental monitoring. With the growing popularity of Internet of Things (IoT) devices, HAR has become even more important since sensor data from wearables, smartphones, and other connected devices can be utilised to monitor human behaviours in real-time.(Mata et al., 2023) HAR can be used in healthcare to monitor patients' daily activities, detect falls, and aid individuals with disabilities. HAR can be used in sports to analyse athletes' motions and optimise training programmes. HAR can also be utilised in security applications to detect and deter criminal activity. (Bansal et al., 2022) propose a big data architecture for network security. Furthermore, IoT sensor data analysis can provide useful insights into human behaviour and activity patterns. In environmental monitoring, for example, IoT sensors may detect air quality, noise pollution, and temperature, among other things, and HAR can be used to identify pollution sources and devise methods to mitigate their impact on human health. (Hashmi et al., 2023) IoT sensors in retail can be used to track client movements and analyse their behaviour in order to enhance store layout and product placement. (Jain et al., 2022) focuses on enhancing the accuracy of machine learning in blockchain-based transaction classification IoT sensors can be utilised in transportation to monitor traffic flow and design efficient transportation systems. The applications of HAR and IoT sensor data analysis are limitless, and they have the potential to revolutionise many facets of our life. (Kaushik et al., 2022) present an innovative approach to cybersecurity using deep learning. Despite the potential benefits of HAR and IoT-based analysis, various problems must be addressed; where the huge amounts of data created by IoT devices, which can be overwhelming and difficult to analyse, are a major concern. Another difficulty is the requirement for accurate and dependable sensor data, as flaws and noise in the data might lead to erroneous activity detection findings. Furthermore, while collecting and analysing personal data from IoT devices, privacy considerations must be addressed. The advantages of HAR and IoT-based analysis outweigh the drawbacks, and more research in this field has the potential to improve many parts of our life.(Çalışkan et al.,2023)

Literature Review

Alashti et al. (2023) suggested an activity detection using lightweight pipeline that uses skeleton data from several viewpoints to improve the perception of human action by an assistive robot. Data sampling, geographical temporal data processing, and representation and classification algorithms are all part of the pipeline. This study compares detecting and classifying human actions using a modified traditional M-LeNet model to a Vision Transformer. Both approaches are tested against a multi-perspective dataset of human activities in the house (RHMHAR-SK), and the results show that integrating camera viewpoints improves recognition accuracy. Furthermore, in the Ambient Assisted Living (AAL) context, where bandwidth and computer resources are frequently constrained, their pipeline provides an efficient and scalable solution.

An et al. (2023) presented a transfer learning HAR paradigm with two components: fine-tuning and representational analysis. The representational analysis step identifies common elements that can be shared by multiple users as well as user-specific features that must be customised. Using this knowledge, the reusable portion of classifier is transferred to new users, while the remaining portion is fine-tuned. When compared to the baseline, the proposed approach improved accuracy by up to 43% and reduced training time by up to 66%.

For identifying human activities from sensor data, Diykh et al. (2023) suggested a hybrid strategy incorporating hierarchical dispersion entropy and AdaB_CNN. On three publicly available datasets, the suggested model outperformed most current approaches in HAR: UCI_HAR 2012, PAMAP2 and WISDM.

Dua et al. (2023) proposed a unique hybrid classifier for Human Activity Recognition (HAR) named "ICGNet" that blends Convolutional Neural Network (CNN) and Gated Recurrent Unit (GRU) architectures. The proposed ICGNet makes use of CNN and GRU strengths to take and store local features and relationships in time series based multivariate data.

Islam et al. (2023) proposed a feature fusion technique that is multi-level for multimodal based human activity recognition that employs a CNN that is multi-level with Convolution Block Attention Module (CBAM) to process visual data and Convolutional Long Short Term Memory (ConvLSTM) to handle time-sensitive multi-source sensor information. On the UP-Fall detection dataset, the suggested method outperformed existing state-of-the-art algorithms across various performance parameters.

Li et al. (2023) proposed an approach for recognising one user's daily behaviour at a time in multitenant smart home scenarios. For a sensor status frequency-inverse, they suggested a a method they called the CSA approach for HAR. The authors then created a distance matrix that is spatial based on the configuration of ambient sensors to aid in context awareness and data noise reduction.

Malik et al. (2023) proposed a CNN-LSTM approach with cascading posture features for multi-view human activity recognition. To extract more robust characteristics, the suggested method combines posture and depth information with a cascading process that use 3D skeleton poses. The approach was tested on two datasets, UWA3DII and NW-UCLA, and outperformed numerous state-of-the-art methods.

Sharma et al. (2023) developed LTRACN, a method for recognising single human activities. To extract features from sequential data, the approach integrates LSTM, TRA, and CNN. On the UCI-HAR dataset, the suggested model outperformed the classic CNN and LSTM models in terms of accuracy.

Soni et al. (2023) developed an improved DL based approach for human activity recognition in IoHT using smartphones. To extract features from accelerometer data, the proposed technique combines CNN and RNN with attention methods. The model outperforms various state-of-the-art approaches when tested on two datasets.

Yi-Fei et al. (2023) suggested a method for recognising human activities based on self-attention. The system employs a self-attention mechanism to find the relationships between various bodily joints, followed by a CNN to extract features. On two datasets, CAD-60 and UCI-HAR, the proposed model achieves state-of-the-art performance.

Zhang et al. (2023) suggested a human activity detection approach based on smartphone sensors that employ feature selection and deep decision fusion. The suggested method identifies the most discriminative features using the ReliefF algorithm and then classifies the activities using a deep decision fusion method. The proposed method performs significantly better than numerous state-of-the-art methods when tested on two datasets.

Zhang et al. (2023) developed a wearable sensor-based human activity detection technique based on the ConvTransformer model. To extract features from sensor data, the proposed technique combines

CNN and transformer. The suggested model outperforms numerous state-of-the-art approaches when tested on three datasets.

Problem Statement

Human activity detection utilising IoT sensor data presents a number of obstacles, including noisy and incomplete data, individual diversity in activity patterns, and the requirement for real-time processing and categorization. To reliably recognise and categorise actions from sensor data, these issues necessitate time series analysis, feature engineering, and deep learning. Because of their large complexity and complicated temporal dynamics, processing and analysing these data streams, particularly time series data, presents substantial hurdles. As a result, reliably recognising human activities from sensor data remains a difficult problem that necessitates advanced signal processing and machine learning techniques. The current work seeks to investigate the use of time series analysis and machine learning approaches for the classification of human activities and the identification of personal qualities in this setting.

Visualisation of the Data and Time Series Analysis

The heatmap, as shown in Figure 1, displays the time series data as a color-coded matrix, with the x-axis representing time and the y-axis representing the various dimensions or variables in the time series. Each cell in the matrix is coloured to represent the value of the time series at a certain time and dimension. Darker colours suggest greater values, whereas lighter colours suggest low values. The heatmap is an effective visualisation technique for investigating patterns and trends in time series data. Figure 2, The figure is a scatter plot that illustrates the singular values produced from the time series data's singular value decomposition (SVD). The scatter plot reveals that the first few single values have a substantially larger magnitude than the remaining singular values. This means that the time series data can be effectively simulated using only a few SVD main components. The SVD can be used to decompose a time series matrix into a set of orthogonal basis functions that represent the most important patterns in the data in the context of time series analysis. Figure 3 shows a scatter plot with a regression line superimposed.

Figure 1. A heatmap visualising time series data

For 100 data points, the scatter plot depicts the connection between two variables, X and Y. Each point on the plot represents a single X and Y observation. Linear regression, a statistical approach for assessing the relationship between variables, is used to calculate the regression line.

Figure 2. A scatterplot of first two components and elementary matrix

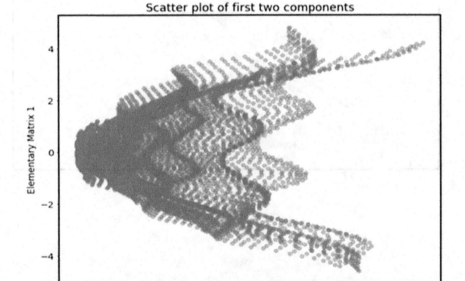

The line indicates the direction and intensity of the linear relationship between X and Y by representing the best-fit line through the data points. The plot also has labels for the X and Y axes as well as a title. Figure 4 shows a toy dataset made up of a sine wave with Gaussian noise introduced, as well as a polynomial fit to the noisy data. The original data is coloured blue, the noisy data is coloured green, and the polynomial fit is coloured orange. This plot's objective is to show how polynomial fitting may be used to approximate the underlying trend of a noisy dataset. A 4th degree polynomial is utilised in this example to fit the noisy data, resulting in a smooth curve that approximates the underlying sine wave. The figure also shows how adding noise to a dataset can make estimating the underlying trend more challenging. Figure 4 The first line of the plot depicts the time series reconstruction using only the first elementary component, designated as F0. The second line depicts the time series reconstruction using the sum of the second, third, and fourth elementary components, indicated as F1+F2+F3. The third line is the time series reconstruction using the sum of the remaining elementary components, designated as F4+...+F19. Finally, the fourth line depicts the time series reconstruction using only the third elementary component, indicated as F3. This type of graphic is excellent for understanding how each elementary component contributes to the overall time series and how different component groupings affect the reconstruction.

Figure 3. A scatterplot with a regression line

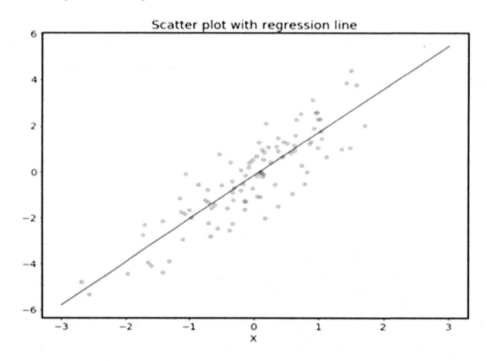

Figure 4. First four components for toy time series

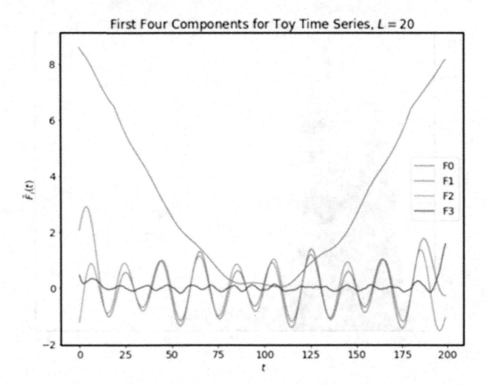

Figure 5 depicts the magnitude of a single subject's acceleration vector while walking from the MotionSense dataset. The x-axis measures time in seconds, and the y-axis measures acceleration in G's. The plot shows that the magnitude of the acceleration vector fluctuates over time, with higher and lower acceleration periods. These acceleration changes are most likely caused by the subject's stride while walking, as different phases of the gait cycle correspond to distinct accelerations. The plot depicts the rhythmic character of walking and explains how this rhythmicity can be analysed using techniques such as Singular Spectrum Analysis (SSA). Figure 6 depicts a walking time series with periodicity where the magnitude of the acceleration vector when walking is measured in units of gravity (G) in the time series. The x-axis measures time in seconds, and the y-axis measures acceleration in G. The plot contains four lines, each indicating a time series segment ranging from 4 to 5 seconds.

Figure 5. Histogram of acceleration magnitudes while walking

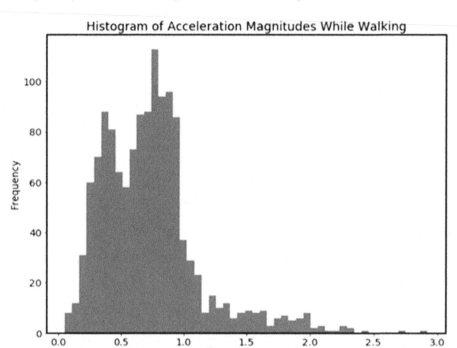

Figure 6. Example of periodicity in walking time series

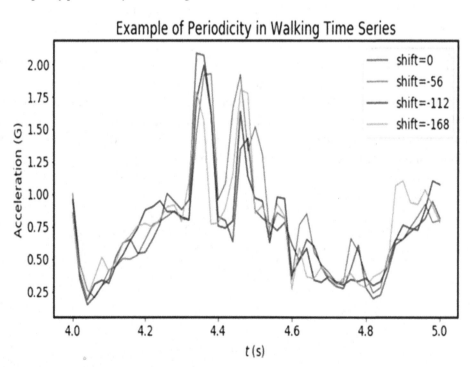

Figure 7 shows a plot depicting a walking time series with periodicity, where the magnitude of the acceleration vector when walking is measured in units of gravity (G) in the time series. The x-axis measures time in seconds, and the y-axis measures acceleration in G. The plot contains four lines, each indicating a time series segment ranging from 4 to 5 seconds. The plot depicted in Figure 8 is a 3D recurrence plot of the Lorenz Attractor system, which is a non-linear, chaotic system used to represent fluid motion. The graphic depicts the system's trajectory in three-dimensional space as it evolves over time. The axes represent the Lorenz system's three variables (X, Y, and Z), while the trajectory depicts a complicated pattern known as a weird attractor. The recurrence plot depicts how the system returns to specific states over time, resulting in a self-similar pattern. This visualisation style is frequently employed in the investigation of complicated systems and time series data. The figure demonstrates the Lorenz attractor's chaotic nature, with the track looping around and crossing itself in a complex manner. The recurrence plot allows us to observe the attractor in a new light, emphasising the system's repeating character and highlighting specific aspects of the attractor that may be difficult to detect from other angles. Furthermore, the plot can be used to investigate the system's dynamics and the link between various factors.

Figure 7. Histogram of magnitude of acceleration vector while walking

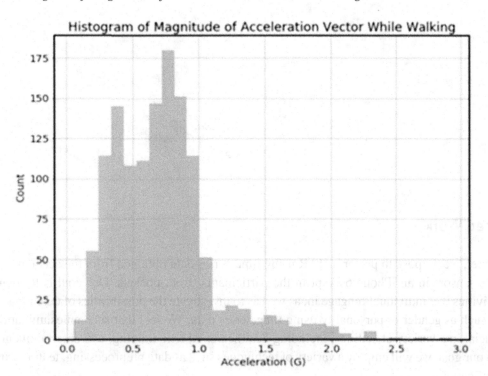

Figure 8. Lorenz attractor recurrence plot

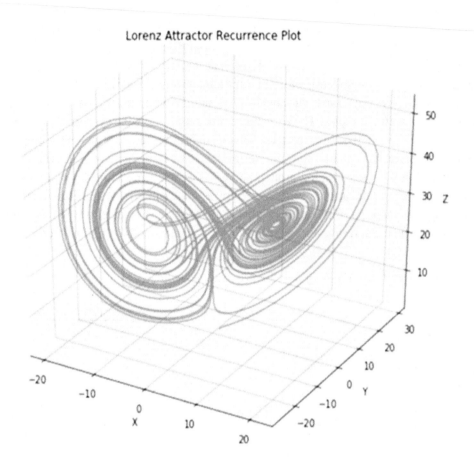

Proposed Work

In this paper, we propose to perform HAR using time-series data obtained from the accelerometer and gyroscope sensors in an iPhone 6s kept in the participants' front pockets. Our goal is to identify the many activities that individuals engage in, as well as to investigate the possibilities of detecting personal qualities such as gender or personality using time-series data. We feel that such an examination could be beneficial in a variety of settings, including healthcare, athletic training, and entertainment. To accomplish our goal, we will employ a variety of techniques such as data preprocessing, feature extraction, and classification.

Result and Discussion

To begin, we will preprocess the raw time-series data to reduce noise and outliers before extracting essential features such as mean, standard deviation, and frequency domain variables. We will next categorise the various activities and personal qualities using machine learning algorithms such as CNNs. CNNs have been proved to be useful in HAR tasks, and we expect they will do well in our proposed analysis.

Figure 9 depicts acceleration data for various activities such as walking, sitting, jogging, and so on. We can see changes in acceleration patterns for different activities by visualising the acceleration data. For each action, the graphic depicts three different acceleration components - x, y, and z - over time.

Figure 9. Acceleration data for various activities such as walking, sitting, jogging, and so on

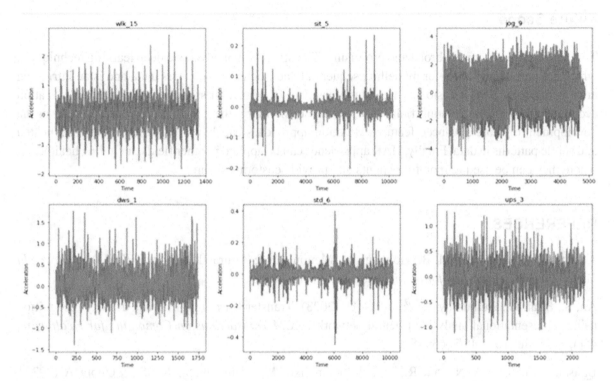

The model has several layers, including convolutional, pooling, and dense layers. The first layer consists of a 2D convolutional layer with 32 3x3 filters and a rectified linear activation function. The next two layers are the same as the first two, but with 64 filters instead of two. The last dense layer comprises 512 units and a dropout rate of 0.5 with a rectified linear activation function. For multi-class classification, the output layer contains three units with a softmax activation function. For training, the model employs the categorical cross-entropy loss function and the Adam optimizer. The dense layer is regularised using L2 regularisation with a regularisation rate of 0.01. The model gets an average accuracy of 96.43% and takes time of 4930 seconds to compile. The average loss is 0.132.

CONCLUSION

The suggested analysis seeks to remove noise and outliers in raw time-series data before extracting critical characteristics and categorising diverse activities and personal qualities with machine learning methods such as Convolutional Neural Networks (CNNs). There are multiple layers in the CNN model, including convolutional, pooling, and dense layers. The first layer of the model is a 2D convolutional layer that has 32 3x3 filters and a rectified linear activation function. To boost generalisation and avoid overfitting, the model additionally incorporates dropout layers and 2D max pooling layers. The last dense layer consists of 512 units with a 0.5 dropout rate and a rectified linear activation function. For multiclass classification, the output layer has three units with a softmax activation function. Model perform really well with an accuracy of 96.43%.

Future Scope

This study's future scope is broad and promising. The utilisation of advanced deep learning techniques,, which are particularly ideal for modelling sequential data such as time-series data, and is an intriguing area of research. Sensor fusion, which combines data from several sensors to increase classification accuracy, is another interesting field of research. Furthermore, there is room for advancement in the development of more advanced feature extraction approaches capable of capturing more complicated and subtle patterns in data. Finally, HAR approaches can be applied in other fields, including healthcare, where they can be used to monitor patients and provide customised care.

REFERENCES

Alashti, M. R. S., Abadi, M. B., Holthaus, P., Menon, C., & Amirabdollahian, F. (2023). *Lightweight human activity recognition for ambient assisted living*. IARIA.

An, S., Bhat, G., Gumussoy, S., & Ogras, U. (2023). Transfer learning for human activity recognition using representational analysis of neural networks. *ACM Transactions on Computing for Healthcare*, *4*(1), 1–21. doi:10.1145/3563948

Bansal, B., Jenipher, V. N., Jain, R., Dilip, R., Kumbhkar, M., Pramanik, S., Roy, S., & Gupta, A. (2022). Big Data Architecture for Network Security. In S. Pramanik, D. Samanta, M. Vinay, & A. Guha (Eds.), *Cyber Security and Network Security*. doi:10.1002/9781119812555.ch11

Çalışkan, A. (2023). Detecting human activity types from 3D posture data using deep learning models. *Biomedical Signal Processing and Control*, *81*, 104479. doi:10.1016/j.bspc.2022.104479

Diykh, M., Abdulla, S., Deo, R. C., Siuly, S., & Ali, M. (2023). Developing a novel hybrid method based on dispersion entropy and adaptive boosting algorithm for human activity recognition. *Computer Methods and Programs in Biomedicine*, *229*, 107305. doi:10.1016/j.cmpb.2022.107305 PMID:36527814

Dua, N., Singh, S. N., Semwal, V. B., & Challa, S. K. (2023). Inception inspired CNN-GRU hybrid network for human activity recognition. *Multimedia Tools and Applications*, *82*(4), 5369–5403. doi:10.100711042-021-11885-x

Gupta, M., Ghatak, S., Gupta, A., & Mukherjee, A. L. (Eds.). (2022). Artificial Intelligence on Medical Data. *Proceedings of International Symposium, ISCMM 2021* (Vol. 37). Springer Nature.

Hashmi, M. F., Kunduru, P. R., Mujavar, S. A., Nandigama, S. S., & Keskar, A. G. (2023). Medical Anomaly Detection Using Human Action Recognition. In Image Processing and Intelligent Computing Systems (pp. 215-222). CRC Press.

Islam, M. M., Nooruddin, S., Karray, F., & Muhammad, G. (2023). Multi-level feature fusion for multimodal human activity recognition in Internet of Healthcare Things. *Information Fusion*, *94*, 17–31. doi:10.1016/j.inffus.2023.01.015

Jain, V., Beram, S. M., Talukdar, V., Patil, T., Dhabliya, D., & Gupta, A. (2022, November). Accuracy enhancement in machine learning during blockchain based transaction classification. In *2022 Seventh International Conference on Parallel, Distributed and Grid Computing (PDGC)* (pp. 536-540). IEEE. 10.1109/PDGC56933.2022.10053213

Kaushik, D., & Garg, M., & Gupta, M. (2022). Utilizing Machine Learning and Deep Learning in Cybesecurity: An Innovative Approach. Cyber Security and Digital Forensics: Challenges and Future Trends, (pp.271-293). Wiley.

Li, Y., Yang, G., Su, Z., Li, S., & Wang, Y. (2023). Human activity recognition based on multienvironment sensor data. *Information Fusion*, *91*, 47–63. doi:10.1016/j.inffus.2022.10.015

Malik, N. U. R., Abu-Bakar, S. A. R., Sheikh, U. U., Channa, A., & Popescu, N. (2023). Cascading Pose Features with CNN-LSTM for Multiview Human Action Recognition. *Signals*, *4*(1), 40–55. doi:10.3390ignals4010002

Mata, O., Méndez, J. I., Ponce, P., Peffer, T., Meier, A., & Molina, A. (2023). Energy Savings in Buildings Based on Image Depth Sensors for Human Activity Recognition. *Energies*, *16*(3), 1078. doi:10.3390/en16031078

Saha, B., Samanta, R., Ghosh, S., & Roy, R. B. (2023, January). BandX: An Intelligent IoT-band for Human Activity Recognition based on TinyML. In *24th International Conference on Distributed Computing and Networking* (pp. 284-285). ACM. 10.1145/3571306.3571415

Sharma, P., Mishra, A., Kashyap, N., Muzamil, M., Rawat, R. S., Abidi, A. I., & Umrao, L. S. (2023). LTRACN: A Method for Single Human Activity Recognition. Intelligent Systems and Smart Infrastructure. *Proceedings of ICISSI, 2022*, 24.

Soni, V., Jaiswal, S., Semwal, V. B., Roy, B., Choubey, D. K., & Mallick, D. K. (2023, January). An Enhanced Deep Learning Approach for Smartphone-Based Human Activity Recognition in IoHT. In *Machine Learning, Image Processing, Network Security and Data Sciences: Select Proceedings of 3rd International Conference on MIND 2021* (pp. 505-516). Singapore: Springer Nature Singapore. 10.1007/978-981-19-5868-7_37

Talukdar, V., Dhabliya, D., Kumar, B., Talukdar, S. B., Ahamad, S., & Gupta, A. (2022, November). Suspicious Activity Detection and Classification in IoT Environment Using Machine Learning Approach. In *2022 Seventh International Conference on Parallel, Distributed and Grid Computing (PDGC)* (pp. 531-535). IEEE. 10.1109/PDGC56933.2022.10053312

Veeraiah, V., Gangavathi, P., Ahamad, S., Talukdar, S. B., Gupta, A., & Talukdar, V. (2022, April). Enhancement of meta verse capabilities by IoT integration. In *2022 2nd International Conference on Advance Computing and Innovative Technologies in Engineering (ICACITE)* (pp. 1493-1498). IEEE. 10.1109/ICACITE53722.2022.9823766

Yi-Fei, T., Soon-Chang, P., Ooi, C. P., & Tan, W. H. (2023). Human activity recognition with self-attention. *Iranian Journal of Electrical and Computer Engineering*, *13*(2).

Zhang, Y., Yao, X., Fei, Q., & Chen, Z. (2023). Smartphone sensors-based human activity recognition using feature selection and deep decision fusion. *IET Cyber-Physical Systems: Theory & Applications*.

Zhang, Z., Wang, W., An, A., Qin, Y., & Yang, F. (2023). A human activity recognition method using wearable sensors based on convtransformer model. *Evolving Systems*, 1–17. doi:10.100712530-022-09480-y

Chapter 19
A Comprehensive Review of IoT Reliability and Its Measures:
Perspective Analysis

Sandeep Bhatia
Galgotias University, India

Neha Goel
ⓘ https://orcid.org/0000-0002-7189-6305
Raj Kumar Goel Institute of Technology, India

Vinay Ahlawat
KIET Group of Institutions, India

Bharat Bhushan Naib
Galgotias University, India

Khushwant Singh
Maharshi Dayanand University, India

ABSTRACT

Due to the limited resources available in the connected IoT devices, it is difficult to assess the reliability of the system. IoT reliability desires favourable issues for measuring analysis with proposing, solving, qualifying considering, besides giving IoT responsiveness arrangements is, accordingly, essential for the enormous preparation of IoT innovation over all areas of society. In this unique situation, the most important reason is to propose and investigate survey responsiveness instruments for IoT outcomes, upholding the use of repetitive courses then gadgets that benefit as much as possible from cell phones, as key parts of IoT. Reliability for IoT has networking and various formula-based evaluations that has magnified node to node connection at various levels. The authors go through reliability measurements and models in this chapter. A detailed investigation for the sake to quantify reliability in the IoT based devices has been discussed alongside issues associated with it.

DOI: 10.4018/978-1-6684-8785-3.ch019

INTRODUCTION

The internet of things (IoT) of having partner degree expansion of the web during that monster quantities of "things", just as sensors, actuators, and processors, moreover towards human clients, square measure arranged and ready to offer high goal data on their air and exercise a level of the executives over it. It is still at partner degree beginning phase of advancement, besides bunches of issues/research difficulties ought to towards be settled before it's wide received. a few of those square measure specialized, just as capacity in addition quantifiability, as billions of heterogeneous gadgets will be associated, anyway choosing the best approach to contribute inside the IoT might be a test for business, in addition there additionally are significant social, legitimate besides moral difficulties, just as security in addition protection of data collection, that ought to be settled. However, the future IoT will have an international, cross-industry, and cross-innovation base, the paper surveys the world normalization endeavors that square measure current towards encourage its overall creation besides reception.

THEORETICAL BACKGROUND

The IoT is viewed as the following stage in the development of the web. It will empower typical gadgets to be associated with the web towards accomplish numerous divergent objectives. With possibly billions of gadgets to be associated, plainly normalization will be required so as to maintain a strategic distance from bedlam.

Although while IoT is crucial to our daily lives now, it is clear that IoT utilization will be essential to the development of technology in the years to come. According to a recent forecast, Figure 1 Hayajneh et al. (2020)., there will be over 75.44 billion linked devices worldwide in 2025. Unfortunately, security sometimes takes a backseat as businesses race to create new IoT devices with innovative uses. If any, businesses may employ outdated security standards.

Internet of Things (IoT) usage has increased, and it has recently attracted more attention due to the wide range of applications and devices it is being used for, including wireless sensors, medical devices, sensitive home sensors, and other related IoT gadgets.

Since it takes time to look into all potential vulnerabilities, security considerations are frequently ignored in the rush to quickly deliver new IoT goods on the market. Security concerns have been raised because IoT devices are internet-based and contain sensitive and secret information. Some academics are looking into ways to make these devices more secure.

One gauge is that solitary 0.6% of items that could be important for the IoT are as of now associated. By 2040, there could be doing 100 billion gadgets associated with the web, far more prominent than the quantity of human clients. Whereas the growth of the number of human customers follows a logarithmic curve, the growth of the Internet of Things follows an exponential trajectory. The three primary factors influencing the development of devices, the price of electronic components, and the IoT in remote mode are undergoing change. These characteristics enable the incorporation of tiny implanted sensors and actuators that can be networked into real things. Sensors and actuators, together with preparation and availability, will be the key components of the Internet of Things. The cloud Savvy items, for example, present day telephones use sensors besides actuators towards connect with this present reality. Implanted handling gives keen articles insight while availability besides the cloud gives the way to convey in addition store information. The IoT will eventually develop into a system of individuals, cycles, information, and physical items that intercommunicate and utilizing remote conventions.

Figure 1. Predictions for the internet of things from 2014 to 2025 (in billions)

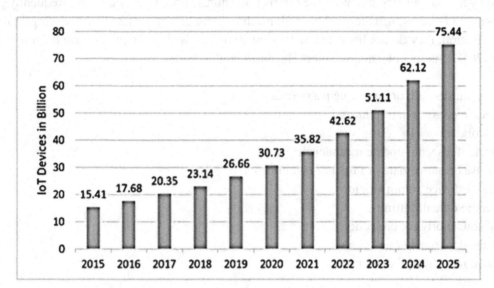

In its Joint Specialized Advisory Group 1, ISO/IEC created an Outstanding Working Gathering (SWG) in the domain of data innovation Maalel et al. (2013) that considered a general understanding of IoT, market requirements, normalisation gaps, and reference structures.

IoT also has the following definition, defined by ISO:

"Their ability that handle data from the real and virtual worlds effectively react depends on the basis of networked objects, people, systems, and data resources."

Including fundamental interfaces and conventions, power needs, and security constraints, a list of more than 400 existing standards with a connection to the Internet of Things was analysed. An official ISO working group has taken the role of the former SWG.

Web of Things (WG10). On the ISO WG10 website, there are nearly 500 active working reports that cover various IoT topics, such as reference structures, interoperability, security, definition, and syntax. IoT Reference Engineering (ISO/IEC N13119) is a draught standard. In the reference engineering, characteristics of IoT areas, a reference model for IoT frameworks, and the interoperability of IoT components are all portrayed. This working group is now a formal subcommittee of three working groups: sensor systems, the internet of things, and wearable innovations.

Examination with too innovation are currently situated to giving network to individuals, things, besides submissions over the Net. The Web of Things (IoT) empowers the improvement at a wide range in smart claims then administrations, consolidating together genuine world in addition virtual world information. As indicated by the World Financial Gathering (WEF), IoT gadgets will develop. The expansion of associated savvy things or gadgets through the Web for an assortment of uses brings a few difficulties. At the point when various keen things are included, dependably sharing data and settling on community-oriented choices in basic applications, for example, e-medical services, home computerization, or modern mechanization, is critical Prasad & kumar (2013).

Attacks on the architect layers and assaults on the data phases are the two basic forms of IoT attacks. Another distinction between IoT and old internet is that content and data on traditional internet are produced by human interactions. With the Internet of Things (IoT), smart machines frequently generate and gather data (sensors, actuators). Although machines can be manipulated into sending or receiving false information, they do not lie. The top 10 security issues with IoT devices have been listed by the OWASP (2018) Internet of Things Project Hayajneh et al. (2020):

- Weak, hardcoded, or guessable passwords
- Insecure ecosystem interfaces
- Network services
- A lack of a secure update mechanism
- The use of out-of-date or insecure components
- Lack of device administration
- Insecure default settings
- Inadequate privacy protection
- Unsafe data transfer and storage
- Inadequate physical hardening

To preserve our privacy, we must make sure that the information gathered and transmitted complies with the following standards: integrity: communicated data is accurate and complete (ii) confidentiality: transferred data, communication among endpoints, sensors, and readers are secured and encrypted (iii) authenticity: transmitted data is verified and comes from approved sensors, endpoints, and readers.

Bhatia et al. (2011), IoT implementation in military research and rescue operations has been discussed. Autonomous robot rescue operations can be carried out to detect alive humans and send data related to human presence remotely. A python language can be used to control vehicular movements.

Jeet et al. (2015), Internet of things can be deployed to control home appliances through voice and head movements. Python programming can be used to control home appliances which can be difficult to control manually by disabled persons.

Ganai et al. (2022), the Internet of things can be used in the healthcare sector to control open access to healthcare and its cyber securities issues has been discussed by author.

Moore et al. (2020), A definition and discussion of IoT reliability were made across the four major architectural layers. A thorough assessment of the literature was presented, examining research on the reliability of devices, networks, and systems as well as cutting-edge techniques for IoT anomaly detection. Lastly, five important research directions for the field of reliability in the Internet of Things have been developed using the results and outputs of this thorough assessment.

Li et al. (2021), The reliability of IoTNT is studied in this paper using the Space Fault Network (SFN). The nodes and edges of IoTNT are identical to the events and connections of SFN in order to do this. The dependability of IoTNT is investigated using a proposed structure analysis method based on SFN. In addition, the impact of potential logical connections between nodes on IoTNT reliability is being researched.

Bhatia et al. (2023), This article compares current wireless protocols and examines how they affect the deployment of smart farming using IoT-WSN. The key elements of the IoT-WSN concept for smart farming are described. It demonstrates many technological characteristics that are used into IoT-based farming approaches. In this study, the role of pertinent technologies and current protocols in smart farming is examined in detail. This research presents a thorough examination of existing wireless protocols utilized in IoT-WSN based smart farming.

Bhatia et al. (2023), Author focus on methods based on real-time human face identification. A webcam can be used to capture a digital image, count the number of people using the gate to enter or exit public transportation, and count the gender of those people to determine the number of seats that are available on the bus based on gender. The webcam can be installed for this purpose at the front of public cars and connected to a Raspberry Pi processing module.

Gautam et al. (2023), Author demonstrate the Framework for LPG Gas Leakage Detection and Weight Monitoring System. Reliability of IoT to sense, detect and transmit data about LPG gas leakage should be maintain to avoid causalities. Author discussed method to detect gas.

Bhatia et al. (2023), Author discussed the benefits and key challenges of current generation 5G and its reliability in current scenario of technological advancement. Author also highlight the importance of 6G and 7G to revolutionize the technological advancement.

Xing (2020), Author initially outlines reliability concerns given by particular supporting technologies at each tier based on the layered IoT architecture. The report then provides a thorough analysis and evaluation of the literature relating to IoT reliability. At four tiers (perception, communication, support, and application), reliability models and solutions are reflected and categorized. Even with a large body of work, IoT reliability research is still in its infancy. Following that, challenging research possibilities and issues are highlighted in regard to recently discovered behaviors and upcoming novel IoT system complexity and dynamics.

Chen et al. (2021), The author evaluates the evaluation method combining direct reliability and indirect reliability, analyses the communication requirements of the emerging services of the power Internet of things, and proposes the reliability incremental routing method of power line communication based on delay constraint. The dependability incremental routing algorithm suggested in this research has some benefits in terms of reliability and time delay, according to simulation. It can satisfy the need for dependable communication in the power Internet of things sector and raise the level of power grid intelligence.

RELIABILITY THEORY

Reliability: The chance that a framework realising its intended presentation may be shown theoretically. A framework's dependability is determined by considering each segment's capability for constant quality. To assess the IoT's dependability, Yong-Fei et al. used the unchanging quality hypothesis. For the application layer, the observation layer, the Internet, mobile devices, satellite communications, and, five dependability capabilities were included.

Based on their suspicions, an estimate of 0.87 was made for the general IoT steady quality. However, as the IoT is still expanding quickly and the authors may not have considered all important factors, this should only be used as a barometer. Other IoT requirements, such Quality of Service and even information for executives, may be managed using dependability theory. How reliable is the IoT data collected?

Li & Tian (2014) stress that factors including information misfortune, clamour, inaccurate information, and information excess where a few sensors have estimated a similar item would affect how reliable the information from IoT sensors is. Considering each of these elements might lead to the creation of a reliable quality model for IoT data.

The dependability of IoT programming itself might be another important problem since programming quality that never wavers can be seen as an extraordinary example of the dependability hypothesis. Segment or material disappointment that prevents a framework from carrying out its intended task controls equipment inflexibility. Since programming may provide unexpected results for many causes, such as erroneous information coming from another source, it is more difficult to assess programming reliability in the IoT. Gadget that was not considered in the structure stage. Outdated nature of implanted programming in IoT frameworks that can't be promptly kept up (for instance, IoT sensors in an atomic reactor presented to radiation) may likewise influence dependability. There are a few bits of effort anywhere the idea of unwavering quality into IoT has dealt with. In addition to discussing architecture guidelines for addressing dependability difficulties, Kempf et al. (2011) discuss their contribution to IoT unwavering quality. Prasad & kumar (2013) present a few approaches to handle IoT dependability and look at prospective energy efficiency in addition to unwavering quality (EER) issues. Because continuing observation requests are required by cutting-edge IoT, the need for EER is satisfied. By the way, Prasad & kumar (2013) simply brands overall considerations after modern demands to refine dependability in IoT exchanges. "In an investigation of dependability for crisis applications in IoT, An Adaptive Joint Protocol based on Implicit ACK (AJIA) for bundle tragedy recovery and course quality evaluation is described by Maalel et al. (2013). The arrangement is merely offered as a hypothetical choice without any approval for a trial.

As a testbed, Zhu, (2015) conducts a full SCALE (Safe Community Awareness and Alerting Network) phase and offers the main challenges and solutions to deal with creating a reliable IoT people group. The creator of SCALE hasn't performed a reliability system study because the system is still in the trial stage.

However, there are currently a few initiatives aimed towards planning and proving IoT reliability. Behera et al. (2015) initial dependability exhibiting strategy for an assistance-arranged IoT arrangement is intended as an example. The article describes a contextual study with two subsystems and is focused on the administration dependability of IoT. Every subsystem receives constant temperature and smoke data from a few sensors, and when those characteristics exceed a certain threshold, an actuator as well as caution is started. However, the paradigm doesn't focus specifically on failure scenarios.

An IoT dependability evaluation tool that considers identification consistency, propagation dependability, and preparation dependability is presented by Li & Tian (2014). Additionally, Ahmad (2014) have developed a method for evaluating the reliability of programming as well as equipment. Most items or articles in the Internet of Things, like many other things in our daily lives, rely on setbacks alone or collectively as a foundation. Without compromising our point of agreement, we might agree that the dissatisfaction structure a particular likelihood appropriation. Hence, when investigating the IoT's dependability and accessibility metrics, it's crucial to consider both the characteristics of IoT devices as a whole and each device individually. "In this instance, we define the IoT device dependability metric as the duration of the operational state of the device. Thus, the chance that the IoT will be in an operational condition within the range [0, t] may be used to construct the reliability metric R(t). Assume that an event E with matching probability Pr reflects the system failure state (E). R(t) = 1-Pr follows from that (E). This provides us R as a function of the failure rate function over a short time (t).

Let h stand for the failure probability over a brief period (t, t+At). Hence, the standard gamma

Probability density function presented in can be used to define the reliability function R(t) as shown in (1).

$$R(t) = (1/ (\alpha-1)!)(\lambda t)^{\alpha-1} e^{-\lambda t} \tag{1}$$

"We consider f(t) to be the probability density function of an event failure in order to determine the expected time to failure or the average failure time for an event. The anticipated value is then explained as stated in (2)

$$PF(t) = \int_0^t t^* f(t.) \, dt \tag{2}$$

"The accessibility measure as well as reliability measurement remain equal when a device does not need repair. If both the repair rate and the failure rate are h, as shown in (3), we could specifically test the availability measure A(t).

$$A(t) = \frac{\mu + \lambda_e}{\lambda + \mu} \frac{-(\lambda + \mu) \cdot t}{\lambda + \mu} \tag{3}$$

"In the event that t, (3) turns into the limiting probability for availability MTR = mean of repair = $1//\mu$ as t→∞.

A user can predict how the IoT system will behave overall after computing the dependability and availability probability that are relevant to the occurrences. The reliability data in the Internet of Things can be interpreted as follows: IoT failure time points and durations should be modelled for future applications, along with the possibility that an IoT device is now operational waiting lines, how to maximise the functionality of the devices, and how to maximise their functionality. To investigate the dependability data in this section, we therefore assume that the standard Gamma function described in (1). The discretization of the gamma function yields two failure concepts: the failure function itself and the survival function.

Figure 2. A broad overview of the internet of things architecture

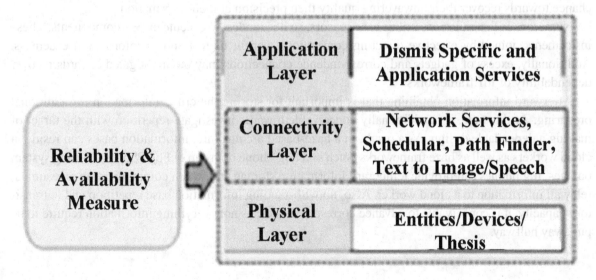

We will make use of the conventional Gamma probability density function model to explore a number of reliability and availability problems that typically occur in IoT. We suggest utilizing a Markov model to represent the failure-causing events in future research.

IOT RELIABILITY

In as of late, with the development of IoT applications in numerous zones, dependability consumes picked up in significance because of the effect that disappointments can have on the exhibition of IoT frameworks. In basic territories, the outcomes could be deplorable. Shortcomings in an IoT system may start in the correspondence framework, vitality framework, mechanical framework, then so forth.

IoT frameworks remain made of a few structure squares. Figure 2 demonstrations an overall construction of an IoT framework, where here are 4principles squares: "Things", "Communication, Computing besides Storage, then Services besides Applications". Our effort has centered around expanding the dependability of the Communication square.

The term unwavering quality can be utilized sufficiently, for instance on account of utilization strength, protection from security issues, versatility, self-design, long haul ease of use, or generally speaking framework dependability Kempf et al. (2011).

IoT arrange framework unwavering quality is vital for the dependability of the entire framework. For instance, crisis claims need the hard broadcast of instrument information by least deferral. Disappointment in conveying sensor information in a solid then convenient way for improving system unwavering quality, for example, parts improvement technique, powerful and inventive plan, utilization of misrepresented segments, and auxiliary excess, in which the last gives dependability through elective courses. The unwavering quality model that we propose in the ensuing segment depends on auxiliary repetition.

In an IoT framework, an object is a brilliant item where information is gathered through sensor hubs. The best possible determination of sensors is vital, then it must be an element of the sort of utilizations to be executed. Besides, the framework may consolidate data from a few sensors, which may speak to a chance towards recover their unwavering quality then precision at their information.

Numerous IoT submissions utilize ease sensors, so it is modest to execute excess components. These instruments, joined by effective directing conventions, permit us to improve information exactness. Additionally, excess of gadgets and correspondence connections may be investigated towards recover dependability of IoT frameworks.

Afterward information obtaining that is important for storing the crude information for additional preparing. At this time, it is additionally conceivable towards investigate repetition with the target of making a dependable design. As a result, in a haze-based architecture, information bases can reside in cloud workers as well as haze frameworks (such as a Fog Phone or group of Fog Phones). During system outages, fog phones can provide temporary information storage and, when communication is resumed, relay all information to a cloud worker. Also, notwithstanding information base repetition and to transitory capacity, it is conceivable to advance correspondences, as not everything information require to be put away halfway.

REDUNDANCY-BASED IOT RELIABILITY FRAMEWORK

That may end in significant expenses, client disappointment, with physical harm to individuals while belongings 3.

"There are two methods regularly used to accomplish WSN dependability: re transmission and excess. A three-dimensional reference model for WSN unwavering quality is proposed by Mahmood et al. (2015) (Figure 3). With this orientation archetypal, unwavering quality can be established scheduled re transmission as well as repetition, using bounce by-jump or start to finish correspondence, as well as functioning at parcel level or event level.

Figure 3. WSN reliability study reference model
Mahmood et al. (2015)

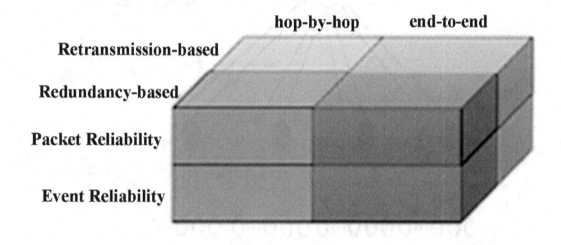

On the other hand, Chaturvedi (2016) offers some crucial methods.

IOT RELIABILITY MODELS BASED ON REDUNDANCY AND ITS IMPLEMENTATION

By incorporating recurrence into IoT frameworks, reliability may continue to improve. In this area, we propose some basic models for expanding dependability, that utilization and join correspondence joins excess and gadgets repetition. These replicas do exclude multi-jump instrument systems.

The fundamental prototypical of an IoT framework remains formed by an IoT arrange wherever every instrument (article) is associated with the focal worker utilizing a passage gadget and a system interface. This classical is portrayed into Figure 4.

As their entryways continue to be fundamental components in IoT frameworks, their overabundance regains overall consistency. In this way, the severance is counted evenly upon admission can be used to develop a direct unchanging quality model. Each gateway is linked to a backup gateway in this scenario, allowing the pair to operate independently as a master-slave pair. Figure 4 depicts that. A 1:1 redundancy, or one backup connection, is simply seen as part of another possible model with each primary connection that joins each gateway to the server.

Figure 4. Reliability model with redundancy of gateway- Implementation

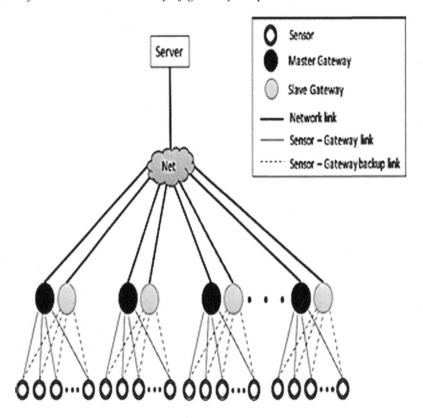

On Then again, into this model introduced at Figure 6, unwavering quality is better through utilizing both correspondence joins excess besides entryway repetition. The model introduced in the Figure 6 may be rearranged for the situation anywhere devices are coordinated for variables, that a serious normal circumstance.

PRECISE DEMONSTRATING OF IOT SYSTEM RELIABILITY

At the point when a framework develops in size and intricacy is more inclined to disappointments, influencing its exhibition Chaturvedi (2016). IoT frameworks should likewise bargain this issue, then one compelling method of doing so is through parts besides additionally way assorted variety.

Applying diagram hypothesis, it is conceivable to show frameworks unwavering quality. The utilization of diagram hypothesis aimed at dependability considers remained planned through Misra & Pao (1970). Using this type of exploratory light has become more important today owing to system.

Figure 5. Model for reliability with additional communication channels

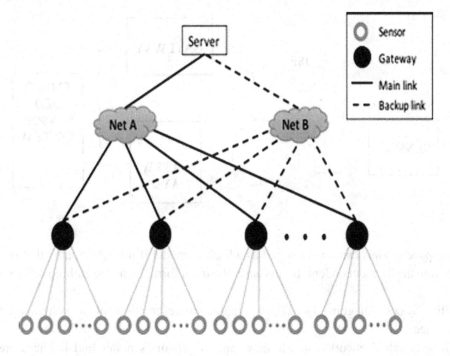

Figure 6. Dependability assessment using a dual redundancy model

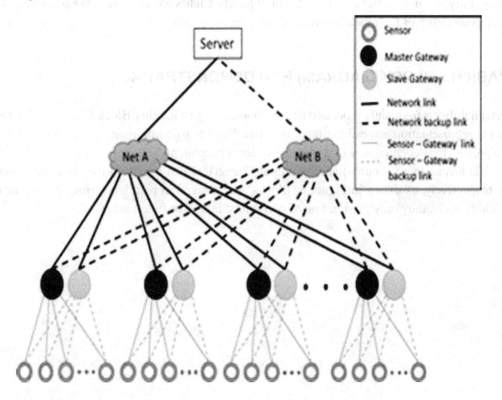

Figure 7. Simple IOT system reliability block diagram with redundant gateway

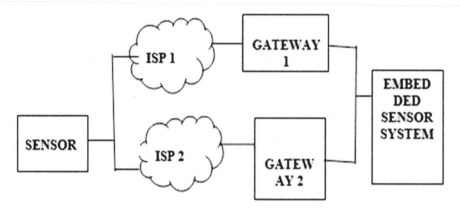

Non-state-space models can be employed as a backup model if it is anticipated that the constituent parts will be statistically independent. In this case, the main formalisms are Bobbio & Trivedi (2017) ":

- Reliability Block Diagram (RBD), a graphic representation of a system's dependability performance.
- Reliability Graphs/Networks, which show how a network's nodes and linkages are connected visually.
- A tree called a fault tree (FT), which was built using deductive reasoning.

In this study, we used RBD and Unwavering Quality Charts/Systems as two earlier formalisms to test the dependability of the repeating models that were suggested.

(RELIABILITY BLOCK DIAGRAM) RBD DEMONSTRATING

Systems reliability is frequently modelled and evaluated using Reliability Block Diagrams (RBDs). RBD enables the representation and estimation of system reliability by combining a set of blocks, where each block expresses the dependability of a specific system component.

An RBD has an input (Source) and an output (Target), which represent the start and finish of the system, respectively, as shown in Figure 8. Blocks are arranged in parallel, series, or combinations of these. Additionally, they only exist in one of two states: failing or functioning.

Figure 8. RBD organized in (a) series, (b) parallel, and (c) combined

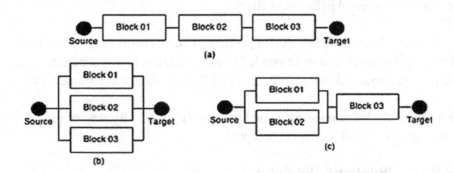

When a system's components are working properly, the related RBD is set up as a list of the system's components. Figure 8(a) illustrates a series structure with n independent components that is used to determine the system reliability.

The system reliability is calculated using Equation 4 below:

$$R(t) = n \prod i=1 \; Ri(t) \tag{4}$$

The block's reliability at time t is known as Ri(t). As a result, the reliability of the individual building blocks that make up a system with n components in sequence is multiplied by time t to determine its reliability. For instance, if each block's reliability is 0.9, an RBD with three blocks has reliability equal to 0.729.

Failures are independent of each other, and each block has a reliability associated. For example, block Bi has reliability Ri(t) associated to "working state" at time t. Meanwhile, 1 − Ri(t) represents the "failure state" at time t.

On the other hand, the related RBD is structured in parallel when a given system, consisting of n components, depends on at least one of them functioning well. Equation (5) can now be used to calculate the reliability at time t of the system depicted in Figure 8(b).

$$R(t) = 1 - n \prod i=1 \; (1 - Ri(t)) \tag{5}$$

In this instance, the reliability of a system with three blocks, each of which has its dependability set to 0.9, is 0.999. It is interesting to consider how the way the blocks are organized impacts the system's overall dependability.

Other combinations, such as series-parallel, parallel-series, bridge, and k-out-of-n, can be made in addition to series and parallel combinations. Equations (4) and (5) are also used to determine the dependability of these extra combinations. Blocks 01 and 02 are shown in parallel in Figure 8(c), while Block 03 is shown in series. In this instance, the series one is calculated once the parallel structure's reliability has been determined. The overall system dependability is 0.891, assuming that each block has reliability of 0.9.

An essential RBD model amasses the disappointment paces all things considered before modules for outline.

We carefully considered three repetition scenarios for our evaluation, which were aligned with the dependability models suggested in Segment III:

- Case 1: The basic Internet of Things (IoT) system, which consists of a worker, an Internet service provider (ISP), a passage, and an embedded sensor system, is free of superfluous parts and is constructed from pre-arranged modules. Example 2: Passage Excess IoT framework.

Example 3: Passage's IoT framework in addition to ISP excess. By way of referenced previously, measurable autonomy of all IoT squares is accepted.

Reliability Charts/Systems Displaying

Similar instances of sub-area and are presently dissected utilizing the Unwavering quality Diagrams/ Systems formalism. Once more, we expect that all components of the system are measurably free. Besides, in this methodology, we accept that solitary the edges or connections have relegated disappointment likelihood.

The methods for a chart G= (N, E), where N is the configuration of hubs and E is the configuration of edges or connections that connect two hubs, can be used to communicate with our system (Chaturvedi, 2016). Connections or associations can be one-way or two-way.

According to the Bernoulli dissemination, the chart deliberation is commonly described as a paired probabilistic system, G= (N, E, P), where P is the likelihood task assigned to each component and p speaks to the likelihood of the up state as well as (1-p) the likelihood of the down state (Bobbio & Trivedi, 2017).

Several methods, such as Aggregate of Disjoint Items (SDP) (Roman et al., 2011) or Parallel Choice Charts (BDDs) (Silva et al., 2013), can be used to evaluate the reliability of the articulation.

An SDP is dependent on the personality of Boolean capacities, where the relationship of two capacities can be expressed as the association of disparate terms. A BDD can effectively encapsulate these capacities because it is a condensed representation of a Boolean capacity.

The BDD approach and its reduction structure, known as Diminished Arranged BDD (ROBDD) (Silva et al., 2013), which is a parallel tree with one level by factor, were used in this work. The preowned computation is dependent on the recursive application of the Shannon deterioration rule agreed to (9) in a probabilistic interpretation.

RELIABILITY ISSUES WITH THE INTERNET OF THINGS (IOT)

IOT Reliability faces a number of different difficulties: -

1) **Insufficient device security:** This is the absence of adequate safeguards against online attacks, hacking, data theft, and unauthorized access on electronic devices like computers, smartphones, and IoT devices. Insecure software, weak passwords, unpatched vulnerabilities, a lack of encryption, and other security problems can all contribute to this. In order to guarantee the security and privacy of sensitive information kept on these devices, it is crucial to routinely update the software and put in place robust security measures. IoT devices frequently have poor security measures that are simple to exploit.

2) **Absence of uniformity:** In a certain field or industry, the lack of accepted requirements or norms is referred to as a lack of standardization. This may lead to incompatibility between various systems, goods, or procedures, which could result in muddle, inefficiency, or lower interoperability. For instance, in the world of technology, a lack of standardization might make it difficult for various systems and devices to communicate and exchange data. This can be avoided by establishing standards and procedures, which also ensure compatibility and uniformity. IoT devices are not standardized, which makes it challenging to consistently protect them.

3) **Network attack susceptibility:** A network, system, or device's susceptibility to being infiltrated or exploited by cybercriminals is referred to as its network attack vulnerability. This may occur as a result of flaws in the network architecture, outdated software, careless password management, or an absence of adequate security precautions. Network attacks can lead to data theft, privacy violations, service interruptions, and monetary losses. Strong security measures, including firewalls, encryption, and frequent software upgrades, should be put in place to lessen vulnerability to network attacks, as well as user education regarding safe internet usage. IoT devices are susceptible to attacks like denial-of-service (DoS) attacks since they rely on networks.

4) **Transmission of unprotected data:** This term describes the flow of data across a network or the internet without the necessary security. As a result, the data may be exposed to malicious actors' interceptions, alteration, or theft. Unsecured data transmission can happen when insecure protocols are utilized or when data is sent via an unencrypted network connection. Using secure protocols like SSL/TLS or VPN and encrypting the data before sending it are crucial for protecting sensitive data during transmission. Even if the data is intercepted during transmission, this can aid in maintaining its confidentiality and integrity. IoT devices frequently send private information that, if improperly secured, could be intercepted or altered.

5) **Privacy difficulties:** Problems with the gathering, storing, using, and sharing of personal information are referred to as privacy issues. Concerns concerning who has access to personal information, how it is used, and whether it is shielded from illegal access or misuse can fall under this category. Since personal data is gathered and stored on an unprecedented scale in the digital era, privacy concerns have risen in importance. In order to resolve privacy concerns, people and organizations must put in place the proper security measures to safeguard personal information, be honest about how it is being used, and respect people's right to privacy control. In order to set guidelines and protect people's personal information, privacy laws and regulations have also been formed. As a result of the potential for unauthorized collection and use of personal data, the enormous amount of data produced by IoT devices poses privacy concerns. In order to set guidelines and protect people's personal information, privacy laws and regulations have also been formed.

6) **Software flaws:** Vulnerabilities in software are weak points or errors in the code that can be used by attackers to get unauthorized access, steal confidential data, or engage in harmful activities. The usage of out-of-date or unsupported software can result in software vulnerabilities as well as flaws or mistakes committed throughout the development process. Attackers can take advantage of these flaws to take over a system, put malware on it, or steal confidential data. It is crucial for software developers to adhere to secure coding principles and for users to maintain their software updated and properly configured in order to lower the risk of software vulnerabilities.

7) **Threats posed by insiders:** Rather than from outside sources like hackers or cybercriminals, insider threats are security problems that originate from within a company. These dangers may come from a variety of sources, including contractors who abuse their access privileges, employees who

harm the organization intentionally or unintentionally, and insiders who are forced to jeopardize its security. Intellectual property theft, data breaches, and reputational harm to the company are all possible outcomes of insider threats. Organizations should also have a strategy in place for identifying insider security issues, responding to them, and recovering from them. If employees or contractors who have access to IoT systems harm others, whether on purpose or accidentally, they could pose a security risk.

Implementing security measures like encryption, secure authentication, and software upgrades is crucial to addressing these issues and ensuring the safe and secure operation of IoT systems and devices.

WAYS TO IMPROVE INTERNET OF THINGS (IOT) RELIABILITY

Enhancing IoT reliability involves making networks more secure against cyber threats and intelligent edge-based data storage and computation. Implementing reliable IoT for smart cities can reduce latency, improve performance and increase energy efficiency.

Over the past decade, smart devices have seemingly taken over the world. As per a 2021 study, there are approximately 175 million smart homes globally. Similarly, automated supply chains can use smart sensors and plugs for asset tracking and inventory management.

The Internet of Things (IoT) powers the networks that connect such smart devices and applications. As you can imagine, manufacturers, hospitals, exporters and businesses in other sectors depend heavily on such applications for critical functions—such as performing a bypass surgery with robotic arms or mixing two explosive chemicals autonomously in just the

right proportion to make a solution essential to business operations. In such circumstances, the reliability factor of IoT devices must be absolutely unwavering.

Here are some ways in which high reliability in IoT for smart cities can be achieved figure 8.:

1) By Boosting Edge IoT Capabilities

Older IoT networks involve edge components with lower computing power. This makes the transfer of data through the various IoT devices slower, thereby making the overall network needlessly laggy. You can prevent this by incorporating computing resources—and, by extension, computing power and storage space at the edge devices. This will serve two purposes in an IoT network. Firstly, this will prevent the stagnation of data in the backend. Instead, improved data processing and storage in the edge devices of a network will ease data flow and prevent congestion. Secondly, the increase in computing power and autonomous decision-making capability within edge devices will make IoT-driven operations faster and more accurate.

2) By Implementing Mesh IoT Networks

Traditional IoT networks use star networks in which all the devices in a network are connected with a common gateway. The gateway acts as a carrier for all the data collected by the sensors and devices. Although star networks are cost-effective to implement, problems such as black spots and limited network coverage are common in such IoT networks. In contrast, mesh networks, which involve sensor devices distributing data with one another through multiple nodes, increase coverage area for data transmission, scalability and self-healing algorithms. This greatly increases the reliability of IoT for smart cities, even when two devices are separated by a large distance.

3)   By Building a Multi-layered Defence for IoT Networks

Another factor that needs to be considered when IoT reliability is considered is cybersecurity. IoT networks may involve several businesses and stakeholders. To make the overall network resistant to DDoS, data breaches and other kinds of cyber threats, such stakeholders must build a multi-layered infrastructure wherein all devices have tight Identity and Access Management (IAM) protocols through Multi-Factor Authentication (MFA) applications. This will make hacking into an IoT network difficult for cyber-criminals.

Such measures will simply prop up the reliability aspect of IoT for smart cities and businesses and ensure that the critical operations that use the technology face zero to minimal downtime.

There are five directions for IoT reliability research, and smart applications can use IoT reliability Moore et al. (2020).

Direction 1: Vertical and real-time measurement
We need to be able to attest to the system's stability in real-time, or as near to real-time as is feasible, if the IoT is to control important infrastructure, such as security and vital traffic systems.

Direction 2: All gadgets, all protocols
This survey has shown the enormous variety of protocols and equipment that are planned to connect to and use IoT services. Numerous research organizations are working to create communication protocols that are lighter and more efficient, and this work is resulting in standards for communication proto-cols that continue to change every day. Additionally, new IoT hardware and devices keep appearing on the consumer market every day. Consequently, the optimal dependability solution must be independent of communication protocols, hardware, and software.

Direction 3: Complete stack awareness
While many researchers had successfully solved a single problem, or subset of problems, in IoT reliability research, it was concluded from the literature review that no study had been conducted with a thorough understanding of end-to-end dependability. This is no simple undertaking given the size and complexity of emerging IoT implementations. This does not imply, however, that researchers should attempt to create a "one size fits all" reliability method, as this would run counter to the first research goal indicated in this paper. Instead, unique dependability solutions that consider the entire IoT architecture should be proposed for each IoT sector.

Direction 4: Gathering knowledge about reliability from anomalies
When abnormalities arise in IoT services, a lot of work is put into identifying and reporting them. Although this activity is both important and valuable, it does not always increase reliability without taking an additional step. The user may not always be able to determine whether an IoT system is less dependable just by being aware of an occurrence. As a result, it is necessary to do research on how to synthesize data regarding emerging abnormalities in IoT systems into data on how the dependability has been impacted. For instance, there might not necessarily be an immediate risk to life if a sensor in a smart house that is watching over an assisted living situation malfunction. In

contrast, if a temperature sensor in a smart factory starts producing false readings, there is a risk that dangerous gear would malfunction.

Direction 5: Anticipate and manage failure before it happens.

The task of measuring reliability is well covered in this work. The task of predictive maintenance can be taken into consideration if the research is to go one step farther than this objective. Can we derive a correct maintenance date from the quantifiable reliability of a system if we are able to do so? Furthermore, is it possible to further categories this at the component level and make it a dynamic process that bases outcomes on real-time reliability data rather than utilizing a history of past failures to predict a failure date? An important step in the investigation of IoT reliability would be the resolution of this research question.

CONCLUSION

Finally, this study looked at the "Internet of Things, a set of systems that are committed to perception and the board of articles in the real world". There are several ways in which the Internet of Things is different from the current global Internet.

For instance, the link layers are tuned for low power consumption, the networks are often unmanaged, many applications are safety-critical, and the majority of nodes must be implementable in a lightweight way. The authors present some architectural guidelines to cope with reliability issues ranging from packet transmission to network lifetime and application behavior by drawing on their experiences designing and deploying such networks.

IoT innovations have captured the interest of researchers and developers all across the world. IoT developers and researchers are working together to scale up the technology for the benefit of society. So, if we consider the numerous difficulties and shortcomings in current technical approaches, advancements are justifiably possible. We've discussed a number of problems and difficulties in this post that IoT designers must overcome if they want to improve reliability with a more powerful architecture.

In this chapter, we discuss dependability models and metrics. The IoT not only offers services but also produces a lot of data. The importance of big data analytics, which can produce precise results that can be used to develop a better IoT system, is consequently also emphasized.

REFERENCES

Ahmad, M. (2014). Reliability Models for the Internet of Things: A Paradigm Shift. *2014 IEEE Int. Symp. Softw. Reliab. Eng. Work.*, (pp. 52–59). IEEE. 10.1109/ISSREW.2014.107

Al Hayajneh, A., Bhuiyan, M. Z. A., & McAndrew, I. (2020). I. Improving Internet of Things (IoT) Security with Software-Defined Networking (SDN). *Computers*, *9*(1), 8. doi:10.3390/computers9010008

Behera, R. K., Reddy, K. H. K., & Roy, D. S. (2015). Reliability modelling of service oriented Internet of Things. In *2015 4th International Conference on Reliability, Infocom Technologies and Optimization: Trends and Future Directions, ICRITO 2015*. IEEE. 10.1109/ICRITO.2015.7359216

Bhatia, S., Dhillon, H. S., & Kumar, N. (2011, December). Alive human body detection system using an autonomous mobile rescue robot. In *2011 Annual IEEE India Conference* (pp. 1-5). IEEE. 10.1109/INDCON.2011.6139388

Bhatia, S., Gautam, D., Kumar, S., & Verma, S. (2023). *Automatic Seat Identification System in Smart Transport using IoT and Image Processing.* 2023 3rd International Conference on Intelligent Communication and Computational Techniques (ICCT), Jaipur, India. 10.1109/ICCT56969.2023.10075664

Bhatia, S., Jaffery, Z. A., & Mehfuz, S. (2023). *A Comparative Study of Wireless Communication Protocols for use in Smart Farming Framework Development.* 2023 3rd International Conference on Intelligent Communication and Computational Techniques (ICCT), Jaipur, India. 10.1109/ICCT56969.2023.10075696

Bhatia, S., Mallikarjuna, B., Gautam, D., Gupta, U., Kumar, S., & Verma, S. (2023). *The Future IoT: The Current Generation 5G and Next Generation 6G and 7G Technologies.* 2023 International Conference on Device Intelligence, Computing and Communication Technologies, (DICCT), Dehradun, India. 10.1109/DICCT56244.2023.10110066

Bobbio & Trivedi, K. (2017). Reliability and Availability Engineering: Modeling, Analysis, and Applications. Cambridge University.

Chaturvedi, S. K. (2016). *Network Reliability: Measures and Evaluation.* Wiley. doi:10.1002/9781119224006

Chen, W., Yuan, J., Luo, A., & Xie, F. (2021). *Reliability Incremental Routing for Power Line Communication Based on Power Internet of Things.* 2021 IEEE 6th International Conference on Cloud Computing and Big Data Analytics (ICCCBDA), Chengdu, China. 10.1109/ICCCBDA51879.2021.9442543

Ganai, P. T., Bag, A., Sable, A., Abdullah, K. H., Bhatia, S., & Pant, B. (2022) *A Detailed Investigation of Implementation of Internet of Things (IOT) in Cyber Security in Healthcare Sector.* 2022 2nd International Conference on Advance Computing and Innovative Technologies in Engineering (ICACITE), Greater Noida, India. 10.1109/ICACITE53722.2022.9823887

Gautam, D., Bhatia, S., Goel, N., & Mallikaijuna, B. G., & Naib, B. (2023). *Development of IoT Enabled Framework for LPG Gas Leakage Detection and Weight Monitoring System.* 2023 International Conference on Device Intelligence, Computing and Communication Technologies, (DICCT), Dehradun, India. 10.1109/DICCT56244.2023.10110294

Jeet, V., Dhillon, H. S., & Bhatia, S. (2015). Radio Frequency Home Appliance Control Based on Head Tracking and Voice Control for Disabled Person. *2015 Fifth International Conference on Communication Systems and Network Technologies*, Gwalior, India. 10.1109/CSNT.2015.189

Kempf, J., Arkko, J., Beheshti, N., & Yedavalli, K. (2011). *Thoughts on reliability in the internet of things*. Interconnecting Smart Objects with Internet Work.

Li, S., Cui, T., & Alam, M. (2021). Reliability analysis of the internet of things using Space Fault Network. *Alexandria Engineering Journal*, *60*(1), 1259–1270. doi:10.1016/j.aej.2020.10.049

Li, Y., & Tian, L. (2014). Comprehensive Evaluation Method of Reliability of Internet of Things. In *IEEE 2014 Ninth International Conference on P2P, Parallel, Grid, Cloud and Internet Computing*, (pp. 262 –266). IEEE.

Maalel, N., Natalizio, E., Bouabdallah, A., Roux, P., & Kellil, M. (2013). Reliability for emergency applications in internet of things. In *Proceedings - IEEE International Conference on Distributed Computing in Sensor Systems, DCoSS 2013*, (pp. 361–366). IEEE. 10.1109/DCOSS.2013.40

Mahmood, M. A., Seah, W. K. G., & Welch, I. (2015). Reliability in wireless sensor networks: A survey and challenges ahead. *Computer Networks*, 79(December), 166–187. doi:10.1016/j.comnet.2014.12.016

Misra, K. B., & Rao, T. S. M. (1970). Reliability Analysis of Redundant NetworksUsingFlowGraphs. *IEEE Transactions on Reliability*, *R-19*(1), 19–24. doi:10.1109/TR.1970.5216374

Moore, S. J. (2022). IoT reliability: a review leading to 5 key research directions. CCF Transactions on Pervasive Computing and Interaction, 2, 147-163.

Moore, S. J., Nugent, C. D., Zhang, S., & Cleland, I. (2020). IoT reliability: A review leading to 5 key research directions. *CCF Trans. Pervasive Comp. Interact.*, *2*(3), 147–163. doi:10.100742486-020-00037-z

Prasad, S. S., & Kumar, C. (2013). A Green and Reliable Internet of Things. *Commun. Netw.*, *5*(1), 44–48. doi:10.4236/cn.2013.51B011

Roman, R., Alcaraz, C., Lopez, J., & Sklavos, N. (2011, March). Key management systems for sensor networks in the context of the Internet of Things. *Computers & Electrical Engineering*, *37*(2), 147–159. doi:10.1016/j.compeleceng.2011.01.009

Silva, M. D., Leandro, R., Batista, D. M., & Guedes, L. A. (2013). A dependability evaluation tool for the Internet of Things. *Computers & Electrical Engineering*, *39*(7), 2005–2018. doi:10.1016/j.compeleceng.2013.04.021

Xing, L. (2020). Reliability in Internet of Things: Current status and future perspectives. *IEEE Internet of Things Journal*, *7*(8), 6704–6721. doi:10.1109/JIOT.2020.2993216

Zhu, Q. (2015). Enhancing Reliability of Community Internet of Things Deployments with Mobility. In *34th International Symposium on Reliable Distributed System*, Montreal - Quebec, Canada

Chapter 20
Sustainable IoT for Smart Environmental Control

Kavitha Rajamohan

iD https://orcid.org/0000-0002-7803-8901

CHRIST University (Deemed), India

Sangeetha Rangasamy

iD https://orcid.org/0000-0002-1850-232X

CHRIST University (Deemed), India

M. Ajai Kumar

CHRIST University (Deemed), India

Bidura Sarkar

CHRIST University (Deemed), India

Jubaraj Mukherjee

CHRIST University (Deemed), India

ABSTRACT

The network of physical items/things/objects, that are implanted with sensors, software, and other networking technologies to communicate and exchange data with other devices and systems through the internet, is referred to as the Internet of Things (IoT). Environmental control in the context of IoT systems refers to the use of connected devices and sensors to manage and regulate various aspects of the environment, such as temperature, lighting, air quality, water quality, and more. The goal is to create an intelligent environment that is more efficient, comfortable, and sustainable.

DOI: 10.4018/978-1-6684-8785-3.ch020

1. INTRODUCTION

1.1. Environment at a Glance

Since the past decade, planet Earth has been facing numerous environmental issues, such as climate change, global warming, industrial pollution, deforestation, resource depletion, and many more. Such environmental issues create a profound impact on the health of the general public. Since the population is growing day by day, the amount of data required to prevent environmental degradation is increasing daily. It is necessary to have an accurate understanding of the environment and to investigate and end the factors that contribute to environmental degradation.

1.2. IoT Devices and Sensors Available

IoT devices and sensors supporting IoT devices are increasingly being used to monitor and track various aspects of our environment, which consist of temperature, humidity, air quality, and water quality. Some of the commonly available IoT sensors for monitoring the environment are temperature sensors, air quality sensors, water quality sensors, soil moisture sensors, humidity sensors, smoke detectors, rain detectors, pH sensors, and more. These devices can be used to monitor the environment in real-time and provide quality data. After filtering the data, it is used to make decisions according to the current prevailing conditions. This data is used to predict the future state of the environment.

1.3. Environment and IoT

The integration of IoT devices and sensors with the environment has the capability to revolutionize the methods of monitoring and tracking environmental data. IoT devices enable environmental monitoring in real-time and data collection from a variety of sources, including weather stations, sensors, and satellites. This information can be used to spot environmental patterns and create plans for dealing with environmental problems. IoT devices can also be used to automate environmental processes like waste management and energy preservation.

The integration of IoT devices and sensors with the environment is an important step towards creating a more sustainable and healthy planet. By monitoring the environment in real-time, we can better understand the factors that contribute to environmental degradation and develop strategies for addressing these challenges.

2. STATUS OF IOT ENABLED ENVIRONMENT ACROSS GLOBE

Globally, the use of IoT devices and sensors in the environment is expanding quickly. Governments and commercial businesses are investing in IoT technology worldwide to enhance environmental management and monitoring. Here are a few instances of how IoT technology is utilized around the globe:

- **Europe:** To promote the use of IoT technology on the continent, the European Commission has created the "Smart Anything Everywhere" programme. In order to follow the movements of wildlife and to check the quality of the air and water, IoT devices and sensors are utilized.

- **United States:** To enhance the monitoring of air and water quality, the Environmental Protection Agency (EPA) in the US has implemented IoT technology. Smart cities are also utilizing IoT gadgets and sensors to improve energy use and reduce waste.
- **Asia:** The use of IoT technology in the environment is expanding quickly throughout Asia. For instance, sensors and IoT devices are being utilized in China to monitor air quality in large cities. IoT technology is being used in India to monitor water quality and boost irrigation system effectiveness.
- **Australia:** The Great Barrier Reef and other environmentally vulnerable places are being monitored in Australia using IoT technology. The movement of fauna is tracked, and coral reef health is monitored, using IoT devices and sensors.

The Internet of Things application has the potential to significantly strengthen and improve the current agricultural industry. According to a Statista analysis, IoT use in agriculture is anticipated to increase globally. According to reports, the IoT market for the agriculture sector was 14.79 billion USD in 2018, but it is anticipated to grow to 30 billion USD by 2030. Globally, the application of IoT technology in the environment is expanding quickly. To enhance environmental monitoring and management and to build more sustainable and wholesome communities, governments and corporate organizations are investing in this technology according to (Deshmukh, 2019.)

3. GOVERNMENT INITIATIVES IN INDIA TOWARDS SUSTAINABLE ENVIRONMENT

The government is aware that these admirable goals are marred by uncertainties. These difficulties are recognised by the Indian government. While aiming for rapid and sustainable economic growth, it is aware that such growth cannot be supported if its social and environmental foundations are unstable. The improvement of human well-being is given priority in the national vision, which includes not only access to an appropriate level of food consumption and other consumer products but also to fundamental social services, particularly those related to education, health, drinking water, and basic sanitation. Additionally, it prioritizes increased access to economic and social possibilities for all people and groups, as well as increased participation in decision-making. The management and conservation of natural resources is a key component of development strategy. The 2030 agenda for sustainable development consists of 17 sustainable development goals which were approved by the world leaders in 2015. These goals focus on the future development of all countries with an emphasis at eradication of poverty, environmental sustainability, and maintenance of prosperity. The entire global society depends on achieving these goals. The value of international cooperation in helping developing countries achieve their climate objectives cannot be overstated. India's development strategy has long been founded on ideas that are similar to those espoused in the 2030 Development Agenda. Global cooperation is needed to combat climate change, and India has been addressing these issues alongside other pressing development needs. The nation has significantly increased the amount of renewable energy capacity it has installed. The demand for natural resources in India has increased due to the country's high growth rate and fast urbanization, putting pressure on the environment and generating questions about sustainability. Resource efficiency can be a crucial instrument for meeting the nation's resource needs with the least amount of negative environmental impact. In India, air pollution has become a significant problem. In order to comprehensively address the issues of air pollution across the entire country, the Indian Government established the National Clean Air Program (NCAP) as a time-bound, national level project. By bringing a variety of strategies into use, India has made significant progress. The Swachh Bharat Mission (SBM), Beti Bachao Beti Padhao (BBBP), Pradhan Mantri Awas Yojana (PMAY), Pradhan Mantri Jan-Dhan Yojana (PMJDY), Deen Dayal Upadhyay Gram Jyoti Yojana (DDUGJY), and Pradhan Mantri Ujjwala Yojana are some of India's current flagship policies and programs that have helped the country advance(Press Information Bureau, 2021.).

3.1. Ganga: The Lifeline Of India

Cleaning up the great River Ganga through the Namami Gange Mission has been a top government priority in accomplishing SDG 6 (Ensure availability and sustainable management of water and sanitation for all). The mission was started as a priority programme with a 20,000 crore budget for the years 2015 to 2020.

Figure 1. Mission ganga budget allocation from 2011-2020
(Details of Funds Spent on Namami Gange Programme, 2021)

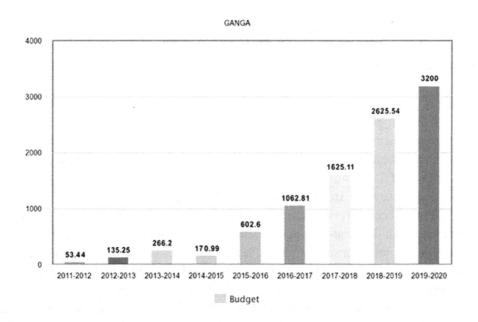

A total of 6,106.25 crore has been spent on the programme from 2014–15 to 2018–19, which is a significant increase above prior such programmes as displayed in the statistics given in Figure 1. The National Mission for Clean Ganga (NMCG) was given authority under the Environment (Protection) Act of 1986 to expedite implementation and develop policies for the long-term sustainability of the Ganga rejuvenation efforts. This was done to ensure effective implementation and proper coordination with the State and Local Bodies(Details of Funds Spent on Namami Gange Programme, 2021).

3.2. Major Components of Namami Gange Mission

- **Sewerage Project Management:** For sewerage projects, the policy decision to use Public-Private Partnership (PPP) approach of Hybrid Annuity Model (HAM) and 15 years long-term Operation & Maintenance (O&M) included in the project cost and improved governance through 'One City One Operator' approach ensured competitive and positive market participation along with synergy in implementation.

- **Urban Sanitation:** A report prepared by the Consortium of 7 IITs identified 10 cities that contributed more than 60 percent pollution load in Ganga. The Mission extended a comprehensive coverage of these cities with construction and rehabilitation of Sewage Treatment Plants (STPs) for a prospective year of 2035, inception and diversion of drains, solid waste management through cleanliness drives on ghats and deployment of skimmers for river surface cleaning.

4. APPLICATIONS OF IOT IN ENVIRONMENTAL MONITORING

IoT device utilization has made it possible to create creative answers to a range of environmental sustainability concerns. Systems for managing energy more intelligently can cut energy costs and carbon emissions. Municipalities can monitor the air quality index and take corrective action to lower air pollution thanks to air quality monitoring devices. Waste collection schedules are optimized through intelligent waste management systems, which also lessen their environmental impact. Systems for fleet management optimize travel patterns and enhance driver performance, cutting back on hazardous emissions. Smart water monitoring systems minimize water waste by detecting water leaks, regulating water quality, and monitoring fill levels. Human interventions that are prone to inaccuracy and inefficiency are minimized with smart agriculture. Wireless IoT sensors enable smart cold chains to check the environment for food.

4.1. Smart Buildings

Sensors of various types are used for building automation and monitoring, and they enable real-time data collection. HEMS-IoT (Home Energy Management Systems—Internet of Things) is a big data and machine learning-based energy management system. They are used for temperature control, fire detection, heating, ventilating, and air-conditioning (HVAC), water quality monitoring, occupant detection, carbon dioxide monitoring, waste management, maintaining sustainability, and so on. ZigBee technologies can be used to minimize the waste of electrical energy. ZigBee is a wireless communication protocol to create personal area networks using small, low-powered digital radios. ZigBee is a wireless ad hoc network. ZigBee can be used for automatic meter readings and in smart grids. The ZigBee Home Area Network architectural model is shown in Figure 2.

Figure 2. ZigBee home area network

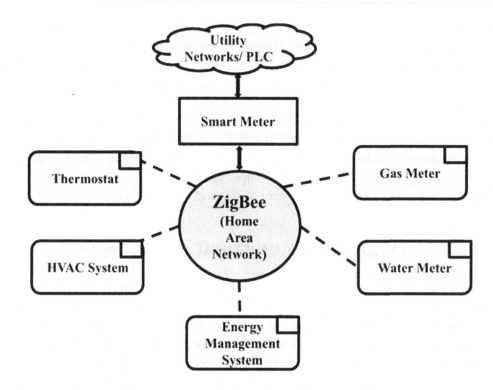

To deploy the smart building sensors, the LoRa (long-range) ecosystem can be used. Long-range transceivers with wireless radio frequency technology that have a long battery life, a programmable bit rate, and a low energy consumption rate are used here. The LoRaWAN technology can be used in place of Bluetooth, Wi-Fi, or cellular networks. There is cost-effective and seamless connectivity between the public, private, and hybrid networks. The Oakland City Center in California is one of the best examples of a smart building, as it uses the VAV (Variable Air Volume) system. Here temperature and humidity data are collected and analyzed by an AI algorithm, and setpoints are forwarded to the HVAC system, which makes decisions to maintain the environment. Air quality monitoring is also available here. If a virus is discovered in the environment, a "decontamination mode" prevents virus transmission by raising the temperature. Digital ceilings connect HVAC, lighting systems, smart meters, video surveillance cameras, sensors, and actuators in a smart building with the help of network infrastructures. A communication protocol for smart buildings is BACnet (Building Automation and Control Network), which allows the equipment to communicate using a PLC (programmable logic controller).

4.2. Smart Agriculture

IoT plays a major role in smart agriculture. Crop and farm monitoring is done with the help of drones, sensors, and video recording cameras. Computer imaging combined with machine learning tools can be used to analyze the collected data. IoT can be used to keep track of water tank levels to make the irrigation process more efficient. IoT-based technologies can be used for frost detection and automated irrigation systems. Drones fitted with HD cameras can be used to detect pests on agricultural crops. They can be used to spray pesticides and use sensors to keep track of the amount of pesticide sprayed. Pollution to the environment is reduced by balancing the amount of pesticides sprayed.

- **IoT-based irrigation system:** SWAMP (Ullah et al., 2021) stands for smart water management platform; it is an IoT-based smart irrigation project for efficient freshwater utilization. It has the ability to automate water management and distribution while also preventing over- and under-irrigation. EEWMP (Energy Efficient Water Management Platform) is an advanced version of SWAMP. It consists of field-deployed sensors, sinks, fusion centers, and open-source clouds. The SWAMP architecture consists of five layers: the device communication layer, the acquisition security management layer, the data management layer, the irrigation layer, and the water application services layer. Several technologies include cloud computing, sensors, drones, semantic computing, communication protocols, etc. But some shortcomings of SWAMP include increased energy consumption due to the sending of redundant reports; thus, an energy-efficient sensor communication model is used, namely EEWMP. EEWMP consumes 30% less energy and increases network stability.

- **Precision Agriculture:** Here IoT can be used for livestock management, inventory management, vehicle and cattle tracking, field monitoring, and so on. One can even monitor soil conditions (such as pH and moisture), nutrient and water level tracking, and irrigation management. IoT can be used in greenhouses to maintain specific climatic conditions. Solar-powered sensors can be used in such situations. These sensors can be used to detect light intensity, pressure, humidity, and temperature. Cloud-based data storage systems integrated with IoT platforms can be used to collect and analyze sensed data. Priva Horticulture is a good example of IoT-based smart agriculture; it has features such as greenhouse climate control, greenhouse sensors, smart irrigation systems for greenhouses, data-driven greenhouses, and crop and labor insights. The central irrigation system maintains the correct balance of nutrients in the water to be distributed among the crops.

4.3. Water Management

Sensors are used for water supply chain management, checking the chemical composition of treated water and acidification, checking the water pressure of pipelines, monitoring water temperature, controlling water leakage in distribution pipelines, monitoring water quality to manage diseases and pollution, and waste-water management. Smart Water Management helps us make reasonable and sustainable use of water. Smart water meters help measure the quantity of water consumed. SWAMP and EEWMP (Efficient Water Management Platform for Smart Irrigation) are used for water management in irrigation systems. The following is a solution to the water management and distribution problem:

- According to (Ali et al., 2022), to assess physical quantities like water pH levels, turbidity, and flow rates, sensors were installed on a water distribution network (WDN). The data was sent to the Firebase platform via a network of sensors. Water leakage was tested using an IoT testbed design.
- According to (Manoj et al., 2022), The quality of the water was determined by variables including pH, dissolved oxygen, nitrogen, and water temperature. Based on those parameters, the sensors collected data, which they then sent to an Arduino or Raspberry Pi for processing. On the basis of a comparative examination of the facts gathered, decisions were made. Through a cellular network

connected to the Arduino, the alerts were delivered to the farmer. The aeration unit was activated if the dissolved oxygen concentration was lower than expected, and the motor was activated if the water level was lower than necessary. Data comparison was conducted on a regular basis and at predetermined intervals.

- According to (Alobaidy et al., 2022), LAPs (low-altitude platforms) linked to wireless technologies were utilized to transport data in harsh or hilly areas. Water quality sensors, LoRa (long-range) wireless equipment, smart utility network (SUN) technology, LAP balloons, and WiFi equipment were all used in that scenario. The architectural model of LAP-AIN is depicted in Figure 3. LAP-AIN stands for low altitude platform based Airborne IoT Network mainly designed for monitoring water quality in harsh climatic conditions.

Figure 3. Architecture model for LAP-AIN

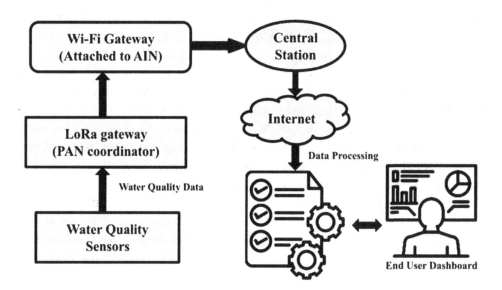

It is possible to optimize water usage, identify and reduce water wastage, enhance water quality, and assure sustainable water resource management by merging AI and IoT approaches in smart water management. Effective water management methods are made possible by the integration of real-time data collecting, advanced analytics, and intelligent decision-making. These strategies benefit both the environment and human communities.

Techniques based on artificial intelligence can allocate water resources more efficiently by taking demand, availability, and quality into account. To optimize water distribution, sophisticated algorithms can evaluate the present water supply, reservoir levels, and usage trends. This includes choosing the water supply route that is the most effective, balancing water demand among various industries, and making plans for future water requirements. AI algorithms analyze the data and provide precise irrigation recommendations, optimizing water usage and promoting sustainable agriculture practices. Water demand and distribution may be made more efficient with AI and IoT. AI algorithms can foresee changes in water demand by looking at past usage trends, weather predictions, and other pertinent information. In order to ensure effective water use, this information can be utilized to dynamically regulate water supply, plan irrigation, or use demand response methods.

4.4. Soil Moisture Monitoring

Keeping track of the soil moisture levels using IoT-based technologies is an important aspect of smart agriculture. Soil moisture sensors help in determining the water content in the soil. For instance, tensiometers are soil moisture sensors that measure the tension between soil particles and water particles. In tensiometers, data collection is direct; no interpretation needs to be done. It doesn't need a high level of maintenance. Another example is solid-state sensors, which are less expensive. The disadvantage of solid-state sensors is that they do not work in arid soils with high salinity. There are also electromagnetic sensors that use time-domain reflectometry to measure the dielectric permittivity of soil, which is calibrated for moisture content. The data read from the sensors can be transmitted to remote locations using cellular or radio links, displayed on the readers, or stored in loggers. Telemetry-based systems can send data to remote locations in real-time. Telemetry-based systems may have the following architecture: The data loggers in the field are connected to a wireless network, and they transmit data to a central controller, which again sends data to a website using cellular links, radio links, and satellite links. The architectural model of telemetry based systems is depicted in figure 4.

The capacitance based soil moisture sensor is widely used. The working principle is that the dielectric constant of soil particles changes with the moisture content of the soil. The sensor consists of multiple electrodes embedded at different depths in the soil and the capacitance between them is measured and is correlated with the soil moisture level. AI algorithms can analyze historical and real-time data from sensors to predict optimal irrigation schedules, detect anomalies and provide suggestions for efficient water management, data from several sensors can be aggregated and analyzed to gain insights into the soil moisture patterns. The scalability, accuracy, sensing depth, and integration capabilities of these sensors enable data collection and analysis.

Figure 4. Architecture of telemetry based systems

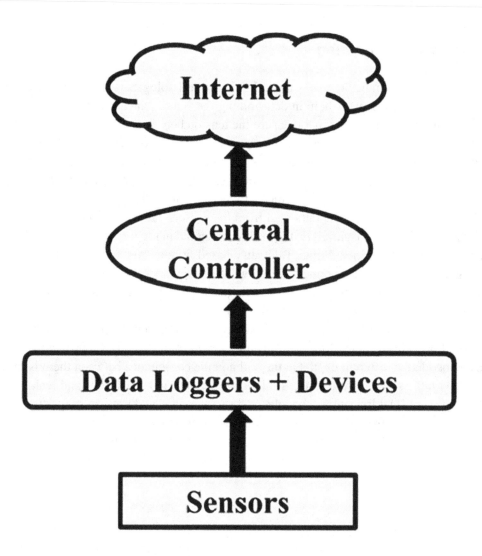

4.5. Smart Energy Management

IoT-based energy management systems can be used in electric supply chains to enable smart metering, real-time monitoring of power consumption, preventing energy waste, lowering carbon emissions from electricity production, increasing the use of renewable energy sources, and lowering costs. According to (Madhuri et al., 2022), Energy management is required for reducing supply chain shortages and for boosting power effectiveness. In the Intelligent Power Management System (IPMS), price optimization techniques are present that depend on detector elements that measure the duration of power use. EMS (Energy Management Systems) regulate and manage electric power usage. This paper uses a three-layered architecture for the integration of a smart grid. The three-layered architecture for a smart energy grid consists of the following:

- A data collection layer consisting of sensors, actuators, and wireless sensor networks;
- A data communication layer consisting of LAN: Bluetooth, ZigBee, Wi-Fi, and LoRaWAN; and
- A data storage layer consisting of fog, clouds, edge computing, and big data analytics.

Along with the three layers of software platforms present here, a collection of components is combined with a microcontroller unit. End devices, in conjunction with the microcontroller unit, can support short-range, low-power communication protocols while conserving energy. Real-time operating systems allow devices to be more productive by preventing buffering. Examples of real-time IoT operating systems are RIOT, FreeRTOS, Nano-RK, LiteOS, and so on. Gateways connect multiple IoT-enabled devices with the cloud for data communication. According to (Xiaoyi et al., 2021), a multi-objective distributed dispatching algorithm was used for green energy management in smart cities. This algorithm focused on converting renewable sources of energy to batteries, heat, and load storage. In that case, the cost was linked to utility functions and energy consumption. Significant energy waste was reduced in that case by utilizing IoT for green energy management.

Smart energy management using IoT and AI techniques involves the integration of advanced technologies to optimize energy consumption, improve efficiency, and enhance sustainability. Large amounts of energy data gathered from IoT devices are subjected to the use of AI techniques, such as machine learning and data analytics. The analysis of both historical and current data by machine learning algorithms can reveal patterns, outliers, and trends in energy consumption. These algorithms can also anticipate future energy demand and consumption using predictive analytics. AI algorithms can reduce energy use by analyzing data from IoT devices and other sources like weather predictions and price signals. Machine learning algorithms are able to identify patterns and automatically modify energy use to prevent wastage. Energy usage can be dynamically modified based on demand, grid circumstances, and pricing using demand response systems.

5. USE CASES FROM INDIAN CONTEXT

5.1. Smart Buildings

According to (Majdi et al., 2022), India metropolitans house millions of human lives and managing buildings whether residential, corporate or shopping malls is an arduous task. Figure 5 depicts an IoT based Smart Building Architecture. Automating the process using AI/ML enabled IoT systems makes it easier to manage in terms of:

- **Energy efficiency:** India faces power shortage in several regions, hence using an IoT enabled system to manage and monitor the power consumption of the building is crucial to maintain a steady flow of electricity.
- **Indoor air quality:** Automatic monitoring of indoor air quality using smart sensors can inform the inhabitants about it. Some air purification system actuators will also help set the air quality to a healthy level.
- **Security:** Unauthorized access can be mitigated using IoT based solutions in public buildings and secure locations thereby preventing theft, damage and trespassing.

- **Comfort:** Automated systems can be employed to monitor and adjust temperature, humidity, lighting and ventilation to ensure occupant comfort, especially in hot temperature regions in India.

Figure 5. IoT based smart building architecture

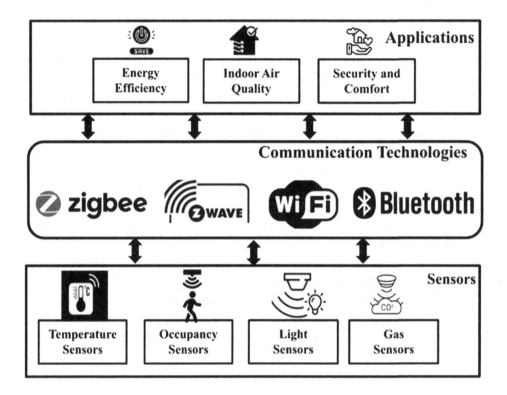

1. **TCS Olympus Centre, Chennai:** One example of a smart building in Chennai is the TCS Olympus Centre, which features intelligent lighting, automated shading, and an advanced building management system for energy efficiency and occupant comfort.
2. **Hiranandani Gardens, Mumbai:** This residential complex has been built with LEED (Leadership in Energy and Environmental Design) certification, making it one of the most energy-efficient buildings in India. It also has energy-efficient lighting and automated climate control systems.
3. **The 42, Kolkata:** The residential skyscraper in Kolkata is one of the tallest smart-buildings in India, boasting features such as safety/security automation, water tank sensors, smart energy meters, automated lighting control among others.
4. **The Central Park, New Delhi:** This residential complex is equipped with a number of smart building features, including automated lighting control, energy-efficient air conditioning, and a water management system.

5.2. Smart Agriculture

India is a primarily agrarian country, it heavily relies on its farmers to enrich the economy to an extent where agriculture contributes to 18.8% of Gross Value Added in the year 2021-2022. Automating the process of farming using specialized IoT devices and optimizing it with the help of AI prediction models can contribute to an exceptionally streamlined production system. This also includes the possibility of detecting calamities and taking preventive action to protect the crops in case of such disasters. The Indian Government has taken several active steps to ensure the field of Smart Agriculture is developed maintaining the efficiency and profitability of the farmers. One of the key use cases for IoT smart agriculture in India is precision farming. With over 58% of the Indian population engaged in agriculture, precision farming can play a significant role in improving the country's agricultural output. By using IoT sensors, farmers can analyze data on crop growth, soil moisture, and nutrient levels, enabling them to optimize their farming practices and reduce costs according to (Pathmudi et al., 2023)

1. **IoT-Based Smart Agriculture:** Farm Automation system that uses sensors and wireless communication to monitor and control various parameters such as soil moisture, temperature, humidity, and light intensity. The system also includes automatic irrigation, fertilization, and pest control, which can help reduce costs and increase yields.
2. **Livestock Monitoring:** By using smart wearable sensors on farm livestock, a plethora of health issues can be easily identified in the early stages and mitigated using proper care. Incorporating a GPS location sensor the whereabouts of livestock can also be located especially when going out for grazing and other activities.
3. **Smart Irrigation System:** IOT-Based Precision Agriculture development and implementation of a smart irrigation system in Indian agriculture. The system uses IoT sensors to monitor soil moisture and weather conditions, and it automatically controls the water supply to crops. This technology helps reduce water usage and improves crop yields.
4. **Pest and Disease Control in Farms:** Presence of pests infecting and damaging plants in large farms can be handled by IoT based recognition systems and can help farmers be informed on the infestation of their crops. Similarly plant diseases which are evident via the condition of the leaves can also be monitored as discussed by (Nayagam et al., 2023).

5.3. Water Management

Smart water management using IoT technology is vital in addressing the water crisis that many countries, including India, are facing. By leveraging IoT sensors and real-time data analysis, it is possible to optimize water usage, monitor the quality of water, and reduce wastage of this precious resource. Figure 6 depicts a smart water meter and leakage detection system. Some of the use cases of IoT smart water management in India include:

*   **Monitoring water quality:** IoT sensors can be installed in water treatment plants and distribution networks to detect changes in water quality. By continuously monitoring water quality, authorities can quickly identify contamination, treat it, and avoid waterborne diseases (Singh et al., 2022).
*   **Leak detection:** With IoT-enabled water meters and sensors, leaks in the water distribution system can be identified early, and the necessary measures can be taken to reduce water loss.

- **Irrigation management:** IoT sensors can be used in agricultural fields to monitor soil moisture levels and provide farmers with information on the best time to irrigate crops. This can help conserve water and reduce the cost of irrigation as discussed by (Parvathi Sangeetha et al., 2022)
- **Water conservation:** Smart water meters can be used to monitor water usage in households, and residents can be alerted when they exceed their daily or monthly water limit. This can help in reducing water wastage and promoting responsible water use.

Figure 6. Smart metering and water leakage detection system

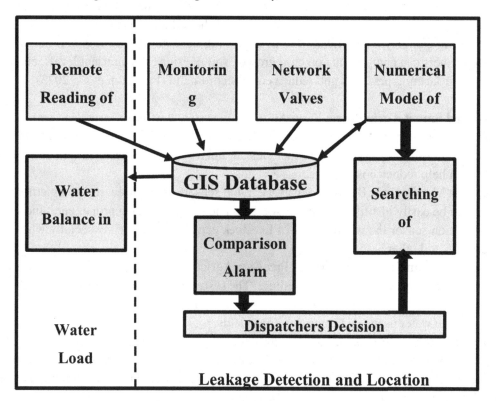

In conclusion, IoT smart water management can play a crucial role in managing India's water crisis by reducing wastage, improving water quality, and promoting responsible water use. Indian rivers like the Yamuna and Ganga are affected with pollutants including sewage and industrial waste. Even though they are some of the primary sources of water for countless citizens it is not managed and controlled properly. Using an IoT enabled smart water management system we can monitor the pollution in these rivers and inform the respective authorities. Several Indian cities use IoT enabled water management systems for example in Gujarat, AMRUT(Atal Mission for Rejuvenation and Urban Transformation) project implemented by RMC (Rajkot Municipal Corporation) in 2018, uses IoT enabled sensors and real-time data monitoring to detect leaks, and identify water wastage.

5.4. Soil Moisture Monitoring

ESP32 Moisture Sensor is an excellent sensor developed by Adafruit technologies and has immense potential. A simple sensor like this can be used for a wide array of applications that are listed below:

- **Precision Irrigation:** It can be used to judge the appropriate time for irrigation, how much water to use when the land needs the irrigation and this inturn reduces the amount of water used in the longer run. It is also useful in curbing the amount of electricity used in the irrigation system. Indian states like Punjab, Haryana, Maharashtra, Tamil Nadu, Karnataka, and Andhra Pradesh use this technology for farming. The Central Soil Salinity Research Institute (CSSRI) and the Indian Institute of Soil Science (IISS) conduct several researches based on the domain and can be referred to for future scope of work according to (Prasanna Lakshmi et al., 2023)
- **Drought Management:** The onset of drought destroys agricultural lands in states across India such as Rajasthan and Gujrat. Farmers with the help of IoT based systems enabled with AI/ML prediction can be informed about the onset of drought in their soil which they can then work to mitigate by using various techniques such as crop rotation, intercropping and other soil protective methods. Apart from Gujarat and Rajasthan, Andhra Pradesh also employs soil moisture management to prevent droughts. The Indian Institute of Remote Sensing (IIRS) and the National Remote Sensing Centre (NRSC) are good sources of information on drought management.
- **Flood management:** Soil moisture monitoring can be used to anticipate floods and help with flood management. Monitoring soil moisture for flood control is a frequent practice in states like Assam, Bihar, and West Bengal. Information about flood control can be found at the National Centre for Medium Range Weather Forecasting (NCMRWF) and the Indian National Centre for Ocean Information Services (INCOIS).
- **Soil Conservation:** By monitoring soil moisture, farmers can choose the best soil conservation practices to enhance the production and health of their soils. States like Madhya Pradesh, Uttar Pradesh, and Bihar frequently adopt soil conservation with soil moisture monitoring. The Central Soil and Water Conservation Research and Training Institute (CSWCRTI) and the National Bureau of Soil Survey and Land Use Planning (NBSS&LUP) are good sources of information on soil conservation.

As discussed above, using IoT and progressing technologies we can manage the risks of drought, flood, soil erosion, and farming irrigation which is quintessential for India and its agrarian nature of livelihood.

5.5. Smart Energy Management

India has set objectives to reach net-zero carbon emissions by 2070, achieve cumulative renewable energy installations of 50% by 2030, and lower the carbon intensity of the country's economy by less than 45% by the end of the decade. To achieve such a gargantuan task our country needs to rely on the effective solutions that Smart Energy Management can provide. It is becoming more and more common to observe our country adopting this method as it leads to lesser emissions, reduces energy consumption and cuts overall cost. Some smart energy management initiatives taken by the government are as follows:

- **Smart Grids:** States in India, notably Gujarat, Delhi, and Maharashtra, are deploying smart grids. Smart networks enhance energy distribution and cut waste by using cutting-edge technology like real-time data analytics and automation.They also make it easier for renewable energy sources to be integrated into the grid, leading to more effective and environmentally friendly energy use. (Archana et al., 2022)
- **Energy-efficient Buildings:** As discussed before in the Smart Buildings section, in places like Bangalore and Mumbai, where high energy consumption and air pollution are important concerns. The necessity for energy-efficient buildings is becoming more prevalent. These structures incorporate energy-saving design elements like insulation, natural ventilation, and effective lighting. These features also enhance indoor air quality. In order to optimize energy consumption based on occupancy and weather conditions, they also combine smart technology such as sensors and automation.
- **Electric Vehicles:** With the advent of electric vehicles in the recent past and its rising popularity pan-India, cities like Bangalore, Delhi and Mumbai are fast adopting the technology for locomotion. E-vehicles are not only environmentally friendly in terms of carbon emissions but also more energy efficient. This results in creating a group IoT situation where all the cars can be connected to each other with the energy consumption being monitored, which will result in smart management of the energy being used by them based on a database of commuters across the city. The Indian government has also introduced schemes such as FAME (Faster Adoption and Manufacturing of Electric Vehicles) for a quick and seamless transfer to the use of e-vehicles in our country.
- **Renewable Energy:** Conventional sources of energy are non-renewable ones. Apart from having a tremendous carbon footprint it is also not a sustainable way of powering our future. With major investment in solar and wind power, India is one of the markets for renewable energy. Solar and wind energy projects now make up a sizable fraction of the energy mix in states like Gujarat, Rajasthan, and Tamil Nadu. These states have achieved tremendous success in deploying renewable energy conservation techniques. These initiatives enhance energy production and cut waste by utilizing cutting-edge technologies like real-time monitoring and predictive analytics.

Smart energy management is being used in several aspects of India since the country's goal is to achieve sustainability in all operations. There is a high chance in the near future that the rising population will demand more deployment of IoT in the aforementioned sectors to ensure that the goal is easily achieved.

6. INFRASTRUCTURE AND TECHNOLOGY REQUIREMENTS

Sustainable IoT (Internet of Things) for smart environmental control is a technology-based solution aimed at promoting sustainability in the management of the environment. To achieve this, it requires several key infrastructure and technology requirements.

6.1. Power-Consumption

As the device is meant to collect continuous data in order to process, The IoT devices need a lot of power and that power cannot be directly connected to the grid. Instead the IoT device will be manufactured with the specific feature that it consumes less power and becomes highly efficient. This will make the devices with power backup of the IoT which will make the device stay long and make it working, collecting and sending data to the cloud uninterruptedly. This can be achieved by using hardware components that are designed to minimize energy consumption and reducing the carbon footprint of the software.

6.2. Cloud

As the need of storing the data and analyzing the data is increasing day – by – day, the method of storing and analyzing the data on the device locally has been reduced because of the evolution of the cloud where all the data is being stored in the cloud by the medium of API communication or any other methods and all the processing being done on the cloud. This enables the IoT device to work effectively and efficiently so that it consumes less power and does only one job of sending the data to the cloud and the cloud keeps the backup of the data and processes faster than the processor which is within the IoT itself.

6.3. Artificial Intelligence (AI)

As AI is developing day by day, the algorithm's used for sensing the data by the IoT are becoming more efficient day by day. The AI detects the environment and decides when to make the IOT to process and when to sense and send data to the cloud where the need of sending data rapidly will be reduced. Due to this procedure algorithm, this benefits the whole architecture in many ways such as reducing the power consumption, reducing the need of consumption of data, reducing the data to be stored, reducing the processing power of the cloud, Increasing the lifespan of the IoT device and its components and sensors. These are some of the benefits and thus, will be enhanced more in the future.

6.4. Internet

As the technology has been developing rapidly, the internet service providers have been enhancing the internet connectivity worldwide at a rapid increasing rate. Due to this the IoT has started evolving as the complexity of sending the data from the iot to cloud has been reduced from low to none. Anyone now can rent a hosting platform and run an IoT device to send data to their hosting platform which is so cheaper compared to the old days. As it's been working through the internet, the data is monitored in real-time with low latency of below 1 second. The introduction of 5G will change the future usage of IoT devices mainly in the medical field.

6.5. Security

The more the use of digital devices, the exposure of data will also be more. As the IoT has increased worldwide, the rate of cyber threats has also increased in the past decade. The data which is sent to the cloud or the data which is present in the cloud can be manipulated easily if the security of the infrastructure is weak. This cyber-attack on the devices or the cloud could lead to loss of data and its security. These can be prevented by maintaining proper firewalls for the cloud, using encryption and SSL(Secure Sockets Layer) still many more to be developed to protect the system.

Efficiency, which includes the utilization of low-power sensors and devices, effective data processing algorithms, and energy-efficient hardware components, is a crucial component of sustainable IoT infrastructure. These elements can help the IoT infrastructure to conserve energy, cut carbon emissions, and increase sustainability in general. The usage of durable and secure hardware and software components that are created to reduce downtime and guarantee that the system is constantly operational is required for reliability in sustainable IoT infrastructure. As a result, the system is more sustainably maintained and repaired less frequently. The use of eco-friendly materials and technologies that limit waste, reduce pollution, and maximize the use of resources is necessary for sustainable IoT infrastructure. This makes the IoT system more environmentally friendly, sustainable, and protects the environment for present and future generations. Environmental management that is environmentally friendly will result in cut down on energy use, and ensure that our environment is conserved for future generations by implementing sustainable IoT solutions.

7. SWOC ANALYSIS ON IOT IN SMART ENVIRONMENT

Figure 7. SWOC analysis of IoT in a smart environment

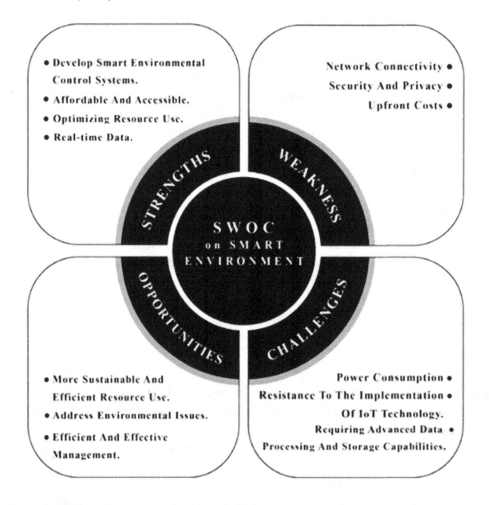

SWOC analysis is a tool used to analyze the Strengths, Weaknesses, Opportunities, and Challenges of a particular situation or context. Figure 7 depicts SWOC analysis on IoT in a smart environment. Here is a SWOC analysis of IoT environment:

Strengths:

- IoT technology provides **real-time data** on environmental conditions, enabling us to monitor and manage environmental factors effectively.
- IoT technology can be used to **develop smart environmental control systems**, optimizing resource use and reducing waste.
- IoT sensors and devices are becoming more **affordable and accessible**, making it easier to implement IoT solutions in different settings.

Weaknesses:

- IoT technology requires internet connectivity, and in some areas, **network connectivity** is weak or not available.
- **Security and privacy** concerns may arise when collecting and sharing data from IoT devices and sensors.
- The implementation of IoT infrastructure can require significant **upfront costs,** which may be a barrier to adoption in some settings.

Opportunities:

- IoT technology has the potential to revolutionize environmental monitoring and management, leading to **more sustainable and efficient resource use.**
- IoT technology can be used to identify and **address environmental issues** such as pollution, climate change, and natural resource depletion.
- IoT solutions can be customized to meet the needs of different environments, enabling more **efficient and effective management.**

Challenges:

- IoT technology can generate vast amounts of data, **requiring advanced data processing and storage capabilities.**
- IoT technology requires significant **power consumption**, which can be a barrier to implementation in remote or off-grid areas.
- There may be **resistance to the implementation of IoT technology** from individuals or organizations that are resistant to change or prefer traditional methods of environmental management.

8. WAYFORWARD

Based on the SWOC analysis, here are some potential ways forward for IoT in the environment:

- **Develop more efficient IoT technologies:** To address the challenges related to power consumption and data processing, efforts should be made to improve the efficiency of IoT sensors and devices, data storage and processing, and communication technologies.
- **Address security and privacy concerns:** Security and privacy should be prioritized in the development and implementation of IoT technologies to ensure that sensitive data is protected and that users have confidence in the system.
- **Reduce the costs of IoT infrastructure:** To increase the adoption of IoT in different settings, efforts should be made to reduce the upfront costs of IoT infrastructure, such as through the development of more affordable sensors and devices.
- **Partner with stakeholders:** Collaboration and partnerships with relevant stakeholders, such as local governments, businesses, and environmental organizations, can help to address environmental issues more effectively through the use of IoT technologies.
- **Educate and train users:** To increase awareness and understanding of IoT technologies and their potential benefits, efforts should be made to educate and train users on how to use and maintain IoT devices and systems.

Overall, the way forward for IoT in the environment is to address the challenges and weaknesses highlighted in the SWOC analysis while leveraging the strengths and opportunities of the technology to develop more efficient and effective solutions for environmental monitoring and management.

9. CONCLUSION

The way the environment is managed and monitored may be completely changed by IoT technology. IoT sensors and devices make it possible to cut waste and optimize resource consumption by delivering real-time data on environmental conditions. This may result in more environmentally friendly and economically viable environmental management measures, as well as better health and wellbeing for people and other living beings. To ensure that IoT technology can be used effectively for sustainable environmental control, there are a few issues and flaws that are presented in this chapter to be fixed. IoT sensors and gadgets require a lot of electricity to operate, which can be a problem in remote or off-grid places. Power consumption is a serious problem. In addition, IoT technology collecting massive data demands sophisticated data processing and storage capabilities, which can be costly and call for substantial infrastructure investments.

Addressing security and privacy issues is also crucial, especially as more sensitive data is gathered and shared through IoT devices. To promote greater adoption and usage of IoT technology, it is essential to guarantee the security of sensitive data and user confidence in the system. Additionally, there may be a barrier to adoption in some environments due to the high upfront expenditures associated with implementing IoT infrastructure. This can be resolved by making an effort to lower the price of IoT infrastructure, for example, by creating more economical sensors and devices. Efforts should be made to develop more effective IoT technologies, address security and privacy concerns, lower the cost of IoT infrastructure, collaborate with pertinent stakeholders, educate and train users, and develop IoT technologies that are more effective in order to overcome these challenges and leverage the strengths and opportunities of IoT technology for sustainable environmental control. By doing this, IoT technology's potential can be optimized to build a more sustainable and healthy planet for everyone.

REFERENCES

Ali, A. S., Abdelmoez, M. N., Heshmat, M., & Ibrahim, K. (2022). A solution for water management and leakage detection problems using IoTs based approach. *Internet of Things*, *18*, 100504. doi:10.1016/j.iot.2022.100504

Alobaidy, H. A. H., Nordin, R., Mandeep, J. S., Abdullah, N. F., Haniz, A., Ishizu, K., Matsumura, T., Kojima, F., & Ramli, N. (2022). Low Altitude Platform-based Airborne IoT Network (LAP-AIN) for Water Quality Monitoring in Harsh Tropical Environment. *IEEE Internet of Things Journal*, *9*(20), 1–1. doi:10.1109/JIOT.2022.3171294

Archana, S., Shankar, R., & Singh, S. (2022). Development of smart grid for the power sector in India. *Cleaner Energy Systems*, *2*, 100011. doi:10.1016/j.cles.2022.100011

Details of Funds Spent on Namami Gange Programme. (2021). Pib.gov.in. https://pib.gov.in/PressReleaseIframePage.aspx?PRID=1739094

Madhuri, N. S., Shailaja, K., Saha, D. P. R., Glory, K. B., & Sumithra, M. (2022). IOT integrated smart grid management system for effective energy management. Measurement. *Sensors (Basel)*, *24*, 100488. doi:10.1016/j.measen.2022.100488

Majdi, A., Dwijendra, N. K. A., Muda, I., Chetthamrongchai, P., Sivaraman, R., & Hammid, A. T. (2022). A smart building with integrated energy management: Steps toward the creation of a smart city. *Sustainable Energy Technologies and Assessments*, *53*, 102663. doi:10.1016/j.seta.2022.102663

Manoj, M., Dhilip Kumar, V., Arif, M., Bulai, E.-R., Bulai, P., & Geman, O. (2022). State of the Art Techniques for Water Quality Monitoring Systems for Fish Ponds Using IoT and Underwater Sensors: A Review. *Sensors (Basel)*, *22*(6), 2088. doi:10.339022062088 PMID:35336256

Namami, G. (2023, March 18). *National Mission for Clean Ganga(NMCG)*. Ministry of Jal Shakti, Department of Water Resources, River Development & Ganga Rejuvenation, Government of India. https://nmcg.nic.in/

Nayagam, M., Vijayalakshmi, B., Somasundaram, K., Mukunthan, M. A., Yogaraja, C. A., & Partheeban, P. (2023). Control of pests and diseases in plants using IOT Technology. Measurement. *Sensors (Basel)*, *26*, 100713. doi:10.1016/j.measen.2023.100713

Parvathi Sangeetha, B., Kumar, N., Ambalgi, A. P., Abdul Haleem, S. L., Thilagam, K., & Vijayakumar, P. (2022). IOT based smart irrigation management system for environmental sustainability in India. *Sustainable Energy Technologies and Assessments*, *52*, 101973. doi:10.1016/j.seta.2022.101973

Pathmudi, V. R., Khatri, N., Kumar, S., Abdul-Qawy, A. S. H., & Vyas, A. K. (2023). A systematic review of IoT technologies and their constituents for smart and sustainable agriculture applications. *Scientific African*, *19*, e01577. doi:10.1016/j.sciaf.2023.e01577

Prasanna Lakshmi, G. S., Asha, P. N., Sandhya, G., Vivek Sharma, S., Shilpashree, S., & Subramanya, S. G. (2023). An intelligent IOT sensor coupled precision irrigation model for agriculture. Measurement. *Sensors (Basel)*, *25*, 100608. doi:10.1016/j.measen.2022.100608

Singh, S., Rai, S., Singh, P., & Mishra, V. K. (2022). Real-time water quality monitoring of River Ganga (India) using internet of things. *Ecological Informatics, 101770*, 101770. doi:10.1016/j.ecoinf.2022.101770

Ullah, R., Abbas, A. W., Ullah, M., Khan, R. U., Khan, I. U., Aslam, N., & Aljameel, S. S. (2021). EEWMP: An IoT-Based Energy-Efficient Water Management Platform for Smart Irrigation. *Scientific Programming, 2021*, 1–9. doi:10.1155/2021/5536884

Xiaoyi, Z., Dongling, W., Yuming, Z., Manokaran, K. B., & Benny Antony, A. (2021). IoT driven framework based on efficient green energy management in smart cities using multi-objective distributed dispatching algorithms. *Environmental Impact Assessment Review, 88*, 106567. doi:10.1016/j.eiar.2021.106567

Chapter 21
Evolutionized Industry With the Internet of Things

Sangeetha Rangasamy
https://orcid.org/0000-0002-1850-232X
CHRIST University (Deemed), India

Kavitha Rajamohan
https://orcid.org/0000-0002-7803-8901
CHRIST University (Deemed), India

V. S. Lavan
CHRIST University (Deemed), India

C. Mayur
CHRIST University (Deemed), India

Mary F. Lalitha
CHRIST University (Deemed), India

ABSTRACT

Technological development has led us to various paths. Though each technological change has evolved in its own space and time, it resulted in amenities and a pragmatic approach to problem-solving. One such technology is the "Internet of Things" - IoT which was coined in the year 1999 by a computer scientist called Kevin Ashton. Despite its start in the late 90s, it has come a long way to achieve the status of 10 Billion active IoT devices by the end of 2021, with the IoT solutions costing an estimated economic value of 4-11 trillion by 2025.

DOI: 10.4018/978-1-6684-8785-3.ch021

1. INDUSTRIAL INTERNET OF THINGS (IIOT)

IIoT refers to the integration of a network of connected things technology in industrial settings, specifically in manufacturing, energy, transportation, and so on. IIoT involves connecting a variety of sensors, devices, and machines to the internet, enabling real-time communication, monitoring, and control of industrial processes. This technology allows for improved efficiency, productivity, and better decision making, enabling new business models.

1.1 Industrial Revolution

The term revolution refers to the set of activities that have taken place to bring changes in the way of doing things differently. From an Industry perspective it is the speeding up of innovation by the means of introduction of new tools that benefit various fields. Some major areas that it helped include optimized production by reducing the use of labor power, production of various goods and services, efficiently managing the time (Wilkinson, 2022). This is the era of the 5th Industrial Revolution but looking back at history, this Industrial Revolution is dated back to the era of 18 hundreds where in The Great Britain saw a sharp incline in the technological advancement that eventually led to merger between industry and technology. This revolution over the centuries has led us to the new phase called Evolution where technology has already provided a platform and people are at the receiving end to make use of it and push it into another level thereby people, process and industry are just evolving within a revolution.

1.2 Evolution of Industrial IoT

The evolution of IIoT is the development from simple machine-to-machine communication to a complex network of interconnected devices and systems (Desjardins, 2018). The key highlights include the growth of data volume, complexity of devices, and the need for integration with other technologies as key drivers of the evolution of IIoT. Fig 1 shows the developmental phase of various technologies within each Industrial evolution.

Figure 1. Industrial evolution road map

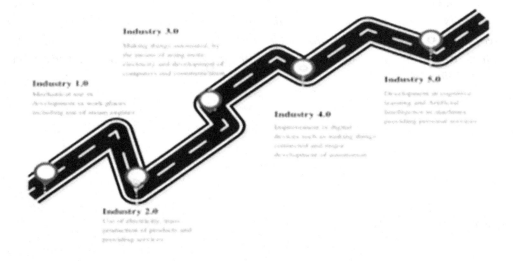

1.3 Benefits of Using IIoT

Based on today's scenario, the Industries have to be nimble in moving forward so as to meet the demands of the market. Fig 2 shows benefits of the Internet of Things in the Industry domain. Considering the adoption of the Internet of Things in the Industry key benefits can be listed as follows (Mendoza, 2020):

Figure 2. Benefits of industrial internet of things

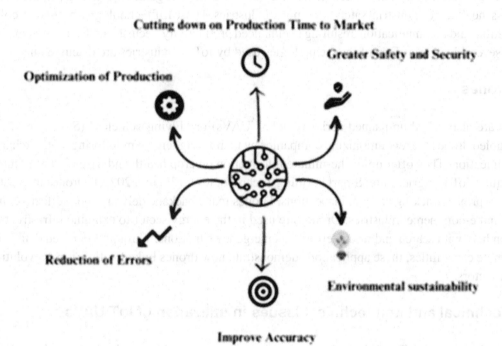

- **Cutting down on Production Time to Market:**By the use of automation in the manufacturing sector, time consumption, which is of very much essence, can be reduced, hence leaving industrialists with more time for other activities.
- **Reduction of Errors:**The use of Internet of Things in Industry can have benefits which leads the product to go through fewer errors so as to keep the product ready for the market.
- **Greater Safety and Security:** Through the usage of sensors, the technology has led to increased security and safety to all the segments of the society.
- **Optimization of Production:** By connecting machines, devices, and sensors to the internet, the IIoT can provide real-time data and insights that can be used to identify inefficiencies and improve productivity
- **Environmental sustainability:** The IIoT can help companies reduce their environmental impact by optimizing energy usage and reducing waste through more efficient processes.

2. IIOT TECHNOLOGIES

IoT innovations like sensors, drones, and wearables have found extensive use in several sectors. By providing real-time monitoring of machinery, ambient conditions, and asset tracking, sensors are essential in sectors including industry, agriculture, healthcare, and smart cities. Drones are used for airborne surveillance, data collecting, and surveying in the construction, agricultural, logistics, and search and rescue sectors. For monitoring vital signs, tracking activity levels, and enhancing general wellbeing, wearable technology, such as smartwatches and fitness trackers, is gaining popularity in the fitness and wellness, healthcare, industrial safety, and sports industries. These IoT technologies boost data collecting, automation, and communication, resulting in enhanced productivity, security, and consumer experiences across several industries.Few of the technologies used by IoT in industries are discussed here.

2.1 Drones

Drones are also called unmanned aerial vehicles (UAVs) or "Flying vehicles" (Stöcker et al., 2017). IoT-enabled drones are revolutionizing companies by facilitating remote monitoring, data exchange, and communication. They offer up-to-the-minute information on crop health and irrigation requirements in agriculture (IoT in Agriculture: Smart Farming Technologies for Future, 2022.). Drones in construction help with project tracking and site inspections. Drones make package delivery more effective in the logistics and e-commerce industries. Drones are used in the energy sector to examine infrastructure, and they can help with search and rescue efforts in emergency situations. Through improved data collecting and remote capabilities, these applications demonstrate how drones linked with IoT are revolutionizing several sectors.

2.2. Technical and non-technical issues in utilization of IoT UAVs

The issue in using the IoT-enabled UAVs is security. Since it is an unmanned aerial vehicle containing an embedded camera, it may affect the privacy of a person around (D. Liao et al. 2017). In most countries there are few restricted regions over which the drones cannot be flown over, therefore the communication between the Ground Control Room (GCR) and UAV should be protected.There are few security strategies used for such scenarios like (Vijayakumar et al., 2016):

- Vehicular Ad hoc Networks (VANETs)
- Wireless LAN (WLAN)
- Voice over IP (VoIP)
- Radio frequency identification (RFID)
- Wireless Sensor Networks (WSN)
- Internet of things (IoT)

2.21.1 Drones in healthcare

Drones equipped with IoT technology provide a variety of uses and advantages in the healthcare sector (Matthew, 2021). Here are a few applications for drones in IoT healthcare:

- **Delivery of medical goods:** Drones can be used to quickly and effectively carry medical supplies, such as medicines, vaccinations, and emergency gear, to hard-to-reach places. Drones may be remotely followed and monitored with IoT connection, resulting in secure and dependable delivery. Drones that can carry essential medical supplies quickly and efficiently to remote locations can improve access to healthcare and perhaps save lives (Scalea et al., 2021). These drones bridge access gaps in healthcare and address inequities by ensuring secure and dependable delivery with real-time monitoring and tracking. Figure 3 explains how drones are used in delivery in the healthcare sector.

Figure 3. Drone carriage for medical goods

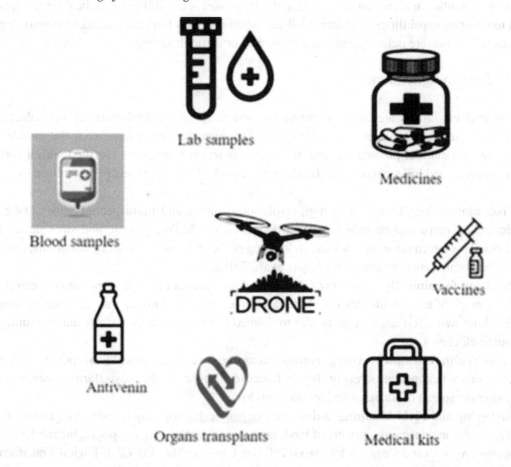

- **Disaster Response:** Drones that are outfitted with medical supplies and sensors can be used to deliver quick medical aid during emergencies or natural disasters. They can evaluate the situation, find survivors, and supply essential supplies, allowing for quicker reaction times and sometimes even lifesaving.
- **Remote patient monitoring:** It is possible using IoT-enabled drones that are fitted with sensors to track vital signs, gather patient data, and send it instantly to healthcare practitioners. This allows

remote patient monitoring at their homes or other remote locations, enhancing patient access to healthcare and facilitating the early identification of health conditions.

- **Support for Telemedicine:** Drones with cameras and Internet of Things (IoT) connection can help in telemedicine consultations. They can give medical staff visual support, enabling remote patient condition evaluation and simplifying virtual exams.
- **Services to medical emergencies:** Drones can be fitted with defibrillators and medical equipment to give urgent emergency medical care in difficult-to-reach areas. They can provide life-saving interventions before conventional medical aid arrives since they can reach accident sites or rural locations faster than typical ambulances (Graboyes et al., 2020).

Drones in healthcare allow quicker reaction times, greater accessibility to medical services, and enhanced monitoring capabilities by utilizing IoT technologies. They have the potential to improve patient outcomes and healthcare delivery, particularly in rural or difficult settings.

2.2.2 Drones in Agriculture

In IoT agriculture, drones are being employed to modernize agricultural methods and enhance crop management. Drones give farmers the tools they need to collect important data and make wise decisions for increased agricultural productivity and resource efficiency. Drones are outfitted with a variety of sensors, cameras, and IoT technologies (Haghi et al., 2017). Drones are used in agriculture for:

- **Crop monitoring:** Drones with high-resolution cameras and multispectral sensors take aerial pictures of crops and provide important details on the health, growth, and stress levels of the plants. This information aids farmers in locating problem areas like insect infestations, nutritional deficits, or irrigation requirements (Ayaz et al., 2019).
- **Precision farming:** By combining GPS technology and data from drones, farmers can produce exact maps of their fields that enable focused interventions. Drones may be used to accurately distribute water, fertilizer, or pesticides to defined locations, reducing waste and maximizing resource efficiency.
- **Crop Health Assessment:** Using thermal sensors on drones, it is possible to spot changes in crop temperature that indicate stress or illness. Because of this early diagnosis, farmers can act quickly to stop the spread of illnesses and reduce crop losses.
- **Surveying and Field Mapping:** Advanced imaging technology-equipped drones can swiftly and precisely scan enormous amounts of land, producing accurate maps, topographic models, and 3D photos (Agricultural Drones - Industrial IoT Use Case Profile | IoT ONE Digital Transformation Advisors, 2020). The design of irrigation systems, soil analysis, and land planning are all aided by this data.
- **Monitoring of livestock:** In large-scale farming operations, drones are employed to monitor cattle. Drones equipped with thermal imaging cameras can detect and follow animals, ensuring their safety, detecting stray animals, or finding missing cattle.

Drones in agriculture provide farmers with real-time data, remote monitoring, and automated procedures by utilizing IoT capabilities. This technological fusion raises production, lowers expenses, and encourages environmentally friendly farming methods. Drone usage in IoT agriculture is revolutionizing the sector by empowering farmers to make data-driven decisions and enhance crop management in general.

2.2.3 Internet of Quantum Drones (IoQDs)

Paul Benioff introduced the concept and principles of quantum information in 1980. Richard Feyman's development to quantum computing and Shor's integer factorization algorithm contributed to quantum computing (Feynman, 1982). **Qubits or quantum bits** are used in quantum computers. It can simultaneously be zero and one whose probability factor always depends on the outcomes. The application of IoT to the drone system has many disadvantages like privacy and authentication issues. Hence in recent years the prototypes of drone- based quantum network communication have been implemented. This has been implemented on a small scale where quantum signals are used over short distances to exchange the quibts. With the feasibility, the information can be easily transferred and can be easily set for any applications. The drone has the advantage of being positioned or re-positioned anywhere at any point of time hence IoD is very useful in rural areas. IoQDs has the following characteristics:

- The message can be transmitted using quantum communication between device and the users.
- Fast communication process is possible using IoQDs used optical fiber based quantum cryptography technique that covered a range of 421 kms (Lucamarini, 2018).
- IoQDs provide full flexibility which can be used in real-time applications such as military, traffic management, internet services, and secure transferring of messages over long distances.

2.1.3.1 Applications of Quantum drones

There are various real time applications with the development of IoQDs networks. Some of the applications are shown in fig 4. Few of the areas where IoQDs are highly functional are (Kumar, et. al 2022):

Figure 4. Applications of IoQDs

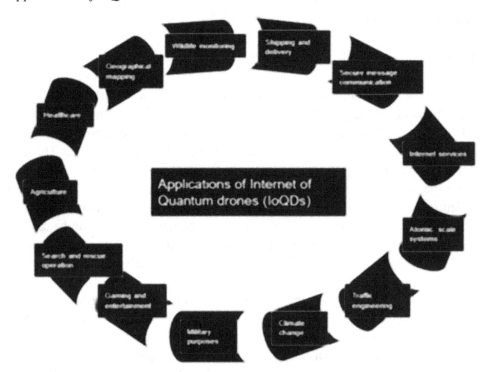

- **IoQDs for radar and traffic management:** IoQDs detect, sense and identify using quantum computers. Quantum drones could be used in radar systems to improve the accuracy and resolution of radar images. They are used in traffic management to help monitor and control the traffic flow.

- **IoQDs for quantum internet services:** Drone-based quantum internet services are much faster than the normal internet services. The data is more secure. Using drones the internet connectivity can be adjusted to get the required bandwidth.

- **IoQDs for climate change:** Quantum computers and quantum computing help in handling the data concerned to climate. These provide data processing and computation much faster hence it plays a prominent role during calamities like hurricanes, tsunamis, earthquakes and heat waves. The data regarding these calamities can be processed in advance so many lives and properties can be protected.

2.3 Sensors

Sensors are a key component of IIoT technology. A sensor is a device that recognizes changes in physical or environmental conditions and converts them into a processing-ready electrical signal by other devices. In the context of the IIoT, sensors are used to collect data from various components of a manufacturing process or an industrial environment. This data can then be analyzed and used to improve processes, increase efficiency, and optimize performance. For example, sensors are used to monitor the performance of machines, equipment, and other assets, allowing manufacturers to detect potential problems and perform predictive maintenance.

2.2.1 Sensors in Healthcare

In the context of healthcare, sensors are used to gather data about patient health and the environment, and then analyze this data to improve patient outcomes, streamline processes, and reduce costs. The use of sensors in healthcare is part of the larger trend towards the integration of IoT technology in healthcare, which seeks to connect medical devices, equipment, and other assets to collect and analyze vast amounts of data.

Some examples of how sensors are being used in healthcare include remote monitoring of patients, wearable IoT devices for self-monitoring, and the use of IoT and sensors to support clinical decision-making. Some important sensors within the healthcare include:

- **Temperature sensors:** These sensors are used to monitor patient temperature and provide early warning for fever or hypothermia.
- **Pressure sensors:** Pressure sensors are used to monitor blood pressure, intraocular pressure, and intracranial pressure.
- **Motion sensors:** Motion sensors can be used to monitor patient activity and detect falls in elderly patients.
- **Pulse oximeters:** These sensors measure blood oxygen levels and heart rate and are commonly used in hospitals and home care settings.
- **Electrocardiogram (ECG) sensors:** ECG sensors are used to monitor heart activity and detect irregular heartbeats.
- **Blood glucose sensors:** These sensors are used to monitor blood glucose levels in patients with diabetes.
- **Accelerometers:** Accelerometers are used to monitor movement and posture in patients, such as in rehabilitation programs.
- **Imaging sensors:** Imaging sensors such as X-ray and MRI machines are used to diagnose and monitor various medical conditions.

2.3.2 Sensors in Sports

The use of sensors in sports is a rapidly growing field, and wearable IoT devices such as fitness trackers and smartwatches are increasingly being used by athletes to monitor their own performance and health. In addition, sensors can be used to monitor a wide range of athletic performance metrics, including speed, acceleration, heart rate, and power output, allowing athletes and coaches to optimize training, identify areas for improvement, and track progress over time (Linnamo, 2023).

Some important domains that make use of sensors in sports include:

- **Performance Monitoring:** Sensors can be used to monitor a wide range of athletic performance metrics, including speed, acceleration, heart rate, and power output. This information can be used to optimize training, identify areas for improvement, and track progress over time.

- **Injury Prevention:** Sensors can be used to monitor movement patterns, balance, and other indicators of injury risk, allowing athletes and coaches to identify and address potential injury risks before they become a problem.
- **Training Optimization:** Sensors can be used to monitor recovery and training progress, helping athletes and coaches to optimize training, reduce the risk of overtraining, and maximize performance.
- **Wearables:** Wearable IoT devices such as fitness trackers and smartwatches allow athletes to monitor their own performance, health, and wellness, and share this information with coaches and trainers.

2.3.3 Sensors in Childcare

The use of sensors in childcare plays a vital role in monitoring the child's actions, providing parents with valuable information and to assess the perceptions of parents towards this technology (Ul Hasan & Negulescu, 2020). It was highlighted in a study that, majority of parents were open to the idea of using sensors to monitor their children's activity levels and health, as long as the technology was easy to use and provided useful information (Won et al., 2022).

However, some sensors that are commonly used in childcare monitoring include:

- **Wearable sensors:** These can be worn by the child and used to monitor their physical activity, heart rate, sleep patterns, and other health metrics.
- **Environmental sensors:** These can be used to monitor the environment in childcare facilities, such as temperature, humidity, and air quality.
- **Location tracking sensors:** These can be used to track the location of children and provide real-time alerts if they wander out of designated areas.
- **Food and drink sensors:** These can be used to monitor the eating and drinking habits of children, providing valuable information for parents and caregivers.
- **Sleep monitoring sensors:** These can be used to monitor the sleep patterns of children, helping parents and caregivers understand their sleep quality and habits.

Overall, the sensors and IoT technology have the potential to be a valuable tool in childcare, providing valuable insights and information for parents and caregivers. However, it is important to carefully consider the feasibility and acceptability of these technologies and address any privacy and security concerns that may arise.

2.4 Wearables

Wearables are a type of IoT device that can be worn or attached to the body to gather data on various physical and physiological parameters (Majumder et al., 2017). Wearables typically include sensors, wireless connectivity, and other advanced technologies that enable them to collect and transmit data to other devices and cloud-based servers and can play a significant role in transforming various industries, including healthcare, fitness, and entertainment. Wearable devices: can be grouped into two categories: passive and active. Passive wearables don't require user interaction and rely on a connected mobile device to control them. They gather and keep track of data such as heart rate, steps, and sleep patterns. Examples of passive wearables include fitness trackers, smartwatches, and wearable sensors. Active wearables, on the other hand, are interactive and can be controlled by touch, voice commands, or gestures. They can perform various functions such as messaging, making calls, playing music, and making payments.

2.4.1 Wearables in Healthcare

In recent years a wide range of wearable IoT healthcare applications have been developed and deployed. The rapid increase in wearable devices allows the transfer of patient personal information between different devices, at the same time personal health and wellness information of patients can be tracked and attacked (Jiang, 2020). There are many techniques that are used for protecting patient information in medical and wearable devices. Some of the ways that IoT technologies are used in wearable devices in healthcare are:

- **Wearable Sensors**: Wearable devices in healthcare typically include sensors, such as accelerometers, gyroscopes, heart rate monitors, and glucose sensors, which can gather data on various physical and physiological parameters. These sensors can provide valuable insights into a person's health and wellness, enabling healthcare providers to identify potential health issues and intervene early.
- **Wireless Connectivity**: Wearable devices in healthcare typically use wireless connectivity technologies, such as Bluetooth, Wi-Fi, and cellular connectivity, to transmit data to other devices, such as smartphones, laptops, and cloud-based servers. This enables healthcare providers to access the data from anywhere, and enables the wearable devices to interact with other devices and applications.
- **Artificial Intelligence**: Artificial intelligence (AI) and machine learning algorithms can be used to analyze the data collected by wearable devices in healthcare, providing personalized insights and recommendations. For instance, wearable devices can use AI to provide users with recommendations on how to improve their sleep or physical activity, based on the data collected from their sensors.
- **Remote Monitoring:** Wearable devices in healthcare can be used for remote monitoring, enabling healthcare providers to monitor patients remotely, and reducing the need for in-person visits. For instance, wearable devices can be used to monitor patients with chronic conditions, such as diabetes or heart disease, and to alert healthcare providers if the patient's condition worsens.
- **Clinical Trials:** Wearable devices in healthcare can be used in clinical trials to gather data on patients in real-time, providing valuable insights into the efficacy and safety of new treatments and medications. (Lee, 2016)

2.4.2 Wearables in Sports

The availability of wearable performance devices and sensors is on the rise for both the general public and athletic teams. Advances in technology enable individual athletes, sports teams, and healthcare providers to track functional movements, workloads, and biometric indicators to enhance performance and prevent injury.Some of the key ways that IoT technologies are used in sports wearables include:

- **Performance Monitoring:** Wearables in sports usually include sensors such as accelerometers, gyroscopes, heart rate monitors, and GPS, which can track physical and physiological parameters. These sensors provide information on the athlete's performance, allowing coaches and trainers to identify areas for improvement and create customized training plans.
- **Wireless Connectivity:** Sports wearables often use wireless connectivity technologies such as Bluetooth, Wi-Fi, and cellular connectivity to transmit data to other devices such as smartphones and laptops. This allows coaches and trainers to access the data from anywhere and enables the wearables to interact with other devices and applications.
- **Data Analysis:** Artificial intelligence (AI) and machine learning algorithms can be used to analyze the data gathered by sports wearables, providing personalized insights and recommendations. For example, wearables can use AI to provide athletes with recommendations on how to improve their sleep or nutrition based on the data collected from their sensors.
- **Injury Prevention:** Wearables in sports can be used to monitor and prevent injuries. For example, wearable devices can detect when an athlete is at risk of injury and provide real-time alerts to the athlete and coach. (Borg et al., 2019)

2.4.3 Wearables in Childcare

Wearable devices are designed to help parents monitor their children's safety and well-being. It provides peace of mind to parents when children are not in close proximity. Some of the ways that IoT technologies are used in wearable devices for childcare are:

- **Location Tracking:** Wearable devices for childcare can be equipped with GPS technology, which allows parents to track their child's location in real-time. This can provide peace of mind for parents who are worried about their child's safety, especially when they are not in close proximity.
- **Health Monitoring:** Some wearable devices for childcare can track various health parameters, such as heart rate, temperature, and activity levels. This can help parents monitor their child's well-being, especially when they are not in close proximity.
- **Communication:** Wearable devices for childcare can be equipped with communication features, such as voice or text messaging. This can allow parents to stay in touch with their child, even when they are not in close proximity.
- **Emergency Alerts:** Some wearable devices for childcare can be programmed to send emergency alerts, such as if the child is in danger or if the device is removed. This can provide peace of mind for parents and help ensure the child's safety.(Rodríguez, P, 2020)

3. APPLICATIONS OF IIOT

Industries including automotive, FMCG, finance, healthcare, hospitality, and construction have been transformed by IoT. IoT makes safety features, vehicle communication, and proactive maintenance possible in the car industry. Real-time inventory management and personalised customer experiences are advantageous for FMCG companies. IoT is used by the financial sector for fraud detection and safe transactions. Remote patient monitoring and individualised treatments are made possible in healthcare through IoT. Smart room management solutions enhance the client experience in the hotel industry.

Project management, equipment tracking, and safety are all improved in construction. Figure 5 explains the applications of Internet of things in various industrial sectors. Overall, IoT fosters innovation, personalisation, and efficiency across various sectors.

Figure 5. Applications of IoT in industry

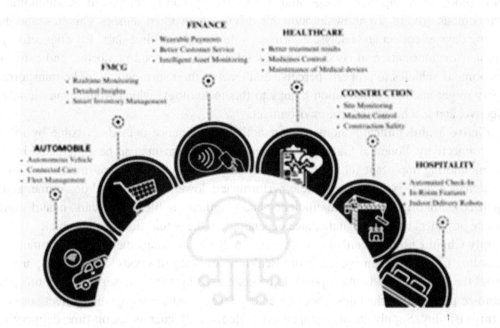

3.1 IIoT in Manufacturing Industry

The Internet of Things (IoT) has completely changed the industrial sector by converting outdated facilities into intelligent, networked spaces. IoT optimizes processes, lowers downtime, and boosts overall efficiency via real-time monitoring, predictive maintenance, and simplified supply chain management (Peranzo, 2022). Manufacturers may obtain important insights into the functioning of their equipment by utilizing sensor data and analytics, enabling proactive maintenance and better decision-making (Kalsoom et al., 2021). IoT also improves logistics and supply chain visibility by enabling real-time inventory tracking. With the idea of the "smart factory," automation, robots, and IoT are all made possible, boosting production and creativity. In conclusion, the use of IoT in manufacturing transforms procedures, boosts competitiveness, and paves the path for the Industry 4.0 revolution.

3.1.1 Automobile Industry

IoT is fundamentally transforming the automotive industry, revolutionizing the way vehicles are manufactured, operated, and experienced (Shrivastava et al., 2020) . Here are some key ways in which IoT is reshaping the automotive industry:

- **Vehicle Connectivity:** IoT enables vehicles to become connected, turning them into smart devices on wheels. Through embedded sensors, IoT connectivity, and telematics systems, vehicles can communicate with each other, infrastructure, and external services. This connectivity enables real-time data exchange, advanced navigation, remote diagnostics, vehicle-to-vehicle (V2V) and vehicle-to-infrastructure (V2I) communication, and over-the-air software updates. Connected cars offer enhanced safety, improved driving experiences, and access to a range of services and applications.

- **Autonomous Driving:** IoT is essential to the growth and development of autonomous cars. Autonomous driving. To allow autonomous driving, networked sensors, cameras, and data processing devices collect and analyse enormous volumes of real-time data. IoT connectivity makes it easier for information to be shared across infrastructure, cloud platforms, and cars, enabling autonomous vehicles to properly perceive and react to their surroundings. The transportation industry might undergo a revolution thanks to this technology, which would make it safer, more effective, and less harmful to the environment.

- **Predictive Maintenance:** Automotive predictive maintenance is made possible by IoT sensors and connectivity. Potential faults and maintenance requirements can be anticipated by continually monitoring important vehicle components and analyzing data in real-time. This preventative maintenance strategy improves vehicle performance, lowers unscheduled downtime, and lowers repair costs. In order to increase efficiency and lengthen the lifespan of cars, manufacturers and service providers can plan maintenance depending on the actual state of the vehicle.

- **Supply Chain Enhancement:** IoT improves supply chain management in the automotive sector, according to supply chain optimisation. Real-time tracking of goods, components, and vehicles across the supply chain is made possible by connected systems and sensors. Manufacturers can optimize production schedules, lower inventory levels, and boost logistical effectiveness thanks to this visibility. Supply chain management fueled by IoT guarantees on-time delivery of parts, shortens lead times, and boosts overall operational effectiveness.

- **Improvements to the Customer Experience:** Personalized and linked experiences for customers are made possible by IoT for automakers. Advanced entertainment systems, seamless connectivity to smartphones and other devices, individualized settings, and access to a variety of digital services are just a few of the features available in connected automobiles. Additionally, remote car monitoring, diagnostics, and vehicle health reports are made possible by IoT connectivity, increasing customer convenience and satisfaction.

- **Security and Safety:** IoT technologies improve the security and safety of vehicles. In order to help drivers make wise judgements, connected cars may offer real-time information on traffic patterns, road conditions, and possible risks. IoT-enabled security systems offer tracking capabilities in case of theft and aid in preventing car theft. Additionally, with the development of cybersecurity, measures are taken to safeguard connected automobiles from online dangers and unauthorized access.

In general, the Internet of Things is revolutionizing the automobile sector by allowing connected, intelligent, and effective cars. It delivers a customized and smooth driving experience, boosts operational effectiveness, optimizes supply chain management, and increases safety. IoT will fuel more innovation and influence the direction of mobility as it develops.

3.1.2 Fast Moving Consumer Goods (FMCG)

The FMCG (Fast-Moving Consumer Goods) businesses' production procedures are significantly being impacted by IoT (Internet of Things). FMCG producers may increase production, improve quality assurance, streamline supply chain management, and provide customers with more individualized goods by implementing IoT technology (Kumar Sagar et al., 2018) . Here are some significant changes caused by IoT in FMCG manufacturing:

- **Monitor smart production:** IoT sensors and linked devices are installed across the production line to monitor various elements of the manufacturing process in real-time. This is known as smart production monitoring. This involves measuring things like machine performance, pressure, humidity, and temperature. Manufacturers can spot production bottlenecks, increase productivity, and proactively deal with any quality or performance concerns by gathering and analyzing this data.
- **Control of quality and traceability:** IoT's capacity to provide real-time data on product performance and quality allows for improved quality control procedures. Critical parameters may be monitored by sensors built into production equipment, assuring constant product quality and lowering faults. IoT also makes it possible for producers to track ingredients, completed goods, and raw materials along the supply chain. This aids in risk mitigation, identifying and resolving quality concerns, and regulatory compliance.
- **Supply Chain Improvement:** IoT is essential to the FMCG supply chain's optimisation as it relates to supply chain management. Manufacturers may see inventory levels, production capacity, and demand trends in real-time by implementing IoT sensors and linked systems. This makes it possible to plan production more efficiently, estimate demand more precisely, and manage inventories more effectively. In order to fulfil customer demand, IoT-powered supply chain optimisation minimises stockouts, cuts waste, and assures on-time delivery.
- **Customization and Personalization:** IoT empowers FMCG producers to give consumers customized and personalized goods and experiences. Manufacturers may customize their products to specific consumers by gathering and analyzing data on consumer preferences, purchase trends, and product usage. This covers distinct product variants depending on consumer preferences, personalized promotions, and customized packaging.
- **Remote Monitoring and Maintenance:** The Internet of Things makes it possible to remotely monitor and maintain manufacturing equipment. Real-time data on equipment performance may be provided through connected sensors, enabling predictive maintenance and minimizing downtime. Manufacturers may more effectively schedule maintenance tasks, remotely check the status of their machines, and get alerts when repair is needed. This helps to assure continuous output, optimize maintenance schedules, and prevent equipment failure.
- **Sustainability and energy efficiency:** IoT technology help the FMCG sector adopt sustainable production procedures. IoT-enabled devices may be used by factories to track energy use, resource use, and waste production. This allows them to spot areas for improvement and adopt energy-saving measures. Systems fueled by the internet of things (IoT) can optimize energy use in real-time, cutting costs and environmental effect.

The Internet of Things (IoT) is, overall, changing how FMCG firms manufacture their products by allowing smart production monitoring, quality control, supply chain optimisation, personalisation, remote monitoring, and sustainability efforts. FMCG producers can increase productivity, boost product quality, and provide customers more individualized experiences by utilizing IoT technology.

3.2 IIoT in Service Industry

The Internet of Things (IoT), which enables smarter, more connected operations and improves consumer experiences, is revolutionizing the service sector. IoT enables service providers to gather and analyze real-time data, allowing proactive decision-making, automation, and customisation. It does this by integrating devices, sensors, and systems. IoT is used in the service sector for a variety of tasks including supply chain optimisation, customer experience improvement, facility management, asset tracking, remote monitoring, predictive maintenance, and safety and security measures. With IoT, service companies may streamline processes, boost productivity, give clients individualized experiences, and offer seamless services.

3.2.1 Financial Services

The Internet of Things (IoT) is reshaping the financial services sector and how financial institutions operate, interact with clients, and provide services. Here are a few significant IoT uses in the financial services sector:

- **Smart Banking:** IoT makes smart banking solutions possible, allowing for the provision of real-time data on client behavior, preferences, and transactional patterns by way of linked devices and sensors. Financial institutions may provide targeted goods and services, issue personalized notifications, and personalize banking experiences with the use of this data. Smart banking enabled by IoT also includes location-based services for increased convenience, wearable devices for transactions, and contactless payments. (D. Liao et al. 2017)
- **Asset tracking and management:** IoT makes it easier for financial organizations to track and manage their assets effectively. Banks can track and manage precious assets like cash, papers, and equipment in real-time by implementing IoT sensors and tags. Processes for inventory management are streamlined, losses are decreased, and security is improved. Optimizing cash logistics and ATM replenishment processes is another benefit of real-time asset tracking.
- **Fraud Detection and Security:** IoT improves security and fraud prevention procedures in the financial services sector. Financial institutions may identify abnormalities and possible fraud by analyzing IoT-generated data from a variety of sources, including transaction patterns, user behavior, and location-based data. IoT-enabled security systems may also keep an eye on physical locations, guard entry points, and send out instant notifications when something seems fishy.
- **Risk management and insurance:** The insurance industry can now analyze and manage risks with greater accuracy thanks to IoT. IoT devices, such wearables or telematics sensors in cars, gather information on user behavior, driving habits, health measurements, and other topics. Insurance providers can utilize this information to provide usage-based insurance, individualized rates, and proactive risk mitigation services. Real-time monitoring of environmental elements,

such as weather conditions or property dangers, that may have an influence on insurance policies is also a part of IoT-powered risk management.

- **Smart ATMs and Branch Automation:** IoT makes it easier for the financial services sector to install smart ATMs and branch automation. Connected ATMs allow extra services beyond cash withdrawal, personalize user experiences, and deliver targeted promotions. IoT sensors can keep an eye on the state of ATMs, allowing for preventative maintenance and reducing downtime. IoT-enabled branch automation improves operational effectiveness and customer happiness by streamlining procedures such as queueing, self-service kiosks, and digital signage.
- **Data analytics and insights:** Financial institutions may learn a lot about consumer behavior, market trends, and risk patterns by combining IoT-generated data with cutting-edge analytics. Targeted marketing techniques, product innovation, and data-driven decision-making are all aided by these insights. IoT data analytics also helps with attempts to combat money laundering, fraud, and regulatory compliance.

Personalized banking experiences, effective asset monitoring, fraud detection, risk management, smart ATMs, and data-driven insights are just a few of the advantages that IoT applications in the financial services sector may provide. Financial institutions may increase customer happiness, boost operational effectiveness, reduce risks, and maintain competitiveness in the continuously changing digital market by using IoT technology.

3.2.2 IoT in Healthcare

The Internet of Things (IoT) is redefining the healthcare sector by boosting operational effectiveness, changing the way healthcare services are provided, and improving patient outcomes (Hochstein et al., 2020). The following are some important IoT uses in the healthcare industry:

- **Monitoring a patient remotely:** Remote patient monitoring is made possible by IoT, allowing medical professionals to keep an eye on patients' vital signs, medication compliance, and general health status from a distance. Wearable sensors and implanted devices that are connected send real-time data to medical specialists allowing for the early identification of health conditions and prompt action. Remote patient monitoring raises the standard of care for managing chronic diseases, minimizes hospital readmissions, and improves patient outcomes.
- **Smart medical equipment:** IoT drives a variety of linked prostheses, continuous glucose monitoring, and smart infusion pumps, among other smart medical equipment. Real-time monitoring, accurate dosage, and individualized treatment regimens are made possible by these devices' data collection and transmission. The accuracy, safety, and effectiveness of healthcare delivery are all improved by IoT-enabled medical devices.
- **Asset and Inventory Management:** IoT makes it easier for healthcare institutions to manage assets and inventories effectively. Medical equipment, supplies, and medications are tracked via connected sensors and tags in terms of their location, status, and use. As a result, inventory management procedures are made more efficient and timely availability is ensured. Healthcare environments benefit from improved operational performance and cost-effectiveness thanks to IoT-powered asset and inventory management.

- **Virtual care and telemedicine:** IoT makes telemedicine and virtual care solutions possible, allowing patients to communicate with medical professionals from a distance. IoT enables virtual diagnosis, remote consultations, and at-home care through video consultations, remote monitoring tools, and secure data transfer. This lowers healthcare costs, increases patient comfort, and improves access to healthcare services, particularly in rural or disadvantaged regions.
- **Real-time Locating Systems (RTLS):** RTLS systems driven by IoT keep track of the whereabouts of assets, personnel, and patients inside healthcare institutions. This enhances the management of patient flow, optimizes staffing, and guarantees resource efficiency. Healthcare professionals may offer timely care, improve patient safety, and increase operational effectiveness with the use of RTLS.
- **Health and Wellness Monitoring using Telemedicine and Virtual Cars:** Health and wellness monitoring is made possible via telemedicine and virtual cars, as well as IoT devices and wearables. These gadgets monitor parameters including heart rate, activity level, sleep schedule, and stress level. Individuals may actively control their health, take part in preventative care, and make wise lifestyle decisions thanks to IoT-powered health and wellness monitoring gadgets.
- **Safety and Smart Facilities:** IoT improves the security and effectiveness of medical facilities. In order to maintain the best conditions for patient care and medicine storage, connected devices continuously monitor environmental factors including temperature, humidity, and air quality. The physical security of healthcare institutions is improved by IoT-powered security solutions, such as surveillance cameras and access control, protecting patients, personnel, and important data.

With tools like telemedicine, smart medical equipment, remote patient monitoring, and real-time locating systems, IoT is revolutionizing the healthcare industry. These innovations promote efficiency, provide individualized treatment, and improve patient outcomes. Healthcare providers may provide patients easily accessible and secure services thanks to IoT.

3.2.3 IoT in Hospitality

The hospitality service business is being significantly impacted by IoT (Rangaiah, 2021), which is changing how hotels and resorts run their operations and improving the visitor experience. Here are a few prevalent IoT uses in the hotel industry:

- **Smart Room Management:** The Internet of Things (IoT) allows smart room management systems, which automate and regulate a variety of room tasks. Voice commands or mobile applications can be used by guests to manage the temperature, lighting, curtains, and entertainment systems in their rooms. IoT-powered room management improves comfort and personalisation for a seamless and relaxing stay for visitors.
- **Sustainability and energy management:** IoT makes energy management easier in hospitality settings by tracking and maximizing energy use. Connected sensors gather information about temperature, lighting, and occupancy to enable sophisticated energy management systems. By automatically adjusting energy use based on visitor presence, these systems maximize efficiency and lower energy costs. By providing real-time monitoring and management of water use and waste management, IoT also helps with sustainability initiatives.

- **Improved visitor Services:** IoT improves visitor services by offering individualized encounters and on-demand support. Personalized messages, offers, and suggestions based on visitor preferences and position on the property are made possible by beacons and location-based sensors. IoT-enabled chatbots and virtual assistants provide round-the-clock concierge services, answering questions from visitors, disseminating knowledge, and expediting requests.

- **Smooth check-in and check-out processes:** The check-in and check-out procedures at hotels are streamlined by IoT. In order to check in quickly, visitors can utilize self-service kiosks and smartphone apps rather than going to the front desk. Keyless entry systems that are IoT-enabled let visitors enter their rooms using smartphones or smart cards instead of traditional keys, increasing security.

- **Inventories and Asset Management:** IoT makes it easier for the hotel sector to manage inventories and assets effectively. Connected sensors monitor the supply and amenity inventories, providing prompt replenishment and reducing stockouts. IoT also facilitates asset tracking and management, optimizing maintenance schedules and lowering operational expenses, for items like linens, furniture, and equipment.

- **Security and Safety:** IoT improves security and safety procedures in lodging facilities. Connected security cameras, alarms, and access control systems keep an eye on and secure the property to ensure the safety of visitors and employees. IoT-powered sensors may identify dangers like fire or leakage and send out right away notifications for quick action. IoT also makes it possible for effective evacuation procedures and emergency response systems.

- **Personalization and data analytics:** IoT-generated data and cutting-edge analytics offer insightful information on visitor behavior, preferences, and patterns. With the use of this information, hotels may better target their advertising and operating plans. Personalized suggestions, loyalty programmes, and targeted promotions are made possible by IoT-powered data analytics, increasing customer pleasure and loyalty.

In conclusion, IoT applications in the hospitality sector have a variety of advantages, such as smart room management, energy efficiency, improved guest services, streamlined check-in and check-out procedures, inventory and asset management, safety and security measures, and data-driven personalization. By utilizing IoT technology, hospitality businesses may enhance operational effectiveness, deliver distinctive guest experiences, and compete in a sector that is always changing.

3.2.4 IoT in Construction Sector

The Internet of Things (IoT) is transforming the construction services sector and bringing about major gains in productivity, efficiency, and safety (Newton, 2023). Here are a few important IoT uses in the building industry:

- **Connected Jobsite Monitoring:** Using sensors and cameras, IoT makes it possible to monitor construction sites in real-time. These interconnected gadgets offer useful information on worker security, equipment use, environmental factors, and progress monitoring. Multiple construction sites may be remotely monitored and managed by project managers, increasing safety procedures, maximizing resource allocation, and ensuring regulatory compliance.

- **Tracking of assets and equipment:** IoT enables effective tracking and management of assets and equipment in the construction industry (Rybakov, 2021). Real-time visibility of the location, use, and maintenance requirements of equipment is made possible through connected sensors and GPS trackers. This aids construction organizations in reducing downtime, proactive scheduling of maintenance, and theft prevention.

- **Remote equipment maintenance and diagnostics:** Construction organizations can remotely monitor and analyze equipment performance thanks to IoT. Connected sensors on equipment and tools gather information on operational circumstances, performance indicators, and maintenance needs. As a result, it is possible to do preventative maintenance, which minimizes project delays and boosts overall effectiveness by decreasing equipment malfunctions and optimizing maintenance schedules.

- **Smart Safety Systems:** The implementation of smart safety systems on construction sites improves safety. Smart helmets and vests and other connected wearables can monitor a worker's vital signs, spot falls or accidents, and send out prompt notifications. Construction employees will work in a safer environment thanks to IoT-powered safety systems that also feature real-time monitoring of dangerous situations, gas detection, and emergency response systems (Martinez et al., 2020).

- **Security at Construction Sites:** IoT enhances security measures at construction sites. Connected security cameras and access control systems keep an eye on and safeguard the space, preventing theft and unauthorized entry. In order to increase site security and lower risks, IoT-powered security systems may connect with other IoT devices for automatic warnings and real-time video surveillance.

- **Construction Supply Chain Optimisation:** The Internet of Things (IoT) improves the construction supply chain by enabling real-time inventory and material visibility and tracking. RFID tags and connected sensors track the flow of commodities, enabling effective inventory management, cutting down on delays, and avoiding shortages. This guarantees prompt material delivery and speeds the building process.

- **Automation of buildings and energy management:** Energy management and building automation are made possible by IoT in the construction sector. Based on occupancy and environmental factors, connected sensors and devices manage lighting, HVAC systems, and energy consumption. By doing this, the built environment becomes more sustainable and comfortable while also improving energy efficiency and lowering utility bills.

The Internet of Things (IoT) applications in the construction services sector provide a wide range of advantages, including real-time jobsite monitoring, equipment tracking, remote diagnostics and maintenance, smart safety systems, site security, supply chain optimisation, and building automation. Construction organizations may increase safety, increase project efficiency, save costs, and deliver projects on time and within budget by using IoT technology.

4. RISK AND CHALLENGES

While the IIoT has the potential to greatly improve efficiency and productivity, it also poses significant security and privacy challenges. The importance of ensuring the safe and secure deployment of IIoT technology cannot be overstated, as security breaches can have significant financial and reputational consequences. To address such security challenges, organizations must implement robust security measures to ensure that without much of a compromising of security and the industrialists should consider risks before moving on to the full fledged implementation of Internet of Things in Industries.

4.1 Risk Considerations

There are several generic risks that are to be considered while organizations are deploying IIoT technology (Pal. S et. al 2021):

- **Threats to confidentiality:** The unauthorized access to sensitive industrial or personal data can compromise the confidentiality of sensitive information.
- **Threats to integrity:** The unauthorized modification of data can compromise the accuracy and reliability of the information.
- **Threats to availability:** Cyberattacks can disrupt the availability of IIoT systems and devices, leading to operational disruptions and potential financial losses.
- **Threats to accountability:** The lack of accountability for security incidents can make it difficult to determine the cause of security breaches and hold those responsible accountable.

Some specific risk include:

- **Cyber Attacks:** One of the biggest risks is the vulnerability of IIoT devices to cyber attacks, as these devices often lack basic security measures such as encryption, authentication, and authorization. Hackers can exploit these vulnerabilities to gain unauthorized access to sensitive data and control systems, leading to data breaches, downtime, and financial losses.
- **Lack of Standardization and Compatibility:** Another risk is the lack of standardization and compatibility between IIoT devices and systems, which can lead to interoperability issues and make it difficult to integrate new devices and systems into existing networks. This can also increase the risk of security breaches, as different devices and systems may have different levels of security.
- **Lack of Knowledge:** Additionally, the lack of understanding about the security risks of IIoT can lead to complacency among organizations, who may overlook the need to implement proper security measures and continuously monitor their systems for potential security threats.

Furthermore, this just does not cover all the risk that has to be considered but the industrialists must keep their open mind while including the technology of Internet of Things or any technology for that matter as a famous saying goes with risk comes the reward. By taking a proactive approach to security, organizations can ensure the safe and secure deployment of IIoT technology. Now to secure an IIoT system there are a number of security issues it should consider. The next section presents the security recommendations for that can be kept in mind while deploying an IIoT system.

4.2 Security Recommendations

It is a major need for organizations to consider security when deploying IIoT systems, as these systems are often critical to the operation of industrial processes and have a direct impact on human safety, environmental protection, and economic sustainability. There is an increasing number of sophistication of cyber criminals and hence there is an importance in implementing strong cybersecurity measures to prevent data breaches and unauthorized access to industrial control systems (Raimundo & Rosário, 2022).

The following are the recommendation to address the security challenges of the IIoT:

- **Implement a comprehensive cybersecurity strategy:** Organizations should develop a comprehensive cybersecurity strategy that includes measures such as secure authentication, network segmentation, and continuous monitoring and threat intelligence.
- **Adopt industry-standard security protocols and technologies:** Organizations should adopt industry-standard security protocols and technologies, such as secure communications protocols, encryption, and firewalls, to ensure the protection of their systems and data.
- **Collaborate with security experts:** Organizations should collaborate with security experts to stay up-to-date on the latest security threats and best practices.
- **Implement secure authentication:** Organizations should implement secure authentication mechanisms, such as multi-factor authentication, to prevent unauthorized access to their systems and data.
- **Network segmentation:** Organizations should segment their networks to isolate critical systems and reduce the risk of system-wide disruption.
- **Continuous monitoring and threat intelligence:** Organizations should implement continuous monitoring and threat intelligence systems to detect and respond to emerging threats in real-time.

By implementing these security measures and staying vigilant against emerging threats, organizations can ensure the safe and secure deployment of IIoT technology and protect their systems and data against cyber attacks. Organizations can reduce their exposure to cybersecurity threats by implementing secure authentication and authorization, network segmentation, and continuous monitoring. These recommendations ensure that organizations are well-prepared to address the security challenges of IIoT and take full advantage of its benefits. Ultimately, a strong focus on cybersecurity is essential to realizing the full potential of IIoT in industrial management.

5. FUTURE OF INDUSTRIAL INTERNET OF THINGS

The future of the IIoT is poised to bring about numerous benefits, including improved efficiency, increased productivity, and reduced costs. With a focus on several key areas, the IIoT is set to play a major role in shaping the future of industry and commerce. Integration of Artificial Intelligence and Machine Learning: AI and machine learning technologies will be used to analyze vast amounts of data generated by connected devices and machines. This will allow organizations to make better and more informed decisions, automate various processes, and improve overall efficiency. For example, AI algorithms can analyze sensor data to predict equipment failures, enabling organizations to schedule maintenance before the failure occurs (Al-khafajiy et al., 2019).

- **Adoption of Edge Computing:** Edge computing involves processing data at the source, rather than in the cloud or data center. This will allow organizations to respond to real-time events and reduce the latency and bandwidth requirements of data transmission. The adoption of edge computing will also help organizations reduce costs associated with data storage and transmission.

- **Deployment of 5G Networks:** The widespread adoption of 5G networks is expected to greatly enhance the capabilities of the IIoT. With faster and more reliable connectivity, organizations will be able to connect more devices and process larger amounts of data, further increasing the efficiency and productivity of industrial processes.
- **Implementation of Robust Cybersecurity Measures:** As the IIoT continues to grow, cybersecurity will become an increasingly important issue. Organizations will need to implement robust security measures to protect their devices, networks, and data from cyber-attacks and other security threats. This includes everything from encryption and firewalls to secure software updates and secure device management.
- **Expansion of the Internet of Medical Things (IoMT):** The Internet of Medical Things (IoMT) is a subcategory of the IIoT that deals specifically with connected medical devices. The IoMT has the potential to revolutionize the healthcare industry by improving patient outcomes, reducing costs, and streamlining processes. For example, wearable medical devices can collect real-time biometric data, allowing physicians to better monitor patients and detect potential issues early (Khan et al., 2016).

The future of the Industrial Internet of Things is poised to bring about significant benefits and efficiencies in various industries. As the IIoT continues to evolve, organizations will need to adapt and embrace these technologies to remain competitive in the rapidly changing technological landscape. The integration of AI and machine learning, the adoption of edge computing, the deployment of 5G networks, and the implementation of robust cybersecurity measures will play a crucial role in shaping the future of the IIoT.

6. STRENGTHS, WEAKNESSES, OPPORTUNITIES, AND CHALLENGES OF IIOT APPLICATIONS

Industrial IoT has a variety of advantages, disadvantages, possibilities, and issues. Its advantages include economic savings, environmental friendliness, and stimulating innovation. IIoT improves operations, lowers costs, promotes sustainability, and encourages cooperation and data-driven decision-making. However, there are obstacles to overcome, such as security issues, data management difficulties, and the integration of various IoT devices. Healthcare, wearables, and investment opportunities present opportunities, while obstacles include mobility concerns, data management, and the heterogeneity of IoT devices (*An IIoT SWOT analysis*, 2020). Addressing these obstacles and capitalizing on IIoT's strengths and prospects are critical to the technology's effective deployment and revolutionary influence across sectors. A SWOC analysis is shown in Figure 6.

Figure 6. SWOC analysis of industrial IoT

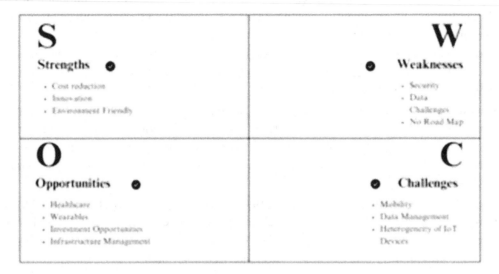

6.1 Strengths of Industrial IoT

Cost savings, environmental friendliness, and innovation are all advantages of Industrial IoT (IIoT). Through real-time monitoring and data analytics, IIoT enables organizations to reduce maintenance costs and optimize resource allocation by employing predictive maintenance tactics. It contributes to environmental sustainability by promoting energy efficiency and better resource management. Furthermore, IIoT promotes innovation by opening up new options and integrating technology such as AI and sophisticated analytics, which drive product creation and process optimisation. Overall, the capacity of IIoT to create cost savings, promote sustainability, and inspire innovation in the industrial sector is one of its key assets.

6.2 Weakness of Industrial IoT

Security flaws, data problems, and the absence of a standardized roadmap are among the shortcomings of Industrial IoT (IIoT). IIoT systems are vulnerable to cyber-attacks and unauthorized access, necessitating stringent security safeguards. Dealing with enormous volumes of data while maintaining data quality and privacy may be difficult. The absence of a standardized roadmap complicates technology selection and system integration. Overcoming these flaws necessitates the implementation of robust security measures, the resolution of data management issues, and the promotion of industry collaboration for standardization and interoperability.

6.3 Opportunities of Industrial IoT

Industrial IoT (IIoT) prospects include breakthroughs in healthcare, wearables, financial opportunities, and infrastructure management. In healthcare, IIoT provides remote patient monitoring, personalized therapies, and real-time health tracking. Wearable technology provides data-driven insights and personalized experiences. IIoT also opens up prospects for investment in sensor technologies, networking platforms, and analytics tools. IIoT allows real-time monitoring, predictive maintenance, and improved decision-making in infrastructure management. These possibilities fuel breakthroughs and improvements in a variety of sectors.

6.4 Challenges of Industrial IoT

Mobility difficulties, data management complications, and the variety of IoT devices are among the challenges in Industrial IoT (IIoT). IIoT systems need seamless communication across several sites, yet collecting and analyzing enormous amounts of data presents difficulties. The wide variety of IoT devices and protocols creates integration and compatibility challenges. Furthermore, security and privacy issues necessitate strong safeguards to preserve data and ensure secure communication. Overcoming these obstacles is critical for realizing the full potential of IIoT and reaping its advantages.

CONCLUSION

As a famous quote from Christian Kubis- **"Industry 4.0 is not really a revolution. It's more of an evolution."**(*changing people government-Local News-3E International Fair showcases jobs and more*, 2021). The quote makes it very clear that the current situation in technology is huge and considerable changes have already been made, but now is the time for evolution in the industry by the use of technological trends that are in exercise, such as the Internet of Things.As the technologies evolve it is likely to see more innovative and impactful applications in the future. In conclusion, the Industrial Internet of Things (IIoT) has the potential to transform the industrial landscape by enabling real-time monitoring, predictive maintenance, and automation. However, the adoption of IIoT also poses significant challenges related to data management, security, scalability, and interoperability. With the continued development and integration of emerging technologies such as edge computing, fog computing, 5G networks, blockchain, and artificial intelligence, the evolution of IIoT is set to accelerate, bringing new opportunities and challenges for industries. As the technology continues to evolve, it will be essential to focus on collaboration, innovation, and standardization to overcome the challenges and unlock the full potential of IIoT for the benefit of society and the economy.

REFERENCES

IoT ONE Digital Transformation Advisors. (2020.). *Industrial IoT Use Case Profile*. IoT ONE. https://www.iotone.com/usecase/agricultural-drones/u54

Al-khafajiy, M., Baker, T., Chalmers, C., Asim, M., Kolivand, H., Fahim, M., & Waraich, A. (2019). Remote health monitoring of elderly through wearable sensors. *Multimedia Tools and Applications*, 78(17), 24681–24706. doi:10.100711042-018-7134-7

Ayaz, M., Ammad-uddin, M., Sharif, Z., Mansour, A., & Aggoune, E.-H. M. (2019). Internet-of-Things (IoT) based Smart Agriculture: Towards Making the Fields Talk. *IEEE Access : Practical Innovations, Open Solutions*, 7, 1–1. doi:10.1109/ACCESS.2019.2932609

Borg, G., Löllgen, H., & Döllinger, M. (2019). Wearables in Sports: Opportunities and Challenges. *Sports Medicine and Health Science*, *1*(2), 62–67. doi:10.1016/j.smhs.2019.02.002

Desjardins, J. (2018, January 9). *Timeline: The History of the Industrial Internet of Things*. Visual Capitalist. https://www.visualcapitalist.com/timeline-industrial-internet-things/

Feynman, R. P. (1982). Simulating physics with computers. *International Journal of Theoretical Physics*, *21*(6-7), 467–488. doi:10.1007/BF02650179

GraboyesR. F.BryanD.CoglianeseJ. (2020, January 1). Overcoming Technological and Policy Challenges to Medical Uses of Unmanned Aerial Vehicles. Papers.ssrn.com. https://papers.ssrn.com/sol3/papers.cfm?abstract_id=3561743 doi:10.2139/ssrn.3561743

Haghi, M., Thurow, K., & Stoll, R. (2017). Wearable Devices in Medical Internet of Things: Scientific Research and Commercially Available Devices. *Healthcare Informatics Research*, *23*(1), 4. doi:10.4258/hir.2017.23.1.4 PMID:28261526

Hochstein, B., Rangarajan, D., Mehta, N., & Kocher, D. (2020). An Industry/Academic Perspective on Customer Success Management. *Journal of Service Research*, *23*(1), 3–7. doi:10.1177/1094670519896422

IoT in agriculture: Smart Farming technologies for Future. (2022). Cropin. Www.cropin.com. https://www.cropin.com/iot-in-agriculture

Jiang, Y. (2020). Combination of wearable sensors and internet of things and its application in sports rehabilitation. *Computer Communications*, *150*, 167–176. doi:10.1016/j.comcom.2019.11.021

Kalsoom, T., Ahmed, S., Rafi-ul-Shan, P. M., Azmat, M., Akhtar, P., Pervez, Z., Imran, M. A., & Ur-Rehman, M. (2021). Impact of IoT on Manufacturing Industry 4.0: A New Triangular Systematic Review. *Sustainability (Basel)*, *13*(22), 12506. doi:10.3390u132212506

Khan, Y., Ostfeld, A. E., Lochner, C. M., Pierre, A., & Arias, A. C. (2016). Monitoring of Vital Signs with Flexible and Wearable Medical Devices. *Advanced Materials*, *28*(22), 4373–4395. doi:10.1002/adma.201504366 PMID:26867696

Kumar Sagar, P., Garg, K., & Dutta, C. (2018, August 1). *Application of Internet of Things in Fast Moving Consumer Goods Sector to increase Business Efficiency*. IEEE Xplore. doi:10.1109/ICG-CIoT.2018.8753033

Liao, D., Sun, G., Li, H., Yu, H., & Chang, V. (2017). The framework and algorithm for preserving user trajectory while using location-based services in IoT-cloud systems. *Cluster Computing*, *20*(3), 2283–2297. doi:10.100710586-017-0986-1

Linnamo, V. (2023). Sensor Technology for Sports Monitoring. *Sensors (Basel)*, *23*(2), 572. doi:10.339023020572 PMID:36679367

Lucamarini, M. (2018). Record Distance for Quantum Cryptography. *Physics (College Park, Md.)*, *11*. https://physics.aps.org/articles/v11/111

Majumder, S., Mondal, T., & Deen, M. (2017). Wearable Sensors for Remote Health Monitoring. *Sensors (Basel)*, *17*(12), 130. doi:10.339017010130 PMID:28085085

Martinez, J. G., Gheisari, M., & Alarcón, L. F. (2020). UAV Integration in Current Construction Safety Planning and Monitoring Processes: Case Study of a High-Rise Building Construction Project in Chile. *Journal of Management Engineering*, *36*(3), 05020005. doi:10.1061/(ASCE)ME.1943-5479.0000761

Matthew, U. O., Kazaure, J. S., Amaonwu, O., Adamu, U. A., Hassan, I. M., & Kazaure, A. A. (2021). Role of Internet of Health Things (IoHTs) and Innovative Internet of 5G Medical Robotic Things (IIo-5GMRTs) in COVID-19 Global Health Risk Management and Logistics Planning. In M. Niranjanamurthy, S. Bhattacharyya, & N. Kumar (Eds.), *Intelligent Data Analysis for COVID-19 Pandemic. Algorithms for Intelligent Systems*. Springer. doi:10.1007/978-981-16-1574-0_2

Mendoza, M. (2020, July 19). *Industrial IoT - The Top 5 Benefits of Industry 4.0*. Hitachi Solutions. https://global.hitachi-solutions.com/blog/industrial-iot-benefits/

Newton, E. (2023, March 29). *5 Unique Applications of IoT in Construction*. IoT For All https://www.iotforall.com/5-unique-applications-of-iot-in-construction#:~:text=Construction%20sites%20can%20also%20use,see%20if%20it%20needs%20repair

Pal, S., & Jadidi, Z. (2021). Analysis of Security Issues and Countermeasures for the Industrial Internet of Things. *Applied Sciences (Basel, Switzerland)*, *11*(20), 9393. doi:10.3390/app11209393

Peranzo, P. (2022, March 18). *IoT in Manufacturing: 8 Use Cases & Upcoming Trends*. Imaginovation. https://imaginovation.net/blog/iot-in-manufacturing/

Raimundo, R. J., & Rosário, A. T. (2022). Cybersecurity in the Internet of Things in Industrial Management. *Applied Sciences (Basel, Switzerland)*, *12*(3), 1598. doi:10.3390/app12031598

Rangaiah, M. (2021, May 14). *9 IoT Applications in the Hospitality Industry*. Analytics Steps. Www.analyticssteps.com. https://www.analyticssteps.com/blogs/9-iot-applications-hospitality-industry

Rybakov, A. (2021, October 22). *Applying IoT in the construction industry: top-7 use cases*. Agile Vision. Www.agilevision.io. https://www.agilevision.io/blog/applying-iot-in-the-construction-industry-top-7-use-cases

Scalea, J. R., Pucciarella, T., Talaie, T., Restaino, S., Drachenberg, C. B., Alexander, C., Qaoud, T. A., Barth, R. N., Wereley, N. M., & Scassero, M. (2021). Successful Implementation of Unmanned Aircraft Use for Delivery of a Human Organ for Transplantation. *Annals of Surgery*, *274*(3), e282–e288. doi:10.1097/SLA.0000000000003630 PMID:31663974

Shrivastava, A., Bhardwaj, A., & Hasteer, N. (2020). IoT in Automobile Sector: State of the Art. *2020 10th International Conference on Cloud Computing, Data Science & Engineering (Confluence)*. IEEE. 10.1109/Confluence47617.2020.9058202

Stöcker, C., Bennett, R., Nex, F., Gerke, M., & Zevenbergen, J. (2017). Review of the Current State of UAV Regulations. *Remote Sensing (Basel)*, *9*(5), 459. doi:10.3390/rs9050459

Ul Hasan, M. N., & Negulescu, I. I. (2020). Wearable technology for baby monitoring: A review. *Journal of Textile Engineering & Fashion Technology*, *6*(4). doi:10.15406/jteft.2020.06.00239

Vijayakumar, P., Chang, V., Deborah, L. J., Balusamy, B., Shyn, P. G., Vijayakumar, P., Chang, V., Deborah, L. J., Balusamy, B., & Shyn, P. G. (2016). *Computationally Efficient Privacy Preserving Anonymous Mutual and Batch Authentication Schemes for Vehicular Ad Hoc Networks.* Future Generation Computer Systems. https://eprints.soton.ac.uk/403454/

Wilkinson, F. (2022, June 2). *Industrial Revolution and Technology.* National Geographic. https://education.nationalgeographic.org/resource/industrial-revolution-and-technology/

Won, H., Lee, H., Song, G., Kim, Y., & Kwak, N. (2022). Reliable Data Collection Methodology for Face Recognition in Preschool Children. *Sensors (Basel), 22*(15), 5842. doi:10.339022155842 PMID:35957397

Chapter 22
Integration of WSN and IoT:
Wireless Networks Architecture and Protocols – A Way to Smart Agriculture

Sandeep Bhatia
Galgotias University, India

Zainul Abdin Jaffery
Jamia Millia Islamia, India

Shabana Mehfuz
Jamia Millia Islamia, India

Neha Goel
Raj Kumar Goel Institute of Technology, India

ABSTRACT

IoT refers to a network of interconnected devices which are able to share data with each other. Each device in a network has a unique address. The placement of devices called nodes in any network has always been a matter of concern associated with its routing, shortest path, power dissipation by nodes and topology etc. The integration of a wireless sensor network with the internet of things is an area of research among the researchers and provides a valid approach for networking in any IoT enabled system. WSN- IoT integration finds applications in smart agriculture. In this chapter the authors will discuss the methods of WSN-IoT integration and its benefits among smart agriculture. The authors will discuss the application using Lora WAN to enhance better crop yield. LoRa WAN is a low cost and long-range device which is suitable for smart agriculture application for sending collected data to a remote location which we cannot communicate through Zigbee, wi-fi, Bluetooth and other short-range devices. The researchers will explain the routing methods and protocols associated with smart agriculture.

DOI: 10.4018/978-1-6684-8785-3.ch022

INTRODUCTION

Wireless Sensor Networks (WSNs) and the Internet of Things (IoT) are two related technologies that are increasingly being integrated to create new applications and services. WSNs are networks of small, battery-powered devices equipped with sensors that can collect data about their environment. The IoT, on the other hand, is a system of interconnected devices that can communicate with each other and with the internet, and is characterized by its ability to collect, analyze and act on data.

The term "Internet of Things" (IoT) refers to the interconnectedness of physical objects, such as vehicles, home appliances, and other goods, which enable these things to communicate and exchange data.

These objects are embedded with electronics, software, sensors, and connectivity. The Internet of Things (IoT) concept entails extending Internet connectivity to a variety of gadgets and common objects in addition to traditional devices like desktop and laptop computers, smartphones, and tablets. Offering advanced device, system, and service connectivity that goes beyond machine-to-machine communications and encompasses a range of protocols, domains, and applications is the ultimate goal of the Internet of Things.

The Internet of Things (IoT) has quickly expanded to play a significant role in how people live, interact, and conduct business. Web-enabled devices are transforming our universal rights into a larger switched-on space to live in all over the world.

Applications

Examples of applications that can be created by integrating WSNs and IoT technologies include:

1. **Smart agriculture:** WSNs can be used to collect data about soil moisture, temperature, and other parameters, which can be integrated with IoT platforms to enable precision agriculture.
2. **Smart cities:** By deploying WSNs throughout a city, it is possible to collect data about traffic, air quality, and other factors, which can be integrated with IoT platforms to enable smarter city planning.
3. **Industrial automation:** WSNs can be used to collect data about equipment performance, which can be integrated with IoT platforms to enable predictive maintenance and other advanced automation applications.
4. **Healthcare monitoring:** WSNs can be used to collect data about patient vital signs, which can be integrated with IoT platforms to enable remote monitoring and other healthcare applications.

Benefits

Integrating WSNs and IoT technologies can provide a number of benefits, such as:

1. **Enhanced sensing capabilities**: WSNs can provide real-time data about physical parameters such as temperature, humidity, and light levels, which can be integrated with IoT platforms to enable smarter decision-making.
2. **Improved energy efficiency:** WSNs are designed to operate on low-power, so they can be integrated with IoT devices to help reduce overall energy consumption.

3. **Increased scalability:** By combining the scalability of IoT devices with the flexibility of WSNs, it is possible to create large-scale systems that can support a wide range of applications.
4. **Better reliability:** The redundancy built into WSNs can improve the reliability of IoT systems, ensuring that critical data is always available.
5. **Enhanced security:** WSNs can be integrated with IoT devices to provide more secure communication channels, ensuring that data is protected from unauthorized access.

IoT-WSN in Smart Agriculture

We are discussing the application of IoT-WSN in smart agriculture in our study.

The following are the several areas of agriculture where the Internet of Things is bringing about changes:

1. Greenhouses utilize sensors to monitor and adjust the temperature conditions in order to maximize production and maintain quality.
2. To produce the compost and avoid fungus and other microbial contaminants, regulate the varied humidity and temperature conditions.
3. Using sensors to track the animals and locate where they are grazing, examine farm air quality, and find dangerous gases coming from animal waste.
4. Constant crop monitoring to minimize spoilage and crop waste.

By 2050, the World Bank predicts that there will be 9 billion people on the planet. 70 percent more food must be produced to meet the demand. Food sustainability is a major concern as a result. With established approaches, profits and productivity are declining. The environment has an impact on agriculture's traditional methods as well. Therefore, traditional agricultural practices face a lot of obstacles.

The present farm revolution will be significantly impacted by smart agriculture.

*   It will sustainably reduce input, labor, and time.
*   It will increase output and revenue.
*   It will ensure sustainability and lessen its negative environmental effects.

Smart Agriculture

"A technology-enabled approach to farming management that observes, measures, and analyses the needs of specific fields and crops."

In Smart Agriculture, an integration of ICT with conventional farming methods to improve the quantity and quality of farm output.ICT will include a variety of products- Drones, sensors, actuators, robotics, and GPS. The newest addition is 5G technology. Smart agricultural sensors, robotic drones, GPS-enabled tools, and satellite imaging are examples of IoT-enabled gadgets. On the soil, the crop, the available equipment, the hyper-local weather forecasts, and other variables, real-time data is gathered. Inputs from IoT and AI/ML-powered predictive analytics software are added to this.

The first section of this article explains the framework of our proposed system using IoT-WSN based approach by deploying specific set of sensors in agriculture land, this is called sensing unit which sense and send data to a remote server using long range networking protocol called LoRa WAN and after processing, the required information assists the farmers in terms of increased soil quality, better production, weather monitoring and livestock management.

The second section of this article provides an explanation of wireless sensor network architecture and LoRa WAN as a networking protocol, and some more protocols like Zigbee, Wi-Fi and Bluetooth and etc. explained under this section. Also, hardware requirements in IoT-WSN discussed in this section.

The third section demonstrates objectives of our study.

The fourth section includes the challenges and issues associated with WSN-IoT enable smart agriculture.

The fifth section explained the routing methods associated with smart agriculture.

The sixth section defines the references preceded by the conclusion and at the end there is a comparative analysis of different agriculture approaches.

RELATED WORK

A solution based on smart agriculture is needed to handle the key difficulties that are arising in agriculture, such as the declining water levels (in rivers and tanks) and environmental factors. A system for monitoring crops utilizing temperature and moisture was presented by Liqiang et al. (2011). In order to limit the amount of water used, sensors Ocampo et al. (2017) created an algorithm based on soil moisture and temperature readings. Their suggested solution uses photovoltaic panels for power and duplex communication lines for data. Their technology makes use of a cellular network interface for data transfer and irrigation scheduling computations. The development of several methods for precision agriculture applications has been aided by the progress in the field of wireless sensor networks. WSNs are used in a system by Chaudhary et al. (2011) to monitor and regulate greenhouse parameters. The examination of several agricultural fields has shown that agricultural outputs are deteriorating daily. Precision agriculture promises to boost production rates while requiring less additional labor as new technology are adopted. There are numerous studies that help farmers by offering an effective strategy built on modern technologies for raising overall output rates Varman et al. (2017). Using distributed wireless sensor networks, Keshtgari and Deljoo (2011) created a system. Their suggested solution is intended to provide controlled irrigation for enhancing productivity while offering real-time monitoring of the agricultural land. efficiently using water resources. The system has a detailed design that uses software-controlled irrigation and real-time field sensing. The system's architecture offers a low-cost, effective method of controlling irrigation via WSNs. WSN-based system for precise field variables including soil moisture, temperature, and humidity levels are crucial to agriculture. In order to collect the parameters, sensor nodes were placed below the soil. These sensor nodes communicate with one another using established protocols and send the measured data to the control station via the sink node. A communication protocol has been created by Zhao et al. (2017) for rapid transmission of sensor data. Their suggested architecture produces extremely few duty cycles while lengthening the monitoring system's lifespan. A system that makes use of sensors and UART (universal asynchronous receiver transmitter) interface has been created by Bachuwar et al. (2018). Hourly sampling, data buffering, transmission, and status validation all take place during the transmission process. Due to sensor deployment under the soil, attenuation in radio frequency signals is seen in the experimental results, and the deployment cost is rather high. Gayatri et al. (2015) offers an approach for using GPS (global positioning system) enabled sensors to monitor agricultural fields and control irrigation using sensors in the ground. Their method offers a practical remedy for greenhouse monitoring by periodically logging and storing data in the cloud.

Heble et al. (2018) have provided a practical farming option for areas with scarce water resources. An autonomous water flow control system was created using simple electronics and inexpensive sensors. Real-time monitoring considers sensing temperature and humidity and showing those values on LCD. According to the current moisture level, their system supplies water effectively to the required plants. A time-dependent, low-cost irrigation scheduling system has been proposed by Reche et al. (2014). Their system is constructed with a sensor network that can measure soil temperature and moisture. The GSM module processes the real-time sensed data and sends it to the consumers through SMS.

Nawandar and Satpute (2019) described an irrigation system made up of irrigation controls and soil moisture sensors. Their technique establishes a wireless connection between the irrigation controller and farmer computer. Where the water quality is poor or scarce, the intended system can be used. They have put in place a decision-support system to maximize irrigation management, fertilizer use, and water utilization for crop development. A smart agriculture system based on cloud computing and the Internet of Things was created by Channe et al. (2015). The use of the cloud platform allows for load balancing and the dynamic distribution of resources. In 2017, (Rajeswari et al.) proposed a system built on an IoT platform and control network. Additionally, they feature an information management system for handling and storing data in their design. Access to the saved data is available to support various agricultural research projects. The comparative analysis of a conducted literature review of various agricultural systems is shown in Table 1. The majority of smart farming and precision agricultural solutions do not offer an economical option for farming. The irrigation system proposed by Reche et al. (2014) is the most effective option for smart agriculture since it offers real-time data analysis. The created concept offers an economical approach to smart agriculture. Real-time field parameters are measured by the suggested system, and the data is stored on a cloud platform for analysis in real-time. The use of IoT in military research and search and rescue operations has been described in Sandeep et al. (2011). Robotic autonomous rescue operations can be used to find alive people and provide information about their whereabouts. Vehicle movement can be managed using the Python programming language.

According to Jeet et al. (2014), the Internet of Things can be used to deploy voice and head motion controls for residential appliances. Python programming can be used to operate home appliances that are challenging for disabled people to operate manually.

Sandeep et al. (2022) discusses how the Internet of Things can be utilized in the healthcare industry to limit public access to healthcare data and increase cyber security.

IoT implementation for automatic seat identification in public transport using image processing discussed by Sandeep et al. (2023). Also discussed IoT-WSN application in smart agriculture using LoRa WAN protocol to communicate data gathered by sensors from agriculture land to a large distance.

Figure 1. IoT-WSN framework for smart agriculture
Yang et al. (2021)

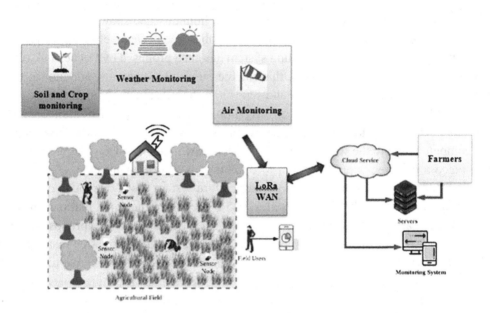

FRAMEWORK OF OUR SYSTEM

Our design's main goal is to demonstrate a smart agricultural system based on IoT and sensor networks that can manage irrigation and fertilizer applications to maximize crop productivity. For the system to be designed without human interaction, soil moisture, temperature, humidity, and air monitoring sensors are needed. The proposed framework for smart agriculture, based on sensor networks and IoT, is shown in Figure 1. Four operational levels make up the suggested design: sensor deployment, middleware, communication, and application layer.

Layer Related to Sensor Placement

The initial layer of operation in the realm of agriculture is the deployment of sensors. For the purpose of gathering field data, the suggested system makes use of a soil moisture sensor, a temperature and humidity sensor, and an air monitoring sensor. Figure 1 displays the sensor nodes that have been deployed. Real-time field data collection and ongoing field monitoring are the responsibilities of these sensor nodes. The soil moisture sensor measures the relative temperature and humidity levels by measuring the soil moisture content, temperature, and humidity. The same air monitoring sensor is employed to determine the direction and speed of the wind. These sensor nodes converse with one another and send the data they have collected to the sink or master node. The sink node, an Arduino Uno-based microcontroller, is in charge of gathering field data and sending it to the Thing Speak cloud platform for analysis and storage. Real-time data collection is done by the experimentation's hardware for wind speed, relative temperature, and humidity. The sensor nodes connect wirelessly with one another and send sensed data to a GSM-capable sink node. This sink node is in charge of collecting field data and sending it to the cloud for analysis and storage.

Middleware Layer

The second operational layer of our suggested system is middleware. This layer is in charge of automating the agricultural process and managing the actuators. The controller receives the gathered values that are above the threshold for ground station analysis. These operational layers keep an eye on the temperature, humidity, wind speed, and soil moisture because these are important variables directly influence crop output. When the soil moisture is seen to be below a threshold value, the first of three configuration criteria is considered, and an alarm message is generated and sent to the user via email. In order to switch on the water motor for soil irrigation, the automated system then generates a question. The water motor is automatically turned off by the system after the desired irrigation level has been reached. When relative temperature and humidity measurements above the threshold, the factor takes this into account. To maintain the ambient atmosphere, the system proposes opening the flap of the greenhouse or polyhouse via an automated generated alert.

The third element is the wind direction and speed, which are measured by air monitoring devices. Systems generate an alarm for covering the area and the targeted plants, protecting them from harm, when the air monitoring value exceeds the threshold. The middleware layer is in charge of the aforementioned three tasks. Based on a real-time analysis of observed field data, these activities are performed in the ground station utilizing the Thing Speak platform. By regulating these three variables in the selected agricultural region, crop production will rise overall while using less resources.

Networking and Cloud Layer

The networking layer of the suggested system is the third operational layer. This layer is in charge of communicating with the deployed sensors and is based on the ESP32 microcontroller.

Because this gateway has a GSM module, the captured data can be transmitted wirelessly to the base station through it. This module has Wi-Fi capability and transmits data wirelessly. The deployed sensor keeps track of field data like humidity, temperature, and wind speed and sends it to the gateway node. The controller gateway is running an IP protocol that transmits the gathered data to the Thing Speak cloud for analysis and decision-making.

Application Layer

The challenges with the use of smart agriculture are effectively resolved by the development in the fields of cloud computing and edge computing. The created model collects field data using the Thing Speak cloud platform and offers effective storage and analysis. Different channels are created in this operational layer for various field characteristics, representing their frequent monitoring and saving of data fields with time and date. The gateway node sends the precise measurement data to the appropriate channels.

In the Thing Speak cloud, the data is analyzed using MATLAB to calculate irregularities and forecast results. For the ground station's time-based analysis, these data values are shown using MATLAB. Data from the agricultural fields can be remotely seen at any time and from any location via a web service.

WIRELESS SENSOR NETWORK ARCHITECTURE

- Main components of a WSN node are

    ○ Controller
    ○ Communication device(s)
    ○ Sensors/actuators
    ○ Memory
    ○ Power supply

Figure 2. WSN architecture for smart agriculture

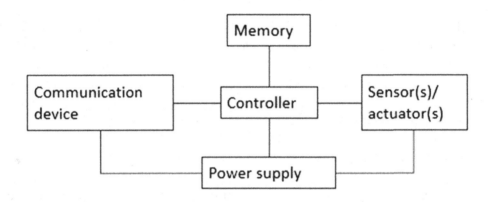

A wireless sensor node (WSN) typically consists of a microcontroller, transceiver, external memory, power source, and one or more sensors. The sensors capture data from the environment and send it to the controller for processing. The controller performs tasks, processes data, and controls the functionality of other components in the WSN node. The microcontroller is a commonly used controller due to its low cost, flexibility, ease of programming, and low power consumption. Additionally, the WSN node may include analog-to-digital converters (ADCs) to convert the analog signals produced by the sensors into digital signals as shown in fig 2.

LoRaWAN

LoRaWAN is a wireless protocol used for managing network connectivity for LoRa sensors, which are designed for simple sensor devices requiring long battery life and long-range connectivity. LoRaWAN is considered one of the best LPWAN standards, making it ideal for IoT applications that involve many low-power devices. LoRaWAN sensors are capable of transferring data across long distances, making them particularly useful for IoT sensors operating in remote areas without public access. Compared to WiFi, Zigbee, and Bluetooth, which have poor range and propagation through objects, LoRaWAN is a better option for IoT sensors needing long-range connectivity. A typical wireless sensor node (WSN) based on LoRaWAN includes a microcontroller, transceiver, external memory, power source, and one or more sensors. These sensors capture data from the environment and send it to the controller for processing as shown in fig 3. Additionally, the WSN node may include analog-to-digital converters (ADCs) to convert analog signals produced by the sensors into digital signals. LoRaWAN sensors are used in various applications, such as environmental monitoring, smart homes, and data center room monitoring Polonelli et al. (2018).

Figure 3. LoRa WAN architecture for smart agriculture
Source- lora-alliance.org

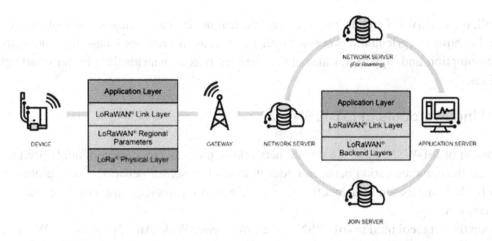

LoRaWAN (Long Range Wide Area Network) is a low-power, wide-area networking technology designed for long-range communication with low data rates. It is particularly well-suited for smart agriculture applications due to its ability to cover large areas, provide long-range connectivity, and operate with low power consumption.

Here's an overview of the LoRaWAN architecture for smart agriculture:

1. **End Devices:** These are the sensor nodes or devices deployed in the field to monitor various parameters in agriculture, such as soil moisture, temperature, humidity, light intensity, and crop health. These devices are typically battery-powered and designed to operate for an extended period.
2. **Gateways:** LoRaWAN gateways act as intermediaries between the end devices and the network server. They receive the sensor data from the end devices over long distances (up to several kilometers) and forward it to the network server using the LoRaWAN protocol. Gateways are usually connected to the internet via Ethernet, Wi-Fi, or cellular networks.
3. **Network Server:** The network server is responsible for managing the entire LoRaWAN network. It receives the data packets forwarded by the gateways and performs several tasks, including data decryption, device authentication, and data routing to the appropriate application server.
4. **Application Server:** The application server receives the data from the network server and processes it according to the specific requirements of the smart agriculture application. It performs data analysis, visualization, and decision-making based on the received sensor data. The application server can also send commands or configuration changes back to the end devices through the network server and gateways.

5. **Cloud Services/Back-end System:** In many cases, the application server may interact with cloud services or a back-end system for further data storage, analytics, and integration with other applications or services. This allows for more advanced data processing, historical analysis, and data sharing with other stakeholders.

6. **User Interface:** The user interface can be a web application or a mobile application that provides a graphical representation of the sensor data, analytics, and control options for the end users. It allows farmers or agricultural experts to monitor the conditions in real-time, receive alerts, and make informed decisions for irrigation, fertilization, pest control, or other farming activities.

Overall, the LoRaWAN architecture for smart agriculture enables efficient and cost-effective monitoring and control of agricultural processes over large areas. It provides long-range connectivity, low power consumption, and supports a wide range of sensors, making it an ideal choice for smart agriculture applications.

Networking Protocols in IoT-WSN

In the context of IoT-WSN, there are several networking protocols that are commonly used to connect and manage the communication between nodes in a wireless sensor network. These protocols help to ensure reliable transmission of data, efficient use of network resources, and effective management of network connectivity.

One popular protocol used in IoT-WSN is the Low-Power Wide-Area Network (LPWAN) protocol, which is designed to enable long-range communication between low-power devices. LPWAN protocols are ideal for IoT applications that require long-range connectivity and low-power consumption, such as environmental monitoring, smart homes, and industrial automation.

Another commonly used protocol in IoT-WSN is the ZigBee protocol, which is a low-power, wireless mesh network protocol that is designed for low-data rate applications. ZigBee is well-suited for home automation and industrial control applications, where it can be used to connect a large number of low-power devices in a mesh network.

In addition to LPWAN and ZigBee, there are several other networking protocols that are commonly used in IoT-WSN, including Bluetooth Low Energy (BLE), Wi-Fi, and cellular networks. Each of these protocols has its own strengths and weaknesses, and the choice of protocol will depend on the specific requirements of the application.

Overall, the choice of networking protocol in IoT-WSN is an important consideration that can have a significant impact on the performance, reliability, and scalability of the network. It is important to carefully evaluate the different options and choose a protocol that is well-suited to the specific needs of the application.

Hardware Requirements in IoT-WSN

In wireless sensor nodes (WSN), the hardware components typically include a microcontroller, transceiver, external memory, power source, and one or more sensors. The microcontroller is responsible for running the software and managing the sensors and the transceiver, which facilitates wireless communication between the node and the gateway or other nodes in the network. The external memory provides storage for data and code, while the power source can be a battery, solar panel, or energy harvesting module.

In terms of networking protocols, LoRaWAN is commonly used for managing network connectivity for LoRa sensors in IoT applications involving many low-power devices. Other commonly used protocols include ZigBee, Bluetooth Low Energy (BLE), Wi-Fi, and cellular networks. It is crucial to carefully evaluate each option to select a protocol that is best suited for the specific needs of the application.

In IoT-WSN (Internet of Things - Wireless Sensor Networks), the hardware requirements depend on the specific application and use case. However, there are some general considerations that can be made regarding the hardware requirements for IoT-WSN:

1. **Sensor Nodes**: The sensor nodes are the fundamental components of WSNs. They should be low-power devices that can operate for extended periods on a limited power supply, such as batteries or energy harvesting systems. The sensors must be capable of measuring the parameters that are essential for the application. For example, temperature, humidity, pressure, light, motion, etc.
2. **Communication**: The communication module is essential for the WSN nodes to communicate with each other and with the gateway or the cloud. The communication module should be low-power and LoRaWAN is a popular protocol in smart agriculture due to its ability to provide long-range connectivity in remote areas. Other commonly used protocols include ZigBee and IEEE 802.11 g (WiFi 2.4 GHz). The requirements for IoT modules based on the LoRaWAN protocol include low cost, low consumption, small dimensions, fast prototyping, and easy programming. Proper selection of hardware components and communication protocols is important for the effective implementation of IoT-WSN in smart agriculture applications.
3. **Processing Unit**- The processing unit is responsible for processing the data gathered by the sensing unit.
4. **Power Unit:** The power unit supplies the required power to the sensor node components.

Since sensor nodes in WSNs are resource-constrained, selecting the right protocol is crucial to ensure low-power consumption and efficient use of resources. Several protocol designs have been proposed, including ZigBee, Bluetooth, and Wi-Fi, each with its strengths and weaknesses. Careful evaluation and selection of components and protocols are essential to ensure effective implementation of IoT-WSN in various applications.

OBJECTIVES OF STUDY

1. To proposed framework for smart agriculture based on IoT-WSN.
2. To demonstrate various networking protocols used in IoT-WSN based smart agriculture.
3. To explain IoT-WSN challenges in smart agriculture.
4. To study routing protocols in WSN.
5. To discuss role of Machine learning in Agriculture.

CHALLENGES ASSOCIATED WITH THE IMPLEMENTATION OF TECHNOLOGIES IN IOT-WSN

The technical hurdles and trade-offs involved in developing connected devices that are both secure and functional are referred to as design issues in the IoT (Internet of Things). The following are some of the major IoT design challenges:

Interoperability

The effective exchange of data between various systems, devices, or components is referred to as interoperability. Interoperability is a major issue in the Internet of Things (IoT) setting since so many different types of devices are being connected to the internet. Lack of standards in the IoT can cause problems with data sharing and communication between devices, creating a disjointed and ineffective system. Organizations and industry groups are striving to create standards and protocols to guarantee interoperability across IoT devices in order to address this difficulty. This includes creating standard data formats, communication protocols, and security guidelines. For the Internet of Things to reach its full potential and for linked devices to cooperate effectively, interoperability is crucial. ensuring effective data exchange and seamless collaboration between various IoT devices.

Security

The security of sensitive data and systems from unwanted access, theft, or damage is a significant problem in the Internet of Things (IoT). IoT devices frequently fall victim to cyber assaults because of their increased internet exposure and their constrained computational power. IoT security issues might include things like:

1. IoT device security is the first step in ensuring that they are shielded from malware and unauthorized access.
2. Network security: Defending against cyberattacks while allowing connectivity between IoT devices and the network.
3. Data security: preventing unauthorizes access to or tampering with the data collected and transmitted by IoT devices.
4. Privacy: safeguarding the privacy of people whose personal data is gathered and sent by IoT devices.

Organizations should use strong security measures like encryption, firewalls, and routine software updates to meet these security challenges.

They should also regularly analyses and audit security to find and fix any potential security concerns. Organizations can protect IoT systems and sensitive data by putting security first and lowering the risk of cyberattacks. Defending IoT devices from online attacks and unauthorized access, as well as the sensitive data they collect and communicate.

Scalability

Scalability is the capacity of a system to manage growing workloads or user populations without noticeably degrading performance. Scalability is a significant issue in the Internet of Things (IoT) setting since the number of connected devices is expanding quickly, resulting in an increase in data and communication traffic. IoT scaling challenges include:

1. Data management: Securing and managing the vast amounts of data produced by IoT devices.
2. Network capacity: Ensuring that networks are able to manage the growing amount of data and communication.
3. Device management: Effectively controlling the expanding number of IoT devices and making sure they are simple to set up and maintain.

Organizations should implement scalable infrastructures, such cloud computing, that can handle the expanding number of IoT devices and the data they produce, in order to overcome these scalability concerns. To deal with the rising volume of data, they should also install effective data management and storage solutions, such as distributed databases and data lakes.

Organizations can guarantee that their IoT systems can handle the increasing number of connected devices and continue to provide great performance and efficiency by giving scalability a high priority. creating systems that can properly handle the data flow caused by a high number of connected devices.

Reliability

The capacity of a system to carry out its intended function repeatedly and successfully over time. Reliability is a crucial issue in the context of the Internet of Things (IoT), as the failure of even a single IoT device can have serious repercussions. Among the difficulties with IoT reliability are:

1. Device failure: Ensure that IoT devices are created and intended to be dependable and operate properly even in challenging settings.
2. Network connectivity: Keeping connections between IoT devices and the network strong and dependable even in the face of hardware or software problems.
3. Data accuracy: Ensuring the accuracy and dependability of the data that IoT devices collect and communicate.

Organizations should use robust and dependable hardware and software designs for IoT devices to handle these dependability difficulties, and they should do routine testing and maintenance to find and fix any problems. To ensure that the system keeps working in the event of a failure, they should also put in place redundant systems and failover procedures. Organizations can ensure that their IoT systems operate consistently and without error and deliver the desired advantages and results by placing a high priority on reliability. Ensuring that, despite hardware or software failures, IoT systems continue to be usable and accessible.

Power Consumption

Power consumption is the measure of how much energy a system or device consumes. Given that many IoT devices are created to be compact, low-power, and battery-operated, power consumption is a significant challenge in the IoT setting. The following are a few issues with IoT power consumption:

1. Battery life: Ensuring that IoT devices have enough battery life to operate between recharges or without constant replacement.
2. Energy efficiency: Ensuring that IoT gadgets are made to use energy wisely and cut down on the system's overall power usage.
3. Power management: Putting into practice efficient power management strategies, like sleep modes, to lower IoT devices' power consumption when not in use.

Privacy

As IoT devices collect, store, and communicate copious amounts of sensitive and personal data, privacy is a crucial issue in the Internet of Things (IoT). IoT presents a number of privacy challenges, including:

1. **Data collection**: Making sure that only the information that is required is gathered and that it is done so in a way that respects people's right to privacy.
2. **Data storage**: Making sure that IoT device data is securely saved and that access to it is tightly limited.
3. **Data sharing**: Limiting who gets access to the information gathered by IoT devices and making sure it isn't distributed without the right authorization.
4. Organizations should develop strong privacy policies and practices, such as data protection, data minimization, and data retention, to solve these privacy challenges. Additionally, consumers should be made aware of the privacy risks associated with IoT devices and encouraged to take precautions to safeguard their privacy. Also, in order to preserve the privacy of people whose information is collected by IoT devices, companies should implement privacy-enhancing technologies like encryption and anonymization. Organizations can contribute to ensuring that people's rights and freedoms are respected and that sensitive information is shielded from illegal access or misuse by putting privacy first. defending the privacy of people whose personal data is gathered and sent by IoT devices.

Figure 4. Measures to make IOT networks more reliable
Source: Allerin.com

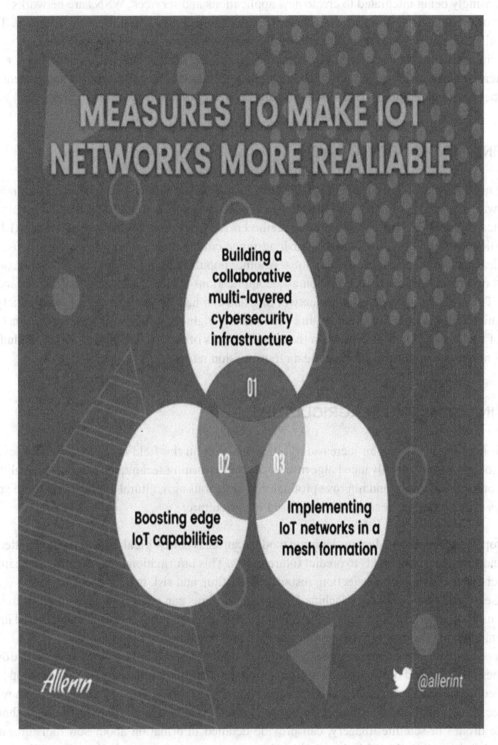

Wireless Sensor Networks (WSNs) and the Internet of Things (IoT) are two related technologies that are increasingly being integrated to create new applications and services. WSNs are networks of small, battery-powered devices equipped with sensors that can collect data about their environment. The IoT, on the other hand, is a system of interconnected devices that can communicate with each other and with the internet, and is characterized by its ability to collect, analyze and act on data. Fig 4 represent parameters to make IoT system more reliable by boosting edge IoT capabilities, IoT implementation in a mesh network and building a collaborative multi-layered cybersecurity infrastructure.

ROUTING METHODS IN IOT-WSN

Routing methods play a crucial role in IoT-WSN by enabling communication between sensor nodes and the gateway. The choice of routing method depends on the network topology, application requirements, and device capabilities. One popular routing method is the Low-Power Wide-Area Network (LPWAN) using LoRaWAN protocol. It is suitable for long-range communication with low power consumption and is ideal for smart agriculture applications. Another popular routing method is the ZigBee protocol, which is optimized for low-power, low-data-rate applications such as home automation and industrial control. The protocol is based on mesh networking and can handle multiple devices and data types. It is also suitable for applications that require low latency, high reliability, and security. Ultimately, the choice of routing method will depend on the specific needs of the IoT-WSN application, including the required range, power consumption, and data transmission rate.

MACHINE LEARNING IN AGRICULTURE

Machine learning is playing an increasingly important role in the field of agriculture. By leveraging large amounts of data and advanced algorithms, machine learning techniques can provide valuable insights, optimize processes, and improve productivity in various agricultural applications. Here are some examples of how machine learning is being used in agriculture:

1. Crop yield prediction: Machine learning models can analyze historical data on weather patterns, soil conditions, and crop yields to predict future yields. This information helps farmers make informed decisions regarding crop selection, resource allocation, and risk management.
2. Disease and pest detection: Machine learning algorithms can analyze images of crops to identify signs of diseases, pests, or nutrient deficiencies. This enables early detection and targeted intervention, allowing farmers to take appropriate measures to minimize crop loss.
3. Weed identification and management: Machine learning models can be trained to recognize different types of weeds and distinguish them from crops. This information can be used to develop targeted weed management strategies, such as precise herbicide application or robotic weed removal.
4. Precision agriculture: Machine learning techniques, combined with remote sensing technologies like drones or satellite imagery, can provide detailed information about soil moisture, nutrient levels, and crop health. This data enables farmers to optimize resource allocation, tailor irrigation and fertilization practices, and improve overall efficiency.

5. Livestock monitoring: Machine learning algorithms can analyze sensor data from wearable devices attached to livestock to monitor their health, activity levels, and behavior. This helps farmers identify early signs of illness, detect estrus in breeding animals, and improve animal welfare.

6. Supply chain optimization: Machine learning can be used to optimize various aspects of the agricultural supply chain, including logistics, storage, and distribution. By analyzing historical data and real-time information, machine learning models can optimize routes, predict demand, and reduce waste.

7. Farm management systems: Machine learning algorithms can be integrated into farm management systems, helping farmers make data-driven decisions. These systems can provide recommendations on planting schedules, fertilization rates, and optimal harvest times based on various factors like weather data, historical trends, and specific farm conditions.

Overall, machine learning in agriculture has the potential to enhance productivity, reduce environmental impact, and improve sustainability by enabling farmers to make more informed decisions and optimize their operations.

SMART AGRICULTURE IN TERMS OF ML, DL ALGORITHMS

Smart agriculture, also known as precision agriculture, is the application of modern technologies, including machine learning (ML) and deep learning (DL) algorithms, to optimize farming practices and improve crop yield, quality, and resource efficiency. ML and DL algorithms play a crucial role in analyzing large volumes of data collected from various sources, such as sensors, satellite imagery, and weather stations, to provide insights and make data-driven decisions in agriculture. Here are some examples of ML and DL algorithms used in smart agriculture:

1. Crop yield prediction: ML algorithms can analyze historical data on weather conditions, soil quality, irrigation, and other relevant factors to predict crop yield. This information helps farmers optimize their planting strategies, manage resources effectively, and plan for future harvests.

2. Disease detection: ML algorithms can be trained on large datasets of images or sensor data to detect and diagnose crop diseases and pests. By analyzing patterns and identifying early signs of diseases, farmers can take preventive measures and apply targeted treatments, reducing the need for broad-spectrum pesticides and minimizing crop losses.

3. Weed detection and control: ML algorithms can analyze images or sensor data to identify and differentiate between crops and weeds. This information enables farmers to target specific areas for weed control, reducing the use of herbicides and minimizing the impact on the environment.

4. Irrigation optimization: ML algorithms can analyze soil moisture data, weather forecasts, and plant physiological parameters to optimize irrigation scheduling. By dynamically adjusting irrigation based on real-time data, farmers can ensure that crops receive the right amount of water, minimizing water wastage and improving water use efficiency.

5. Nutrient management: ML algorithms can analyze soil nutrient data and crop requirements to optimize fertilizer application. By providing precise recommendations on the type and quantity of fertilizers needed, farmers can minimize nutrient wastage, reduce environmental impact, and improve crop productivity.

6.   Crop quality assessment: DL algorithms can analyze images or sensor data to assess the quality and ripeness of crops. By automating the assessment process, farmers can make informed decisions regarding harvest timing, grading, and sorting, ensuring higher-quality produce and minimizing post-harvest losses.

7.   Pest and disease forecasting: ML algorithms can analyze historical weather data, crop growth stages, and pest/disease occurrence data to forecast the risk of pest outbreaks or disease epidemics. This information helps farmers implement preventive measures, such as timely spraying or adjusting planting dates, to mitigate risks and protect their crops.

These are just a few examples of how ML and DL algorithms are being applied in smart agriculture. The field is rapidly evolving, and new algorithms and techniques are constantly being developed to address various challenges in agriculture and improve farming practices.

COMPARATIVE ANALYSIS

Table 1 shows comparative analysis of various agriculture approaches in terms of parameters, devices, cloud storage and management and microcontroller etc.

Table 1. Comparative study of different agriculture approaches

References	Parameters	Smart Devices	Cloud System	Storage	Microcontroller
(Liqiang et al., 2011)	Soil Moisture	No	No	No	Nil
Channe et al., 2015)	Soil Moisture, Temp., Humidity	Yes	Yes	Yes	Arduino UNO
(Chaudhary et al., 2011)	Soil Moisture, Temp.	Yes	Yes	Yes	PIC16F887
(Reche et al., 2014)	Wind Speed, Soil Moisture, Temp., Humidity	Yes	Yes	Yes	Irrigation Controller
Heble et al., 2018) Microcontroller	Temp. Soil Moisture, Wind Speed	Yes	No	No	PIC Microcontroller

Yang et al. (2021)

CONCLUSION AND FUTURE SCOPE

In this paper, a framework for agricultural field monitoring and irrigation control is proposed using the concept of IoT. Internet of Things provides an efficient solution for smart agriculture and proves its importance for helping farmers by increasing the overall productivity. This framework is designed using sensor networks and cloud platforms for storing and analyzing data in real-time. The system senses different field information like soil moisture, temperature, humidity, and air quality and transmits this information to the cloud using a gateway. The cloud platform is responsible for storing data and performing data aggregation, analyses, and visualization tasks. Based on the sensed information system predict the upcoming weather and amount of dryness present in the soil. The system schedules the irrigation process based on the real-time collected data and weather prediction. This system generates alerts that suggest farmers or users whether soil irrigation is required or not along with the percentage of soil moisture content. The implementation of this system is the agricultural field can benefits users by improving crop yields and production rates. The work can be extended for diagnosing crop diseases, insects, and weeds that occur in crops. Artificial neural network and image processing can be adopted for the detection and diagnosis of crop diseases by installing digital cameras in the agricultural field.

IoT-WSN can help farmers make more informed decisions, improve their productivity and profitability, and reduce waste and environmental impact. As these technologies continue to evolve, we can expect to see more and more innovative applications of IoT and WSN in the agricultural industry.

In addition, the selection of a suitable wireless protocol is crucial for achieving long-range connectivity and improving power consumption. LoRaWAN is a popular protocol in smart agriculture due to its ability to provide long-range connectivity in remote areas. Other commonly used protocols include ZigBee and IEEE 802.11 g (Wi-Fi 2.4 GHz). The requirements for IoT modules based on the LoRaWAN protocol include low cost, low consumption, small dimensions, fast prototyping, and easy programming. Proper selection of hardware components and communication protocols is important for the effective implementation of IoT-WSN in smart agriculture applications.

In conclusion, the integration of WSNs and IoT technologies can provide a number of benefits and enable the creation of new applications and services in a wide range of fields.

REFERENCES

Bhatia, S., Dhillon, H. S., & Kumar, N. (2011). Alive human body detection system using an autonomous mobile rescue robot. *2011 Annual IEEE India Conference*, Hyderabad, India. 10.1109/INDCON.2011.6139388

Bhatia, S., Gautam, D., Kumar, S., & Verma, S. (2023). Automatic Seat Identification System in Smart Transport using IoT and Image Processing. 2023 3rd International Conference on Intelligent Communication and Computational Techniques (ICCT), Jaipur, India. 10.1109/ICCT56969.2023.10075664

Bhatia, S., Jaffery, Z. A., & Mehfuz, S. (2023). *A Comparative Study of Wireless Communication Protocols for use in Smart Farming Framework Development*. 2023 3rd International Conference on Intelligent Communication and Computational Techniques (ICCT), Jaipur, India. 10.1109/ICCT56969.2023.10075696

Channe, H., Kothari, S., & Kadam, D. (2015). Multidisciplinary model for smart agriculture using internet-ofthings (IoT), sensors, cloud-computing, mobile-computing & big-data analysis. *Int. J. Computer Technology and Application*, 6(3), 374–382.

Chaudhary, D. D., Nayse, S. P., & Waghmare, L. M. (2011). Application of wireless sensor networks for greenhouse parameter control in precision agriculture. *International Journal of Wireless & Mobile Networks*, 3(1), 140–149. doi:10.5121/ijwmn.2011.3113

Dan, L. I. U., Xin, C., Chongwei, H., & Liangliang, J. (2015). Intelligent agriculture greenhouse environment monitoring system based on IOT technology. In *2015 International Conference on Intelligent Transportation, Big Data and Smart City* (pp. 487-490). IEEE. 10.1109/ICITBS.2015.126

de Ocampo, A. L. P., & Dadios, E. P. (2017). Energy cost optimization in irrigation system of smart farm by using genetic algorithm. In *2017IEEE 9th International Conference on Humanoid, Nanotechnology, Information Technology, Communication and Control, Environment and Management (HNICEM)* (pp. 1-7). IEEE. 10.1109/HNICEM.2017.8269497

Elijah, O., Rahman, T. A., Orikumhi, I., Leow, C. Y., & Hindia, M. N. (2018). An overview of Internet of Things (IoT) and data analytics in agriculture: Benefits and challenges. *IEEE Internet of Things Journal*, *5*(5), 3758–3773. doi:10.1109/JIOT.2018.2844296

Ganai, P. T., Bag, A., Sable, A., Abdullah, K. H., Bhatia, S., & Pant, B. (2022). A Detailed Investigation of Implementation of Internet of Things (IOT) in Cyber Security in Healthcare Sector. *2022 2nd International Conference on Advance Computing and Innovative Technologies in Engineering (ICACITE)*, Greater Noida, India. 10.1109/ICACITE53722.2022.9823887

Gayatri, M. K., Jayasakthi, J., & Mala, G. A. (2015). *Providing Smart Agricultural solutions to farmers for better yielding using IoT. In 2015 IEEE Technological Innovation in ICT for Agriculture and Rural Development (TIAR)*. IEEE.

Heble, S., Kumar, A., Prasad, K. V. D., Samirana, S., Rajalakshmi, P., & Desai, U. B. (2018). A low power IoT network for smart agriculture. In *2018 IEEE 4th World Forum on Internet of Things (WF-IoT)* (pp. 609-614). IEEE. 10.1109/WF-IoT.2018.8355152

Jeet, V., Dhillon, H. S., & Bhatia, S. (2015). Radio Frequency Home Appliance Control Based on Head Tracking and Voice Control for Disabled Person. *2015 Fifth International Conference on Communication Systems and Network Technologies*, Gwalior, India. 10.1109/CSNT.2015.189

Keshtgari, M., & Deljoo, A. (2011). *A wireless sensor network solution for precision agriculture based on zigbee technology*. Academic Press.

Khanna, A., & Kaur, S. (2019). Evolution of Internet of Things (IoT) and its significant impact in the field of Precision Agriculture. *Computers and Electronics in Agriculture, 157*, 218–231. doi:. compag.2018.12.039 doi:10.1016/j

Khatri-Chhetri, A., Aggarwal, P. K., Joshi, P. K., & Vyas, S. (2017). Farmers' prioritization of climate-smart agriculture (CSA) technologies. *Agricultural Systems, 151*, 184–191. doi:10.1016/j.agsy.2016.10.005

Lakshmisudha, K., Hegde, S., Kale, N., & Iyer, S. (2016). Smart precision based agriculture using sensors. *International Journal of Computer Applications, 146*(11), 36–38. doi:10.5120/ijca2016910916

Liqiang, Z., Shouyi, Y., Leibo, L., Zhen, Z., & Shaojun, W. (2011). A crop monitoring system based on wireless sensor network. *Procedia Environmental Sciences, 11*, 558–565. doi:10.1016/j.proenv.2011.12.088

Mekala, M. S., & Viswanathan, P. (2017). A novel technology for smart agriculture based on IoT with cloud computing. In *2017 International Conference on I-SMAC (IoT in Social, Mobile, Analytics and Cloud) (I-SMAC)* (pp. 75-82). IEEE. 10.1109/I-SMAC.2017.8058280

Mekala, M. S., & Viswanathan, P. A. (2017). *Survey: Smart agriculture IoT with cloud computing. In 2017 international conference on microelectronic devices, circuits and systems (ICMDCS).* IEEE.

Nawandar, N. K., & Satpute, V. R. (2019). IoT based low cost and intelligent module for smart irrigation system. *Computers and Electronics in Agriculture, 162,* 979–990. doi:10.1016/j.compag.2019.05.027

Patil, K. A., & Kale, N. R. (2016). A model for smart agriculture using IoT. In *2016 International Conference on Global Trends in Signal Processing, Information Computing and Communication (ICGTSPICC)* (pp. 543-545). IEEE. 10.1109/ICGTSPICC.2016.7955360

Polonelli, T., Brunelli, D., Bartolini, A., & Benini, L. (2018). A LoRaWAN Wireless Sensor Network for Data Center Temperature Monitoring. *Applications in Electronics Pervading Industry, Environment and Society.*

Rajeswari, S., Suthendran, K., & Rajakumar, K. (2017). A smart agricultural model by integrating IoT, mobile and cloud-based big data analytics. In *2017 International Conference on Intelligent Computing and Control (I2C2)* (pp. 1-5). IEEE. 10.1109/I2C2.2017.8321902

Reche, A., Sendra, S., Díaz, J. R., & Lloret, J. (2014). A smart M2M deployment to control the agriculture irrigation. In *International conference on ad-hoc networks and wireless* (pp. 139-151). Springer.

Srbinovska, M., Gavrovski, C., Dimcev, V., Krkoleva, A., & Borozan, V. (2015). Environmental parameters monitoring in precision agriculture using wireless sensor networks. *Journal of Cleaner Production, 88,* 297–307. doi:10.1016/j.jclepro.2014.04.036

TongKe. (2013). Smart agriculture based on cloud computing and IOT. *Journal of Convergence Information Technology, 8*(2).

Varman, S. A. M., Baskaran, A. R., Aravindh, S., & Prabhu, E. (2017). Deep learning and IoT for smart agriculture using WSN. In *2017 IEEE International Conference on Computational Intelligence and Computing Research (ICCIC)* (pp. 1-6). IEEE.

Veena, S., Mahesh, K., Rajesh, M., & Salmon, S. (2018). The survey on smart agriculture using IOT. *Int J Innov Res EngManag, 5*(2), 63–66.

Yang, J. W., Sharma, A., & Kumar, R. (2021). IoT-Based Framework for Smart Agriculture. *International Journal of Agricultural and Environmental Information Systems, 12*(2), 1–14. doi:10.4018/IJAEIS.20210401.oa1

Zhao, W., Lin, S., Han, J., Xu, R., & Hou, L. (2017). *Design and implementation of smart irrigation system based on LoRa. In 2017 IEEE Globecom Workshops (GC Wkshps).* IEEE. doi:10.1109/GLOCOMW.2017.8269115

Chapter 23
The Current Generation 5G and Evolution of 6G to 7G Technologies:
The Future IoT

Sandeep Bhatia
Galgotias University, India

Neha Goel
Raj Kumar Goel Institute of Technology, India

Soniya Verma
KIET Group of Institutions, India

ABSTRACT

Broadband cellular networking technologies 5G/6G/7G for establishing connections between users and sending data over the internet are sometimes referred to as the "internet of things" or embedded items. IoT brings about a huge amount of traffic as a result of the strong interaction among the millions of connected things available at the time of deployment to particular applications. The increase in the popularity of the internet of things has been likened to an increase in the number of barriers. In this study, the current use of 5G edge network infrastructure with IoT-enabled and hybrid and multi-cloud deployments need 6G and 7G technologies, and the authors examine the current state of IoT as well as potential conditions and challenges that may influence IoT acquisition. In this chapter there is a comparison of different communication technology in the context of IoT. At the end of the chapter, the impact of 5G, 6G, and beyond 7G technologies have been discussed in the preface of IoT.

DOI: 10.4018/978-1-6684-8785-3.ch023

INTRODUCTION

A new generation of wireless mobile telecommunications technology appears every ten years or so. The usage of additional frequency bands, faster data rates, and new services characterize this evolution as it moves us closer to being able to connect essentially the entire physical universe. A tool that sends real-time data across geographic, cultural, linguistic, and temporal barriers is known as the Internet as explained by Guevara and AuatCheein (2020). By connecting numerous devices to the internet, people can easily connect with each other online and improve their quality of life Prinima et al. (2016). In the past, 4G had a lot of problems with mobile internet and users had to deal with ambiguities and Internet Protocol instability. The subscribers are not satisfied with 4G, and the industry looking to improve the network latency with 5G and 6G expands the industries in all things in terms of the low network latency and high reliability. As when the physical world would be connected through the internet it would ease the life of people. Also, people could commute using the Internet Gupta and Jha (2015). Several aspects of our daily lives have been made better thanks to the internet, including health monitoring, home appliance control, R&D, science, smart manufacturing, tourism, entertainment, mass communication, and more. IoT technology is used in almost 200 nations for data, information, news, and opinion exchange Sawanobori (2016). Automating and improving human existence is the ultimate purpose of IoT. Figure 1 depicts a rough layout of the digital transition from 6G to 7G and the rapid growth of telecommunications technology across all industrial partners and IT marketing in the telecommunications sector.

The global scope of the Internet of Things' applications is expanding. Now, a wide range of devices, including smartphones, intelligent vehicles, and a variety of other devices, can all share data through the Internet 5G-PPP (2020). Wireless communication is supported by 3G, which also improved 2G and 2. After 3G is upgraded to 4G and 5G, it is utilized for edge networks, fast data transport, and improved voice control. For video conferencing and long-term testing (LTE), 4G is upgraded to 5G, which is currently operational. The personal and social dimensions encompass the following application domains: An essential component of the internet is networking Samsung 5G Vision (2015). For example, a vehicle outfitted with the Internet of Things (IoT) technology will not only assist drivers by giving maps, but will also provide real-time in-formation, but will also provide instructions and propose the best routes to reach the location Zanella et al. (2014). In the year 1999/2000 once SIM card inserted, then 3G running inside mobile and it supported MAN, it became a popular network service in IoT and provides a more prominent way to establish calls and this technology enables people to communicate with each other as well to access data in limited access with more charges levied by the service providers, all these things in IoT recognizedas the 'thing/object' Nassar et al. (2017). In Yan et al. (2020), the author discussed 4G, 5G, 6G and 7G impact on IoT enabled applications. The IoT revolution with emerging communication technologies has been discussed by the author in Figure 1.

Figure 1. Evolution from 5G to 6G and jump to 7G

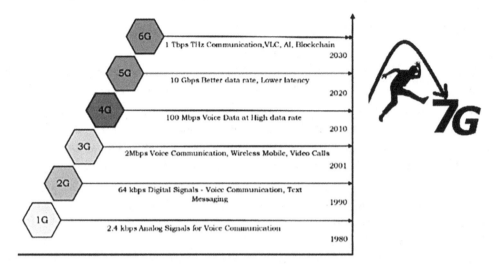

The second section of this article provides an explanation as the related work of the integrated environment of emerging technologies. The third section demonstrates the methodology and objectives. The fourth section gives the explanation. The fifth section defines the references preceded by the conclusion. Wireless technology has advanced significantly during the past decade. Wireless technology reaches its growth and revolution at 7.5G. Wireless technology FG (Future Generation) mobile communications will offer higher data transmission speeds than 6G and 7G. Advanced techniques are always evolving in all areas of mobile and wireless communications, making wireless technology one of the domains that is constantly developing at a fast rate of pace.

The rollout of 5G mobile communication technology is now in its infancy. Although we currently have a wide range of technologies, each of which is capable of carrying out specific tasks like supporting voice traffic using voice over IP (VoIP), broadband data access in mobile environments, etc., there is a great need to introduce new technologies that can combine all of these systems into a single, unified system. Since 8G is all about smoothly integrating terminals, networks, and applications, it offers a solution to this issue Yan et al. (2020). In this work an attempt has been made to present a study of several cellular technologies notably 4G, 5G, 6G, 7G, and Future G (FG) correspondingly and detail comparison among them.

The major goal of this study is to determine how mobile wireless communication will develop in the future.

RELATED WORK

IoT is considered to be used across various regions and platforms. Healthcare, agriculture, smart buildings (schools, hospitals, and homes), supply chain management, transportation, and defense are among areas where technology is being used Lee et al. (2017). The current 5g technology includes SDN, IoT and AI it supports for core edge technologies and based application Srinivasan et al. (2022). For the integrated environment 5G not sufficient minimum 6G required, 6G is 1000 times faster than 5G and improves the

SDN, AI and IoT within the relevant areas of security and privacy Ziegler et al. (2013). The network slicing said that multi-access and edge computing (MEC) shows the integrated environment of end-to-end network service capability with artificial intelligence (AI), IoT, and Software Defined Network (SDN), these technologies guaranteed to transfer the data with a high service rate and provided the service level agreement (SLA) as shown in Figure 2.

Figure 2. Upcoming technologies integrated environment

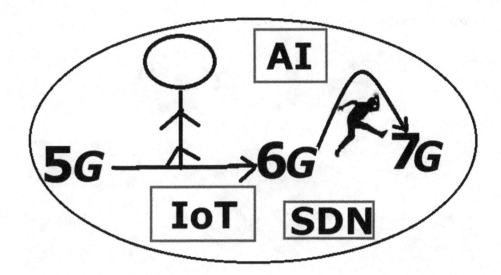

The network services will be enhanced from 6G to 7G. IoT changed the way we want to live upon and automate our lifestyle. IoT large spread usage has the potential to change our daily lives. The IoT is very useful for boosting city smartness and infrastructure in general Yadav et al. (2021). 6G operates a higher energy radio spectrum than 5G, 6G is sufficient for "smart house" as per security and privacy 6G also not sufficient, network-enabled medical gadgets looking for 6G/7G gadgets find applications in healthcare Mihret et al. (2021). IoT optimizes the connected devices in terms of energy consumption and low cost. IoT assists elderly people and hearing-impaired persons by providing smart gadgets which facilitate voice recognition. Under IoT enabled smart transportation, connected cars, bridges embedded with smart sensors give us new ways to deal with daily life problems like huge traffic and insufficient space for parking by providing real time information about the movement of traffic to peoples Adebusola et al. (2020). IoT faces security issues as all the things are vulnerable and anyone can hack it for illegal purposes. The evolution and concept of smart things required for 7G, as well as the issues associated with IoT required for 7G, will be discussed in the paper Altameem et al. (2022). Ganai et al. (2022) explained IoT 4G implementation in cyber security in healthcare sector. Author explained the detailed analysis of cyber security issues in healthcare sector using IoT 4G. Qadir et al. (2021) explained 6G internet of Things architecture, issues and its detailed investigation towards IoT applications. Nguyen et al. (2022) discussed 6G implementation, future scope and its use cases in IoT applications. Sandeep et al. (2023) discussed current generation 4G and 5G to 6G analysis covered in this paper. Salameh et al. (2022) discussed 5G to 6G challenges, technologies and its applications. Sandeep et al. (2023)

discussed implementation of IoT 4G in automatic seat identification system in smart transport using image processing. Sandeep et al. (2023) explained the importance of LoRa WAN in IoT 4G, 5G and 6G communication. Gautam et al. (2023) discussed IoT 3G communication in gas leakage detection to reduce causalities caused by gas leakage.

MOBILE COMMUNICATION EVOLUTION 1G TO 5G

Figure 3. Mobile communication evolution 1G to 5G

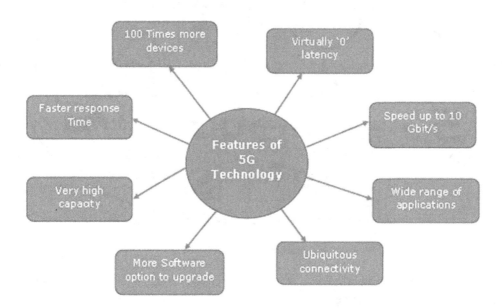

In the initial part of the 1980s, the first generation, or 1G, was released. It was distinguished by its ability to transmit voice over analogue technology. It had limitations even though it was a big advancement at the time. For instance, it lacked data service to turn voice into digital signals, had subpar voice quality, and lacked worldwide roaming service Prinima et al. (2016).

The second generation, or 2G, was launched in the late 1990s, and it improved speech quality and increased data rate capacity. Global System for Mobile Communications (GSM) was a digital standard used during this generation that included services like Short Message Service (SMS), Multimedia Message Service (MMS), which was fueled by the introduction of devices with color screens, and Wireless Application Protocol (WAP), which permitted Internet access services using a mobile device. Although though such multimedia programs used a lot of energy, a notable benefit of 2G mobiles was that the battery lasted longer than it would on modern smartphones because radio transmissions actually use very little power.

In the late 2000s, the third generation, or 3G, first emerged. It introduced the first truly wireless data, enabling users to access the entire Internet. High data transmission speeds, a key component of 3G technology, enabled the creation of sophisticated multimedia applications. Also, the disruption of services previously inaccessible to mobile devices, such as web surfing, email access, TV streaming, video conferencing, and GPS (Global Positioning System), among others, was made possible by the new frequency bands and position data Prinima et al. (2016). The 3G generation was notable for the consumer market because of its wide range of uses, but it also increased the cost of 3G devices and increased their energy consumption, making them more power-hungry than the majority of 2G models Gupta and Jha (2015).

The fourth generation, or 4G, was released in 2010 and is still in use today. It is completely based on the Internet Protocol (IP). The primary objective of 4G technology is to deliver high-quality, secure, affordable services, multimedia, and Internet via IP at significantly higher data rates than previous generations Prinima et al. (2016). In particular, 4G provides the widely adopted high-speed wireless broadband currently in use, enabling mobile video and cloud applications like video games, high-definition mobile streaming, and 3D television to reach their full potential.

Today's 4G technology supports the development of the Internet of Things by providing consumer data rates in megabytes order, latency in milliseconds order, and device density for about 2000 connected devices per square kilometer globally (IoT). Despite these capabilities, 4G would be replaced by the next generation (5G) by the beginning of the next decade due to an exponential rise in demand, new mobile telecommunication breakthroughs, and other factors, according to Sawanobori et al. (2016) as shown in Figure 3.

New network and service capabilities will be introduced with the advent of 5G. It will guarantee continuity, greater data rates, lower latency, a massive number of simultaneous connections, and network ubiquity everywhere in the world Prinima et al. (2016), even in environments where current 4G technology struggles, like high mobility (for example, in trains) and extremely dense or sparsely populated areas (e.g., stadiums, shopping malls).

Moreover, 5G will be a crucial facilitator for a true Internet of Things, offering a platform to connect an enormous number of sensors and actuators with strict energy efficiency and transmission restrictions 5G-PPP (2020).

Figure 4. 5G features
Source: Mihret et al. (2021)

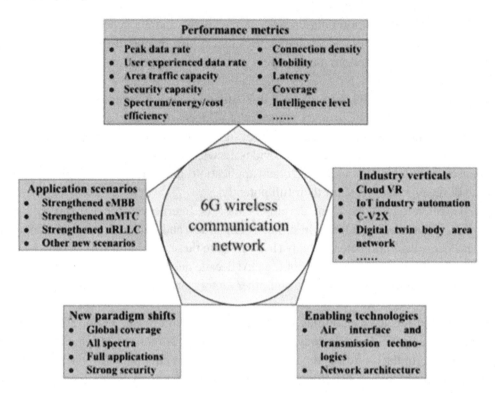

Industry and academic efforts are concentrated on defining the specifications for 5G services, herald-ing the beginning of the 5G era Sawanobori et al. (2016). This effort is motivated by the unprecedented growth in the number of connected devices, mobile data traffic, and the limitations of the 4G technolo-gies to address this enormous data demand. The ability of a 5G device to retain network connectivity at all times and locations will allow for the connection of all the network's devices. In order to do this, the fundamental 5G system design is anticipated to allow up to one million simultaneous connections per square kilometer, enabling the launch of a number of cutting-edge IoT service concepts Samsung 5G Vision (2015). 5G have more than 100 times devices to connect, have almost no latency, 10Gbit/s speed, quick response, easy upgradation and diverse applications as shown in Figure 4. Through 5G, IoT applications can achieve higher degree of communication.

The Internet of Things (IoT) is a recent digital communication paradigm that enables everyday things to connect with one another and with users Samsung 5G Vision (2015). As a result, the IoT seeks to broaden the concept of the Internet and make it more immersive by facilitating simple interaction with a variety of objects, including household appliances, security cameras, industrial actuators, traffic lights, and cars, among others Samsung 5G Vision (2015). Data are being generated and gathered in this con-text from the enormous number of connected devices. The management of various forms of data, in accordance with needs, and the creation of more valuable services are both significantly aided by the integration of Cloud Computing and Big Data technology Yan et al. (2020).

Such technologies are essential to ensuring the Smart City IoT paradigm in urban settings. It satisfies the demand for information and communications technologies (ICT) solutions to be adopted by the majority of national governments for the management of public affairs Yan et al. (2020).

Instead of settling for average speeds, the 6G mobile network/Internet uses an air fiber technology with masts and transceivers on towering structures and even lampposts to form a local network that can offer astounding speeds. The optimum way to transfer highly secure information from transmitters to destinations will be via an air fiber combination. The integrated network of 6G offers locals a really all-in-one solution: 6G can handle all of your business's telecommunications demands, including uploads, downloads, super-fast broadband Internet, multiple line telephones, CCTV monitoring, and video conferencing.

Satellites will be integrated into 6th generation wireless mobile communication networks to provide worldwide coverage. Four nations have worked together to construct the systems for worldwide coverage. America developed the global positioning system (GPS), China developed the COMPASS system, EU developed the Galileo system, and Russia developed the GLONASS system. These autonomous mechanisms make space travel challenging. 7th generation (7G) wireless mobile communication networks will collaborate on a single mission.

In order to support smart communities, this article analyses new 6G connection technologies, including holographic beamforming, edge computing, backscatter communications, and IoT networks powered by artificial intelligence. However, a number of potential future research avenues for 6G-based IoT networks are also mentioned.

By conducting an in-depth analysis of the 6G and IoT convergence, we investigate the new potential presented by 6G technologies in IoT networks and applications in this article. We first discuss some of the most fundamental 6G technologies, such as edge intelligence, reconfigurable intelligent surfaces, space-air-ground-underwater communications, Terahertz communications, massive ultra-reliable and low-latency communications, and blockchain, which are anticipated to power future IoT networks.

We provide an in-depth discussion of the roles of 6G in a wide range of potential IoT applications via five key domains, namely Healthcare Internet of Things, Vehicular Internet of Things and Autonomous Driving, Unmanned Aerial Vehicles, Satellite Internet of Things, and Industrial Internet of Things. This is especially true when compared to the other related survey papers.

Mobile call rates will be relatively expensive in 6G, but they will be lower in 7G/7.5G/8G, which will assist lower-level users. With the aid of 6G technology, fly sensors will be used to paint the globe. These fly sensors will transmit data to their remote observer stations, which will also keep an eye out for any activity in a particular region, such as that of intruders or terrorists as shown in Figure 5. 6G expected to launch in 2030, some European countries supposed to launch 6G first to increased network speed.

6G will support latest IoT technologies like Artificial intelligence, blockchain, Robotics, Machine learning and cyber security.

Figure 5. 6G architecture

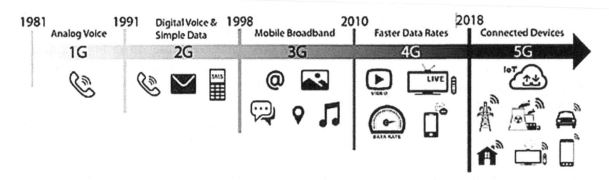

6G networks will rely on new enabling technologies, such as air interface and transmission technologies and novel network architecture, such as waveform design, multiple access, channel coding schemes, multi-antenna technologies, network slicing, cell-free architecture, and cloud/fog/edge computing, to meet efficiency, better intelligence level, and security requirements as shown in Figure 6.

For global coverage, the 7G mobile network is similar to the 6G. The next generation of mobile communication will be the most advanced, but there will still be research on challenging topics like using a mobile phone while travelling from one country to another because satellites are also moving at a constant speed and in a specific orbit, as well as the standards and protocols for cellular to satellite and satellite to satellite communication systems.

Only once all standards and protocols have been established can the 7G vision become a reality. This might be achievable in the generation that comes after 7G and might be called 7.5G or 8G. There won't be any problems with data capacity coverage or hand-off when 7G fixes all of its flaws. The cost of a mobile phone call and its services will be the sole demand from the user at that time. This problem will once more spark an evolution in standards and technology and open up new directions for computing research. This new technological advancement in mobile phone call rates and services will be known as 7.5G or 8G.

The Internet of things can find applications in rescue operations through deployment of autonomous military robots for the detection of the human body. This robot is equipped with the dedicated sensors to perform the operation of sensing the human presence and simultaneously send its current location to remote servers through RFID antenna/ sim card. So, a more specific application of IoT can be found in military operations where smart robots can do rescue operations Mondal et al. (2015).

In Swamy et al. (2023), the Internet of things can find its applications in home automation through voice control and head movement disabled person can control the home appliances and this makes it easier to use devices which can't be controlled manually.

In Sandeep et al. (2011), the Internet of things can be used in the healthcare sector to protect more service security. In medical hospitals, healthcare related information is vulnerable, always there is a risk of data leak. In this paper the author performs a detailed investigation of the internet of things execution in the healthcare industry to increase cyber security and protect data leakage.

In IoT the captured data from sensors need to be communicated to a remote distance via some networking technologies like 3G and 4G onwards. With the increasing number of applications and connected devices in IoT there is a demand for faster communication technology like 5G, 6G and beyond to achieve higher throughput and low latency Sandeep et al. (2015).

CHALLENGES ASSOCIATED WITH THE IMPLEMENTATION OF TECHNOLOGIES IN IoT

How 5G Will Affect the Internet of Things

A speedier communication channel will be provided by 5G; you may anticipate rates of up to a few gigabits per second. Your gadgets may work together and complete tasks more quickly as a consequence. Also, it will offer a network with extremely low latency; early 5G deployment, according to Verizon, demonstrated a latency of 30 ms. That will make it easier to do delicate procedures like surgery using IoT devices. Now, you may connect more devices to 5G without experiencing quality reduction due of its tremendous bandwidth.

5G IoT Challenges

In fact, 5G will have a significant impact on how various businesses operate. Before completely deploying them, however, researchers need to overcome a number of issues.

1. Because 5G uses short waves, you require towers close to one another. Thus, a 5G network requires more cellular towers.
2. There are heightened security threats due to the fact that many 5G components will be virtual. A network will also require more stringent security measures as its user base grows.
3. Installing new network gear will be quite expensive. You need more apparatus, and it must be able to operate in the high-frequency band. But, if vendors share hardware, you might be able to save these expenses.

Figure 6. 6G network paradigm

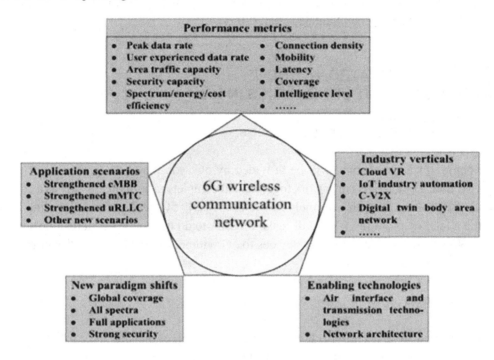

Uses for 5G Internet of Things

There will be many uses for 5G and IoT:

- Instead of going to a parking space, people can utilize a 5G IoT network to park their cars. Similar to that, you can call it directly.
- Farmers may remotely control equipment and monitor crops and livestock via the IoT network.
- Remote surgery can be carried out by doctors using low latency technology.

Your amusement may also be enhanced by the minimal latency. You could play AAA games without installing them, for instance.

- Homeowners can keep an eye on and manage their properties while they are away. You could monitor your robotic vacuum while lying on the beach in the sun, for instance as shown in Figure 7.

Figure 7. 6G network
Source: Telecom.com

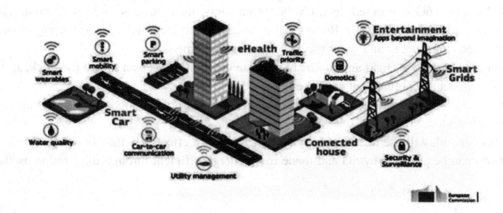

6G Internet of Things and Effect

The Internet of Things (IoT) and sixth generation (6G) wireless communication networks are expected to transform customer services and applications and pave the way for completely intelligent and autonomous systems in the future as shown in Figure 8.

Figure 8. Wireless network development leading to future 6G-IoT
Source: Nguyen et al. (2022)

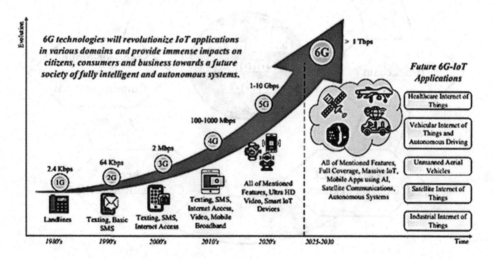

The potential for 6G to change how people connect with the physical, digital, and human worlds. 6G is expected to offer artificial intelligence (AI), augmented reality (AR), metaverse, and virtual reality (VR) by 2030. The improved information that sensors, AI, machine learning (ML), and digital twins will be able to convey to us in real-time will have a significant impact on how we experience and function in everyday life.

The Internet of Things (IoT) will manage home appliances, lower electricity use, and improve automated manufacturing, connected cars, drone agriculture, and other areas.

With the help of 6G, connected electric vehicles, cameras, and roadways could all communicate with one another to improve traffic flow. Robots and connected equipment will manage supply chains more effectively, reducing energy and water use as well as carbon emissions.

Sensors can be used in smart agriculture to regulate water flow, keep an eye on livestock, and enable precise pesticide application to cut carbon emissions.

Smart grids and 6G could help with the transition to renewable energy sources and improve how energy is distributed.

The 6G network will use less electricity and operate more effectively than the 5G network. Future applications can be powered by 6G and made more energy-efficient through digitization as shown in Figure 9.

Figure 9. 6G IoT enabling technologies
Source: Nguyen et al. (2022)

With the exceptional characteristics listed below, 6G may raise network quality to a new level.

- Obtaining a very high data throughput starting at 1 Tb/s to handle huge IoT connectivity where seamless mobility, spectrum availability, and mobile traffic coexist.

- Raising mobile traffic capacity to 1 Gb/s/m2 in order to meet needs for extremely high throughput and IoT device density.
- Reaching a very high device connectivity density of 107 devices/km2, which is beneficial for the construction of exceptionally dense IoT networks.
- Obtaining incredibly low network latencies (10–100 s) to meet the demands of haptic applications like e-health and autonomous driving. Figure 9 shows 6G IoT enabling technologies like Blockchain, Edge Intelligence, Intelligent surfaces and THz communication.

6G Impact on Communication

- We can be more productive at work and at home since 6G will be able to transport even more data from IoT.
- Phones could eventually replace our keys and cash.
- Voice or movement can take the place of typing.
- Healthcare could change as a result of wearables, telesurgery, and implanted sensors.
- Online conference calls using holograms might be successful.
- Connected cars will be able to intervene in a collision.
- Network signals have the ability to detect our location and the environment around us. When paired with AI, ML, and digital twinning, they will also be able to detect radio frequency (RF) signals.
- By using satellites to reach rural areas that do not currently have Internet connection, 6G could ultimately provide global connectivity while reducing the need to build cell towers.

7G INTERNET OF THINGS

The unavoidable sophisticated cellular technology (7G), which will successor 5G and 6G and be able to offer significantly better capacity, higher frequencies, and significantly lower latency in communications, will take the place of 5G and 6G. 7G will be able to satisfy the requirements of extremely high bandwidth, almost zero latency, and universal integration. Society always demands more in terms of digital capacity, speed, connectivity, and total capability for wireless and mobile computing.

The next-generation (7G) networks will provide the norms for data collection and communication and are anticipated to significantly advance imaging, data analytics, location awareness, AGI (Artificial General Intelligence), and quantum and HPC-based next-generation computation capabilities. Also, the 7G computational infrastructure will be able to decide on its own where is the best place for computing to take place. This will primarily be the mobile device itself, but it will also be possible for informational computing to occur through other electronic devices. The degree of capacity in 7G networks will be unparalleled, and it will expand the capabilities and scope of 7G applications to allow ever-smarter and more inventive wireless comprehension, sensing, and imaging applications.

Higher frequencies provided by 7 G will significantly increase throughput.

By the time 7G networks are deployed, edge and core computing will be considerably more smoothly linked as a part of a combined communications/computation infrastructure framework. Once 7G technology is put into use, this will offer a number of benefits, including better accessibility to AGI capabilities. with a significant increase in data produced by 7G networks.

7G will have significant effects on public safety and tackling important concerns like:

- Threat recognition
- Crime prevention
- Reading minds
- Tracking one's health
- Preparing for emergencies
- Gas and toxicity sensing
- IoT device administration

METHODOLOGY

Smart items are causing our society to have an autonomous and intelligent nature. IoT-based smart linked objects are a group of interconnected devices that can be monitored and managed from a central processing workstation, or hub site, and they offer online services to apps Gawas et al. (2015). Mobile edge networks help develop a 6G network for IoT-enabled smart devices, which enhances access to IoT, SDN, and AI capabilitiesMassive URLL communications as shown in Figure 10. 6G is anticipated to be available by 2030 Gawas et al. (2015). In the event that intelligent items automate the objects, study will begin in 6G 2020. Smart shoes and phones, as well as smart refrigerators, washing machines, and other household items, are all on the horizon, but security breaches are also a worry with IoT enabled apps Vaigandla et al. (2021).

Capabilities Massive URL in Future IoT and Current 5G and Next 6G and 7G

In the context of IoT and future wireless technologies, longer and more complex URLs may be used to convey detailed information, transmit sensor data, or control and manage IoT devices as shown in Figure 10. However, the actual capabilities and specific usage of URLs will depend on the evolution of internet protocols, communication standards, and the requirements of the applications built on top of these technologies.

It's important to note that the evolution of wireless technologies is a complex and ongoing process, and the specific features and capabilities of 6G and 7G are still being researched and defined.

Figure 10. The future IoT and current 5G and next 6G and 7G technology
Source: tutorialspoint.com (2023)

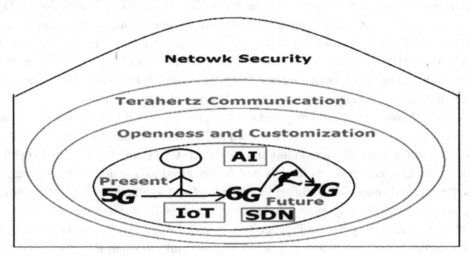

Table 1. 1G to 8G comparative analysis

Technology Generations	Features								
	Year	Data Rate	Enabled Technology and services	Type of Switching	Multiplexing	Core Network	Frequency	Handoff	Location of First Commercialization
1G	1970-1980	2.4Kbps	Cellular Analog, Voice only and AMPS	Circuit	FDMA	PSTN	Few thousand Kbps	Horizontal	Tokyo
2G	1980-1990	64Kbps	GSM, EDGE, GPRS, Digital Voice and short messaging and Cellular Digital	Circuit and packet	TDMA and CDMA	PSTN and packet Network	Few thousand Kbps	Horizontal	Finland
3G	1990-2000	2Mbps	Integrated high quality audio, video and Broadband IP, CDMA 2000, WCDMA, UMTS	Packet except circuit for air interface	CDMA	Packet Network	Few Mbps	Horizontal and Vertical	Japan
4G	2000-2015	200Mbps	Unified IP and combination of LAN, WAN, PAN, WLAN, LTE, VoIP and WiMAX, Wearable devices	Packet Switching	OFDMA	Internet	2-6 GHz	Horizontal and Vertical	South Korea
5G	2015-2026 (Predicted)	1Gbps and higher	4G+WWWW, LAS-CDMA, OFDM, MC-CDMA and IPv6, Wearable devices with AI	IPv6 but still up-dation pending	MIMO, CDMA	Internet	4G Frequency	Horizontal and Vertical	Mexico
6G	2027-2035 (Predicted)	10-11Gbps	GPS, Galileo systems, 5G+ Satellite, Ultra-fast internet access	All Packet	CDMA	Internet	90 GHZ-3.5THz	Horizontal and Vertical	South Korea by 2028
7G	Beyond 2035 (Predicted)	About 12+ Gbps	Secured and global cellular services, Ultra-fast internet access	All packet	CDMA	Internet	90 GHZ-3.5THz	Horizontal and Vertical	Norway, USA, China and Japan by 2035-2040
8G	Beyond 2040 (Predicted)	About 20 Gbps	Secured and global cellular services, Ultra-fast internet access	All Packet	CDMA	Internet	Beyond 4THz	Horizontal and Vertical	

Source: Sandeep et al. (2011)

As shown in Table 1, with an integrated environment of IoT, AI, and SDN with the layer of openness and customization, we are currently moving from 5G to 6G to 6G to 7G in the near future. This technology is becoming more and more innovative for all industries, and openness and customization support agile services and API interfaces Khutey et al. (2015). It is surrounded by terahertz communication in the 100 GHz to 10 THz frequency range that was used in the 6G to 7G period. This communication has no restrictions and has never been used before. Quantum key distribution (QKD) and post-quantum cryptography are provided via network security, as demonstrated in Figure 10.

Examples of IoT interaction and integration in the global Internet include the integration of IPv6 and global interoperability, as well as IoT and Cloud integration. IoT links billions of intelligent objects worldwide while respecting each one's individual constraints ÖZTÜRK et al. (2021). In the future years, new products, services, and business models will be produced as a result of the interaction between businesses, people, and smart things. Smart things are defined as autonomous physical or digital entities having sensing, analysis, responding, and networking capabilities Guevara et al. (2020). Our lives are made more remarkable by the intelligence that is added to everyday objects in our environment. The world of connected devices is now at your fingertips thanks to smart things Prinima et al. (2016). Traditional gadgets are even less intelligent, convenient, secure, safe, and efficient than smart things. These make it simple to connect tangible objects to the internet to automate, monitor, control, and have fun with them from any position, at any time, above any network, and by anybody. Sensors, GPS, Wi-Fi, Bluetooth, ambient intelligence, EPC, barcode, and telemedicine are some of the embedded Internet of Things technologies that can help with this smartness Gupta et al. (2015). Smart refers to any physical entity that is connected to the internet that has certain sensing capabilities. Its primary objectives are as follows:

- Identify users and their social relationships Sawanobori et al. (2016).

  Gain access to the info of the user 5G-PPP (2020).

- Determine the user's social context by evaluating into their framework, interests, and attributes Samsung 5G Vision (2015).
- Infer social aspirations focuses on the cultural environment Zanella et al. (2014).
- Behavior coordination Nassar et al. (2017).

As we advance year by year, we are entering a smarter world. A group of interconnected devices known as "Smart Things" can be monitored and managed online and by a central CPU 5G-PPP (2020). In other words, Big Data Analytics (BDA) is assisting in the processing of better decisions as smart things transform the physical world into a massive information system Samsung 5G Vision (2015). Notwithstanding the fact that it wasn't acknowledged until 1999 Zanella et al. (2014). One of the very first internet-connected devices to exist was the Coke machine. A server application was built by programmers working several feet above the vending machine to keep track of how long a storage column had been empty Nassar et al. (2017). The programmers could use the Internet to connect to the machine, check its condition, and see whether there was a cold drink waiting for them if they went down to it. Now in the current situation in the world, IoT is gathering momentum and people look to it as a future Yan et al. (2020). Various forecasts around smart gadgets or innertubes things have been created based on numerous evaluations. Ac-cording to the "Internet of Things" paper in 2025, there might be up to 16 billion linked devices, on average, each person on the earth owns about 6 gadgets Lee et al. (2017). IMS research estimates that by 2028, there will be 22 billion Internet-connected gadgets Srinivasan et al. (2022).

FINAL DISCUSSION

The IoT poses substantial hurdles that could prevent it from attaining its full potential Ziegler et al. (2013). As per the network security aspects 5G not sufficient 6G/7G leads the integrated environment, as well as developing economies and development, are among the obstacles.

Security: Many IoT deployments face new and distinct security problems Yadav et al. (2021). These obstacles should be resolved in terahertz communication, as should ensuring the security of IoT goods and services. 6G/7G provides security components of the IoT Mihret et al. (2021).

Privacy: Approaches that respect individual privacy preferences over a wide range of expectations are necessary for the IoT to reach its full potential. IoT consumers can benefit greatly in openness customization Adebusola et al. (2020).

In the world of IoT applications, one of the most overused buzzwords is "smart connected things". These applications have a direct impact on our daily lives right from the wake up until we go to bed late at night. Smart Phones have invaded even manmade structures, enabling these applications to be used. Most of us are connected with technological equipment like servers, computers, tablets, phones, and most of the time have access to the internet, so we need to have faster technology driven by 5G and beyond communication technology Altameem et al. (2022).

Integrated environment: The capacity of two or more systems to share data and subsequently use that data is referred to as interoperability. Because the IoT will contain a variety of heterogeneous devices, interoperability is essential, networks and systems must all function together to build an IoT Gawas et al. (2015). Various integrated applications like threat detection, facial recognition, knowledge based and decision-making systems require a 6G/7G approach Vaigandla et al. (2021). There are various aspects also to look up to for things to move successfully ahead. If developing countries are to prosper from IoT, they will need to respond as well Khutey et al. (2015).

The Internet of things will be driven by advanced cellular networks 4G, 5G and beyond 6G. There is a need to have faster communication technology which has low latency, higher throughput and low cost ÖZTÜRK et al. (2021).

Given the pace of technological advancements, it is expected that wireless technologies will continue to evolve beyond 6G. However, at this point, it is challenging to predict the specific features and capabilities of 7G as it is still a subject of ongoing research and development.

Some potential areas of focus for 7G technology could include even higher data rates, extremely low latency, further improvements in network reliability and capacity, enhanced energy efficiency, and the ability to support advanced applications that may include futuristic technologies such as advanced artificial intelligence, immersive virtual reality experiences, massive-scale IoT deployments, and highly reliable autonomous systems.

It's worth noting that the development and deployment of new wireless generations take time, typically spanning several years or even decades. Extensive research, standardization efforts, and advancements in underlying technologies are required before a new generation can be realized and commercially deployed. Therefore, it will likely be some time before we see concrete details and specifications for 7G wireless technology.

IMPACT OF 4G, 5G, 6G, AND 7G ON IoT

The Internet of Things (IoT) has been greatly impacted by 4G, 5G, 6G, and 7G since they make it easier to transport data from sensors to far-off locations than it was with old transmission methods. Using new technology generation higher data rate with low latency and high throughput can be achieved as mentioned in Table 1 and Table 2 shows merits and de merits of all technology upgrade.

Table 2. 4G, 5G, 6G, and 7G comparison on the basis of benefits and limitations

Technology Generations	4G	5G	6G	7G
Merits	High speed, High Handoff, MIMO based Technology, Global Technology	Large coverage area, low power consumption, more spectral efficiency and high security	Global coverage	Less cost of call, Less handoff and no issue of data handling capacity
Demerits	Complicated and more hardware requirement and hard to execute	Still under observation, trials are going on, hard to say as incompetent technological support across the countries, infrastructure development high cost, security breaches and privacy issues need to work upon	High cost of call and similar 5G disadvantages	Similar 5G and 6G issues

Source: Khutey et al. (2015)

The evolution of wireless communication technologies such as 4G, 5G, 6G, and potentially 7G will have a significant impact on the Internet of Things (IoT) ecosystem. Here's a breakdown of how each generation may influence IoT:

1. 4G (LTE)

4G brought considerable improvements in terms of network speed, capacity, and latency compared to its predecessor, 3G. It enabled the growth of IoT applications by providing reliable and faster connections. However, 4G has certain limitations in terms of supporting massive numbers of connected devices, especially in dense IoT deployments.

2. 5G

5G is designed to be a game-changer for IoT. It offers significantly higher data rates, reduced latency, increased capacity, and improved network reliability. These features make 5G well-suited for supporting a vast number of IoT devices simultaneously. It enables real-time communication, enhances the performance of IoT applications, and opens up possibilities for new use cases such as autonomous vehicles, smart cities, and industrial automation.

3. 6G (Expected)

While 6G is still in the early stages of development, it is expected to further revolutionize IoT. Some potential features and advancements being explored for 6G include terahertz frequencies, enhanced data rates, ultra-low latency, improved energy efficiency, and advanced device-to-device communication capabilities. These improvements could enable even more demanding IoT applications, such as immersive augmented reality (AR)/virtual reality (VR), holographic communication, and ubiquitous sensing.

4. 7G (Potential)

As of my knowledge cutoff in September 2021, 7G is a speculative concept that extends beyond current research and development. It is difficult to predict the specific impact of 7G on IoT accurately. However, it is conceivable that 7G might continue to push the boundaries of network performance, offering even higher data rates, lower latency, and seamless connectivity. It could potentially enable transformative IoT applications that require unprecedented levels of speed, reliability, and efficiency.

In summary, each generation of wireless technology brings advancements that enhance the capabilities of IoT. While 4G laid the foundation, 5G is a significant enabler for IoT, and 6G and beyond hold the promise of unlocking new frontiers and supporting even more advanced and immersive IoT use cases.

CONCLUSION AND FUTURE SCOPE

The new diversity applications 6G/7G support digital manufacture, operation, and administration, it provides flexibility, redundancy, self-healing, network reliability, and stability. Smart Things are playing an extremely integral part of everyday lives, and the possibilities are endless. The best thing about the Internet of Things is that it improves people's quality of life, increases operational efficiency, and handles circumstances where human involvement is impossible. The Internet of Things, on the other hand, confronts severe difficulties that could stymie its potential benefits. We will need to place more emphasis on this.

The future IoT and 6G/7G technologies provide low-carbon transformation towards the ICT industry. It improves the network throughput and less resource consumption. And provide end-to-end AI, IoT, and SDN integrated capability with open-source cloud infrastructure.

Mobile wireless communication is an emerging concept and over many years it has fast growth. Efforts are going on to reduce the various technologies to one standard resulting in 5G to 7G revolution. 5G looking towards wireless world with no limitation and 6G integrates 5G to provide faster communication with higher data rate, low latency and higher throughput. With 6G handoff is the matter of concern leading the emergence of 7G.

Since last year, wireless technology has grown significantly. At 7.5G, wireless technology has attained its progression and revolution. Future Generation (FG) wireless technology will feature faster data transmission speeds than 6G and 7G. Advanced techniques are constantly evolving in all sectors of mobile and wireless communications, making wireless technology one of the industries that is growing at a rapid rate. The rollout of 5G mobile communication technology is now in its infancy.

The world tends to move towards wireless completely where uninterrupted and unlimited access to internet and data resources with good quality, faster speed, higher bandwidth and less cost.

In the future, the terminal will modify the air interface to accommodate the available radio access technology; now, the infrastructure handles this. Also, the lack of competent technological support (ancient gadgets) throughout the majority of the world is the biggest obstacle to the success of future technologies.

So, new ones must be installed in place of all of them. In general, building infrastructure for 5G, 6G, and 7G is expensive, and there are still problems with security and privacy. Finally, it should be mentioned that after the introduction of the 1G cellular network in 1981, the 2G generation followed in 1992, the 3G generation followed in 2001, and the 4G generation entered the market in 2012-2013. Hence, 5G will launch in 2020, followed by 6G and 7G in 2030 and 2040, respectively.

The current generation of wireless technology, 5G, has brought significant advancements in terms of data transfer speeds, low latency, and the ability to connect a massive number of devices simultaneously. These capabilities have facilitated the growth and expansion of the Internet of Things (IoT) by enabling seamless communication between IoT devices and supporting emerging technologies like autonomous vehicles, smart cities, and augmented reality.

Looking ahead, the evolution from 5G to 6G and beyond holds even greater potential for the future of IoT. While 6G is still in its early stages of development, it aims to provide even higher data rates, lower latency, and improved reliability compared to its predecessor. 6G is expected to support new use cases such as holographic communication, advanced virtual reality experiences, and remote tactile interactions. It may incorporate technologies like terahertz frequencies and advanced antenna systems to achieve these goals.

Beyond 6G, the concept of 7G is still speculative, and there is limited information available. However, it is anticipated that 7G will surpass the capabilities of 6G, offering even faster speeds, ultra-low latency, and advanced features that we can only imagine at this point.

In the realm of IoT, these advancements in wireless technologies, from 5G to 6G and potentially 7G, will likely enable more sophisticated and interconnected IoT systems. Longer and more complex URLs may be utilized to convey detailed information, transmit sensor data, or control and manage IoT devices.

However, it's important to remember that the evolution of wireless technologies is an ongoing process, and the specific features and capabilities of 6G and 7G are still being researched and defined. As these technologies continue to develop, they will shape the future of IoT, revolutionizing industries, enhancing connectivity, and enabling innovative applications that were previously unimaginable.

REFERENCES

Adebusola, J. A. (2020). An Overview of 5G Technology. In *2020 International Conference in Mathematics, Computer Engineering and Computer Science (ICMCECS)*. IEEE. 10.1109/ICMCECS47690.2020.240853

Altameem, A., Mallikarjuna, B., Saudagar, A. K. J., Sharma, M., & Poonia, R. C. (2022). Improvement of Automatic Glioma Brain Tumor Detection Using Deep Convolutional Neural Networks. *Journal of Computational Biology*, 29(6), 530–544. doi:10.1089/cmb.2021.0280 PMID:35235381

Bhatia, S., Dhillon, H. S., & Kumar, N. (2011). Alive human body detection system using an autonomous mobile rescue robot. In *2011 Annual IEEE India Conference*. IEEE 10.1109/INDCON.2011.6139388

Bhatia, S., Gautam, D., Kumar, S., & Verma, S. (2023). Automatic Seat Identification System in Smart Transport using IoT and Image Processing. *2023 3rd International Conference on Intelligent Communication and Computational Techniques (ICCT),* 1-6. 10.1109/ICCT56969.2023.10075664

Bhatia, S., Jaffery, Z. A., & Mehfuz, S. (2023). A Comparative Study of Wireless Communication Protocols for use in Smart Farming Framework Development. *2023 3rd International Conference on Intelligent Communication and Computational Techniques (ICCT),* 1-7. 10.1109/ICCT56969.2023.10075696

Bhatia, S., Mallikarjuna, B., Gautam, D., Gupta, U., Kumar, S., & Verma, S. (2023). The Future IoT: The Current Generation 5G and Next Generation 6G and 7G Technologies. *2023 International Conference on Device Intelligence, Computing and Communication Technologies (DICCT),* 212-217. 10.1109/DICCT56244.2023.10110066

Ercüment, Ö., & Tuğrul, Ç. (2021). A Survey on The Future of Communication Technologies: IoT, 6G, 7G and Beyond. In *1st International Conference on Electrical-Electronics and Computer Engineering.* Avrasya University.

Ganai, P. T., Bag, A., Sable, A., Abdullah, K. H., Bhatia, S., & Pant, B. (2022). A Detailed Investigation of Implementation of Internet of Things (IOT) in Cyber Security in Healthcare Sector. *2022 2nd International Conference on Advance Computing and Innovative Technologies in Engineering (ICACITE),* 1571-1575. 10.1109/ICACITE53722.2022.9823887

Gautam, D., Bhatia, S., Goel, N., & Mallikaijuna, B. (2023). Development of IoT Enabled Framework for LPG Gas Leakage Detection and Weight Monitoring System. *2023 International Conference on Device Intelligence, Computing and Communication Technologies (DICCT),* 182-187. 10.1109/DICCT56244.2023.10110294

Gawas, A. U. (2015). An overview on the evolution of mobile wireless communication networks: 1G-6G. *International Journal on Recent and Innovation Trends in Computing and Communication, 3*(5), 3130–3133.

Guevara & AuatCheein. (2020). The Role of 5G Technologies: Challenges in Smart Cities and Intelligent Transportation Systems. *Sustainability, 12*(16), 6469. doi:10.3390/su12166469

Gupta, A., & Jha, R. K. (2015). A Survey of 5G Network: Architecture and Emerging Technologies. *IEEE Access : Practical Innovations, Open Solutions, 3,* 1206–1232. doi:10.1109/ACCESS.2015.2461602

Jeet, V., Dhillon, H. S., & Bhatia, S. (2015). Radio Frequency Home Appliance Control Based on Head Tracking and Voice Control for Disabled Person. *2015 Fifth International Conference on Communication Systems and Network Technologies,* 559-563. 10.1109/CSNT.2015.189

Khutey, R. (2015). Future of wireless technology 6G & 7G. *International Journal of Electrical and Electronics Research, 3*(2), 583–585.

Lee, S. K., Bae, M., & Kim, H. (2017). Future of IoT networks: A survey. *Applied Sciences (Basel, Switzerland), 7*(10), 1072. doi:10.3390/app7101072

Mihret & Haile. (2021). 4G, 5G, 6G, 7G and Future Mobile Technologies. *J Comp Sci Info Technol, 9*(2), 75.

Mondal, S., Sinha, A., & Routh, J. (2015). A survey on evolution of wireless generations 0G to 7G. *International Journal of Advance Research in Science and Engineering*, *1*(2), 5–10.

Nassar, A. S., Montasser, A. H., & Abdelbaki, N. A. (2017). Survey on Smart Cities' IoT. *Proceedings of the International Conference on Advanced Intelligent Systems and Informatics*, 9–11.

Nguyen, D. C., Ding, M., Pathirana, P. N., Seneviratne, A., Li, J., Niyato, D., Dobre, O., & Poor, H. V. (2022). 6G Internet of Things: A Comprehensive Survey. *IEEE Internet of Things Journal*, *9*(1), 359–383. doi:10.1109/JIOT.2021.3103320

Prinima, D., & Pruthi. (2016). Evolution of Mobile Communication Network: From 1G to 5G. *Int. J. Innov. Res. Comput. Commun. Eng.*, *4*, 224–227.

Qadir, Z., Munawar, H. S., Saeed, N., & Le, K. N. (2021). *Towards 6G Internet of Things: Recent Advances*. Use Cases, and Open Challenges.

Salameh, A. I., & El Tarhuni, M. (2022, April). From 5G to 6G—Challenges, Technologies, and Applications. *Future Internet*, *14*(4), 117. doi:10.3390/fi14040117

Samsung. (2015). *5G Vision* (White Paper). Samsung Electronics Co.

Sawanobori, T. K. (2016). *The Next Generation of Wireless: 5G Leadership in the U.S.* (White Paper). CTIA Everything Wireless. Available online: https://5g-ppp.eu/wp-content/uploads/2015/02/5G-Vision-Brochure-v1.pdf

Srinivasan, R. (2022). A Comparative Study: Wireless Technologies in Internet of Things. In *2022 2nd International Conference on Advance Computing and Innovative Technologies in Engineering (ICACITE)*. IEEE. 10.1109/ICACITE53722.2022.9823664

Swamy, A. (2023). Advance Cellular Networks (4G, 5G, 6G). *International Journal of Health Sciences*, *2*, 10955–10966.

Vaigandla, K. K., Azmi, N., Podila, R., & Karne, R. K. (2021). A Survey on Wireless Communications: 6g And 7g. *International Journal Of Science, Technology & Management*, *2*(6), 2018–2025. doi:10.46729/ijstm.v2i6.379

Yadav, P. (2021). Evolution of Wireless Communications with 3G, 4G, 5G, and Next Generation Technologies in India. In Advances in Electronics, Communication and Computing. Springer. doi:10.1007/978-981-15-8752-8_35

Yan, J., Liu, J., & Tseng, F. M. (2020). An evaluation system based on the self-organizing system framework of smart cities: A case study of smart transportation systems in China. *Technological Forecasting and Social Change*, *153*, 119371. doi:10.1016/j.techfore.2018.07.009

Ziegler, S. (2013). Iot6–moving to an ipv6-based future iot. In The Future Internet Assembly. Springer. doi:10.1007/978-3-642-38082-2_14

Compilation of References

A., R., S., K., N., N., & P., P. (2022). *Collaborative Edge Data Caching In Online.* 2022 Second International Conference on Advanced Technologies in Intelligent Control, Environment, Computing & Communication Engineering (ICATIECE), Bangalore, India. 10.1109/ICATIECE56365.2022.10047075

Abbas, H., Al-Fuqaha, A., Guizani, M., Rayes, A., & Mohammadi, M. (2020). IoT Applications in Industrial Scenarios: A Comprehensive Survey. *IEEE Communications Surveys and Tutorials, 22,* 2595–2620. doi:10.1109/COMST.2020.2994304

Abbasi, S., Rahmani, A. M., Balador, A., & Sahafi, A. (2021). Internet of Vehicles: Architecture, services, and applications. *International Journal of Communication Systems, 34*(10). doi:10.1002/dac.4793

Abdalla, R., Samara, H., Perozo, N., Carvajal, C. P., & Jaeger, P. (2022). Machine Learning Approach for Predictive Maintenance of the Electrical Submersible Pumps (ESPs). *ACS Omega, 7*(21), 17641–17651. doi:10.1021/acsomega.1c05881 PMID:35664599

Abdulla, A. I., Abdulraheem, A. S., Salih, A. A., Sadeeq, M. A., Ahmed, A. J., Ferzor, B. M., & Mohammed, S. I. (2020). Internet of things and smart home security. *Technol. Rep. Kansai Univ, 62*(5), 2465–2476.

Abiodun, O. I., Jantan, A., Omolara, A. E., Dada, K. V., Umar, A. M., Linus, O. U., & Kiru, M. U. (2019). Comprehensive review of artificial neural network applications to pattern recognition. *IEEE Access : Practical Innovations, Open Solutions, 7,* 158820–158846. doi:10.1109/ACCESS.2019.2945545

Abraham, A., Dash, S., Rodrigues, J. J. P. C., Acharya, B., & Pani, S. K. (2022). A.I., Edge, and IoT-based Smart Agriculture. Londres (Royaume-Uni): Academic Press. doi:10.1016/C2020-0-00516-5

Abraham, G., & Nithya, M. (2021). Smart Agriculture Based on IoT and Machine Learning. *2021 5th International Conference on Computing Methodologies and Communication*, (pp. 414-419). IEEE. 10.1109/ICCMC51019.2021.9418392

Acharya, B., Garikapati, K. (2022). Internet of things (IoT) and Data Analytics in Smart Agriculture: Benefits and challenges. In A. Abraham, S. Dash, J. Rodrigues, B. Acharya, & S. Pani (eds.) Intelligent Data-Centric Systems, AI, Edge, and IoT-based Smart Agriculture, 3-16. Academic Press. doi:10.1016/B978-0-12-823694-9.00013-X

Adballat, A. J. (2021, November 2). *How AI is helping oil and gas companies achieve net zero.* Energy Connects. https://www.energyconnects.com/opinion/thought-leadership/2021/november/how-ai-is-helping-oil-and-gas-companies-achieve-net-zero/

Adebusola, J. A. (2020). An Overview of 5G Technology. In *2020 International Conference in Mathematics, Computer Engineering and Computer Science (ICMCECS).* IEEE. 10.1109/ICMCECS47690.2020.240853

Adhya, D., Chatterjee, S., & Chakraborty, A. K. (2022). Performance assessment of selective machine learning techniques for improved PV array fault diagnosis. *Sustainable Energy. Grids and Networks, 29,* 100582.

Adner, R., & Kapoor, R. (2010). Value creation in innovation ecosystems: How the structure of technological interdependence affects firm performance in new technology generations. *Strategic Management Journal*, *31*(3), 306–333. doi:10.1002mj.821

Advanced Driver Assistance Systems – Overview of ADAS Applications. (2023). Synopsys. https://www.synopsys.com/automotive/what-is-adas.html

Agafonov, A., & Myasnikov, V. (2021). *Traffic Signal Control: a Double Q-learning Approach. Annals of Computer Science and Information Systems.* Fed CSIS. doi:10.15439/2021F109

Agarwal, K., & Guo, Y. X. (2015), May. Interaction of electromagnetic waves with humans in wearable and biomedical implant antennas. In *2015 Asia-Pacific Symposium on Electromagnetic Compatibility (APEMC)* (pp. 154-157). IEEE. 10.1109/APEMC.2015.7175377

Aggarwal, C. C., Ashish, N., & Sheth, A. (2013). *The internet of things: A survey from the data-centric perspective. InManaging and mining sensor data.* Springer.

Ahmad, M. (2014). Reliability Models for the Internet of Things: A Paradigm Shift. *2014 IEEE Int. Symp. Softw. Reliab. Eng. Work.*, (pp. 52–59). IEEE. 10.1109/ISSREW.2014.107

Ahmad, Z., Shahid Khan, A., Wai Shiang, C., Abdullah, J., & Ahmad, F. (2021). Network intrusion detection system: A systematic study of machine learning and deep learning approaches. *Transactions on Emerging Telecommunications Technologies*, *32*(1), e4150. doi:10.1002/ett.4150

Ahmed, M., Xiao, Z., & Shen, Y. (2022). Estimation of Ground PM2.5 Concentrations in Pakistan Using Convolutional Neural Network and Multi-Pollutant Satellite Images. *Remote Sensing (Basel)*, *14*(7), 1735. doi:10.3390/rs14071735

Ahmed, N., Rafiq, J. I., & Islam, M. R. (2020). Enhanced human activity recognition based on smartphone sensor data using hybrid feature selection model. *Sensors (Basel)*, *20*(1), 317. doi:10.339020010317 PMID:31935943

Akash, S. (2021). Edge Computing: Reshaping the Agricultural Sector with Smart Farming. *Analytical Insight.* https://www.analyticsinsight.net/edge-computing-reshaping-the-agricultural-sector-with-smart-farming/

Akhter, R. & Sofi, R. (2022). Precision agriculture using IoT data analytics and machine learning. *Journal of King Saud University - Computer and Information Sciences, 34* (8). . doi:10.1016/j.jksuci.2021.05.013

Al Hayajneh, A., Bhuiyan, M. Z. A., & McAndrew, I. (2020). I. Improving Internet of Things (IoT) Security with Software-Defined Networking (SDN). *Computers*, *9*(1), 8. doi:10.3390/computers9010008

Al Islam, N., & Arifin, F. (2016). Performance analysis of a miniaturized implantable PIFA antenna for WBAN at ISM band. In *2016 3rd International Conference on Electrical Engineering and Information Communication Technology (ICEEICT)* (pp. 1-5). IEEE. 10.1109/CEEICT.2016.7873145

Al Rasyid, M. U., Lee, B. H., & Sudarsono, A. (2015, January 1). Implementation of body temperature and pulseoximeter sensors for wireless body area network. *Sensors and Materials*, *27*(8), 727–732.

Al Rasyid, M. U., Yuwono, W., Al Muharom, S., & Alasiry, A. H. (2016). Building platform application big sensor data for e-health wireless body area network. In *2016 International Electronics Symposium (IES)*, (pp. 409-413). IEEE. 10.1109/ELECSYM.2016.7861041

Ala'raj, M., Majdalawieh, M., & Abbod, M. F. (2020). Improving binary classification using filtering based on k-NN proximity graphs. *Journal of Big Data*, *7*(1), 1–18. doi:10.118640537-020-00297-7

Alalade, E. D. (2020, June). Intrusion detection system in smart home network using artificial immune system and extreme learning machine hybrid approach. In *2020 IEEE 6th World Forum on Internet of Things (WF-IoT)* (pp. 1-2). IEEE. 10.1109/WF-IoT48130.2020.9221151

Alam, M. R., Reaz, M. B. I., & Ali, M. A. M. (2012). A review of smart homes—Past, present, and future. *IEEE transactions on systems, man, and cybernetics, part C (applications and reviews), 42*(6), 1190-1203.

Alam, M. M., & Hamida, E. B. (2014, May). Surveying wearable human assistive technology for life and safety critical applications: Standards, challenges and opportunities. *Sensors (Basel), 14*(5), 9153–9209. doi:10.3390140509153 PMID:24859024

Alarcon, F., & Tovar, E. (2021). Ant Colony Optimization for Multi-Robot Systems. *IEEE Robotics and Automation Letters, 6*, 7163–7169. doi:10.1109/LRA.2021.3093659

Alashti, M. R. S., Abadi, M. B., Holthaus, P., Menon, C., & Amirabdollahian, F. (2023). *Lightweight human activity recognition for ambient assisted living.* IARIA.

Albulayhi, K., Abu Al-Haija, Q., Alsuhibany, S. A., Jillepalli, A. A., Ashrafuzzaman, M., & Sheldon, F. T. (2022). IoT intrusion detection using machine learning with a novel high performing feature selection method. *Applied Sciences (Basel, Switzerland), 12*(10), 5015. doi:10.3390/app12105015

Aldhafeeri, T., Tran, M. K., Vrolyk, R., Pope, M., & Fowler, M. (2020). A review of methane gas detection sensors: Recent developments and future perspectives. *Inventions (Basel, Switzerland), 5*(3), 1–18. doi:10.3390/inventions5030028

Al-Fuqaha, A., Guizani, M., Mohammadi, M., Aledhari, M., & Ayyash, M. (2015). Internet of things: A survey on enabling technologies, protocols, and applications. *IEEE Communications Surveys and Tutorials, 17*(4), 2347–2376. doi:10.1109/COMST.2015.2444095

Ali, A. S., Abdelmoez, M. N., Heshmat, M., & Ibrahim, K. (2022). A solution for water management and leakage detection problems using IoTs based approach. *Internet of Things, 18*, 100504. doi:10.1016/j.iot.2022.100504

Aliar, A. (2022). A Comprehensive Analysis on IoT-based Smart Farming Solutions using Machine Learning Algorithms. *Bulletin of Electrical Engg., and Informatics, 11*(3), 1550-1557. . doi:10.11591/eei.v11i3.3310

Ali, M. H., Jaber, M. M., Abd, S. K., Alkhayyat, A., & Albaghdadi, M. F. (2022). Big data analysis and cloud computing for smart transportation system integration. *Multimedia Tools and Applications*, 1–18. doi:10.100711042-022-13700-7

Ali, P. J. M., Faraj, R. H., Koya, E., Ali, P. J. M., & Faraj, R. H. (2014). Data normalization and standardization: A technical report. *Mach Learn Tech Rep, 1*(1), 1–6.

Al-khafajiy, M., Baker, T., Chalmers, C., Asim, M., Kolivand, H., Fahim, M., & Waraich, A. (2019). Remote health monitoring of elderly through wearable sensors. *Multimedia Tools and Applications, 78*(17), 24681–24706. doi:10.100711042-018-7134-7

Almomani, O., Almaiah, M. A., Alsaaidah, A., Smadi, S., Mohammad, A. H., & Althunibat, A. (2021, July). Machine learning classifiers for network intrusion detection system: comparative study. In *2021 International Conference on Information Technology (ICIT)* (pp. 440-445). IEEE. 10.1109/ICIT52682.2021.9491770

Almusallam, NY, Tari, Z, Bertok, P, & ... (2016). *Dimensionality reduction for intrusion detection systems in multi-data streams-A review and proposal of unsupervised feature selection scheme.* Emerg Comput.

Almutairi, S., Manimurugan, S., & Aborokbah, M. (2019). A new securetransmission scheme between senders and receivers using HVCHC withoutany loss. *EURASIP Journal on Wireless Communications and Networking*, (88), 1–15.

Alobaidy, H. A. H., Nordin, R., Mandeep, J. S., Abdullah, N. F., Haniz, A., Ishizu, K., Matsumura, T., Kojima, F., & Ramli, N. (2022). Low Altitude Platform-based Airborne IoT Network (LAP-AIN) for Water Quality Monitoring in Harsh Tropical Environment. *IEEE Internet of Things Journal*, *9*(20), 1–1. doi:10.1109/JIOT.2022.3171294

Al-Omari, M., Rawashdeh, M., Qutaishat, F., Alshira'H, M., & Ababneh, N. (2021). An intelligent tree-based intrusion detection model for cyber security. *Journal of Network and Systems Management*, *29*(2), 1–18. doi:10.100710922-021-09591-y

Aloraini, F., Javed, A., Rana, O., & Burnap, P. (2022). Adversarial machine learning in IoT from an insider point of view. *Journal of Information Security and Applications*, *70*, 103341. doi:10.1016/j.jisa.2022.103341

Alreshid, E. (2019). Smart Sustainable Agriculture Solution Underpinned by Internet of Things and Artificial Intelligence. *International Journal of Advanced Computer Science and Applications*, *10*(5), 93–102. doi:10.14569/IJACSA.2019.0100513

Al-Rubaie, A. J., Aziz, A. A., Yousif, A. A., Hadi, S. S., & Bader, A. H. (2021). Challenges and Opportunities in IoT-Based Smart Agriculture: A Comprehensive Review. *IEEE Access : Practical Innovations, Open Solutions*, *9*, 127071–127106. doi:10.1109/ACCESS.2021.3105695

Alsheikh, M. A., Niyato, D., Lin, S., Tan, H. P., & Han, Z. (2016, May 20). Mobile big data analytics using deep learning and apache spark. *IEEE Network*, *30*(3), 22–29. doi:10.1109/MNET.2016.7474340

Altameem, A., Mallikarjuna, B., Saudagar, A. K. J., Sharma, M., & Poonia, R. C. (2022). Improvement of Automatic Glioma Brain Tumor Detection Using Deep Convolutional Neural Networks. *Journal of Computational Biology*, *29*(6), 530–544. doi:10.1089/cmb.2021.0280 PMID:35235381

Al-Turjman, F. (2018). Challenges and Opportunities in IoT-Based Healthcare Systems: A Review. *IEEE Access : Practical Innovations, Open Solutions*, *6*, 40027–40043. doi:10.1109/ACCESS.2018.2858027

Alzubi, J., Nayyar, A., & Kumar, A. (2018, November). Machine learning from theory to algorithms: An overview. *Journal of Physics: Conference Series*, *1142*, 012012. doi:10.1088/1742-6596/1142/1/012012

Amitha, I. C., & Narayanan, N. K. (2021). Object Detection Using YOLO Framework for Intelligent Traffic Monitoring. *Lecture Notes in Electrical Engineering*, (pp. 405–412). IEEE. doi:10.1007/978-981-16-5078-9_34

Anoushee, M., Fartash, M., & Akbari Torkestani, J. (2023). An intelligent resource management method in SDN based fog computing using reinforcement learning. *Computing*, 1–30. doi:10.100700607-022-01141-x

An, S., Bhat, G., Gumussoy, S., & Ogras, U. (2023). Transfer learning for human activity recognition using representational analysis of neural networks. *ACM Transactions on Computing for Healthcare*, *4*(1), 1–21. doi:10.1145/3563948

Aqib, M., Kumar, D., & Tripathi, S. (2023). Machine Learning for Fog Computing: Review, Opportunities and a Fog Application Classifier and Scheduler. *Wireless Personal Communications*, *129*(2), 853–880. doi:10.100711277-022-10160-y

Aravind, A. R., Chakravarthi, R., & Natraj, N. A. (2020, September). Optimal mobility based data gathering scheme for life time enhancement in wireless sensor networks. In *2020 4th International Conference on Computer, Communication and Signal Processing (ICCCSP)* (pp. 1-5). IEEE. 10.1109/ICCCSP49186.2020.9315275

Archana, S., Shankar, R., & Singh, S. (2022). Development of smart grid for the power sector in India. *Cleaner Energy Systems*, *2*, 100011. doi:10.1016/j.cles.2022.100011

Arvind, R., Shukla, A., & Yadav, S. (2023). Security Challenges in IoT-based Healthcare Applications: A Review. *Journal of Medical Systems*, *47*(2), 27. doi:10.100710916-022-01803-5

Arya Omnitalk Wireless Solutions. (2021, January 30). *Electronic Toll Collection System*. Fastag. Arya Omnitalk. https://aryaomnitalk.com/electronic-toll-collection-system/

Aryal, A., Becerik-Gerber, B., Anselmo, F., Roll, S. C., & Lucas, G. M. (2019). Smart Desks to Promote Comfort, Health, and Productivity in Offices: A Vision for Future Workplaces. *Frontiers in Built Environment*, *5*, 76. doi:10.3389/fbuil.2019.00076

Arzoumanian, Z., Holmberg, J., & Norman, B. (2005). An astronomical pattern-matching algorithm for computer-aided identification of whale sharks Rhincodon typus. *Journal of Applied Ecology*, *42*(6), 999–1011. doi:10.1111/j.1365-2664.2005.01117.x

Asharf, J., Moustafa, N., Khurshid, H., Debie, E., Haider, W., & Wahab, A. (2020). A review of intrusion detection systems using machine and deep learning in internet of things: Challenges, solutions and future directions. *Electronics (Basel)*, *9*(7), 1177. doi:10.3390/electronics9071177

Ashifuddin Mondal, M., & Rehena, Z. (2019). Intelligent Traffic Congestion Classification System using Artificial Neural Network. *Companion Proceedings of the 2019 World Wide Web Conference*. ACM. 10.1145/3308560.3317053

Ashish Choudhary, A. (2021). IoT-based Smart Agriculture Monitoring System. *Circuit Digest*. https://circuitdigest.com/microcontroller-projects/iot-based-smart-agriculture-moniotring-system

Asim, Y., Azam, M. A., Ehatisham-ul-Haq, M., Naeem, U., & Khalid, A. (2020). Context-aware human activity recognition (CAHAR) in-the-Wild using smartphone accelerometer. *IEEE Sensors Journal*, *20*(8), 4361–4371. doi:10.1109/JSEN.2020.2964278

Aslahi-Shahri, B. M., Rahmani, R., Chizari, M., Maralani, A., Eslami, M., Golkar, M. J., & Ebrahimi, A. (2016). A hybrid method consisting of GA and SVM for intrusion detection system. *Neural Computing & Applications*, *27*(6), 1669–1676. doi:10.100700521-015-1964-2

Asuncion, A., & Newman, D. (2020). *UCI machine learning repository*.

Awan, N., Khan, S., Rahmani, M., Tahir, M., Alturki, R., & Ullah, I. (2021). Machine Learning-Enabled Power Scheduling in IoT-Based Smart Cities, Computers, Materials & Continua. *Tech Science Press*. doi:10.32604/cmc.2021.014386

Ayaz, M., Ammad-uddin, M., Sharif, Z., Mansour, A., & Aggoune, E.-H. M. (2019). Internet-of-Things (IoT) based Smart Agriculture: Towards Making the Fields Talk. *IEEE Access : Practical Innovations, Open Solutions*, *7*, 1–1. doi:10.1109/ACCESS.2019.2932609

Aydin, G., Hallac, I.R., & Karakus, B. (2015). Architecture and implementation of a scalable sensor data storage and analysis system using cloud computing and big data technologies. *Journal of Sensors*.

Azzawi, M. A., Hassan, R., & Bakar, K. A. A. (2016). A review on Internet of Things (IoT) in healthcare. *International Journal of Applied Engineering Research: IJAER*, *11*(20), 10216–10221.

Baballe, M. A. (2022b, July 26). *Accident Detection and Alerting Systems: A Study*. Research Gate. https://www.researchgate.net/publication/362264272_Accident_Detection_and_Alerting_Systems_A_Study

Badidi, E. (2022). Edge AI and Blockchain for Smart Sustainable Cities: Promise and Potential. *Sustainability (Basel)*, *14*(13), 7609. doi:10.3390u14137609

Badillo, S., Banfai, B., Birzele, F., Davydov, I. I., Hutchinson, L., Kam-Thong, T., & Zhang, J. D. (2020). An introduction to machine learning. *Clinical Pharmacology and Therapeutics*, *107*(4), 871–885. doi:10.1002/cpt.1796 PMID:32128792

Badri, H., Bahreini, T., Grosu, D., & Yang, K. (2018). Risk-Based Optimization of Resource Provisioning in Mobile Edge Computing. *2018 IEEE/ACM Symposium on Edge Computing (SEC)*, Seattle, WA, USA. 10.1109/SEC.2018.00033

Bakthavatchalam, K., Karthik, B., Thiruvengadam, V., Muthal, S., Jose, D., Kotecha, K., & Varadarajan, V. (2022). IoT Framework for Measurement and Precision Agriculture: Predicting the Crop Using Machine Learning Algorithms. *Technologies*, *10*(13), 13. doi:10.3390/technologies10010013

Balducci, F., Impedovo, D., & Pirlo, G. (2018). Machine Learning Applications on Agricultural Datasets for Smart Farm Enhancement. *Machines*, *6*(3), 38. doi:10.3390/machines6030038

Band, S. S., Ardabili, S., Sookhak, M., Chronopoulos, A. T., Elnaffar, S., Moslehpour, M., Csaba, M., Torok, B., Pai, H.-T., & Mosavi, A. (2022). When Smart Cities Get Smarter via Machine Learning: An In-Depth Literature Review. *IEEE Access : Practical Innovations, Open Solutions*, *10*, 60985–61015. doi:10.1109/ACCESS.2022.3181718

Banga, A., Gupta, D., & Bathla, R. (2019). Towards A Taxonomy of CyberAttacks on ScadaSystem. *International Conference on Intelligent Computing and Control Systems (ICCS)*, (pp. 343-347). IEEE.

Bank, R. K., Priyadarshini, R., Dubey, H., Kumar, V., & Mankodiya, K. (2019). FogLearn: leveraging fog-based machine learning for smart system bigdata analytics. In *Geospatial Intelligence: Concepts, Methodologies,Tools, and Applications* (pp. 1225–1241). IGI Global.

Bansal, B., Jenipher, V. N., Jain, R., Dilip, R., Kumbhkar, M., Pramanik, S., Roy, S., & Gupta, A. (2022). Big Data Architecture for Network Security. In S. Pramanik, D. Samanta, M. Vinay, & A. Guha (Eds.), *Cyber Security and Network Security*. doi:10.1002/9781119812555.ch11

Baranwal, J., Barse, B., Gatto, G., Broncova, G., & Kumar, A. (2022). Electrochemical Sensors and Their Applications: A Review. *Chemosensors (Basel, Switzerland)*, *10*(9), 363. doi:10.3390/chemosensors10090363

Barbedo, J. G. A. (2019). A review on the use of unmanned aerial vehicles and imaging sensors for monitoring and assessing plant stresses. *Drones (Basel)*, *3*(2), 40. doi:10.3390/drones3020040

Baştanlar, Y., & Özuysal, M. (2014). Introduction to machine learning. *miRNomics: MicroRNA biology and computational analysis*, 105-128.

Batra, G., Jacobson, Z., Madhav, S., Queirolo, A., & Santhanam, N. (2019). *Artificial-intelligence hardware: New opportunities for semiconductor companies*. McKinsey and Company.

Batri, K., & Anbukaruppusamy, S. (2019). Improving TCP Performance in AdHoc Networks. *European Journal of Scientific Research*, *65*(2), 237–245.

Beaulieu, M., Foucher, S., Haberman, D., & Stewart, C. (2018, July). Deep image-to-image transfer applied to resolution enhancement of sentinel-2 images. In *IGARSS 2018-2018 IEEE International Geoscience and Remote Sensing Symposium* (pp. 2611-2614). IEEE. 10.1109/IGARSS.2018.8517655

Beaver, J. M., Borges-Hink, R. C., & Buckner, M.A. (2020). An Evaluationof Machine Learning Methods to Detect Malicious SCADA Communications. *International Conferenceon Machine Learning & Applications*. IEEE.

Beckett-Camarata, J. (2022). Smart and connected cities in Pennsylvania: A multi-case study. Smart Cities and Regional Development (SCRD). *Journal*, *6*(1), 67–78. doi:10.25019crd.v6i1.121

Behera, R. K., Reddy, K. H. K., & Roy, D. S. (2015). Reliability modelling of service oriented Internet of Things. In *2015 4th International Conference on Reliability, Infocom Technologies and Optimization: Trends and Future Directions, ICRITO 2015*. IEEE. 10.1109/ICRITO.2015.7359216

Bhatia, S., Gautam, D., Kumar, S., & Verma, S. (2023). *Automatic Seat Identification System in Smart Transport using IoT and Image Processing.* 2023 3rd International Conference on Intelligent Communication and Computational Techniques (ICCT), Jaipur, India. 10.1109/ICCT56969.2023.10075664

Bhatia, S., Jaffery, Z. A., & Mehfuz, S. (2023). *A Comparative Study of Wireless Communication Protocols for use in Smart Farming Framework Development.* 2023 3rd International Conference on Intelligent Communication and Computational Techniques (ICCT), Jaipur, India. 10.1109/ICCT56969.2023.10075696

Bhatia, S., Mallikarjuna, B., Gautam, D., Gupta, U., Kumar, S., & Verma, S. (2023). *The Future IoT: The Current Generation 5G and Next Generation 6G and 7G Technologies.* 2023 International Conference on Device Intelligence, Computing and Communication Technologies, (DICCT), Dehradun, India. 10.1109/DICCT56244.2023.10110066

Bhatia, S., Dhillon, H. S., & Kumar, N. (2011, December). Alive human body detection system using an autonomous mobile rescue robot. In *2011 Annual IEEE India Conference* (pp. 1-5). IEEE. 10.1109/INDCON.2011.6139388

Bhattacharyya, S., Bose, S., Majumder, S., Sengupta, S., & Bhattacharyya, S. P. (2019). Challenges and opportunities in IoT-based energy management: A review. *Journal of Ambient Intelligence and Humanized Computing, 10*(2), 717–734. doi:10.100712652-018-1006-9

Bhoi, A., Nayak, R. P., Bhoi, S. K., Sethi, S., Panda, S. K., Sahoo, K. S., & Nayyar, A. (2021). IoT-IIRS: Internet of Things based intelligent-irrigation Recommendation System Using Machine Learning Approach for Efficient Water Usage. *PeerJ. Computer Science, 7*, 578. doi:10.7717/peerj-cs.578 PMID:34239972

Bhosale, SV. (2014). Outlier detection in straming data using clustering approached. *International Journal of Computer Science and Information Technologies, 5*(5), 6050–6053.

Bhushan, M., Iyer, S., Kumar, A., Choudhury, T., & Negi, A. (n.d.). *Artificial Intelligence for Smart Cities and Villages: Advanced Technologies.*

Bhushan, M., Kumar, A., Samant, P., Bansal, S., Tiwari, S., & Negi, A. (2021). Identifying Quality Attributes of FODA and DSSA Methods in Domain Analysis using a Case Study. *2021 10th International Conference on System Modeling & Advancement in Research Trends (SMART), 562–567.* 10.1109/SMART52563.2021.9676289

Bhushan, M., Ángel Galindo Duarte, J., Samant, P., Kumar, A., & Negi, A. (2021). Classifying and resolving software product line redundancies using an ontological first-order logic rule based method. *Expert Systems with Applications, 168*, 114167. doi:10.1016/j.eswa.2020.114167

Bhushan, M., Negi, A., Samant, P., Goel, S., & Kumar, A. (2020). A classification and systematic review of product line feature model defects. *Software Quality Journal, 28*(4), 1507–1550. doi:10.100711219-020-09522-1

Bhushan, M., Pandit, A., & Garg, A. (2023). Machine learning and deep learning techniques for the analysis of heart disease: A systematic literature review, open challenges and future directions. *Artificial Intelligence Review, 1–52.* doi:10.100710462-023-10493-5

Bianchi, C., Tuzovic, S., & Kuppelwieser, V. G. (2022). Investigating the drivers of wearable technology adoption for healthcare in South America. *Information Technology & People.*

Bílek, J., Maršolek, P., Bílek, O., & Buček, P. (2022). Field Test of Mini Photoionization Detector-Based Sensors—Monitoring of Volatile Organic Pollutants in Ambient Air. *Environments - MDPI, 9*(4). doi:10.3390/environments9040049

Bishop, C. M., & Nasrabadi, N. M. (2006). *Pattern recognition and machine learning.* Springer.

Bobbio & Trivedi, K. (2017). Reliability and Availability Engineering: Modeling, Analysis, and Applications. Cambridge University.

Bogue, R. (2015). Detecting gases with light: A review of optical gas sensor technologies. *Sensor Review*, *35*(2), 133–140. doi:10.1108/SR-09-2014-696

Bonomi, F. (2021). Connected vehicles, the internet of things, and fogcomputing. In *The eighth ACM international workshop on vehicularinter-networking (VANET)*, (pp. 13-15). ACM.

Borg, G., Löllgen, H., & Döllinger, M. (2019). Wearables in Sports: Opportunities and Challenges. *Sports Medicine and Health Science*, *1*(2), 62–67. doi:10.1016/j.smhs.2019.02.002

Boughaci, D., & Alkhawaldeh, A. A. (2020). Appropriate machine learning techniques for credit scoring and bankruptcy prediction in banking and finance: A comparative study. *Risk and Decision Analysis*, *8*(1-2), 15–24. doi:10.3233/RDA-180051

Boutaba, R., Salahuddin, M. A., Limam, N., Ayoubi, S., Shahriar, N., Estrada-Solano, F., & Caicedo, O. M. (2018). A comprehensive survey on machine learning for networking: Evolution, applications and research opportunities. *Journal of Internet Services and Applications*, *9*(1), 1–99. doi:10.118613174-018-0087-2

Brelsford, R. (2018, June 6). Repsol launches Big Data, AI project at Tarragona Refinery. *Oil & Gas Journal*. https://www.ogj.com/refining-processing/refining/operations/article/17296578/repsol-launches-big-data-ai-project-at-tarragona-refinery

Brown, B. (2022, November 21). *High IQ it meets ESG at Laredo Petroleum: Cover story: Magazine*. The American Oil & Gas Reporter Magazine. https://www.aogr.com/magazine/cover-story/high-iq-it-meets-esg-at-laredo-petroleum

Burd, B., Barker, L., Divitini, M., Perez, F. A. F., Russell, I., Siever, B., & Tudor, L. (2018, January). Courses, content, and tools for internet of things in computer science education. In *Proceedings of the 2017 ITiCSE conference on working group reports* (pp. 125-139). 10.1145/3174781.3174788

Cabanac, M. (1971). Physiological Role of Pleasure: A stimulus can feel pleasant or unpleasant depending upon its usefulness as determined by internal signals. *Science*, *173*(4002), 1103–1107. doi:10.1126cience.173.4002.1103 PMID:5098954

Cai, S., Ma, Z., Skibniewski, M. J., & Bao, S. (2019). Construction automation and robotics for high-rise buildings over the past decades: A comprehensive review. *Advanced Engineering Informatics*, *42*, 100989. doi:10.1016/j.aei.2019.100989

Cai, Z., Xiong, Z., Xu, H., Wang, P., Li, W., & Pan, Y. (2021). Generative adversarial networks: A survey toward private and secure applications. *ACM Computing Surveys*, *54*(6), 1–38. doi:10.1145/3459992

Çalışkan, A. (2023). Detecting human activity types from 3D posture data using deep learning models. *Biomedical Signal Processing and Control*, *81*, 104479. doi:10.1016/j.bspc.2022.104479

Caner, S., & Bhatti, F. (2020). A conceptual framework on defining businesses strategy for artificial intelligence. *Contemporary Management Research*, *16*(3), 175–206. doi:10.7903/cmr.19970

Cao, C., Shi, Y., & Su, Y. (2021). A Distributed Task Allocation Algorithm for Heterogeneous Robot Teams. *IEEE Transactions on Automation Science and Engineering*, *18*, 666–680. doi:10.1109/TASE.2020.2991522

Cao, X., Sun, G., Yu, H., & Guizani, M. (2022). PerFED-GAN: Personalized federated learning via generative adversarial networks. *IEEE Internet of Things Journal*. doi:10.1109/JIOT.2022.3172114

Carcano, A., Coletta, A., Guglielmi, M., Masera, M., Nai Fovino, I., & Trombetta, A. (2021). andA. Trombetta, "A Multidimensional Critical State Analysis for Detecting Intrusions in SCADA Systems,". *IEEE Transactions on Industrial Informatics*, *7*(2), 179–186. doi:10.1109/TII.2010.2099234

Cassisi, C., Ferro, A., Giugno, R., Pigola, G., & Pulvirenti, A. (2013). Enhancing densitybased clustering: Parameter reduction and outlier detection. *Information Systems*, *38*(3), 317–330. doi:10.1016/j.is.2012.09.001

Cena, C. G., Cardenas, P. F., Pazmino, R. S., Puglisi, L., & Santonja, R. A. (2013). A cooperative multi-agent robotics system: Design and modelling. *Expert Systems with Applications*, *40*(12), 4737–4748. doi:10.1016/j.eswa.2013.01.048

Chacko, A., & Hayajneh, T. (2018). Security and privacy issues with IoT in healthcare. *EAI Endorsed Transactions on Pervasive Health and Technology*, *4*(14), 155079. doi:10.4108/eai.13-7-2018.155079

Chalmers, D., MacKenzie, N. G., & Carter, S. (2021). Artificial intelligence and entrepreneurship: Implications for venture creation in the fourth industrial revolution. *Entrepreneurship Theory and Practice*, *45*(5), 1028–1053. doi:10.1177/1042258720934581

Channe, H., Kothari, S., & Kadam, D. (2015). Multidisciplinary model for smart agriculture using internet-ofthings (IoT), sensors, cloud-computing, mobile-computing & big-data analysis. *Int. J. Computer Technology and Application*, *6*(3), 374–382.

Chaturvedi, S. K. (2016). *Network Reliability: Measures and Evaluation*. Wiley. doi:10.1002/9781119224006

Chaudhary, D. D., Nayse, S. P., & Waghmare, L. M. (2011). Application of wireless sensor networks for greenhouse parameter control in precision agriculture. *International Journal of Wireless & Mobile Networks*, *3*(1), 140–149. doi:10.5121/ijwmn.2011.3113

Chen, P., Xu, H., Fan, X., Hu, J., & Song, T. (2022). Multi-Slot Dynamic Computing Resource Optimization in Edge Computing. *2022 14th International Conference on Wireless Communications and Signal Processing (WCSP)*, Nanjing, China. 10.1109/WCSP55476.2022.10039042

Chen, W., Yuan, J., Luo, A., & Xie, F. (2021). *Reliability Incremental Routing for Power Line Communication Based on Power Internet of Things*. 2021 IEEE 6th International Conference on Cloud Computing and Big Data Analytics (ICCCBDA), Chengdu, China. 10.1109/ICCCBDA51879.2021.9442543

Chen, X., Liu, W., Chen, J., & Zhou, J. (2020). An Edge Server Placement Algorithm in Edge Computing Environment. 2020 12th International Conference on Advanced Infocomm Technology (ICAIT), Macao, China. 10.1109/ICAIT51223.2020.9315526

Chen, C. H. (Ed.). (2015). *Handbook of pattern recognition and computer vision*. World Scientific.

Chen, D., Yue, L., Chang, X., Xu, M., & Jia, T. (2021). NM-GAN: Noise-modulated generative adversarial network for video anomaly detection. *Pattern Recognition*, *116*, 107969. doi:10.1016/j.patcog.2021.107969

Cheng, B., Dong, M., Zheng, Y., Zhang, X., & Wu, X. (2021). An intelligent robot system for steel beam transportation and assembly. *Journal of Intelligent & Robotic Systems*, *103*, 249–259. doi:10.100710846-020-01322-3

Chen, H., Li, X., Zhang, S., & Li, Y. (2021a). Intelligent Control Strategy for Smart Manufacturing Systems: A Comprehensive Review. *IEEE Transactions on Industrial Informatics*, *17*, 1973–1990. doi:10.1109/TII.2020.3020355

Chen, J., Zhao, X., & Yan, G. (2021b). Robotic manipulation of large-scale structures in civil engineering. *Automation in Construction*, *121*, 103471. doi:10.1016/j.autcon.2020.103471

Chen, M., Li, W., Hao, Y., Qian, Y., & Humar, I. (2018, September 1). Edge cognitive computing based smart healthcare system. *Future Generation Computer Systems*, *86*, 403–411. doi:10.1016/j.future.2018.03.054

Chen, Q., Wang, W., Huang, K., De, S., & Coenen, F. (2021). Multi-modal generative adversarial networks for traffic event detection in smart cities. *Expert Systems with Applications*, *177*, 114939. doi:10.1016/j.eswa.2021.114939

Chen, S., Mihara, K., & Wen, J. (2018). Time series prediction of CO2, TVOC and HCHO based on machine learning at different sampling points. *Building and Environment, 146*(July), 238–246. doi:10.1016/j.buildenv.2018.09.054

Chen, X., & Huang, J. (2019). Odor source localization algorithms on mobile robots: A review and future outlook. *Robotics and Autonomous Systems, 112*, 123–136. doi:10.1016/j.robot.2018.11.014

Chen, Z., Feng, X., & Zhang, S. (2022). Emotion detection and face recognition of drivers in autonomous vehicles in IoT platform. *Image and Vision Computing, 128*, 104569. doi:10.1016/j.imavis.2022.104569

Chkirbene, Z., Eltanbouly, S., Bashendy, M., AlNaimi, N., & Erbad, A. (2020, February). Hybrid machine learning for network anomaly intrusion detection. In *2020 IEEE International Conference on Informatics, IoT, and Enabling Technologies (ICIoT)* (pp. 163-170). IEEE. 10.1109/ICIoT48696.2020.9089575

Cho, G., Yim, J., Choi, Y., Ko, J., & Lee, S. H. (2019). Review of machine learning algorithms for diagnosing mental illness. *Psychiatry Investigation, 16*(4), 262–269. doi:10.30773/pi.2018.12.21.2 PMID:30947496

Choubey, S., & Karmakar, G. (2021). Artificial intelligence techniques and their application in oil and gas industry. *Artificial Intelligence Review, 54*(5), 3665–3683. doi:10.100710462-020-09935-1

Chugh, N, Chugh, M, & Agarwal, A. (2014). Outlier detection in streaming data: a research perspective. *2014 Int Conf Parall Distrib Grid Comput.* (pp. 429–4). IEEE.

Chui, K. T., Alhalabi, W., Pang, S. S., Pablos, P. O., Liu, R. W., & Zhao, M. (2017, December). Disease diagnosis in smart healthcare: Innovation, technologies, and applications. *Sustainability (Basel), 9*(12), 2309. doi:10.3390u9122309

Chui, M., Manyika, J., & Miremadi, M. (2018). What AI can and can't do (yet) for your business. [What-AI-can-and-cant-do-yet-for-your-business.pdf] [mckinsey.com]. *The McKinsey Quarterly, 1*, 97–108.

Chu, J., Li, W., Yang, X., Wu, Y., Wang, D., Yang, A., Yuan, H., Wang, X., Li, Y., & Rong, M. (2021). Identification of gas mixtures via sensor array combining with neural networks. *Sensors and Actuators. B, Chemical, 329*, 129090. doi:10.1016/j.snb.2020.129090

Cifarelli, L., Wagner, F., & Balossino, P. (2020). The oil & gas upstream cycle: Development and production. *EPJ Web of Conferences, 246*, 1–12. doi:10.1051/epjconf/202024600009 doi:10.1051/epjconf/202024600001

Colombo-Mendoza, L. O., Paredes-Valverde, M. A., Salas-Zárate, M. P., & Valencia-García, R. (2022). Internet of Things-Driven Data Mining for Smart Crop Production Prediction in the Peasant Farming Domain. *Applied Sciences (Basel, Switzerland), 12*(4), 1940. doi:10.3390/app12041940

Columbus, L. (2023). 10 Ways AI has The Potential To Improve Agriculture In 2021. *Forbes.* https://www.forbes.com/sites/louiscolumbus/2021/02/17/10-ways-ai-has-the-potential-to-improve-agriculture-in-2021/?sh=68ca39aa7f3b

Cong, Y., Wang, J., & Li, X. (2016). Traffic Flow Forecasting by a Least Squares Support Vector Machine with a Fruit Fly Optimization Algorithm. *Procedia Engineering, 137*, 59–68. doi:10.1016/j.proeng.2016.01.234

Connected Autonomous Vehicles. (2020, April 21). Ferrovial. https://www.ferrovial.com/en/innovation/technologies/connected-autonomous-vehicles/

Cort'es, R., Bonnaire, X., Marin, O., & Sens, P. (2015, January 1). Stream processing of healthcare sensor data: Studying user traces to identify challenges from a big data perspective. *Procedia Computer Science, 52*, 1004–1009. doi:10.1016/j.procs.2015.05.093

Costanzo, F., Lorenzo, P. D., & Barbarossa, S. (2020). *Dynamic Resource Optimization and Altitude Selection in Uav-Based Multi-Access Edge Computing*. ICASSP 2020 - 2020 IEEE International Conference on Acoustics, Speech and Signal Processing (ICASSP), Barcelona, Spain. 10.1109/ICASSP40776.2020.9053594

Craveiro, F., Duarte, J. P., Bartolo, H., & Bartolo, P. J. (2019). Additive manufacturing as an enabling technology for digital construction: A perspective on Construction 4.0. *Automation in Construction*, *103*, 251–267. doi:10.1016/j.autcon.2019.03.011

DAC. (2022, September 27). *Predictive and preventive maintenance in smart cities*. DAC .Digital. https://dac.digital/predictive-and-preventive-maintenance-in-smart-cities/

Dahane, A., Benameur, R., Kechar, B., & Benyamina, A. (2020). An IoT-Based Smart Farming System Using. *Machine Learning*, 1–6. doi:10.1109/ISNCC49221.2020.9297341

Dakheel, A. H., Dakheel, A. H., & Abbas, H. H. "Intrusion detectionsystem in gas-pipeline industry using machine learning," Periodicals ofEngineering and Natural Sciences, vol. 7, pp. 1030-1040, 2019.

Dang, X., Liu, G., Tang, X., Wang, S., Wang, T., & Zou, M. (2022). Motor Imagery EEG Recognition Based on Generative and Discriminative Adversarial Learning Framework and Hybrid Scale Convolutional Neural Network. *IAENG International Journal of Applied Mathematics*, *52*(4), 1–9.

Dan, L. I. U., Xin, C., Chongwei, H., & Liangliang, J. (2015). Intelligent agriculture greenhouse environment monitoring system based on IOT technology. In *2015 International Conference on Intelligent Transportation, Big Data and Smart City* (pp. 487-490). IEEE. 10.1109/ICITBS.2015.126

Daponte, P., De Vito, L., Glielmo, L., Iannelli, L., Liuzza, D., Picariello, F., & Silano, G. (2019). A review on the use of drones for precision agriculture. *IOP Conference Series. Earth and Environmental Science*, *275*(1), 12022. doi:10.1088/1755-1315/275/1/012022

Darshan, K. R., & Anandakumar, K. R. (2015, December). A comprehensive review on usage of Internet of Things (IoT) in healthcare system. In *2015 International Conference on Emerging Research in Electronics, Computer Science and Technology (ICERECT)* (pp. 132-136). IEEE. 10.1109/ERECT.2015.7499001

Daryabar, F., Dehghantanha, A., & Udzir, N. I. (2021). Towards secure modelfor SCADA systems. International Conferenceon Cyber Security, CyberWarfare and Digital Forensic. CyberSec.

Das, R., & Inuwa, M. M. (2023). A review on fog computing: issues, characteristics, challenges, and potential applications. *Telematics and Informatics Reports,* 100049.

Datta, S. K., Bonnet, C., Gyrard, A., Da Costa, R. P. F., & Boudaoud, K. (2015, October). Applying Internet of Things for personalized healthcare in smart homes. In *2015 24th Wireless and Optical Communication Conference (WOCC)* (pp. 164-169). IEEE. 10.1109/WOCC.2015.7346198

Datta, S. K., Bonnet, C., & Haem, J. (2020). Fog computing architecture to enableconsumer centric internet of things services. In *In 2020InternationalSymposium on Consumer Electronics* (pp. 1–2). ISCE.

De Almeida, J. P. L. S., Nakashima, R. T., Neves-Jr, F., & de Arruda, L. V. R. (2019). Bio-inspired on-line path planner for cooperative exploration of unknown environment by a Multi-Robot System. *Robotics and Autonomous Systems*, *112*, 32–48. doi:10.1016/j.robot.2018.11.005

de Ocampo, A. L. P., & Dadios, E. P. (2017). Energy cost optimization in irrigation system of smart farm by using genetic algorithm. In *2017IEEE 9th International Conference on Humanoid, Nanotechnology, Information Technology, Communication and Control, Environment and Management (HNICEM)* (pp. 1-7). IEEE. 10.1109/HNICEM.2017.8269497 de Souza, P. S. S., Rubin, F. P., Hohemberger, R., Ferreto, T. C., Lorenzon, A. F., Luizelli, M. C., & Rossi, F. D. (2020). Detecting abnormal sensors via machine learning: An IoT farming WSN-based architecture case study. *Measurement*, *164*, 108042. doi:10.1016/j.measurement.2020.108042

Dean, J., & Ghemawat, S. (2008). MapReduce: Simplified data processing on large clusters. *Communications of the ACM*, *51*(1), 107–113. doi:10.1145/1327452.1327492

Demirkan, D. C., Duzgun, H. S., Juganda, A., Brune, J., & Bogin, G. (2022). Real-Time Methane Prediction in Underground Longwall Coal Mining Using AI. *Energies*, *15*(17), 1–12. doi:10.3390/en15176486

Deng, L., Cheng, F., Gao, X., Yu, W., Shi, J., Zhou, L., Zhang, L., Li, M., Wang, Z., Zhang, Y.-D., & Lv, Y. (2023). Hospital crowdedness evaluation and in-hospital resource allocation based on image recognition technology. *Scientific Reports*, *13*(1), 299. doi:10.103841598-022-24221-6 PMID:36609446

Deng, Y., Hua, Y., Napp, N., & Petersen, K. (2019). A Compiler for Scalable Construction by the TERMES Robot Collective. *Robotics and Autonomous Systems*, *121*, 103240. doi:10.1016/j.robot.2019.07.010

Denning, D. E. (1987). An intrusion-detection model. *IEEE Transactions on Software Engineering*, *SE-13*(2), 222–232. doi:10.1109/TSE.1987.232894

Desjardins, J. (2018, January 9). *Timeline: The History of the Industrial Internet of Things*. Visual Capitalist. https://www.visualcapitalist.com/timeline-industrial-internet-things/

Details of Funds Spent on Namami Gange Programme. (2021). Pib.gov.in. https://pib.gov.in/PressReleaseIframePage.aspx?PRID=1739094

Dey, A. (2018). Semiconductor metal oxide gas sensors: A review. *Materials Science and Engineering B: Solid-State Materials for Advanced Technology*, *229*(December 2017), 206–217. doi:10.1016/j.mseb.2017.12.036

Dey, N., Hassanien, A. E., Bhatt, C., Ashour, A., & Satapathy, S. C. (Eds.). (2018). *Internet of things and big data analytics toward next-generation intelligence*. Springer. doi:10.1007/978-3-319-60435-0

Dhanaraju, M., Chenniappan, P., Ramalingam, K., Pazhanivelan, S., & Kaliaperumal, R. (2022). Smart Farming: Internet of Things (IoT)-Based Sustainable Agriculture. *Agriculture*, *12*(10), 1745. doi:10.3390/agriculture12101745

Diao, X., Zheng, J., Cai, Y., Wu, Y., & Anpalagan, A. (2019, December). Fair Data Allocation and Trajectory Optimization for UAV-Assisted Mobile Edge Computing. *IEEE Communications Letters*, *23*(12), 2357–2361. doi:10.1109/LCOMM.2019.2943461

DiMaggio, P. J., & Powell, W. W. (1983). The iron cage revisited: Institutional isomorphism and collective rationality in organizational fields. *American Sociological Review*, *48*(2), 147–160. doi:10.2307/2095101

Ding, S., Wu, F., Qian, J., Jia, H., & Jin, F. (2015). Research on data stream clustering algorithms. *Artificial Intelligence Review*, *43*(4), 593–600. doi:10.100710462-013-9398-7

Din, I. U., Guizani, M., Rodrigues, J. J., Hassan, S., & Korotaev, V. V. (2019). Machine learning in the Internet of Things: Designed techniques for smart cities. *Future Generation Computer Systems*, *100*, 826–843. doi:10.1016/j.future.2019.04.017

Directions, F., & Reading, F. (n.d.). Gas Chromatography and Gas Chromatography } Mass Overview of Derivatization of Sugars for GC, GC-MS or GC-MS-MS Analysis. *Journal of Capillary Electrophoresis*, 2211–2223.

Diro, A., Mahmood, A., & Chilamkurti, N. (2021). Collaborative Intrusion Detection Schemes in Fog-to-Things Computing. *Fog/Edge Computing For Security, Privacy, and Applications*, 93-119.

Diykh, M., Abdulla, S., Deo, R. C., Siuly, S., & Ali, M. (2023). Developing a novel hybrid method based on dispersion entropy and adaptive boosting algorithm for human activity recognition. *Computer Methods and Programs in Biomedicine*, *229*, 107305. doi:10.1016/j.cmpb.2022.107305 PMID:36527814

Domingo, M. C. (2012). An overview of the Internet of Things for people with disabilities. *journal of Network and Computer Applications, 35*(2), 584-596.

Domingos, P. (2012). A few useful things to know about machine learning. *Communications of the ACM*, *55*(10), 78–87. doi:10.1145/2347736.2347755

Dong, R. H., Yan, H. H., & Zhang, Q. Y. (2020). An Intrusion Detection Model for Wireless Sensor Network Based on Information Gain Ratio and Bagging Algorithm. *International Journal of Network Security*, *22*(2), 218–230.

Dua, N., Singh, S. N., Semwal, V. B., & Challa, S. K. (2023). Inception inspired CNN-GRU hybrid network for human activity recognition. *Multimedia Tools and Applications*, *82*(4), 5369–5403. doi:10.100711042-021-11885-x

Duan, X., Xu, F., & Sun, Y. (2020). Research on Offloading Strategy in Edge Computing of Internet of Things. *2020 International Conference on Computer Network, Electronic and Automation (ICCNEA)*, Xi'an, China. 10.1109/ICCNEA50255.2020.00050

Du, H., Xu, W., Yao, B., Zhou, Z., & Hu, Y. (2019). Collaborative optimization of service scheduling for industrial cloud robotics based on knowledge sharing. *Procedia CIRP*, *83*, 132–138. doi:10.1016/j.procir.2019.03.142

Dutta, G., & Goswami, P. (2020). Application of Drone in Agriculture: A Review. *International Journal of Chemical Studies*, *SP-8*(5), 181–187. doi:10.22271/chemi.2020.v8.i5d.10529

Ehatisham-ul-Haq, M., & Azam, M. A. (2020). Opportunistic sensing for inferring in-the-wild human contexts based on activity pattern recognition using smart computing. *Future Generation Computer Systems*, *106*, 374–392. doi:10.1016/j.future.2020.01.003

Ehatisham-ul-Haq, M., Azam, M. A., Naeem, U., Amin, Y., & Loo, J. (2018). Continuous authentication of smartphone users based on activity pattern recognition using passive mobile sensing. *Journal of Network and Computer Applications*, *109*, 24–35. doi:10.1016/j.jnca.2018.02.020

Elahi, M. (2008). Efficient clustering-based outlier detection algorithm for dynamic data stream. *IEEE Fifth Int Conf Fuzzy Syst Knowl Discov*. (pp. 298–304). IEEE.

Elijah, O., Rahman, T. A., Orikumhi, I., Leow, C. Y., & Hindia, M. N. (2018). An overview of Internet of Things (IoT) and data analytics in agriculture: Benefits and challenges. *IEEE Internet of Things Journal*, *5*(5), 3758–3773. doi:10.1109/JIOT.2018.2844296

Elsagheer Mohamed, S. A., & AlShalfan, K. A. (2021). Intelligent Traffic Management System Based on the Internet of Vehicles (IoV). *Journal of Advanced Transportation*, *2021*, 1–23. doi:10.1155/2021/4037533

Ercüment, Ö., & Tuğrul, Ç. (2021). A Survey on The Future of Communication Technologies: IoT, 6G, 7G and Beyond. In *1st International Conference on Electrical-Electronics and Computer Engineering*. Avrasya University.

Fan, J., Lan, W., Geng, S., & Zhao, X. (2022). *Task Caching and Computation Offloading for Muti-User Mobile Edge Computing Network*. 2022 4th International Conference on Communications, Information System and Computer Engineering (CISCE), Shenzhen, China. 10.1109/CISCE55963.2022.9851119

FAO. (2015). Climate Change and Food Security: Risks and Responses. Food and Agriculture Organization of the United Nations Rome.

FAO. (2017). *The future of food and agriculture – Trends and challenges*. Food and Agriculture Organization of the United Nations Rome.

Fatima, N., Memon, K. F., & Ahmed, J. (2021). Precision Agriculture using Internet of thing with Artificial Intelligence: A Systematic Literature Review. *University of Sindh Journal of Information and Communication Technology, 5*(2), 101-110. https://sujo.usindh.edu.pk/index.php/USJICT/article/view/2682

Fayyazi, M., Shahbazmoradi, S., Afshar, Z., & Shahbazmoradi, M. R. (2015). Investigating the barriers of the green human resource management implementation in oil industry. *Management Science Letters, 5*(1), 101–108. doi:10.5267/j.msl.2014.12.002

Feng, X., & Li, P. (2019). A tree species mapping method from UAV images over urban area using similarity in tree-crown object histograms. *Remote Sensing (Basel), 11*(17), 1982. doi:10.3390/rs11171982

Ferozkhan, A. B., & Anandharaj, G. (2021). The Embedded Framework for Securing the Internet of Things. *Journal of Engineering Research, 9*(2).

Feynman, R. P. (1982). Simulating physics with computers. *International Journal of Theoretical Physics, 21*(6-7), 467–488. doi:10.1007/BF02650179

Fortino, G., Giannantonio, R., Gravina, R., Kuryloski, P., & Jafari, R. (2012, December 24). Enabling effective programming and flexible management of efficient body sensor network applications. *IEEE Transactions on Human-Machine Systems, 43*(1), 115–133. doi:10.1109/TSMCC.2012.2215852

Fossen, M. (2022, May 11). *Machine learning can reduce the environmental impact of oil production.* #SINTEFblog. https://blog.sintef.com/sintefenergy/gas-technology/machine-learning-can-reduce-the-environmental-impact-of-oil-production/

Fournier-Viger, P., Nkambou, R., & Tseng, S. M. (2021). RuleGrowth: Mining sequential rules common to several sequences by pattern-growth. Symposium on Applied Computing. ACM.

Francisco, M., Mezquita, Y., Revollar, S., Vega, P., & de Paz, J. F. (2019). Multi-agent distributed model predictive control with fuzzy negotiation. *Expert Systems with Applications, 129*, 68–83. doi:10.1016/j.eswa.2019.03.056

Friha, O., Ferrag, M. A., Shu, L., Maglaras, L., & Wang, X. (2021). Internet of Things for the Future of Smart Agriculture: A Comprehensive Survey of Emerging Technologies. In IEEE/CAA Journal of Automatica Sinica, 8(4), 718-752. doi:10.1109/JAS.2021.1003925

Gago, J., Douthe, C., Coopman, R. E., Gallego, P. P., Ribas-Carbo, M., Flexas, J., Escalona, J., & Medrano, H. (2015). UAVs challenge to assess water stress for sustainable agriculture. *Agricultural Water Management, 153*, 9–19. doi:10.1016/j.agwat.2015.01.020

Ganai, P. T., Bag, A., Sable, A., Abdullah, K. H., Bhatia, S., & Pant, B. (2022) *A Detailed Investigation of Implementation of Internet of Things (IOT) in Cyber Security in Healthcare Sector.* 2022 2nd International Conference on Advance Computing and Innovative Technologies in Engineering (ICACITE), Greater Noida, India. 10.1109/ICACITE53722.2022.9823887

Gandomi, A., & Haider, M. (2015). Beyond the hype: Big data concepts, methods, and analytics. *International Journal of Information Management, 35*(2), 137–144. doi:10.1016/j.ijinfomgt.2014.10.007

Ganz, F., Puschmann, D., Barnaghi, P., & Carrez, F. (2015). A practical evaluation of information processing and abstraction techniques for the internet of things. *IEEE Internet of Things Journal, 2*(4), 340–354. doi:10.1109/JIOT.2015.2411227

Gao, D., Cheng, H., Han, Z., & Yang, S. (2021). *Resource Optimization for the Multi-user MIMO Systems Assisted Edge Cloud Computing.* 2021 IEEE 6th International Conference on Signal and Image Processing (ICSIP), Nanjing, China. 10.1109/ICSIP52628.2021.9688949

Gao, W., Morris, T., Reaves, B., & Richey, D. (2020). On SCADA controlsystem command and response injection and intrusion detection. 2010 eCrime Researchers Summit, Dallas, TX.

Gao, T., Tang, Q., Li, J., Zhang, Y., Li, Y., & Zhang, J. (2022). A Particle Swarm Optimization With Lévy Flight for Service Caching and Task Offloading in Edge-Cloud Computing. *IEEE Access : Practical Innovations, Open Solutions*, *10*, 76636–76647. doi:10.1109/ACCESS.2022.3192846

Gaurav, A., Psannis, K., & Peraković, D. (2022). Security of cloud-based medical internet of things (miots): A survey. [IJSSCI]. *International Journal of Software Science and Computational Intelligence*, *14*(1), 1–16. doi:10.4018/IJSSCI.285593

Gautam, D., Bhatia, S., Goel, N., & Mallikaijuna, B. G., & Naib, B. (2023). *Development of IoT Enabled Framework for LPG Gas Leakage Detection and Weight Monitoring System*. 2023 International Conference on Device Intelligence, Computing and Communication Technologies, (DICCT), Dehradun, India. 10.1109/DICCT56244.2023.10110294

Gawas, A. U. (2015). An overview on the evolution of mobile wireless communication networks: 1G-6G. *International Journal on Recent and Innovation Trends in Computing and Communication*, *3*(5), 3130–3133.

Gayatri, M. K., Jayasakthi, J., & Mala, G. A. (2015). *Providing Smart Agricultural solutions to farmers for better yielding using IoT. In 2015 IEEE Technological Innovation in ICT for Agriculture and Rural Development (TIAR)*. IEEE.

GDS Corp. (n.d.). *Understanding How A Catalyst Sensor Works*. GDS Corp.

Gebeyehu, M. N. (2019). Remote Sensing and GIS Application in Agriculture and Natural Resource Management. *Int J Environ Sci Nat Res.*, *19*(2), 556009. doi:10.19080/IJESNR.2019.19.556009

Genda, K., Abe, M., & Kamamura, S. (2020). *Video Communication Optimization Using Distributed Edge Computing*. 2020 21st Asia-Pacific Network Operations and Management Symposium (APNOMS), Daegu, Korea (South). 10.23919/APNOMS50412.2020.9237044

Getty Images. (2019). *What is a Gas Sensor ? Micron ships 232-layer NAND SSD for PCs, laptops*. Getty Images.

Ghamari, M., Janko, B., Sherratt, R. S., Harwin, W., Piechockic, R., & Soltanpur, C. (2016, June). A survey on wireless body area networks for healthcare systems in residential environments. *Sensors (Basel)*, *16*(6), 831. doi:10.339016060831 PMID:27338377

Ghavate, S. & Joshi, H. (2021). Smart Farming using IoT and Machine Learning with Image Processing. *International Journal of Current Engineering and Technology*, 15-19.

Gielis, J., Shankar, A., & Prorok, A. (2022). A Critical Review of Communications in Multi-robot Systems. *Current Robotics Reports*, *3*(4), 213–225. doi:10.100743154-022-00090-9 PMID:36404913

Gillis, A. (2022). Internet of Things. *Tech Target*. https://internetofthingsagenda.techtarget.com/definition/Internet-of-Things-IoT

GlobalData Thematic Research. (2022, February 14). *Ai is key for O&G companies targeting sustainability*. Offshore Technology. https://www.offshore-technology.com/comment/ai-for-og-sustainability/

Goel, R., Jain, A., Verma, K., Bhushan, M., Kumar, A., & Negi, A. (2020). Mushrooming trends and technologies to aid visually impaired people. *2020 International Conference on Emerging Trends in Information Technology and Engineering (Ic-ETITE)*, (pp. 1–5). IEEE. 10.1109/ic-ETITE47903.2020.437

Goel, S. (2012). Transformation from LEL to UML. *International Journal of Computer Applications*, *48*(12), 888–975.

Gokhale, P., Bhat, O., & Bhat, S. (2018). Introduction to IOT. *International Advanced Research Journal in Science. Engineering and Technology, 5*(1), 41–44.

Gomes, C. M. A., Lemos, G. C., & Jelihovschi, E. G. (2020). Comparing the predictive power of the CART and CTREE algorithms. *Avaliação Psicológica, 19*(1), 87–96.

Goodfellow, I., Pouget-Abadie, J., Mirza, M., Xu, B., Warde-Farley, D., Ozair, S., Courville, A., & Bengio, Y. (2020). Generative adversarial networks. *Communications of the ACM, 63*(11), 139–144. doi:10.1145/3422622

Gopalakrishnan, K. (2018). Deep Learning in Data-Driven Pavement Image Analysis and Automated Distress Detection: A Review. *Data, 3*(3), 28. doi:10.3390/data3030028

Goswami, A., & Banerjee, B. (2021). A Swarm Intelligence Inspired Framework for Cooperative Multi-Robot Systems. *IEEE Transactions on Systems, Man, and Cybernetics. Systems, 51*, 1697–1709. doi:10.1109/TSMC.2019.2952262

Goyal, S., & Singh, A. (2018). Challenges of IoT-Based Supply Chain Management: A Review. *International Journal of Supply Chain Management, 7*(6), 223–231.

GraboyesR. F.BryanD.CoglianeseJ. (2020, January 1). Overcoming Technological and Policy Challenges to Medical Uses of Unmanned Aerial Vehicles. Papers.ssrn.com. https://papers.ssrn.com/sol3/papers.cfm?abstract_id=3561743 doi:10.2139/ssrn.3561743

Guan, K., Rupp, M., Kurner, T., Briso, C., Matolak, D. W., Takada, J.-I., & Wang, W. (2020). IEEE Access Special Section Editorial: 5G and Beyond Mobile Wireless Communications Enabling Intelligent Mobility. *IEEE Access : Practical Innovations, Open Solutions, 8*, 208892–208897. doi:10.1109/ACCESS.2020.3037635

Gubbi, J., Buyya, R., Marusic, S., & Palaniswami, M. (2013). Internet of Things (IoT): A vision, architectural elements, and future directions. *Future Generation Computer Systems, 29*(7), 1645–1660. doi:10.1016/j.future.2013.01.010

Guevara & AuatCheein. (2020). The Role of 5G Technologies: Challenges in Smart Cities and Intelligent Transportation Systems. *Sustainability, 12*(16), 6469. doi:10.3390/su12166469

Gupta, A., Kaushik, D., Garg, M., & Verma, A. (2020). Machine Learning model for Breast Cancer Prediction. *Fourth International Conference on I-SMAC (IoT in Social, Mobile, Analytics and Cloud) (I-SMAC),* Palladam, India.

Gupta, A., Nagda, D., Nikhare, P., & Sandbhor, A. (2021). Smart Crop Prediction using IoT and Machine Learning, *International Journal of Engineering Research & Technology (IJERT), 9*(3), 18-21.

Gupta, M., Ghatak, S., Gupta, A., & Mukherjee, A. L. (Eds.). (2022). Artificial Intelligence on Medical Data. *Proceedings of International Symposium, ISCMM 2021* (Vol. 37). Springer Nature.

Gupta, A., & Jha, R. K. (2015). A Survey of 5G Network: Architecture and Emerging Technologies. *IEEE Access : Practical Innovations, Open Solutions, 3*, 1206–1232. doi:10.1109/ACCESS.2015.2461602

Gupta, B. B., & Quamara, M. (2020). An overview of Internet of Things (IoT): Architectural aspects, challenges, and protocols. *Concurrency and Computation, 32*(21), e4946. doi:10.1002/cpe.4946

Gupta, D., & Shah, M. (2022). A comprehensive study on artificial intelligence in oil and gas sector. *Environmental Science and Pollution Research International, 29*(34), 50984–50997. doi:10.100711356-021-15379-z PMID:34378133

Gupta, N., Janani, S., Dilip, R., Hosur, R., Chaturvedi, A., & Gupta, A. (2022). Wearable Sensors for Evaluation Over Smart Home Using Sequential Minimization Optimization-based Random Forest. *International Journal of Communication Networks and Information Security, 14*(2), 179–188. doi:10.17762/ijcnis.v14i2.5499

Gupta, S., & Singh, N. (2023). Toward intelligent resource management in dynamic Fog Computing-based Internet of Things environment with Deep Reinforcement Learning: A survey. *International Journal of Communication Systems*, *36*(4), e5411. doi:10.1002/dac.5411

Gutierrez-Martinez, V. J., Moreno-Bautista, C. A., Lozano-Garcia, J. M., Pizano-Martinez, A., Zamora-Cardenas, E. A., & Gomez-Martinez, M. A. (2019). A Heuristic Home Electric Energy Management System Considering Renewable Energy Availability. *Energies*, *12*(4), 671. doi:10.3390/en12040671

Habson, K. (2021, October 31). *Cognitive AI to boost sustainability for the oil and gas sector*. Gulf Business. https://gulfbusiness.com/cognitive-ai-to-boost-sustainability-for-the-oil-and-gas-sector/

HaddadPajouh, H., Dehghantanha, A., Parizi, R. M., Aledhari, M., & Karimipour, H. (2021). A survey on internet of things security: Requirements, challenges, and solutions. *Internet of Things*, *14*, 100129. doi:10.1016/j.iot.2019.100129

Haghi, M., Thurow, K., & Stoll, R. (2017). Wearable Devices in Medical Internet of Things: Scientific Research and Commercially Available Devices. *Healthcare Informatics Research*, *23*(1), 4. doi:10.4258/hir.2017.23.1.4 PMID:28261526

Hamilton & Charalambous. (2013). *Leak Detection Technology and Implementation*. IWA Publishing.

Han, L., Yu, C., Xiao, K., & Zhao, X. (2019). A new method of mixed gas identification based on a convolutional neural network for time series classification. *Sensors (Basel)*, *19*(9), 1960. doi:10.339019091960 PMID:31027348

Han, T., Nag, A., Chandra Mukhopadhyay, S., & Xu, Y. (2019). Carbon nanotubes and its gas-sensing applications: A review. *Sensors and Actuators. A, Physical*, *291*, 107–143. doi:10.1016/j.sna.2019.03.053

Han, Z. (2022). *Research on Big Data Mining Application of Internet of Things Based on Artificial Intelligence Technology*. 2022 International Conference on Computing, Robotics and System Sciences (ICRSS), Macau, China. 10.1109/ICRSS57469.2022.00025

Hariri, R. H., Fredericks, E. M., & Bowers, K. M. (2019). Uncertainty in big data analytics: Survey, opportunities, and challenges. *Journal of Big Data*, *6*(1), 1–16. doi:10.118640537-019-0206-3

Harvey, D. J. (2019). Gas chromatography l Gas chromatography/mass spectrometry. Encyclopedia of Analytical Science, 169–179. Science Direct. doi:10.1016/B978-0-12-409547-2.14103-4

Hashmi, M. F., Kunduru, P. R., Mujavar, S. A., Nandigama, S. S., & Keskar, A. G. (2023). Medical Anomaly Detection Using Human Action Recognition. In Image Processing and Intelligent Computing Systems (pp. 215-222). CRC Press.

Hassan, S. R., & Rashad, M. (2023). *Cloud Computing to Fog Computing: A Paradigm Shift*.

Hassler, S. C., & Baysal-Gurel, F. (2019). Unmanned aircraft system (UAS) technology and applications in agriculture. *Agronomy (Basel)*, *9*(10), 618. doi:10.3390/agronomy9100618

Havens, K. J., & Sharp, E. J. (2015). Thermal Imaging Techniques to Survey and Monitor Animals in the Wild: A Methodology. In Thermal Imaging Techniques to Survey and Monitor Animals in the Wild: A Methodology.

Hazra, A., Rana, P., Adhikari, M., & Amgoth, T. (2023). Fog computing for next-generation internet of things: Fundamental, state-of-the-art and research challenges. *Computer Science Review*, *48*, 100549. doi:10.1016/j.cosrev.2023.100549

Heble, S., Kumar, A., Prasad, K. V. D., Samirana, S., Rajalakshmi, P., & Desai, U. B. (2018). A low power IoT network for smart agriculture. In *2018 IEEE 4th World Forum on Internet of Things (WF-IoT)* (pp. 609-614). IEEE. 10.1109/WF-IoT.2018.8355152

Hichri, B., Fauroux, J. C., Adouane, L., Doroftei, I., & Mezouar, Y. (2019). Design of cooperative mobile robots for co-manipulation and transportation tasks. *Robotics and Computer-integrated Manufacturing*, *57*, 412–421. doi:10.1016/j.rcim.2019.01.002

Hidalgo-Mompeán, F., Fernández, J. F. G., Cerruela-García, G., & Márquez, A. C. (2021). Dimensionality analysis in machine learning failure detection models. A case study with LNG compressors. *Computers in Industry*, *128*, 103434. doi:10.1016/j.compind.2021.103434

Hido, S., Tsuboi, Y., Kashima, H., Sugiyama, M., & Kanamori, T. (2011). Statistical outlier detection using direct density ratio estimation. *Knowledge and Information Systems*, *26*(2), 309–336. doi:10.100710115-010-0283-2

Hochstein, B., Rangarajan, D., Mehta, N., & Kocher, D. (2020). An Industry/Academic Perspective on Customer Success Management. *Journal of Service Research*, *23*(1), 3–7. doi:10.1177/1094670519896422

Hoffmann, H. (2020). Kernel PCA for novelty detection. *Pattern Recognition*, *40*(3), 863–874. doi:10.1016/j.patcog.2006.07.009

Holzinger, A., Saranti, A., Angerschmid, A., Retzlaff, C. O., Gronauer, A., Pejakovic, V., Medel-Jimenez, F., Krexner, T., Gollob, C., & Stampfer, K. (2022). (2022). Digital Transformation in Smart Farm and Forest Operations Needs Human-Centered AI: Challenges and Future Directions. *Sensors (Basel)*, *22*(8), 3043. doi:10.339022083043 PMID:35459028

Hong, S., Kim, S. H., & Kwon, M. (2022). Determinants of digital innovation in the public sector. *Government Information Quarterly*, *39*(4), N.PAG. doi:10.1016/j.giq.2022.101723

Hossein, MK, Ibrahim, S, & Hosseinkhani, J. (2013). Outlier detection in stream data by clustering method. *Int J Adv Comp Sci Inform Technol.*, *2*(3), 25–34.

Huo, Z. h., Wang, P., Zhang, S.-j., Wang, D., & Kong, Z. (2020). *A Two-Step Multi-objective Optimization Frame-work for Microgrid Scheduling Problem Based on Cloud-edge Computing*. 2020 IEEE 4th Conference on Energy Internet and Energy System Integration (EI2), Wuhan, China. 10.1109/EI250167.2020.9347092

Hussain, F., Hussain, R., Hassan, S. A., & Hossain, E. (2020). Machine learning in IoT security: Current solutions and future challenges. *IEEE Communications Surveys and Tutorials*, *22*(3), 1686–1721. doi:10.1109/COMST.2020.2986444

Hussain, M., Al-Fuqaha, A., Guizani, M., & Mohammadi, M. (2021). Challenges and Opportunities of IoT-Based Healthcare Systems: A Comprehensive Review. *IEEE Access : Practical Innovations, Open Solutions*, *9*, 44622–44654. doi:10.1109/ACCESS.2021.3078443

Huuskonen, J., & Oksanen, T. (2018). Soil sampling with drones and augmented reality in precision agriculture. *Computers and Electronics in Agriculture*, *154*, 25–35. doi:10.1016/j.compag.2018.08.039

Hu, Z., Tang, J., Wang, Z., Zhang, K., Zhang, L., & Sun, Q. (2018). Deep learning for image-based cancer detection and diagnosis– A survey. *Pattern Recognition*, *83*, 134–149. doi:10.1016/j.patcog.2018.05.014

Hwerbi, K., Benalaya, N., Amdouni, I. (2022). A Survey on the Opportunities of Blockchain and UAVs in Agriculture. *IEEE 11th IFIP International Conference on Performance Evaluation and Modeling in Wireless and Wired Networks (PEMWN)*. IEEE. 10.23919/PEMWN56085.2022.9963871

I., P. T. (n.d.). *India's first electronic toll collection system launched on Ahmedabad-Mumbai Highway*. Indian Tollways - an e-News Magazine on BOT Road Projects. https://www.indiantollways.com/indias-first-electronic-toll-collection-system-launched-on-ahmedabad-mumbai-highway/

Ibrahim, N. M., Alharbi, A. A., Alzahrani, T. A., Abdullah, M. A., Ibrahim, A. A., Hameed, A. M., Albabtain, A. S., Alqahtani, D. A., Alsawwaf, M. K., & Almuqhim, A. A. (2022). Well performance classification and prediction: Deep learning and machine learning long term regression experiments on oil, gas, and water production. *Sensors (Basel)*, *22*(14), 1–22. doi:10.339022145326 PMID:35891005

Ifrim, C., Pintilie, A. M., Apostol, E., Dobre, C., & Pop, F. (2017). The art of advanced healthcare applications in big data and IoT systems. In *Advances in mobile cloud computing and big data in the 5G Era* (pp. 133–149). Springer. doi:10.1007/978-3-319-45145-9_6

Iliev, Y., & Ilieva, G. (2022). A Framework for Smart Home System with Voice Control Using NLP Methods. *Electronics (Basel)*, *12*(1), 116. doi:10.3390/electronics12010116

IndustryTap. (2022). *The Impact of IoT on The Agricultural Sector*. Industry Tap. https://www.industrytap.com/the-impact-of-iot-on-the-agricultural-sector/64411

Ingre, B., Yadav, A., & Soni, A. K. (2018). Decision tree based intrusion detection system for NSL-KDD dataset. In Information and Communication Technology for Intelligent Systems (ICTIS 2017)-Volume 2 2 (pp. 207-218). Springer International Publishing. doi:10.1007/978-3-319-63645-0_23

Ioannou, C., & Vassiliou, V. (2020, May). Experimentation with local intrusion detection in IoT networks using supervised learning. In *2020 16th International Conference on Distributed Computing in Sensor Systems (DCOSS)* (pp. 423-428). IEEE. 10.1109/DCOSS49796.2020.00073

IoT in agriculture: Smart Farming technologies for Future. (2022). Cropin. Www.cropin.com. https://www.cropin.com/iot-in-agriculture

IoT ONE Digital Transformation Advisors. (2020.). *Industrial IoT Use Case Profile*. IoT ONE. https://www.iotone.com/usecase/agricultural-drones/u54

Islam, M. M., Nooruddin, S., Karray, F., & Muhammad, G. (2023). Multi-level feature fusion for multimodal human activity recognition in Internet of Healthcare Things. *Information Fusion*, *94*, 17–31. doi:10.1016/j.inffus.2023.01.015

Iwuanyanwu, C. C. (2021). Determinants and impact of artificial intelligence on organizational competitiveness: A study of listed American companies. *Journal of Service Science and Management*, *14*(5), 502–529. doi:10.4236/jssm.2021.145032

Iyer, R., & Sharma, A. (2019). IoT based home automation system with pattern recognition. *International Journal of Recent Technology and Engineering*, *8*(2), 3925–3929. doi:10.35940/ijrte.B2060.078219

Jafari, R., Razvarz, S., Gegov, A., & Vatchova, B. (2020). Deep Learning for Pipeline Damage Detection: An Overview of the Concepts and a Survey of the State-of-the-Art. *2020 IEEE 10th International Conference on Intelligent Systems, IS 2020 - Proceedings*, (pp. 178–182). IEEE. 10.1109/IS48319.2020.9200137

Jagtap, I., & Babbar, N. (2021). Predicting Air Pollutant using Data Mining and Machine Learning Algorithms. *Journal of Science and Technology*, *06*, 25–30. doi:10.46243/jst.2021.v6.i04.pp25-30

Jain, V., Beram, S. M., Talukdar, V., Patil, T., Dhabliya, D., & Gupta, A. (2022, November). Accuracy enhancement in machine learning during blockchain based transaction classification. In *2022 Seventh International Conference on Parallel, Distributed and Grid Computing (PDGC)* (pp. 536-540). IEEE. 10.1109/PDGC56933.2022.10053213

Jaiswal, A., Sodhi, H. S., Muzamil, H. M., Chandhok, R. S., Oore, S., & Sastry, C. S. (2021). Controlling BigGAN image generation with a segmentation network. In *Discovery Science: 24th International Conference, DS 2021,* (pp. 268-281). Springer International Publishing. https://doi.org/10.1007/978-3-030-88942-5_21

Jang, I., Kim, D., Lee, D., & Son, Y. (2018). An Agent-Based Simulation Modeling with Deep Reinforcement Learning for Smart Traffic Signal Control. *2018 International Conference on Information and Communication Technology Convergence (ICTC)*. IEEE. 10.1109/ICTC.2018.8539377

Javaid, A., Niyaz, Q., Sun, W., & Alam, M. (2016, May). A deep learning approach for network intrusion detection system. In *Proceedings of the 9th EAI International Conference on Bio-inspired Information and Communications Technologies (formerly BIONETICS)* (pp. 21-26). 10.4108/eai.3-12-2015.2262516

Javaid, M., & Khan, I. H. (2021). Internet of Things (IoT) enabled healthcare helps to take the challenges of COVID-19 Pandemic. *Journal of Oral Biology and Craniofacial Research*, *11*(2), ·209–214. doi:10.1016/j.jobcr.2021.01.015 PMID:33665069

Javanmardi, S., Shojafar, M., Mohammadi, R., Alazab, M., & Caruso, A. M. (2023). An SDN perspective IoT-Fog security: A survey. *Computer Networks*, *229*, 109732. doi:10.1016/j.comnet.2023.109732

Jeet, V., Dhillon, H. S., & Bhatia, S. (2015). Radio Frequency Home Appliance Control Based on Head Tracking and Voice Control for Disabled Person. *2015 Fifth International Conference on Communication Systems and Network Technologies*, Gwalior, India. 10.1109/CSNT.2015.189

Jha, K., Doshi, A., Patel, P., & Shah, M. (2019). A comprehensive review on automation in agriculture using artificial intelligence. *Artificial Intelligence in Agriculture*, *2*, 1–12. doi:10.1016/j.aiia.2019.05.004

Jia, M., Komeily, A., Wang, Y., & Srinivasan, R. S. (2019). Adopting Internet of Things for the development of smart buildings: A review of enabling technologies and applications. *Automation in Construction*, *101*, 111–126. doi:10.1016/j.autcon.2019.01.023

Jiang, Y. (2020). Combination of wearable sensors and internet of things and its application in sports rehabilitation. *Computer Communications*, *150*, 167–176. doi:10.1016/j.comcom.2019.11.021

Jian, Y., & Li, Y. (2019). Research on intelligent cognitive function enhancement of intelligent robot based on ant colony algorithm. *Cognitive Systems Research*, *56*, 203–212. doi:10.1016/j.cogsys.2018.12.014

Jia, X., Roels, J., Baets, R., & Roelkens, G. (2019). On-chip non-dispersive infrared CO_2 sensor based on an integrating cylinder†. *Sensors (Basel)*, *19*(19), 1–14. doi:10.339019194260 PMID:31575053

Jijo, B. T., & Abdulazeez, A. M. (2021). Classification based on decision tree algorithm for machine learning. *Evaluation*, *6*, 7.

Jindal, A., Goyal, S., Singh, A., & Singh, P. (2022). Challenges and Opportunities in IoT-Based Smart Manufacturing: A Comprehensive Review. *IEEE Access : Practical Innovations, Open Solutions*, *10*, 27035–27050. doi:10.1109/ACCESS.2022.3071264

Ju, C., & Son, H. Il. (2018). Performance evaluation of multiple UAV systems for remote sensing in agriculture. *Proceedings of the Workshop on Robotic Vision and Action in Agriculture at the IEEE International Conference on Robotics and Automation (ICRA)*, Brisbane, Australia.

Jyotir, M. & Vishal, J. (2021). Internet of Things and Machine Learning in Agriculture. *Internet of Things and Machine Learning*. Nova Publishers. doi:10.52305/MTXX5116

Kaarlela, A., & Halonen, R. (2019). Challenges of IoT for Service Innovation in Healthcare. *IEEE Access : Practical Innovations, Open Solutions*, *7*, 161426–161439. doi:10.1109/ACCESS.2019.2953786

Kabassi, K., & Virvou, M. (2020). Challenges and opportunities in IoT-based education: A review. *Computers & Education*, *144*, 103698. doi:10.1016/j.compedu.2019.103698

Kaddoum, G., Diab, H., & Gani, A. (2019). Challenges and Opportunities in IoT-Based Agriculture: A Review. *IEEE Access : Practical Innovations, Open Solutions*, *7*, 174373–174390. doi:10.1109/ACCESS.2019.2954717

Kadhim, W. (2019). Case study of Dubai as a Smart City. *International Journal of Computer Applications*, *178*(40), 35–37. doi:10.5120/ijca2019919291

Kaggle. (n.d.). *Home*. Kaggle. https://www.kaggle.com/

Kaiwartya, O., Abdullah, A. H., Cao, Y., Altameem, A., Prasad, M., Lin, C. T., & Liu, X. (2016). Internet of Vehicles: Motivation, Layered Architecture, Network Model, Challenges, and Future Aspects. *IEEE Access : Practical Innovations, Open Solutions*, *4*, 5356–5373. doi:10.1109/ACCESS.2016.2603219

Kakandwar, S., Bhushan, B., & Kumar, A. (2023). Integrated machine learning techniques for preserving privacy in Internet of Things (IoT) systems. In *Blockchain Technology Solutions for the Security of Iot-Based Healthcare Systems* (pp. 45–75). Academic Press. doi:10.1016/B978-0-323-99199-5.00012-4

Kalsoom, T., Ahmed, S., Rafi-ul-Shan, P. M., Azmat, M., Akhtar, P., Pervez, Z., Imran, M. A., & Ur-Rehman, M. (2021). Impact of IoT on Manufacturing Industry 4.0: A New Triangular Systematic Review. *Sustainability (Basel)*, *13*(22), 12506. doi:10.3390u132212506

Kaluvan, H., Baskar, P. K., Ramanathan, S., Sundar, S., & Jeyaram, S. (2021). Intelligent transportation system and smart traffic flow with IOT. *Indian Journal of Radio & Space Physics*, *50*. http://op.niscpr.res.in/index.php/IJRSP/article/view/62096/4 65480553

Kaplan, A., & Haenlein, M. (2019). Siri, Siri, in my hand: Who's the fairest in the land? On the interpretations, illustrations, and implications of artificial intelligence. *Business Horizons*, *62*(1), 15–25. doi:10.1016/j.bushor.2018.08.004

Karim, S. M., Habbal, A., Chaudhry, S. A., & Irshad, A. (2022). Architecture, Protocols, and Security in IoV: Taxonomy, Analysis, Challenges, and Solutions. *Security and Communication Networks*, *2022*, 1–19. doi:10.1155/2022/1131479

Kashani, M. H., Madanipour, M., Nikravan, M., Asghari, P., & Mahdipour, E. (2021). A systematic review of IoT in healthcare: Applications, techniques, and trends. *Journal of Network and Computer Applications*, *192*, 103164. doi:10.1016/j.jnca.2021.103164

Katal, A., Wazid, M., & Goudar, R. H. (2013). Big data: issues, challenges, tools and good practices. In *Proceedings of the 6th International Conference on Contemporary Computing* (IC3'13), (pp. 404–409). IEEE. 10.1109/IC3.2013.6612229

Katal, A. (2023). Leveraging Fog Computing for Healthcare. In *Deep Learning Technologies for the Sustainable Development Goals: Issues and Solutions in the Post-COVID Era* (pp. 51–68). Springer Nature Singapore. doi:10.1007/978-981-19-5723-9_4

Kaushik, D., & Garg, M., & Gupta, M. (2022). Utilizing Machine Learning and Deep Learning in Cybesecurity: An Innovative Approach. Cyber Security and Digital Forensics: Challenges and Future Trends, (pp.271-293). Wiley.

Kaygusuz, K. (2009). Energy and environmental issues relating to greenhouse gas emissions for sustainable development in Turkey. *Renewable & Sustainable Energy Reviews*, *13*(1), 253–270. doi:10.1016/j.rser.2007.07.009

Kazy Noor-e-Alam Siddiquee, S. (2022). Development of Algorithms for an IoT-Based Smart Agriculture Monitoring System. Wireless Communications and Mobile Computing. doi:10.1155/2022/7372053

Kedia, S., & Bhushan, M. (2022). Prediction of mortality from heart failure using machine learning. *2022 2nd International Conference on Emerging Frontiers in Electrical and Electronic Technologies (ICEFEET)*, 1–6.

Kekuda, A., Anirudh, R., & Krishnan, M. (2021). Reinforcement Learning based Intelligent Traffic Signal Control using n-step SARSA. *2021 International Conference on Artificial Intelligence and Smart Systems (ICAIS)*. IEEE. 10.1109/ICAIS50930.2021.9395942

Kempf, J., Arkko, J., Beheshti, N., & Yedavalli, K. (2011). *Thoughts on reliability in the internet of things*. Interconnecting Smart Objects with Internet Work.

Keshtgari, M., & Deljoo, A. (2011). *A wireless sensor network solution for precision agriculture based on zigbee technology*. Academic Press.

Khalid, A., Iftikhar, M. A., Almogren, A., Khalid, R., Afzal, M., & Javaid, N. (2021). A blockchain based incentive provisioning scheme for traffic event validation and information storage in VANETs. *Information Processing & Management, 58*(2), 102464. doi:10.1016/j.ipm.2020.102464

Khan, N., Yaqoob, I., Hashem, I.A., Inayat, Z., Ali, M., Kamaleldin, W., Alam, M., Shiraz, M., & Gani, A. (2014). Big data: survey, technologies, opportunities, and challenges. *The Scientific World Journal*.

Khanna, A., & Kaur, S. (2019). Evolution of Internet of Things (IoT) and its significant impact in the field of Precision Agriculture. *Computers and Electronics in Agriculture, 157*, 218–231. doi:. compag.2018.12.039 doi:10.1016/j

Khan, Y., Ostfeld, A. E., Lochner, C. M., Pierre, A., & Arias, A. C. (2016). Monitoring of Vital Signs with Flexible and Wearable Medical Devices. *Advanced Materials, 28*(22), 4373–4395. doi:10.1002/adma.201504366 PMID:26867696

Khatri-Chhetri, A., Aggarwal, P. K., Joshi, P. K., & Vyas, S. (2017). Farmers' prioritization of climate-smart agriculture (CSA) technologies. *Agricultural Systems, 151*, 184–191. doi:10.1016/j.agsy.2016.10.005

Khazai, S., Homayouni, S., Safari, A., & Mojaradi, B. (2021). Anomaly detection in hyperspectral images based on an adaptive support vector method. *IEEE Geoscience and Remote Sensing Letters, 8*(4), 646–650. doi:10.1109/LGRS.2010.2098842

Khetre, A., & Gupta, P. (2021). Path planning of mobile robots for efficient transportation of heavy loads using a novel modified differential evolution algorithm. *Applied Soft Computing, 104*, 107194. doi:10.1016/j.asoc.2021.107194

Khutey, R. (2015). Future of wireless technology 6G & 7G. *International Journal of Electrical and Electronics Research, 3*(2), 583–585.

Kibria, M. G., Nguyen, K., Villardi, G. P., Zhao, O., Ishizu, K., & Kojima, F. (2018, May 17). Big data analytics, machine learning, and artificial intelligence in next-generation wireless networks. *IEEE Access : Practical Innovations, Open Solutions, 6*, 32328–32338. doi:10.1109/ACCESS.2018.2837692

Kimani, K., Oduol, V., & Langat, K. (2019). Cyber security challenges for IoT-based smart grid networks. *International Journal of Critical Infrastructure Protection, 25*, 36–49. doi:10.1016/j.ijcip.2019.01.001

Kim, J., Ha, T., Yoo, W., & Chung, J. M. (2019). Task Popularity based Energy Minimized Computation Offloading for Fog Computing Wireless Networks. *IEEE Wireless Communications Letters, 8*(4), 1200–1203. doi:10.1109/LWC.2019.2911521

Kim, J., Kim, S., Ju, C., & Son, H. (2019). Unmanned aerial vehicles in agriculture: A review of perspective of platform, control, and applications. *IEEE Access : Practical Innovations, Open Solutions, 7*, 105100–105115. doi:10.1109/ACCESS.2019.2932119

Kim, Y. K., Wang, H., & Mahmud, M. S. (2016). Wearable body sensor network for health care applications. In *Smart textiles and their applications* (pp. 161–184). Woodhead Publishing. doi:10.1016/B978-0-08-100574-3.00009-6

Kintzlinger, M., & Nissim, N. (2019). Keep an eye on your personal belongings! The security of personal medical devices and their ecosystems. *Journal of Biomedical Informatics, 95*, 103233. doi:10.1016/j.jbi.2019.103233 PMID:31201966

Kitani, K. M., Ziebart, B. D., Bagnell, J. A., & Hebert, M. (2012, October). Activity forecasting. In *European conference on computer vision* (pp. 201-214). Springer, Berlin, Heidelberg.

Kocejko, T., & Wtorek, J. (2012). Gaze pattern lock for elders and disabled. In *Information Technologies in Biomedicine* (pp. 589–602). Springer. doi:10.1007/978-3-642-31196-3_59

Kodali, R. Jain, V., & Karagwal, S. (2016). *IoT-based Smart Greenhouse*. IEEE. . doi:10.1109/R10-HTC.2016.7906846

Koupaie, HM, Ibrahim, S, & Hosseinkhani, J. (2013). *Outlier detection in stream data by clustering method*. Int J Adv Comput Sci Inf Technol.

Krakowski, S., Luger, J., & Raisch, S. (2022). Artificial intelligence and the changing sources of competitive advantage. *Strategic Management Journal, 1,* 1–28. doi:10.1002mj.3387

Kulbacki, M., Segen, J., Knieć, W., Klempous, R., Kluwak, K., Nikodem, J., Kulbacka, J., & Serester, A. (2018). Survey of drones for agriculture automation from planting to harvest. *2018 IEEE 22nd International Conference on Intelligent Engineering Systems (INES),* 353–358.

Kumar Sagar, P., Garg, K., & Dutta, C. (2018, August 1). *Application of Internet of Things in Fast Moving Consumer Goods Sector to increase Business Efficiency.* IEEE Xplore. doi:10.1109/ICGCIoT.2018.8753033

Kumar, A., Bhushan, M., Galindo, J. A., Garg, L., & Hu, Y.-C. (2023). *Machine Intelligence, Big Data Analytics, and IoT in Image Processing: Practical Applications.* John Wiley & Sons. doi:10.1002/9781119865513

Kumar, D, Bezdek, JC, Rajasegarar, S, & ... (2016). Adaptive cluster tendency visualization and anomaly detection for streaming data. *ACM Transactions on Knowledge Discovery from Data, 11*(2), 24–40.

Kumari, S., & Kaur, A. (2020). Towards Cloud Robotics: Survey of Trends and Technologies. *IEEE Access : Practical Innovations, Open Solutions, 8,* 156800–156819. doi:10.1109/ACCESS.2020.3019383

Kumari, S., & Singh, S. K. (2022). Machine learning-based time series models for effective CO_2 emission prediction in India. *Environmental Science and Pollution Research International, 0123456789.* Advance online publication. doi:10.100711356-022-21723-8 PMID:35780266

Kumar, S., & Bhatt, A. (2013). Foundations of artificial intelligence. *International Journal of Technical Research and Applications, 1*(4), 52–56.

Kumar, S., Singh, R., Kaur, R., & Kumar, S. (2022). Challenges and Opportunities in IoT-Based Smart Grids: A Comprehensive Review. *IEEE Transactions on Industrial Informatics, 18*(2), 1176–1190. doi:10.1109/TII.2021.3072124

Kuppusamy, P., Shanmugananthan, S., & Tomar, P. (2021). Emerging Technological Model to Sustainable Agriculture. In P. Tomar & G. Kaur (Eds.), Artificial Intelligence and IoT-Based Technologies for Sustainable Farming and Smart Agriculture (pp. 101-122). IGI Global. doi:10.4018/978-1-7998-1722-2.ch007

Kurniawan, J., Syahra, S. G., Dewa, C. K., & Afiahayati. (2018). Traffic Congestion Detection: Learning from CCTV Monitoring Images using Convolutional Neural Network. *Procedia Computer Science, 144,* 291–297. doi:10.1016/j.procs.2018.10.530

Kwon, D., Jung, G., Shin, W., Jeong, Y., Hong, S., Oh, S., Bae, J. H., Park, B. G., & Lee, J. H. (2021). Low-power and reliable gas sensing system based on recurrent neural networks. *Sensors and Actuators, B: Chemical, 340*(September 2020), 129258. doi:10.1016/j.snb.2020.129258

Lai, P. (2022). Dynamic User Allocation in Stochastic Mobile Edge Computing Systems. *2022 IEEE World Congress on Services (SERVICES),* Barcelona, Spain. 10.1109/SERVICES55459.2022.00030

Lakshmanna, K., Kaluri, R., Gundluru, N., Alzamil, Z. S., Rajput, D. S., Khan, A. A., Haq, M. A., & Alhussen, A. (2022). A review on deep learning techniques for IoT data. *Electronics (Basel)*, *11*(10), 1604. doi:10.3390/electronics11101604

Lakshmi, J. V. N., & Naresh, G. V. N. (2018). *A Review on Developing Tech-Agriculture using Deep Learning Methods by Applying UAVs*. IJSRCSEIT.

Lakshmisudha, K., Hegde, S., Kale, N., & Iyer, S. (2016). Smart precision based agriculture using sensors. *International Journal of Computer Applications*, *146*(11), 36–38. doi:10.5120/ijca2016910916

Lambertini, A., Mandanici, E., Tini, M. A., & Vittuari, L. (2022). Technical Challenges for Multi-Temporal and Multi-Sensor Image Processing Surveyed by UAV for Mapping and Monitoring in Precision Agriculture. *Remote Sensing (Basel)*, *14*(19), 4954. doi:10.3390/rs14194954

Lamiae, E., Hicham, G. T., Fatiha, E., & Mohammed, B. (2023, March). Smart Home and Machine Learning as a Sustainable Healthcare Solution: Review and Perspectives. In *Innovations in Smart Cities Applications Volume 6: The Proceedings of the 7th International Conference on Smart City Applications* (pp. 145-155). Cham: Springer International Publishing.

Lan, L., You, L., Zhang, Z., Fan, Z., Zhao, W., Zeng, N., Chen, Y., & Zhou, X. (2020). Generative adversarial networks and its applications in biomedical informatics. *Frontiers in Public Health*, *8*, 164. doi:10.3389/fpubh.2020.00164 PMID:32478029

Larranaga, P., Poza, M., Yurramendi, Y., Murga, R. H., & Kuijpers, C. M. H. (1996). Structure learning of Bayesian networks by genetic algorithms: A performance analysis of control parameters. *IEEE Transactions on Pattern Analysis and Machine Intelligence*, *18*(9), 912–926. doi:10.1109/34.537345

LeCun, Y., Bengio, Y., & Hinton, G. (2015). Deep learning. *Nature*, *521*(7553), 436–444. doi:10.1038/nature14539 PMID:26017442

Lee, I. (2022). How the IoT and AI are impacting Agriculture. *Relevant Software*. https://relevant.software/blog/iot-ai-agriculture/

Lee, J. S., Su, Y. W., & Shen, C. C. (2007, November). A comparative study of wireless protocols: Bluetooth, UWB, ZigBee, and Wi-Fi. In *IECON 2007-33rd Annual Conference of the IEEE Industrial Electronics Society* (pp. 46-51). IEEE.

Lee, S. K., Bae, M., & Kim, H. (2017). Future of IoT networks: A survey. *Applied Sciences (Basel, Switzerland)*, *7*(10), 1072. doi:10.3390/app7101072

Lee, S., & Lee, S. (2020). Resource Allocation for Vehicular Fog Computing Using Reinforcement Learning Combined With Heuristic Information. *IEEE Internet of Things Journal*, *7*(10), 10450–10464. doi:10.1109/JIOT.2020.2996213

Lellis, C. (2022, February 8). *10 most regulated industries in the U.S.* Perillon. https://www.perillon.com/blog/10-most-regulated-industries-in-the-us

Li B. (2022). READ: Robustness-Oriented Edge Application Deployment in Edge Computing Environment. *IEEE Transactions on Services Computing, 15*(3), 1746-1759. . doi:10.1109/TSC.2020.3015316

Li, Q., Wu, J. T., Liu, Y., Qi, X. M., Jin, H. G., Yang, C., Liu, J., Li, G. L., & He, Q. G. (2021). Recent advances in black phosphorus-based electrochemical sensors: A review. *Analytica Chimica Acta, 1170*(xxxx), 338480. doi:10.1016/j.aca.2021.338480

Li, Y., & Tian, L. (2014). Comprehensive Evaluation Method of Reliability of Internet of Things. In *IEEE 2014 Ninth International Conference on P2P, Parallel, Grid, Cloud and Internet Computing*, (pp. 262 –266). IEEE.

Liao, D., Sun, G., Li, H., Yu, H., & Chang, V. (2017). The framework and algorithm for preserving user trajectory while using location-based services in IoT-cloud systems. *Cluster Computing*, *20*(3), 2283–2297. doi:10.100710586-017-0986-1

Library of Congress. (n.d.). *Oil and Gas Industry: A Research Guide: U.S. Regulatory Agencies*. Library of Congress. https://guides.loc.gov/oil-and-gas-industry/laws/agencies

Li, G., Liu, H., Huang, T., Han, J., & Xiao, J. (2023). An effective approach for non-singular trajectory generation of a 5-DOF hybrid machining robot. *Robotics and Computer-integrated Manufacturing*, *80*, 102477. doi:10.1016/j.rcim.2022.102477

Li, J., Yu, F. R., Deng, G., Luo, C., Ming, Z., & Yan, Q. (2020). IndustrialInternet: A Survey on the Enabling Technologies, Applications, andChallenges. *IEEE Communications Surveys and Tutorials*, 1504–1526.

Li, M., Li, Y., & Li, X. (2019). Challenges and opportunities in IoT-based transportation systems: A review. *IEEE Internet of Things Journal*, *6*(5), 8249–8261. doi:10.1109/JIOT.2019.2937666

Lin, T., Lv, X., Hu, Z., Xu, A., & Feng, C. (2019). Semiconductor metal oxides as chemoresistive sensors for detecting volatile organic compounds. In Sensors (Switzerland), 19(2). doi:10.339019020233

Linnamo, V. (2023). Sensor Technology for Sports Monitoring. *Sensors (Basel)*, *23*(2), 572. doi:10.339023020572 PMID:36679367

Li, Q., Zhao, J., Gong, Y., & Zhang, Q. (2019). Energy-efficient computationoffloading and resource allocation in fog computing for Internet ofEverything. *China Communications*, *16*(3), 32–41.

Liqiang, Z., Shouyi, Y., Leibo, L., Zhen, Z., & Shaojun, W. (2011). A crop monitoring system based on wireless sensor network. *Procedia Environmental Sciences*, *11*, 558–565. doi:10.1016/j.proenv.2011.12.088

Li, R., Wang, N., Feng, F., Zhang, G., & Wang, X. (2020). Exploring global and local linguistic representations for text-to-image synthesis. *IEEE Transactions on Multimedia*, *22*(12), 3075–3087. doi:10.1109/TMM.2020.2972856

Li, S., Cui, T., & Alam, M. (2021). Reliability analysis of the internet of things using Space Fault Network. *Alexandria Engineering Journal*, *60*(1), 1259–1270. doi:10.1016/j.aej.2020.10.049

Liu, J., Xiong, K., Fan, P., Zhong, Z., & Letaief, K. B. (2019). Optimal Design ofSWIPT-Aware Fog Computing Networks. arXiv preprintarXiv:1901.08997.

Liu, W., Kim, S., Marczuk, K. A., & Ang, M. H. (2014). *Vehicle motion intention reasoning using cooperative perception on urban road*. IEEE. doi:10.1109/ITSC.2014.6957727

Liu, J., & Wu, W. (2021). Automatic Image Annotation Using Improved Wasserstein Generative Adversarial Networks. *IAENG International Journal of Computer Science*, *48*(3).

Liu, Y., Xiao, M., Chen, S., Bai, F., Pan, J., & Zhang, D. (2021). An intelligent edge-chain-enabled access control mechanism for IoV. *IEEE Internet of Things Journal*, *8*(15), 12231–12241. doi:10.1109/JIOT.2021.3061467

Livelihoods. (2021). Regenerative Agriculture: from the key principles to the practice. *Livelihoods*. https://livelihoods.eu/regenerative-agriculture-from-key-principles-to-practice/

Li, Y., & Wang, S. (2018). An Energy-Aware Edge Server Placement Algorithm in Mobile Edge Computing. *2018 IEEE International Conference on Edge Computing (EDGE)*, San Francisco, CA, USA. 10.1109/EDGE.2018.00016

Li, Y., Xia, J., Zhang, S., Yan, J., Ai, X., & Dai, K. (2012). An efficient intrusion detection system based on support vector machines and gradually feature removal method. *Expert Systems with Applications*, *39*(1), 424–430. doi:10.1016/j.eswa.2011.07.032

Li, Y., Yang, G., Su, Z., Li, S., & Wang, Y. (2023). Human activity recognition based on multienvironment sensor data. *Information Fusion*, *91*, 47–63. doi:10.1016/j.inffus.2022.10.015

Lu, J., Liu, X., Yue, T., Wang, W., & Zhao, L. (2020). *Research on Edge Computing-oriented Storm Edge Node Scheduling Optimization Strategy*. 2020 IEEE Sustainable Power and Energy Conference (iSPEC), Chengdu, China. 10.1109/iSPEC50848.2020.9350984

Lu, C. (2019). Design of a Simple Aerial Photo UAV for Agriculture. *Revista de la Facultad de Agronomía*, *36*(3), 781–792.

Lucamarini, M. (2018). Record Distance for Quantum Cryptography. *Physics (College Park, Md.)*, *11*. https://physics.aps.org/articles/v11/111

Lu, H., Du, M., Qian, K., He, X., & Wang, K. (2021). GAN-based data augmentation strategy for sensor anomaly detection in industrial robots. *IEEE Sensors Journal*, *22*(18), 17464–17474. doi:10.1109/JSEN.2021.3069452

Lu, H., & Ma, X. (2020). Hybrid decision tree-based machine learning models for short-term water quality prediction. *Chemosphere*, *249*, 126169. doi:10.1016/j.chemosphere.2020.126169 PMID:32078849

Lu, N., Fan, S., Zhao, Y., Yang, B., Hua, Z., & Wu, Y. (2021). A selective methane gas sensor with printed catalytic films as active filters. *Sensors and Actuators. B, Chemical*, *347*(August), 130603. doi:10.1016/j.snb.2021.130603

Lu, Y., & Xu, X. (2019). Cloud-based manufacturing equipment and big data analytics to enable on-demand manufacturing services. *Robotics and Computer-integrated Manufacturing*, *57*, 92–102. doi:10.1016/j.rcim.2018.11.006

Lyons, M., Appathurai, S., Vasquez, M., Bolikowski, L., Alcalá, P., Carducci, F., & Tarabelloni, N. (2022, October 14). *The AI angle in solving the oil and gas emissions challenge*. BCG Global. https://www.bcg.com/publications/2021/ai-in-oil-and-gas-emissions-challenge

Maalel, N., Natalizio, E., Bouabdallah, A., Roux, P., & Kellil, M. (2013). Reliability for emergency applications in internet of things. In *Proceedings - IEEE International Conference on Distributed Computing in Sensor Systems, DCoSS 2013*, (pp. 361–366). IEEE. 10.1109/DCOSS.2013.40

Madhuri, N. S., Shailaja, K., Saha, D. P. R., Glory, K. B., & Sumithra, M. (2022). IOT integrated smart grid management system for effective energy management. Measurement. *Sensors (Basel)*, *24*, 100488. doi:10.1016/j.measen.2022.100488

Maduranga, M. W. P., & Abeysekera, R. (2020). Machine Learning Applications in IoT Based Agriculture and Smart Farming: A Review. *International Journal of Engineering Applied Sciences and Technology*, *4*(12), 24–27. doi:10.33564/IJEAST.2020.v04i12.004

Mahdavinejad, M. S., Rezvan, M., Barekatain, M., Adibi, P., Barnaghi, P., & Sheth, A. P. (2018). Machine learning for Internet of Things data analysis: A survey. *Digital Communications and Networks*, *4*(3), 161–175. doi:10.1016/j.dcan.2017.10.002

Mahmood, M. A., Seah, W. K. G., & Welch, I. (2015). Reliability in wireless sensor networks: A survey and challenges ahead. *Computer Networks*, *79*(December), 166–187. doi:10.1016/j.comnet.2014.12.016

Majdi, A., Dwijendra, N. K. A., Muda, I., Chetthamrongchai, P., Sivaraman, R., & Hammid, A. T. (2022). A smart building with integrated energy management: Steps toward the creation of a smart city. *Sustainable Energy Technologies and Assessments*, *53*, 102663. doi:10.1016/j.seta.2022.102663

Majumder, S., Mondal, T., & Deen, M. (2017). Wearable Sensors for Remote Health Monitoring. *Sensors (Basel)*, *17*(12), 130. doi:10.339017010130 PMID:28085085

Malathi, C., & Padmaja, I. N. (2021). Identification of cyber attacks using machine learning in smart IoT networks. *Materials Today: Proceedings.*

Maleš, L., Marčetić, D., & Ribarić, S. (2019). A multi-agent dynamic system for robust multi-face tracking. *Expert Systems with Applications, 126,* 246–264. doi:10.1016/j.eswa.2019.02.008

Malik, A. J., & Khan, F. A. (2018). A hybrid technique using binary particle swarm optimization and decision tree pruning for network intrusion detection. *Cluster Computing, 21*(1), 667–680. doi:10.100710586-017-0971-8

Malik, N. U. R., Abu-Bakar, S. A. R., Sheikh, U. U., Channa, A., & Popescu, N. (2023). Cascading Pose Features with CNN-LSTM for Multiview Human Action Recognition. *Signals, 4*(1), 40–55. doi:10.3390ignals4010002

Manasi, S. (2015). Smart Technologies in Smart Cities – Insights into Bengaluru City'sinitiatives. *Journal of Development, Management and Communication.*

Manimurugan, S. & Al- Mutari, S. (2020). A Novel Secret Image HidingTechnique for Secure Transmission. *J. Theo. Appl. Inf. Tech, 95* (1), 166-176.

Manimurugan, S. Al-Mutairi, S., Aborokbah, M., Chilamkurti, N., Ganesan, S., & Patan, R. (2020). Effective Attack Detection in Internet of MedicalThings Smart Environment Using a Deep Belief Neural Network. IEEE.

Manimurugan, S., & Narmatha, C. (2020). A New Approach for IRIS Imageidentification using modified contour segmentation. IEEE international conference on green computing, communication and electricalengineering. IEEE.

Manimurugan, S., & Narmatha, C. (2020). Secure and Efficient Medical ImageTransmission by New Tailored Visual Cryptography Scheme with LSCompressions. *International Journal of Digital Crime and Forensics, 7*(1), 26–50. doi:10.4018/IJDCF.2015010102

Manimurugan, S., Porkumaran, K., & Narmatha, C. (2020). The New Block PixelSort Algorithm for TVC Encrypted Medical Image. *Imaging Science Journal, 62*(8), 403–414. doi:10.1179/1743131X14Y.0000000078

Manogaran, G., Varatharajan, R., Lopez, D., Kumar, P. M., Sundarasekar, R., & Thota, C. (2018, May 1). A new architecture of the Internet of Things and big data ecosystem for securing smart healthcare monitoring and alerting systems. *Future Generation Computer Systems, 82,* 375–387. doi:10.1016/j.future.2017.10.045

Manoj, M., Dhilip Kumar, V., Arif, M., Bulai, E.-R., Bulai, P., & Geman, O. (2022). State of the Art Techniques for Water Quality Monitoring Systems for Fish Ponds Using IoT and Underwater Sensors: A Review. *Sensors (Basel), 22*(6), 2088. doi:10.339022062088 PMID:35336256

Marathe, S. (2019). Leveraging drone based imaging technology for pipeline and RoU monitoring survey. *Society of Petroleum Engineers - SPE Symposium: Asia Pacific Health, Safety, Security, Environment and Social Responsibility 2019.* One Petro. 10.2118/195427-MS

Marouf, M., Salam, Z. A., & Al-Fuqaha, A. (2021). Challenges and Opportunities in IoT-Based Intelligent Transportation Systems: A Review. *IEEE Internet of Things Journal, 8*(4), 2564–2585. doi:10.1109/JIOT.2020.3042138

Martinez, J. G., Gheisari, M., & Alarcón, L. F. (2020). UAV Integration in Current Construction Safety Planning and Monitoring Processes: Case Study of a High-Rise Building Construction Project in Chile. *Journal of Management Engineering, 36*(3), 05020005. doi:10.1061/(ASCE)ME.1943-5479.0000761

Martinka, J., Nečas, A., & Rantuch, P. (2022). The recognition of selected burning liquids by convolutional neural networks under laboratory conditions. *Journal of Thermal Analysis and Calorimetry, 147*(10), 5787–5799. doi:10.100710973-021-10903-2 PMID:34177362

Mata, O., Méndez, J. I., Ponce, P., Peffer, T., Meier, A., & Molina, A. (2023). Energy Savings in Buildings Based on Image Depth Sensors for Human Activity Recognition. *Energies*, *16*(3), 1078. doi:10.3390/en16031078

Mathew, P. S., Pillai, A. S., & Palade, V. (2018). Applications of IoT in healthcare. *Cognitive Computing for Big Data Systems Over IoT: Frameworks, Tools and Applications*, 263-288.

Mathews, N., Christensen, A. L., Stranieri, A., Scheidler, A., & Dorigo, M. (2019). Supervised morphogenesis: Exploiting morphological flexibility of self-assembling multirobot systems through cooperation with aerial robots. *Robotics and Autonomous Systems*, *112*, 154–167. doi:10.1016/j.robot.2018.11.007

Mathurkar, S. S., Patel, N. R., Lanjewar, R. B., & Somkuwar, R. S. (2014). Smart Sensors-based Monitoring System for Agriculture using Field Programmable Gate Array. *2014 International Conference on Circuits, Power and Computing Technologies*, (pp. 339-344). IEEE. 10.1109/ICCPCT.2014.7054914

Matthew, U. O., Kazaure, J. S., Amaonwu, O., Adamu, U. A., Hassan, I. M., & Kazaure, A. A. (2021). Role of Internet of Health Things (IoHTs) and Innovative Internet of 5G Medical Robotic Things (IIo-5GMRTs) in COVID-19 Global Health Risk Management and Logistics Planning. In M. Niranjanamurthy, S. Bhattacharyya, & N. Kumar (Eds.), *Intelligent Data Analysis for COVID-19 Pandemic. Algorithms for Intelligent Systems*. Springer. doi:10.1007/978-981-16-1574-0_2

Mehra, S. (2020, December 20). AI integration for enhancing road safety in India: Three case studies. *Indiaai.Gov.In*. https://indiaai.gov.in/article/ai-integration-for-enhancing-road-safety-in-india-three-case-studies

Mekala, M. S., & Viswanathan, P. (2017). A novel technology for smart agriculture based on IoT with cloud computing. In *2017 International Conference on I-SMAC (IoT in Social, Mobile, Analytics and Cloud) (I-SMAC)* (pp. 75-82). IEEE. 10.1109/I-SMAC.2017.8058280

Mekala, M. S., & Viswanathan, P. A. (2017). *Survey: Smart agriculture IoT with cloud computing. In 2017 international conference on microelectronic devices, circuits and systems (ICMDCS)*. IEEE.

Mekonnen, Y., Namuduri, S., Burton, L., Sarwat, A., & Bhansali, S. (2020). Review - Machine Learning Techniques in Wireless Sensor Network Based Precision AgricultureJ. *Journal of the Electrochemical Society*, *167*(3), 037522. doi:10.1149/2.0222003JES

Melo, R. O., Costa, M. G. F., & Costa Filho, C. F. F. (2020). Applying convolutional neural networks to detect natural gas leaks in wellhead images. *IEEE Access : Practical Innovations, Open Solutions*, *8*, 191775–191784. doi:10.1109/ACCESS.2020.3031683

Mendoza, M. (2020, July 19). *Industrial IoT - The Top 5 Benefits of Industry 4.0*. Hitachi Solutions. https://global.hitachi-solutions.com/blog/industrial-iot-benefits/

Meng, W., Tischhauser, E. W., Wang, Q., Wang, Y., & Han, J. (2018). When intrusion detection meets blockchain technology: A review. *IEEE Access : Practical Innovations, Open Solutions*, *6*, 10179–10188. doi:10.1109/ACCESS.2018.2799854

Menter, Z., Tee, W. Z., & Dave, R. (2021). Application of Machine Learning-Based Pattern Recognition in IoT Devices. In *Proceedings of International Conference on Communication and Computational Technologies* (pp. 669-689). Springer, Singapore. 10.1007/978-981-16-3246-4_52

Merenda, M., Porcaro, C., & Iero, D. (2020). Edge machine learning for ai-enabled iot devices: A review. *Sensors (Basel)*, *20*(9), 2533. doi:10.339020092533 PMID:32365645

Meshram, V., Patil, K., Meshram, V., Hanchate, D., & Ramkteke, S. D. (2021). S.D. Ramkteke, Machine Learning in Agriculture Domain: A state-of-art survey. *Artificial Intelligence in the Life Sciences, 1*, 100010. doi:10.1016/j. ailsci.2021.100010

Microsoft. (2019, August 20). *Predictive Analytics for Traffic.* Microsoft Research. https://www.microsoft.com/en-us/research/project/predictive-analytics-for-traffic/

Mihalache, A. (2017, June 30). Wireless Home Automation System using IoT. Informatica Economica, 21(2/2017), 17–32. doi:10.12948/issn14531305/21.2.2017.02

Mihret & Haile. (2021). 4G, 5G, 6G, 7G and Future Mobile Technologies. *J Comp Sci Info Technol, 9*(2), 75.

Miler, M., Medak, D., & Odobasic, D. (2011). Two-tier architecture for web mapping with NoSQL database couch DB. *Geospatial Crossroads GI Forum, 11*, 62–71.

Ministry of Urban Development Government of India. (2015). *Smart Cities.* MUDGI. https://smartcities.gov.in/themes/habikon/files/SmartCityGuidelines.pdf

Miorandi, D., Sicari, S., De Pellegrini, F., & Chlamtac, I. (2012). Internet of things: Vision, applications and research challenges. *Ad Hoc Networks, 10*(7), 1497–1516. doi:10.1016/j.adhoc.2012.02.016

Mishra, L. (2021, September 6). *Smart lighting the road to sustainability.* Tata Communications New World. https://www.tatacommunications.com/blog/2020/08/smart-lighting-the-road-to-sustainability/

Misra, K. B., & Rao, T. S. M. (1970). Reliability Analysis of Redundant NetworksUsingFlowGraphs. *IEEE Transactions on Reliability, R-19*(1), 19–24. doi:10.1109/TR.1970.5216374

Mogili, U. M. R., & Deepak, B. (2018). Review on application of drone systems in precision agriculture. *Procedia Computer Science, 133*, 502–509. doi:10.1016/j.procs.2018.07.063

Mohammadi, M., Al-Fuqaha, A., Sorour, S., & Guizani, M. (2018). Deep learning for IoT big data and streaming analytics: A survey. *IEEE Communications Surveys and Tutorials, 20*(4), 2923–2960. doi:10.1109/COMST.2018.2844341

Mohammed, M. N., & Ahmed, M. M. (2019). Data Preparation and Reduction Technique in Intrusion Detection Systems: ANOVA-PCA. *International Journal of Computer Science and Security, 13*(5), 167–183.

Mondal, S., Sinha, A., & Routh, J. (2015). A survey on evolution of wireless generations 0G to 7G. *International Journal of Advance Research in Science and Engineering, 1*(2), 5–10.

Moon, D., Im, H., Kim, I., & Park, J. H. (2017). DTB-IDS: An intrusion detection system based on decision tree using behavior analysis for preventing APT attacks. *The Journal of Supercomputing, 73*(7), 2881–2895. doi:10.100711227-015-1604-8

Moore, S. J. (2022). IoT reliability: a review leading to 5 key research directions. CCF Transactions on Pervasive Computing and Interaction, 2, 147-163.

Moore, S. J., Nugent, C. D., Zhang, S., & Cleland, I. (2020). IoT reliability: A review leading to 5 key research directions. *CCF Trans. Pervasive Comp. Interact., 2*(3), 147–163. doi:10.100742486-020-00037-z

Morris, T. H., Jones, B. A., Vaughn, R. B., & Dandass, Y. S. (2021). Deterministic Intrusion Detection Rules for MODBUS Protocols. *Hawaii International Conference on System Sciences.* IEEE.

Mousavi, S. S., Schukat, M., & Howley, E. (2017). Traffic light control using deep policy-gradient and value-function-based reinforcement learning. *IET Intelligent Transport Systems, 11*(7), 417–423. doi:10.1049/iet-its.2017.0153

Moyano-Fuentes, J., & Martínez-Jurado, P.-J. (2016). The influence of competitive pressure on manufacturer internal information integration. *International Journal of Production Research*, *54*(22), 6683–6692. doi:10.1080/00207543.2015.1131866

Mozaffar, M., Liao, S., Xie, X., Saha, S., Park, C., Cao, J., & Gan, Z. (2021). Mechanistic artificial intelligence (mechanistic-AI) for modeling, design, and control of advanced manufacturing processes: Current state and perspectives. *Journal of Materials Processing Technology*, *117485*. doi:10.1016/j.jmatprotec.2021.117485

Muiruri, L. (2021, November 23). *What You Need to Know About Waymo's New Self-Driving Taxi Service*. Makeuseof. https://www.makeuseof.com/waymos-new-self-driving-taxi-service/

Mukhamediev, R. I., Yakunin, K., Aubakirov, M., Assanov, I., Kuchin, Y., Symagulov, A., Levashenko, V., Zaitseva, E., Sokolov, D., & Amirgaliyev, Y. (2023). Coverage path planning optimization of heterogeneous UAVs group for precision agriculture. *IEEE Access : Practical Innovations, Open Solutions*, *11*, 5789–5803. doi:10.1109/ACCESS.2023.3235207

Mukhopadhyay, S. C. (2014). Wearable sensors for human activity monitoring: A review. *IEEE Sensors Journal*, *15*(3), 1321–1330. doi:10.1109/JSEN.2014.2370945

Muna, M. A., Moustafa, N., & Sitnikova, E. (2020). Identification of maliciousactivities in industrial internet of things based on deep learning models. *Journal of Information Security and Applications*, *41*, 1–11.

Nakhkash, M. R., Gia, T. N., Azimi, I., Anzanpour, A., Rahmani, A. M., & Liljeberg, P. Analysis of Performance and Energy Consumption ofWearable Devices and Mobile Gateways in IoT Applications. *COINS '19: Proceedings of the International Conference on Omni-Layer IntelligentSystems*, (pp.68-73). ACM. 10.1145/3312614.3312632

Nalavade, A., Bai, A., & Bhushan, M. (2020). Deep learning techniques and models for improving machine reading comprehension system. *IJAST*, *29*(04), 9692–9710.

Namami, G. (2023, March 18). *National Mission for Clean Ganga(NMCG)*. Ministry of Jal Shakti, Department of Water Resources, River Development & Ganga Rejuvenation, Government of India. https://nmcg.nic.in/

Namiot, D. (2015). On big data stream processing. *Int J Open Inform Technol.*, *3*(8), 48–51.

Narkhede, P., Walambe, R., Mandaokar, S., Chandel, P., Kotecha, K., & Ghinea, G. (2021). Gas detection and identification using multimodal artificial intelligence based sensor fusion. *Applied System Innovation*, *4*(1), 1–14. doi:10.3390/asi4010003

Narmatha, C., Manimegalai, P., & Manimurugan, S. (2020, October). A LosslessCompression Scheme for Grayscale Medical Images Using a P2-Bit ShortTechnique. *Journal of Medical Imaging and Health Informatics*, *7*(6), 1196–1204. doi:10.1166/jmihi.2017.2212

Nassar, A. S., Montasser, A. H., & Abdelbaki, N. A. (2017). Survey on Smart Cities' IoT. *Proceedings of the International Conference on Advanced Intelligent Systems and Informatics*, 9–11.

Natraj, N. A., Kamatchi Sundari, V., Ananthi, K., Rathika, S., Indira, G., & Rathish, C. R. (2022, July). Security Enhancement of Fog Nodes in IoT Networks Using the IBF Scheme. In *Third International Conference on Image Processing and CapsuleNetworks: ICIPCN 2022* (pp. 119-129). Cham: Springer International Publishing. 10.1007/978-3-031-12413-6_10

Nawandar, N. K., & Satpute, V. R. (2019). IoT based low cost and intelligent module for smart irrigation system. *Computers and Electronics in Agriculture*, *162*, 979–990. doi:10.1016/j.compag.2019.05.027

Nayagam, M., Vijayalakshmi, B., Somasundaram, K., Mukunthan, M. A., Yogaraja, C. A., & Partheeban, P. (2023). Control of pests and diseases in plants using IOT Technology. Measurement. *Sensors (Basel)*, *26*, 100713. doi:10.1016/j.measen.2023.100713

Negi, A., & Kaur, K. (2017). *Method to resolve software product line errors.* Information, Communication and Computing Technology: Second International Conference, ICICCT 2017, New Delhi, India.

Nejati, M., Rabiei, S., & Jabbour, C. J. C. (2017). Envisioning the invisible: Understanding the synergy between green human resource management and green supply chain management in manufacturing firms in Iran in light of the moderating effect of employees' resistance to change. *Journal of Cleaner Production*, *168*, 163–172. doi:10.1016/j.jclepro.2017.08.213

Nesa, N., & Banerjee, I. (2017). IoT-based sensor data fusion for occupancy sensing using Dempster–Shafer evidence theory for smart buildings. *IEEE Internet of Things Journal*, *4*(5), 1563–1570. doi:10.1109/JIOT.2017.2723424

Newton, E. (2023, March 29). *5 Unique Applications of IoT in Construction.* IoT For All https://www.iotforall.com/5-unique-applications-of-iot-in-construction#:~:text=Construction%20sites%20can%20also%20use,see%20if%20it%20needs%20repair

Nguyen, T. D., Marchal, S., Miettinen, M., Fereidooni, H., Asokan, N., & Sadeghi, A. R. (2019, July). DÏoT: A federated self-learning anomaly detection system for IoT. In *2019 IEEE 39th International conference on distributed computing systems (ICDCS)* (pp. 756-767). IEEE. 10.1109/ICDCS.2019.00080

Nguyen, D. C., Ding, M., Pathirana, P. N., Seneviratne, A., Li, J., Niyato, D., Dobre, O., & Poor, H. V. (2022). 6G Internet of Things: A Comprehensive Survey. *IEEE Internet of Things Journal*, *9*(1), 359–383. doi:10.1109/JIOT.2021.3103320

Nguyen, T., Gosine, R. G., & Warrian, P. (2020). A systematic review of big data analytics for oil and gas industry 4.0. *IEEE. IEEE Access : Practical Innovations, Open Solutions*, *8*(1), 61183–61201. doi:10.1109/ACCESS.2020.2979678

Niebla-Montero, Á., Froiz-Míguez, I., Fraga-Lamas, P., & Fernández-Caramés, T. M. (2022). Practical Latency Analysis of a Bluetooth 5 Decentralized IoT Opportunistic Edge Computing System for Low-Cost SBCs. *Sensors (Basel)*, *22*(21), 8360. doi:10.339022218360 PMID:36366060

Niennattrakul V, Keogh E, Chotirat AR. (2013). Data editing techniques to allow the application of distance-based outlier detection to streams. *IEEE 13th Int Conf Data Min.* (pp. 947–952). IEEE.

Niua. (2019). ENABLE : REAL-TIME TRAFFIC MONITORING TOOL FOR EFFECTIVE AND EFFICIENT DECISION MAKING. *Niua.* https://www.niua.org/iscfip/compendium/project/enable-real-time-traffic-monitoring-tool-effective-and-efficient-decision
-making

Niu, H., Zhao, T., Wang, D., & Chen, Y. (2019). Estimating evapotranspiration with UAVs in agriculture: A review. *2019 ASABE Annual International Meeting*, 1. ASABE.

Nivash, J. P., Raj, E. D., Babu, L. D., Nirmala, M., & Manoj, K. V. (2014). Analysis on enhancing storm to efficiently process big data in real time. In *Fifth International Conference on Computing, Communications and Networking Technologies (ICCCNT),* (pp. 1-5). IEEE. 10.1109/ICCCNT.2014.7093076

Norasma, C. Y. N., Fadzilah, M. A., Roslin, N. A., Zanariah, Z. W. N., Tarmidi, Z., & Candra, F. S. (2019). Unmanned aerial vehicle applications in agriculture. *IOP Conference Series. Materials Science and Engineering*, *506*(1), 12063. doi:10.1088/1757-899X/506/1/012063

Nord, J. H., Koohang, A., & Paliszkiewicz, J. (2019). The Internet of Things: Review and theoretical framework. *Expert Systems with Applications*, *133*, 97–108. doi:10.1016/j.eswa.2019.05.014

Nori, F., Deypir, M., & Sadreddini, M. H. (2013). A sliding window-based algorithm for frequent closed itemset mining over data streams. *Journal of Systems and Software*, *86*(3), 615–623. doi:10.1016/j.jss.2012.10.011

Nweke, H. F., Teh, Y. W., Mujtaba, G., & Al-Garadi, M. A. (2019). Data fusion and multiple classifier systems for human activity detection and health monitoring: Review and open research directions. *Information Fusion*, *46*, 147–170. doi:10.1016/j.inffus.2018.06.002

O'SheaK.NashR. (2015). *An Introduction to Convolutional Neural Networks*. 1–11. https://arxiv.org/abs/1511.08458

Obeidat, S. M., Al Bakri, A. A., & Elbanna, S. (2020). Leveraging 'green' human resource practices to enable environmental and organizational performance: Evidence from the Qatari oil and gas industry. *Journal of Business Ethics*, *164*(2), 371–388. doi:10.100710551-018-4075-z

Obinikpo, A. A., & Kantarci, B. (2017, December). Big sensed data meets deep learning for smarter healthcare in smart cities. *Journal of Sensor and Actuator Networks.*, *6*(4), 26. doi:10.3390/jsan6040026

OECD. (2021). *Adoption of Technologies for Sustainable Farming Systems, Wageningen Workshop Proc.* Organisation for Economic Co-Operation, and Development.

Ogino, T. (2021). Simplified Multi-objective Optimization for Flexible IoT Edge Computing. 2021 4th International Conference on Information and Computer Technologies (ICICT), HI, USA. 10.1109/ICICT52872.2021.00035

Oh, S., Byon, Y. J., & Yeo, H. (2016). Improvement of Search Strategy With K-Nearest Neighbors Approach for Traffic State Prediction. *IEEE Transactions on Intelligent Transportation Systems*, *17*(4), 1146–1156. doi:10.1109/TITS.2015.2498408

Olaizola, I. G., Quartulli, M., Unzueta, E., Goicolea, J. I., & Flórez, J. (2022). Refinery 4.0, a review of the main challenges of the industry 4.0 paradigm in oil & gas downstream. *Sensors (Basel)*, *22*(23), 9164. doi:10.339022239164 PMID:36501863

Ortiz-Garcés, I., Andrade, R. O., Sanchez-Viteri, S., & Villegas-Ch, W. (2023). Prototype of an Emergency Response System Using IoT in a Fog Computing Environment. *Computers*, *12*(4), 81. doi:10.3390/computers12040081

Osanaiye, O., Cai, H., Choo, K. K. R., Dehghantanha, A., Xu, Z., & Dlodlo, M. (2016). Ensemble-based multi-filter feature selection method for DDoS detection in cloud computing. *EURASIP Journal on Wireless Communications and Networking*, *2016*(1), 1–10. doi:10.118613638-016-0623-3

Oteyo, I. N. (2020). Developing Smart Agriculture Applications: Experiences and Lessons Learnt. *African Conf. on Software Engineering*. Research Gate.

Outchakoucht, A., Hamza, E. S., & Leroy, J. P. (2017). Dynamic access control policy based on blockchain and machine learning for the internet of things. *International Journal of Advanced Computer Science and Applications*, *8*(7). Advance online publication. doi:10.14569/IJACSA.2017.080757

Padhy, S., Alowaidi, M., Dash, S., Alshehri, M., Malla, P. P., Routray, S., & Alhumyani, H. (2023). *AgriSecure: A Fog Computing-Based Security Framework for Agriculture 4*. AgriSecure.

Padhy, S., Alowaidi, M., Dash, S., Alshehri, M., Malla, P. P., Routray, S., & Alhumyani, H. (2023). AgriSecure: A Fog Computing-Based Security Framework for Agriculture 4.0 via Blockchain. *Processes (Basel, Switzerland)*, *11*(3), 757. doi:10.3390/pr11030757

Padvi, M. N., Moholkar, A. V., Prasad, S. R., & Prasad, N. R. (2021). A critical review on design and development of gas sensing materials. *Engineered Science*, *15*, 20–37. doi:10.30919/es8d431

Pal, S., & Jadidi, Z. (2021). Analysis of Security Issues and Countermeasures for the Industrial Internet of Things. *Applied Sciences (Basel, Switzerland)*, *11*(20), 9393. doi:10.3390/app11209393

Panda, P., Sengupta, A., & Roy, K. (2016, March). Conditional deep learning for energy-efficient and enhanced pattern recognition. In 2016 Design, Automation & Test in Europe Conference & Exhibition (DATE) (pp. 475-480). IEEE. doi:10.3850/9783981537079_0819

Pang, LX, Chawla, S, Wei, L, & ... (2013). On detection of emerging anomalous traffic patterns using GPS data. *Data & Knowledge Engineering*, 87(9), 357–373.

Pang, S., Wang, N., Wang, M., Qiao, S., Zhai, X., & Xiong, N. (2021). A Smart Network Resource Management System for High Mobility Edge Computing in 5G Internet of Vehicles. *IEEE Transactions on Network Science and Engineering*, 8(4), 3179–3191. doi:10.1109/TNSE.2021.3106955

Pan, X., Zhang, H., Ye, W., Bermak, A., & Zhao, X. (2019). A Fast and Robust Gas Recognition Algorithm Based on Hybrid Convolutional and Recurrent Neural Network. *IEEE Access : Practical Innovations, Open Solutions*, 7, 100954–100963. doi:10.1109/ACCESS.2019.2930804

Park, Y. S., & Lek, S. (2016). Artificial Neural Networks: Multilayer Perceptron for Ecological Modeling. In Developments in Environmental Modelling (Vol. 28). Elsevier. doi:10.1016/B978-0-444-63623-2.00007-4

Parvathi Sangeetha, B., Kumar, N., Ambalgi, A. P., Abdul Haleem, S. L., Thilagam, K., & Vijayakumar, P. (2022). IOT based smart irrigation management system for environmental sustainability in India. *Sustainable Energy Technologies and Assessments*, 52, 101973. doi:10.1016/j.seta.2022.101973

Pashami, S., Lilienthal, A. J., & Trincavelli, M. (2012). Detecting changes of a distant gas source with an array of MOX gas sensors. *Sensors (Basel)*, 12(12), 16404–16419. doi:10.3390121216404 PMID:23443385

Patel, H., Prajapati, D., Mahida, D., & Shah, M. (2020). Transforming petroleum downstream sector through big data: A holistic review. *Journal of Petroleum Exploration and Production Technology*, 10(6), 2601–2611. doi:10.100713202-020-00889-2

Pathmudi, V. R., Khatri, N., Kumar, S., Abdul-Qawy, A. S. H., & Vyas, A. K. (2023). A systematic review of IoT technologies and their constituents for smart and sustainable agriculture applications. *Scientific African*, 19, e01577. doi:10.1016/j.sciaf.2023.e01577

Patil, A., & Rane, M. (2021). Convolutional Neural Networks: An Overview and Its Applications in Pattern Recognition. *Smart Innovation. Systems and Technologies*, 195, 21–30. doi:10.1007/978-981-15-7078-0_3

Patil, K. A., & Kale, N. R. (2016). A model for smart agriculture using IoT. In *2016 International Conference on Global Trends in Signal Processing, Information Computing and Communication (ICGTSPICC)* (pp. 543-545). IEEE. 10.1109/ICGTSPICC.2016.7955360

Paul, S. KP Bhaumik,. (2016). *AIDCOR:artificial immunity inspired density based clustering with outlier removal*. Int J Mach Learn Cybernet.

Pawar, S., Bhushan, M., & Wagh, M. (2020). The plant leaf disease diagnosis and spectral data analysis using machine learning–A review. *International Journal of Advanced Science and Technology*, 29(9s), 3343–3359.

PennEHRS. (2020). *SOP: Hazardous and Highly Toxic Gases*. PennEHRS. https://ehrs.upenn.edu/health-safety/lab-safety/chemical-hygiene-plan/standard-operating-procedures/sop-hazardous-and

Peranzo, P. (2022, March 18). *IoT in Manufacturing: 8 Use Cases & Upcoming Trends*. Imaginovation. https://imaginovation.net/blog/iot-in-manufacturing/

Perez, A. J., & Zeadally, S. (2017). Privacy issues and solutions for consumer wearables. *IT Professional*, 20(4), 46–56. doi:10.1109/MITP.2017.265105905

Pham, X. Q., Huynh-The, T., & Kim, D.-S. (2021). *A QoE-based Optimization Approach to Computation Offloading in Vehicle-assisted Multi-access Edge Computing.* International Conference on Information and Communication Technology Convergence (ICTC), Jeju Island, Korea. 10.1109/ICTC52510.2021.9621000

Pham, S. T., Vo, P. S., & Nguyen, D. N. (2021). Effective electrical submersible pump management using machine learning. *Open Journal of Civil Engineering, 11*(1), 70–80. doi:10.4236/ojce.2021.111005

Phasinam, K., & Kassanuk, T. (2022). Machine Learning and Internet of Things (IoT) for Real-Time Image Classification in Smart Agriculture. *ECS Transactions, 107*(1), 3305–3311. doi:10.1149/10701.3305ecst

Phuoc, N. V. (2022). The critical factors impacting artificial intelligence applications adoption in Vietnam: A structural equation modeling analysis. *Economies, 10*(129), 1–16. doi:10.3390/economies10060129

Pike, M., Mustafa, N. M., Towey, D., & Brusic, V. (2019). Sensor Networks and Data Management in Healthcare: Emerging Technologies and New Challenges. In *2019 IEEE 43rd Annual Computer Software and Applications Conference (COMPSAC),* (pp. 834-839). IEEE.

Pires, J. N., Faria, B. M., & Neto, P. (2020). Cloud robotics: A survey on current trends and future challenges. *Journal of Intelligent & Robotic Systems, 100*, 1–19. doi:10.100710846-019-01077-4

Polonelli, T., Brunelli, D., Bartolini, A., & Benini, L. (2018). A LoRaWAN Wireless Sensor Network for Data Center Temperature Monitoring. *Applications in Electronics Pervading Industry, Environment and Society.*

Popescul, A., Ungar, L. H., Pennock, D. M., & Lawrence, S. (2013). Probabilistic models for unified collaborative and content-based recommendation in sparse-data environments. *arXiv preprint arXiv:1301.2303.*

Portilla, J., Mujica, G., Lee, J. S., & Riesgo, T. (2019). The extreme edge at the bottom of the Internet of Things: A review. *IEEE Sensors Journal, 19*(9), 3179–3190. doi:10.1109/JSEN.2019.2891911

Prabhakar, P., Arora, S., Khosla, A., Beniwal, R. K., Arthur, M. N., Arias-Gonzáles, J. L., & Areche, F. O. (2022). Cyber Security of Smart Metering Infrastructure Using Median Absolute Deviation Methodology. *Security and Communication Networks, 2022*, 2022. doi:10.1155/2022/6200121

Prajwal, V., & Guru, R. (2022). Smart Greenhouse Farming Using IOT and Machine learning. *Journal of Emerging Technologies and Innovative Research, 9*(8), b708–b718.

Pramanik, M. I., Lau, R. Y., Demirkan, H., & Azad, M. A. (2018, September 1). Smart health: Big data enables a health paradigm within smart cities. Expert Systems with Applications. 2017 Nov 30;87:370-83. .

Pramod, A., Naicker, H. S., & Tyagi, A. K. (2021). Machine learning and deep learning: Open issues and future research directions for the next 10 years. *Computational analysis and deep learning for medical care: Principles, methods, and applications*, 463-490.

Prasad, S. S., & Kumar, C. (2013). A Green and Reliable Internet of Things. *Commun. Netw., 5*(1), 44–48. doi:10.4236/cn.2013.51B011

Prasanna Lakshmi, G. S., Asha, P. N., Sandhya, G., Vivek Sharma, S., Shilpashree, S., & Subramanya, S. G. (2023). An intelligent IOT sensor coupled precision irrigation model for agriculture. Measurement. *Sensors (Basel), 25*, 100608. doi:10.1016/j.measen.2022.100608

Predictive Analytics for Traffic . (2019b, August 19). Microsoft Research. https://www.microsoft.com/en-us/research/project/predictive-analytics-for-traffic/

Prinima, D., & Pruthi. (2016). Evolution of Mobile Communication Network: From 1G to 5G. *Int. J. Innov. Res. Comput. Commun. Eng., 4*, 224–227.

Property, P., Sensors, G., Flow, S. G., Conductivity, T., Sensors, S., Sensors, G. C., Micro, M., Analyzers, G., Absorption, O. S., Spectrometry, M., Chromatography, G., Analysis, M., Positives, F., & Reading, F. (2001). Gas sensors. In *Semiconductor International, 24*(6). doi:10.1016/B978-0-12-803581-8.00548-8

Puliafito, C., Mingozzi, E., Longo, F., Puliafito, A., & Rana, O. (2019). Fog computing for the internet of things: A survey. *ACM Transactions on Internet Technology, 19*(2), 1–41. doi:10.1145/3301443

Qadir, Z., Munawar, H. S., Saeed, N., & Le, K. N. (2021). *Towards 6G Internet of Things: Recent Advances.* Use Cases, and Open Challenges.

Qiu, S., Chen, B., Wang, R., Zhu, Z., Wang, Y., & Qiu, X. (2018). Atmospheric dispersion prediction and source estimation of hazardous gas using artificial neural network, particle swarm optimization and expectation maximization. *Atmospheric Environment, 178*, 158–163. doi:10.1016/j.atmosenv.2018.01.056

Quy, V. K., Hau, N. V., Anh, D. V., Quy, N. M., Ban, N. T., Lanza, S., Randazzo, G., & Muzirafuti, A. (2022). IoT-Enabled Smart Agriculture: Architecture, Applications, and Challenges. *Applied Sciences (Basel, Switzerland), 12*(7), 3396. doi:10.3390/app12073396

R., S., & R., K. (2022). *Survey or Review on the Deep Learning Techniques for Retinal Image Segmentation in Predicting/Diagnosing Diabetic Retinopathy.* IGI Global. doi:10.4018/978-1-6684-4405-4.ch010

Rahamathulla, M. P. (2020, March 1). Cloud-based Healthcare data management Framework. *KSII Transactions on Internet and Information Systems, 14*(3).

Rahman, A., Abbas, S., Gollapalli, M., Ahmed, R., Aftab, S., Ahmad, M., Khan, M. A., & Mosavi, A. (2022). Rainfall Prediction System Using Machine Learning Fusion for Smart Cities. *Sensors (Basel), 22*(9), 3504. doi:10.339022093504 PMID:35591194

Rai, K., Devi, M. S., & Guleria, A. (2016). Decision tree based algorithm for intrusion detection. *International Journal of Advanced Networking and Applications, 7*(4), 2828.

Raimundo, R. J., & Rosário, A. T. (2022). Cybersecurity in the Internet of Things in Industrial Management. *Applied Sciences (Basel, Switzerland), 12*(3), 1598. doi:10.3390/app12031598

Raj, E. D., Nivash, J. P., Nirmala, M., & Babu, L. D. (2014). A scalable cloud computing deployment framework for efficient MapReduce operations using Apache YARN. In *International Conference on Information Communication and Embedded Systems (ICICES2014),* (pp. 1- 6). IEEE.

Rajeev, G. L., & Nancy, R. S, M., John, J., & John, N. M. J. N. M. (2021). Traffic Flow Prediction using Random Forest and Bellman Ford for Best Route Detection. *International Journal of Engineering Research and Technology, 9*(13). https://www.ijert.org/research/traffic-flow-prediction-using-random-forest-and-bellman-ford-for-best-route-detection-IJERTCONV9IS13021.pdf

Rajeswari, S., Suthendran, K., & Rajakumar, K. (2017). A smart agricultural model by integrating IoT, mobile and cloud-based big data analytics. In *2017 International Conference on Intelligent Computing and Control (I2C2)* (pp. 1-5). IEEE. 10.1109/I2C2.2017.8321902

Raj, M. P., Swaminarayan, P. R., Saini, J. R., & Parmar, D. K. (2015). Applications of pattern recognition algorithms in agriculture: A review. *International Journal of Advanced Networking and Applications, 6*(5), 2495.

Rajyaguru, N., Vyas, S., & Vyas, K. (2021). 9 Internet of Things platform for smart Farming. In J. Chatterjee, A. Kumar, P. Rathore, & V. Jain (Eds.), *Internet of Things and Machine Learning in Agriculture: Technological Impacts and Challenges* (pp. 169–202). De Gruyter. doi:10.1515/9783110691276-009

Ramanujam, E., Perumal, T., & Padmavathi, S. (2021). Human activity recognition with smartphone and wearable sensors using deep learning techniques: A review. *IEEE Sensors Journal*, *21*(12), 13029–13040. doi:10.1109/JSEN.2021.3069927

Rana, M. M., Ibrahim, D. S., Asyraf, M. R. M., Jarin, S., & Tomal, A. (2017). A review on recent advances of CNTs as gas sensors. *Sensor Review*, *37*(2), 127–136. doi:10.1108/SR-10-2016-0230

Rana, M., & Bhushan, M. (2022). Machine learning and deep learning approach for medical image analysis: Diagnosis to detection. *Multimedia Tools and Applications*, 1–39. doi:10.100711042-022-14305-w PMID:36588765

Rangaiah, M. (2021, May 14). *9 IoT Applications in the Hospitality Industry*. Analytics Steps. Www.analyticssteps.com. https://www.analyticssteps.com/blogs/9-iot-applications-hospitality-industry

Ranka, S., Rangarajan, A., Elefteriadou, L., Srinivasan, S., Poasadas, E., Hoffman, D., Ponnulari, R., Dilmore, J., & Byron, T. (2020). A Vision of Smart Traffic Infrastructure for Traditional, Connected, and Autonomous Vehicles. *2020 International Conference on Connected and Autonomous Driving (MetroCAD)*. IEEE. 10.1109/MetroCAD48866.2020.00008

Ransbotham, S., Kiron, D., Gerbert, P., & Reeves, M. (2017). Reshaping business with artificial intelligence. *MIT Sloan Management Review*, *59*(1), 1–17.

Rao, V. P. S., & Rao, G. S. (2019). Design and modelling of anaffordable uav based pesticide sprayer in agriculture applications. *2019 Fifth International Conference on Electrical Energy Systems (ICEES)*, (pp. 1–4). IEEE.

Raparelli, E., & Bajocco, S. (2019). A bibliometric analysis on the use of unmanned aerial vehicles in agricultural and forestry studies. *International Journal of Remote Sensing*, *40*(24), 9070–9083. doi:10.1080/01431161.2019.1569793

Ravi, D., Wong, C., Lo, B., & Yang, G. Z. (2016, December 23). A deep learning approach to on-node sensor data analytics for mobile or wearable devices. *IEEE Journal of Biomedical and Health Informatics*, *21*(1), 56–64. doi:10.1109/JBHI.2016.2633287 PMID:28026792

Rayhana, R., Xiao, G., & Liu, Z. (2020). Internet of Things Empowered Smart Greenhouse Farming. *IEEE Journal of Radio Frequency Identification*, *4*(3), 195–211. doi:10.1109/JRFID.2020.2984391

Ray, J. K., Biswas, A. S., Sil, S., Bera, R., Shome, S., Biswas, P., & Mitra, M. (2022). Realization of 5G V2V communication system at 28 GHz for smart vehicle. *Innovations in Systems and Software Engineering*. doi:10.100711334-022-00435-9

Raza, M., Hussain, M., & Ali, A. (2020). Challenges and Opportunities of IoT for Smart Grid: A Systematic Review. *IEEE Access : Practical Innovations, Open Solutions*, *8*, 108704–108723. doi:10.1109/ACCESS.2020.3002175

Razavi-Far, R., Ruiz-Garcia, A., Palade, V., & Schmidhuber, J. (2022). *Generative adversarial learning: architectures and applications*. Springer. doi:10.1007/978-3-030-91390-8

Reche, A., Sendra, S., Díaz, J. R., & Lloret, J. (2014). A smart M2M deployment to control the agriculture irrigation. In *International conference on ad-hoc networks and wireless* (pp. 139-151). Springer.

Reddy, K. S. P. (2020). IoT-based Smart Agriculture using Machine Learning. *2nd Inter. Conf. on Inventive Research in Computing Applications*, (pp. 130-134). IEEE.

Rehman, A., Saba, T., Kashif, M., Fati, S. M., Bahaj, S. A., & Chaudhry, H. (2022). A Revisit of Internet of Things Technologies for Monitoring and Control Strategies in Smart Agriculture. *Agronomy (Basel)*, *12*(1), 127. doi:10.3390/agronomy12010127

Remesh, A., Muralidharan, D., Raj, N., Gopika, J., & Binu, P. K. (2020). Intrusion Detection System for IoT Devices. *2020 International Conference on Electronics and Sustainable Communication Systems (ICESC)*, Coimbatore, India. 10.1109/ICESC48915.2020.9155999

Ren, J, Wu, Q, Zhang, J, & ... (2009). *Efficient outlier detection algorithm for heterogeneous data streams.* Six Int Conf Fuzzy Syst Knowl Discov.

Ridhawi, I. A., Aloqaily, M., Boukerche, A., & Jararweh, Y. (2021). Enabling Intelligent IoCV Services at the Edge for 5G Networks and Beyond. *IEEE Transactions on Intelligent Transportation Systems*, 22(8), 5190–5200. doi:10.1109/TITS.2021.3053095

Rifat, A., Patel, P., & Babu, B. S. (2022). The Internet of Things (IoT) in Smart Agriculture Monitoring. *European Journal of Information Technologies and Computer Science*, 2(1), 14–18. doi:10.24018/compute.2022.2.1.49

Rijwan, K., Indrajeet, K., Jyoti, R., Noor, M., & Shahnawaz, H. (2021). *Opportunities of Artificial Intelligence and Machine Learning in the Food Industry, Journal of Food Quality.* Hindawi. doi:10.1155/2021/4535567

Rizzoli, A. (2023). *8 Practical Applications of AI in Agriculture.* v7Labs. https://www.v7labs.com/blog/ai-in-agriculture

Robin, Y., Amann, J., Goodarzi, P., Schneider, T., Schütze, A., & Bur, C. (2022). Deep Learning Based Calibration Time Reduction for MOS Gas Sensors with Transfer Learning. *Atmosphere (Basel)*, 13(10), 1614. doi:10.3390/atmos13101614

Rodriguez, J. A., Fernandez, F. J., & Arboleya, P. (2018). Study of the Architecture of a Smart City. *The 2nd International Research Conference on Sustainable Energy, Engineering, Materials and Environment.* MDPI. 10.3390/proceedings2231485

Rojas Soares, F. D., Secchi, A. R., & Bezerra de Souza, M. Jr. (2022). Development of a nonlinear model predictive control for stabilization of a gas-lift oil well. *Industrial & Engineering Chemistry Research*, 61(24), 8411–8421. doi:10.1021/acs.iecr.1c04728

Roman, R., Alcaraz, C., Lopez, J., & Sklavos, N. (2011, March). Key management systems for sensor networks in the context of the Internet of Things. *Computers & Electrical Engineering*, 37(2), 147–159. doi:10.1016/j.compeleceng.2011.01.009

Routray, S. K., & Anand, S. (2017, February). Narrowband IoT for healthcare. In *2017 International Conference on Information Communication and Embedded Systems (ICICES)* (pp. 1-4). IEEE.

Rybakov, A. (2021, October 22). *Applying IoT in the construction industry: top-7 use cases.* Agile Vision. Www.agilevision.io. https://www.agilevision.io/blog/applying-iot-in-the-construction-industry-top-7-use-cases

Ryo, M. (2022). Explainable artificial intelligence and interpretable machine learning for agricultural data analysis. *Artificial Intelligence in Agriculture*, 6, 257–265. doi:10.1016/j.aiia.2022.11.003

Sadhukhan, P., & Gazi, F. (2018b). *An IoT based Intelligent Traffic Congestion Control System for Road Crossings.* IEEE. doi:10.1109/IC3IoT.2018.8668131

Sadik S. & Gruenwald L. (2013). Research issues in outlier detection for data streams. *ACM SIGKDD Explor Newslett arch, 15*(1), 33–40.

Sætra, H. S. (2022). The AI ESG protocol: Evaluating and disclosing the environment, social, and governance implications of artificial intelligence capabilities, assets, and activities. *Sustainable Development*, 1. Advance online publication. doi:10.1002d.2438

Saha, B., Samanta, R., Ghosh, S., & Roy, R. B. (2023, January). BandX: An Intelligent IoT-band for Human Activity Recognition based on TinyML. In *24th International Conference on Distributed Computing and Networking* (pp. 284-285). ACM. 10.1145/3571306.3571415

Saha, R., Tariq, M. T., Hadi, M., & Xiao, Y. (2019). Pattern recognition using clustering analysis to support transportation system management, operations, and modeling. *Journal of Advanced Transportation, 2019*, 2019. doi:10.1155/2019/1628417

Saiz-Rubio, V., & Rovira-Más, F. (2020). From Smart Farming towards Agriculture 5.0: A Review on Crop Data Management. *Agronomy (Basel), 10*(2), 207. doi:10.3390/agronomy10020207

Sakr, S., & Elgammal, A. (2016, June 1). Towards a comprehensive data analytics framework for smart healthcare services. *Big Data Research., 4*, 44–58. doi:10.1016/j.bdr.2016.05.002

Salameh, A. I., & El Tarhuni, M. (2022, April). From 5G to 6G—Challenges, Technologies, and Applications. *Future Internet, 14*(4), 117. doi:10.3390/fi14040117

Saleh, S. N., & Fathy, C. (2023). A Novel Deep-Learning Model for Remote Driver Monitoring in SDN-Based Internet of Autonomous Vehicles Using 5G Technologies. *Applied Sciences (Basel, Switzerland), 13*(2), 875. doi:10.3390/app13020875

Salem, O., Serhrouchni, A., Mehaoua, A., & Boutaba, R. (2018, May 31). Event detection in wireless body area networks using Kalman filter and power divergence. *IEEE Transactions on Network and Service Management, 15*(3), 1018–1034. doi:10.1109/TNSM.2018.2842195

Samant, P., Bhushan, M., Kumar, A., Arya, R., Tiwari, S., & Bansal, S. (2021). Condition monitoring of machinery: A case study. *2021 6th International Conference on Signal Processing, Computing and Control (ISPCC)*, 501–505.

Samara, M. A., Bennis, I., Abouaissa, A., & Lorenz, P. (2022). A Survey of Outlier Detection Techniques in IoT: Review and Classification. *J. Sens. Actuator Netw., 11*(1), 4. doi:10.3390/jsan11010004

Samsung. (2015). *5G Vision* (White Paper). Samsung Electronics Co.

Saravana Balaji, B., Salih Mohammed, A., & Al-Atroshi, C. (2020). Adaptability of SOA in IoT Services-An Empirical Survey. *Int. J. Comput. Appl, 182*(31), 25–28.

Sarker, I. H. (2021). Deep Learning: A Comprehensive Overview on Techniques, Taxonomy, Applications, and Research Directions. *SN Computer Science, 2*(6), 420. doi:10.100742979-021-00815-1 PMID:34426802

Sarker, I. H., Abushark, Y. B., Alsolami, F., & Khan, A. I. (2020). Intrudtree: A machine learning based cyber security intrusion detection model. *Symmetry, 12*(5), 754. doi:10.3390ym12050754

Sarker, I. H., Furhad, M. H., & Nowrozy, R. (2021). Ai-driven cybersecurity: An overview, security intelligence modeling and research directions. *SN Computer Science, 2*(3), 1–18. doi:10.100742979-021-00557-0

Sarker, I. H., Kayes, A. S. M., Badsha, S., Alqahtani, H., Watters, P., & Ng, A. (2020). Cybersecurity data science: An overview from machine learning perspective. *Journal of Big Data, 7*(1), 1–29. doi:10.118640537-020-00318-5

Sarwar, A., Hasan, S., Khan, W. U., Ahmed, S., & Marwat, S. N. K. (2022, March). Design of an advance intrusion detection system for IoT networks. In *2022 2nd International Conference on Artificial Intelligence (ICAI)* (pp. 46-51). IEEE. 10.1109/ICAI55435.2022.9773747

Sasi Bhanu, J., Sunitha, D., & Bigul, P. A. (2021). Agricultural Internet of Things using Machine Learning. AIP Conference Proceedings. AIP Publishing. doi:10.1063/5.0058012

Sawanobori, T. K. (2016). *The Next Generation of Wireless: 5G Leadership in the U.S.* (White Paper). CTIA Everything Wireless. Available online: https://5g-ppp.eu/wp-content/uploads/2015/02/5G-Vision-Brochure-v1.pdf

Scalea, J. R., Pucciarella, T., Talaie, T., Restaino, S., Drachenberg, C. B., Alexander, C., Qaoud, T. A., Barth, R. N., Wereley, N. M., & Scassero, M. (2021). Successful Implementation of Unmanned Aircraft Use for Delivery of a Human Organ for Transplantation. *Annals of Surgery, 274*(3), e282–e288. doi:10.1097/SLA.0000000000003630 PMID:31663974

Scholkopf, B., Smola, A., & Müller, K.-R. (1998). Nonlinear Component Analysis as a Kernel Eigenvalue Problem. *Neural Computation, 2*(5), 1299–1319. doi:10.1162/089976698300017467

Schubert, E, Zimek, A, & Kriegel, HP. (2015). Fast and scalable outlier detection with approximate nearest neighbor ensembles. *Database Syst Adv Appl., 90*(50), 19–36.

Schwartz, T., Stevens, G., Jakobi, T., Denef, S., Ramirez, L., Wulf, V., & Randall, D. (2015). What people do with consumption feedback: A long-term living lab study of a home energy management system. *Interacting with Computers, 27*(6), 551–576. doi:10.1093/iwc/iwu009

Selvaraj, S., & Sundaravaradhan, S. (2020). Challenges and opportunities in IoT healthcare systems: A systematic review. *SN Applied Sciences, 2*(1), 139. doi:10.100742452-019-1925-y

Semeraro, F., Griffiths, A., & Cangelosi, A. (2023). Human–robot collaboration and machine learning: A systematic review of recent research. *Robotics and Computer-integrated Manufacturing, 79*, 102432. doi:10.1016/j.rcim.2022.102432

Sen, A., Roy, R., & Dash, S. R. (2021). Smart Farming using Machine Learning and IoT. Wiley. doi:10.1002/9781119769231.ch2

Senthil, K. S. D., & Shamili, M. D. (2022). Smart Farming using Machine Learning and Deep Learning Techniques. *Decision Analytics Journal, 3*. doi:10.1016/j.dajour.2022.100041

Serbanescu, A. V., Obermeier, S., & Yu, D. Y. (2020). ICS threat analysisusing a large-scale honeynet. *3rd International Symposium for ICS &SCADA Cyber Security Research 2020 (ICS-CSR 2020) 3*, (pp. 20-30). ACM.

Sethi, D., & Anand, J. (2019, September 12). Big data and WBAN: Prediction and analysis of the patient health condition in a remote area. *Engineering and Applied Science Research., 46*(3), 248–255.

Sfar, A. R., Natalizio, E., Challal, Y., & Chtourou, Z. (2018). A roadmap for security challenges in the Internet of Things. *Digital Communications and Networks, 4*(2), 118–137. doi:10.1016/j.dcan.2017.04.003

Shafiq, M., Tian, Z., Sun, Y., Du, X., & Guizani, M. (2020). Selection of effective machine learning algorithm and Bot-IoT attacks traffic identification for internet of things in smart city. *Future Generation Computer Systems, 107*, 433–442. doi:10.1016/j.future.2020.02.017

Shahad, P. (2021). Challenges in Streaming Data Analysis for Building an Adaptive Model for Handling Concept Drifts. In *2021 International Conference on System, Computation, Automation and Networking (ICSCAN)*, (pp. 1-6). IEEE.

Shah, K. J., Pan, S., Lee, I., Kim, H., You, Z., Zheng, J., & Chiang, P. (2021). Green transportation for sustainability: Review of current barriers, strategies, and innovative technologies. *Journal of Cleaner Production, 326*, 129392. doi:10.1016/j.jclepro.2021.129392

Shah, S. F. A., Iqbal, M., Aziz, Z., Rana, T. A., Khalid, A., Cheah, Y. N., & Arif, M. (2022). The role of machine learning and the internet of things in smart buildings for energy efficiency. *Applied Sciences (Basel, Switzerland), 12*(15), 7882. doi:10.3390/app12157882

Shaikh, T. A., Mir, W. A., Rasool, T., & Sofi, S. (2022) Machine Learning for Smart Agriculture and Precision Farming: Towards Making the Fields Talk. In: Archives of Computational Methods in Engineering. doi:10.100711831-022-09761-4

Shanmuganathan, M., Almutairi, S., Aborokbah, M. M., & Ganesan, S., & Ramachandran, V. (2020, August). Review of advanced computational approaches onmultiple sclerosis segmentation and classification. *IET Signal Processing, 14*(6), 333–341. doi:10.1049/iet-spr.2019.0543

Sharif, A., Li, J., Khalil, M., Kumar, R., Sharif, M., & Sharif, A. (2017). *Internet of things — smart traffic management system for smart cities using big data analytics.* Active Media Technology. doi:10.1109/ICCWAMTIP.2017.8301496

Sharma, H., Haque, A., &Blaabjerg, F. (2021). Machine Learning in Wireless Sensor Networks for Smart Cities: A Survey. *Electronics, 10*(9), 1012. MDPI AG. doi:10.3390/electronics10091012

Sharma, A., & Singh, S. (2019). Challenges and Opportunities in IoT-Based Agriculture: A Review. In P. Kumar, P. Singh, H. Le Duc, & S. Pandey (Eds.), *Internet of Things: Challenges and Opportunities* (pp. 85–94). Springer. doi:10.1007/978-981-13-3014-7_9

Sharma, O., Rathee, G., Kerrache, C. A., & Herrera-Tapia, J. (2023). Two-Stage Optimal Task Scheduling for Smart Home Environment Using Fog Computing Infrastructures. *Applied Sciences (Basel, Switzerland), 13*(5), 2939. doi:10.3390/app13052939

Sharma, P., Mishra, A., Kashyap, N., Muzamil, M., Rawat, R. S., Abidi, A. I., & Umrao, L. S. (2023). LTRACN: A Method for Single Human Activity Recognition. Intelligent Systems and Smart Infrastructure. *Proceedings of ICISSI, 2022*, 24.

Shetty, S., & Smitha, A. B. (2021). Smart Agriculture Using IoT and Machine Learning. In A. Choudhury, A. Biswas, T. P. Singh, & S. K. Ghosh (Eds.), *Smart Agriculture Automation Using Advanced Technologies. Transactions on Computer Systems and Networks.* Springer. doi:10.1007/978-981-16-6124-2_1

Shiblee Sadik, M, & Gruenwald, L. (2010). DBOD-DS: Distance-based outlier detection for data streams. *Database Expert Syst Appl., 62*(61), 122–136.

Shone, N., Ngoc, T. N., Phai, V. D., & Shi, Q. (2018). A deep learning approach to network intrusion detection. *IEEE Transactions on Emerging Topics in Computational Intelligence, 2*(1), 41–50. doi:10.1109/TETCI.2017.2772792

Shrivastava, A., Bhardwaj, A., & Hasteer, N. (2020). IoT in Automobile Sector: State of the Art. *2020 10th International Conference on Cloud Computing, Data Science & Engineering (Confluence).* IEEE. 10.1109/Confluence47617.2020.9058202

Shruthi, U., Nagaveni, V., & Raghavendra, B. K. (2019). A review on machine learning classification techniques for plant disease detection. *2019 5th International Conference on Advanced Computing & Communication Systems (ICACCS),* (pp. 281–284). IEEE.

Siddavatam, I. A., Satish, S., Mahesh, W., & Kazi, F. (2020). An ensemblelearning for anomaly identification in SCADA system. *2020 7th International Conference on Power Systems (ICPS),* (pp. 457-462). ACM.

Sihare, S. (2022). Future Digital Marketing Revolutionizing E-Commerce. *Towards Excellence, 14*(1).

Sihare, S. R. (2017a). Image-based digital marketing. *International Journal of Information Engineering and Electronic Business, 9*(5), 10. doi:10.5815/ijieeb.2017.05.02

Sihare, S. R. (2017b). Role of m-Banking for Indian Rural Consumers, its Adaptation Strategies, and Challenges: Consumer Behavior Analysis. *International Journal of Information Engineering & Electronic Business, 9*(6). doi:10.5815/ijieeb.2017.06.05

Sihare, S. R. (2018). Roles of E-content for E-business: Analysis. *International Journal of Information Engineering & Electronic Business, 10*(1). Advance online publication. doi:10.5815/ijieeb.2018.01.04

Silva, A., & Antunes, C. (2015). Multi-relational pattern mining over data streams. *Data Mining and Knowledge Discovery*, *29*(6), 1783–1814. doi:10.100710618-014-0394-6

Silva, M. D., Leandro, R., Batista, D. M., & Guedes, L. A. (2013). A dependability evaluation tool for the Internet of Things. *Computers & Electrical Engineering*, *39*(7), 2005–2018. doi:10.1016/j.compeleceng.2013.04.021

Singh, A. K., & Srivastava, S. (2020). Challenges and Opportunities in IoT-Based Supply Chain Management: A Review. In Internet of Things and Big Data Analytics Toward Next-Generation Intelligence (pp. 223-247). Springer. doi:10.1007/978-981-15-4702-6_11

Singh, S. (2022). The Definitive Guide to Enterprise Digital Transformation. *App Inventive.* https://appinventiv.com/guide/digital-transformation-for-business/

Singh, S. N., & Bhushan, M. (2022). Smart ECG monitoring and analysis system using machine learning. *2022 IEEE VLSI Device Circuit and System (VLSI DCS)*, (pp. 304–309). IEEE.

Singh, V. K., Piryani, R., Uddin, A., & Waila, P. (2013, January). Sentiment analysis of textual reviews; Evaluating machine learning, unsupervised and SentiWordNet approaches. In *2013 5th international conference on knowledge and smart technology (KST)* (pp. 122-127). IEEE.

Singh, A., Gutub, A., & Nayyar, A. (2022). *Redefining food safety traceability system through Blockchain: Findings, Challenges, and Open Issues*. Multimed Tools Appl. doi:10.100711042-022-14006-4

Singh, K, & Upadhyaya, S. (2012). Outlier detection: Applications and techniques. *IJCSI Int J Comp Sci Iss.*, *9*(3), 307–323.

Singh, S., Rai, S., Singh, P., & Mishra, V. K. (2022). Real-time water quality monitoring of River Ganga (India) using internet of things. *Ecological Informatics*, *101770*, 101770. doi:10.1016/j.ecoinf.2022.101770

Siregar, R. R. A., Seminar, K. B., Wahjuni, S., & Santosa, E. (2022). Vertical Farming Perspectives in Support of Precision Agriculture Using Artificial Intelligence: A Review. *Computers*, *11*(9), 135. doi:10.3390/computers11090135

Sivakumar, R., & Prabadevi, B. (2022). *Internet of Things and Machine Learning Applications for Smart Precision Agriculture*. IoT Applications Computing. doi:10.5772/intechopen.97679

Skourletopoulos, G., Mavromoustakis, C. X., Mastorakis, G., Batalla, J. M., Dobre, C., Panagiotakis, S., & Pallis, E. (2017). Data and cloud computing: a survey of the state-of-the-art and research challenges. In Advances in mobile cloud computing and big data in the 5G Era, (pp. 23-41). Springer.

Slevi, S. T., & Visalakshi, P. (2021, November). A survey on Deep Learning based Intrusion Detection Systems on Internet of Things. In *2021 Fifth International Conference on I-SMAC (IoT in Social, Mobile, Analytics and Cloud)(I-SMAC)* (pp. 1488-1496). IEEE. 10.1109/I-SMAC52330.2021.9641050

Smrithy, GS, Munirathinam, S, & Balakrishnan, R. (2016). Online anomaly detection using non-parametric technique for big data streams in cloud collaborative environment. [Big Data]. *IEEE Int Conf Big Data*, *2016*, 1950–1955.

Somvanshi, M. (2016).. . *Iccubea.*, *2016*, 7860040.

Song, C., Ma, W., Li, J., Qi, B., & Liu, B. (2022). Development Trends in Precision Agriculture and Its Management in China Based on Data Visualization. *Agronomy (Basel)*, *12*(11), 2905. doi:10.3390/agronomy12112905

Song, D., Yuan, L., Zhao, W., Yu, Q., Du, J., & Pan, K. (2021). Cloud-Edge Computing Resource Collaborative Optimization Method for Power Distribution Fault Analysis Service. *2021 China International Conference on Electricity Distribution (CICED)*, Shanghai, China. 10.1109/CICED50259.2021.9556733

Songhorabadi, M., Rahimi, M., MoghadamFarid, A. M., & Haghi Kashani, M. (2023). Fog computing approaches in IoT-enabled smart cities. *Journal of Network and Computer Applications*, *211*, 103557. doi:10.1016/j.jnca.2022.103557

Song, S., Chang, K., Yun, K., Jun, C., & Baek, J. G. (2022). Defect Synthesis Using Latent Mapping Adversarial Network for Automated Visual Inspection. *Electronics (Basel)*, *11*(17), 2763. doi:10.3390/electronics11172763

Song, S., Zhan, Z., Long, Z., Zhang, J., & Yao, L. (2011). Comparative study of SVM methods combined with voxel selection for object category classification on fMRI data. *PLoS One*, *6*(2), e17191. doi:10.1371/journal.pone.0017191 PMID:21359184

Song, Y., Yau, S. S., Yu, R., Zhang, X., & Xue, G. (2017). An Approach to QoS-based Task Distribution in Edge Computing Networks for IoT Applications. *2017 IEEE International Conference on Edge Computing (EDGE)*, Honolulu, HI, USA. 10.1109/IEEE.EDGE.2017.50

Soni, V., Jaiswal, S., Semwal, V. B., Roy, B., Choubey, D. K., & Mallick, D. K. (2023, January). An Enhanced Deep Learning Approach for Smartphone-Based Human Activity Recognition in IoHT. In *Machine Learning, Image Processing, Network Security and Data Sciences: Select Proceedings of 3rd International Conference on MIND 2021* (pp. 505-516). Singapore: Springer Nature Singapore. 10.1007/978-981-19-5868-7_37

Spinelle, L., Gerboles, M., Kok, G., Persijn, S., & Sauerwald, T. (2017). Review of portable and low-cost sensors for the ambient air monitoring of benzene and other volatile organic compounds. *Sensors (Basel)*, *17*(7), 1520. doi:10.339017071520 PMID:28657595

Srbinovska, M., Gavrovski, C., Dimcev, V., Krkoleva, A., & Borozan, V. (2015). Environmental parameters monitoring in precision agriculture using wireless sensor networks. *Journal of Cleaner Production*, *88*, 297–307. doi:10.1016/j.jclepro.2014.04.036

Sreevidya, SS. (2015). Detection of outliers in data stream using clustering method. *Int J Sci Eng Technol Res.*, *4*(3), 559–563.

Srinivasan, R. (2022). A Comparative Study: Wireless Technologies in Internet of Things. In *2022 2nd International Conference on Advance Computing and Innovative Technologies in Engineering (ICACITE)*. IEEE. 10.1109/ICACITE53722.2022.9823664

Stöcker, C., Bennett, R., Nex, F., Gerke, M., & Zevenbergen, J. (2017). Review of the Current State of UAV Regulations. *Remote Sensing (Basel)*, *9*(5), 459. doi:10.3390/rs9050459

Su, Z., He, G., & Li, Z. (2022). *Using Grasshopper Optimization Algorithm to Solve 0-1 Knapsack Computation Resources Allocation Problem in Mobile Edge Computing*. 2022 34th Chinese Control and Decision Conference (CCDC), Hefei, China. 10.1109/CCDC55256.2022.10034236

Suárez, P., Iglesias, A., & Gálvez, A. (2019). Make robots be bats: Specializing robotic swarms to the Bat algorithm. *Swarm and Evolutionary Computation*, *44*, 113–129. doi:10.1016/j.swevo.2018.01.005

Su, J. (2019, February 12). Venture capital funding for artificial intelligence startups hit record high in 2018. *Forbes*.

Sullivan, N., Grainger, S., & Cazzolato, B. (2019). Sequential single-item auction improvements for heterogeneous multi-robot routing. *Robotics and Autonomous Systems*, *115*, 130–142. doi:10.1016/j.robot.2019.02.016

Sun, D., Li, X., Zhao, L., Wang, Y., Cao, X., & Yang, J. (2020). *Study of Demand-Side Resources Collaborative Optimization System Based on the Cloud & Edge Computing*. 2020 IEEE 5th International Conference on Cloud Computing and Big Data Analytics (ICCCBDA), Chengdu, China. 10.1109/ICCCBDA49378.2020.9095622

Sun, D. W., Zhang, G. Y., & Zheng, W. M. (2014). Big data stream computing: Technologies and instances. *Journal of Software*, 25(4), 839–862.

Suri, R. S., Dubey, V., Kapoor, N. R., Kumar, A., & Bhushan, M. (2022). Optimizing the Compressive Strength of Concrete with Altered Compositions Using Hybrid PSO-ANN. *Information Systems and Management Science: Conference Proceedings of 4th International Conference on Information Systems and Management Science (ISMS) 2021*, (pp. 163–173). IEEE.

Swamy, A. (2023). Advance Cellular Networks (4G, 5G, 6G). *International Journal of Health Sciences*, 2, 10955–10966.

Swamy, S. N., Jadhav, D., & Kulkarni, N. (2017, February). Security threats in the application layer in IOT applications. In *2017 International conference on i-SMAC (iot in social, mobile, analytics and cloud)(i-SMAC)* (pp. 477-480). IEEE. 10.1109/I-SMAC.2017.8058395

Szulczyński, B., & Gębicki, J. (2017). Currently commercially available chemical sensors employed for detection of volatile organic compounds in outdoor and indoor air. *Environments - MDPI, 4*(1), 1–15. doi:10.3390/environments4010021

Taddy, M. (2019). The technological elements of artificial intelligence. In A. Agrawal, J. Gans, & A. Goldfarb (Eds.), *The Economics of Artificial Intelligence: An Agenda* (61–87). National Bureau of Economic Research Conference Report series. Chicago: University of Chicago Press. 10.7208/chicago/9780226613475.003.0002

Taher, N. C., Mallat, I., Agoulmine, N., & El-Mawass, N. (2019). An IoT-Cloud Based Solution for Real-Time and Batch Processing of Big Data: Application in Healthcare. In *2019 3rd International Conference on Bio-engineering for Smart Technologies (BioSMART)*, (pp. 1-8). IEEE.

Tahsien, S. M., Karimipour, H., & Spachos, P. (2020). Machine learning based solutions for security of Internet of Things (IoT): A survey. *Journal of Network and Computer Applications, 161*, 102630. doi:10.1016/j.jnca.2020.102630

Tajudin, K., & Shahina Begam, I. (2020). Intelligent Crop Growth Management System using Internet of Things. *European Journal of Molecular and Clinical Medicine, 7*(9), 2211–2222.

Takeda, A., Kimura, T., & Hirata, K. (2020). *Joint optimization of edge server and virtual machine placement in edge computing environments*. 2020 Asia-Pacific Signal and Information Processing Association Annual Summit and Conference (APSIPA ASC), Auckland, New Zealand.

Talaviya, T., Shah, D., Patel, N., Yagnik, H., & Shah, M. (2020). Implementation of Artificial Intelligence in Agriculture for Optimization of Irrigation and Application of Pesticides and Herbicides. *Artificial Intelligence in Agriculture, 4*, 58–73. doi:10.1016/j.aiia.2020.04.002

Talukdar, S. B., Dhabliya, D., Ahamad, S., & Gupta, A. (2022). Machine Learning for Anomaly Detection in IoT. In Emerging Trends and Applications of Machine Learning in Data Science (pp. 335-352). IGI Global.

Talukdar, V., Dhabliya, D., Kumar, B., Talukdar, S. B., Ahamad, S., & Gupta, A. (2022, November). Suspicious Activity Detection and Classification in IoT Environment Using Machine Learning Approach. In *2022 Seventh International Conference on Parallel, Distributed and Grid Computing (PDGC)* (pp. 531-535). IEEE. 10.1109/PDGC56933.2022.10053312

Tan, A. (2021). How IoT and Machine Learning are Automating Agriculture, Tech Target. *Computer Weekly*. https://www.computerweekly.com/news/252504285/How-IoT-and-machine-learning-are-automating-agriculture

Tang, S., Shelden, D. R., Eastman, C. M., Pishdad-Bozorgi, P., & Gao, X. (2019). A review of building information modeling (BIM) and the internet of things (IoT) devices integration: Present status and future trends. *Automation in Construction, 101*, 127–139. doi:10.1016/j.autcon.2019.01.020

Tariq, N. (2018). Impact of cyberattacks on financial institutions. *Journal of Internet Banking and Commerce, 23*(2), 1–11.

Tarun, K., Sreelakshmi, K., & Peeyush, K. P. (2019). Segregation of Plastic and Non-plastic Waste using Convolutional Neural Network. *IOP Conference Series. Materials Science and Engineering*, *561*(1), 1–7. doi:10.1088/1757-899X/561/1/012113

Ta, V. D., Liu, C. M., & Nkabinde, G. W. (2016). Big data stream computing in healthcare real-time analytics. In *2016 IEEE International Conference on Cloud Computing and Big Data Analysis (ICCCBDA)*, (pp. 37-42). IEEE.

Tavakoli, M., Carriere, J., & Torabi, A. (2020). Robotics, Smart Wearable Technologies, and Autonomous Intelligent Systems for Healthcare During the COVID-19 Pandemic: An Analysis of the State of the Art and Future Vision. *Advanced Intelligent Systems*, *2*(7), 2000071. doi:10.1002/aisy.202000071

Tavallaee, M., Bagheri, E., Lu, W., & Ghorbani, A. A. (2009, July). *A detailed analysis of the KDD CUP 99 data set. In 2009 IEEE symposium on computational intelligence for security and defense applications.* IEEE.

Tax, D. M. J., & Duin, R. P. W. (2020). Data domain description using supportvectors. European Symposium on Artificial Neural Networks. IEEE.

Tefera, H. A., Huang, D., & Njagi, K. (2020). Implementation of IoT and Machine Learning for Smart Farming Monitoring System [IJSBAR]. *International Journal of Sciences, Basic and Applied Research*, *52*(1), 67–77.

Teruel, E., Aragues, R., & López-Nicolás, G. (2019). A distributed robot swarm control for dynamic region coverage. *Robotics and Autonomous Systems*, *119*, 51–63. doi:10.1016/j.robot.2019.06.002

Thakkar P, Vala J, & Prajapati V. (2016). Survey on outlier detection in data stream. *Int J Comput Appl., 136*(2), 0975–8887.

Thakor, V. A., Razzaque, M. A., & Khandaker, M. R. (2021). Lightweight cryptography algorithms for resource-constrained IoT devices: A review, comparison and research opportunities. *IEEE Access : Practical Innovations, Open Solutions*, *9*, 28177–28193. doi:10.1109/ACCESS.2021.3052867

Tian, Y., & Pan, L. (2015). Predicting Short-Term Traffic Flow by Long Short-Term Memory Recurrent Neural Network. *2015 IEEE International Conference on Smart City/SocialCom/SustainCom (SmartCity).* IEEE. 10.1109/SmartCity.2015.63

TongKe. (2013). Smart agriculture based on cloud computing and IOT. *Journal of Convergence Information Technology*, *8*(2).

Torres-Martínez, L. M., Kharissova, O. V., & Kharisov, B. I. (2019). Handbook of ecomaterials. Handbook of Ecomaterials, 1(December 2020), 1–3773. doi:10.1007/978-3-319-68255-6

Tran, M. Q., Amer, M., Abdelaziz, A. Y., Dai, H. J., Liu, M. K., & Elsisi, M. (2023). Robust fault recognition and correction scheme for induction motors using an effective IoT with deep learning approach. *Measurement*, *207*, 112398. doi:10.1016/j.measurement.2022.112398

Triantafyllou, A., Sarigiannidis, P., & Bibi, S. (2019). Article Precision Agriculture: A Remote Sensing Monitoring System Architecture. *Information (Basel)*, *10*(11), 348. doi:10.3390/info10110348

Trivedi, P., Purohit, D., Soju, A., & Tiwari, R. R. (2014). *Major Industrial Disasters in India.*, *9*(4), 7.

Trovato, V., Sfameni, S., Rando, G., Rosace, G., Libertino, S., Ferri, A., & Plutino, M. R. (2022). A Review of Stimuli-Responsive Smart Materials for Wearable Technology in Healthcare: Retrospective, Perspective, and Prospective. *Molecules (Basel, Switzerland)*, *27*(17), 5709. doi:10.3390/molecules27175709 PMID:36080476

Tu, X., Ma, O., Zhang, L., Yang, J., Wang, Y., & Cai, H. (2021). Distributed Multi-Robot System for Planetary Surface Exploration. *IEEE Access : Practical Innovations, Open Solutions*, *9*, 12987–12996. doi:10.1109/ACCESS.2021.3053461

Tyagi, A. K., & Goyal, D. (2020, June). A survey of privacy leakage and security vulnerabilities in the internet of things. In *2020 5th International conference on communication and electronics systems (ICCES)* (pp. 386-394). IEEE. 10.1109/ICCES48766.2020.9137886

Tyralis, H., Papacharalampous, G., & Langousis, A. (2019). A brief review of random forests for water scientists and practitioners and their recent history in water resources. *Water (Basel)*, *11*(5), 910. doi:10.3390/w11050910

Uddin, M. A., Ayaz, M., Mansour, A., Aggoune, H. M., Sharif, Z., & Razzak, I. (2021). Cloud-connected flying Edge Computing for Smart Agriculture. *Peer-to-Peer Networking and Applications*, *14*(6), 3405–3415. doi:10.100712083-021-01191-6

Uganya, G., Rajalakshmi, D., Teekaraman, Y., Kuppusamy, R., & Radhakrishnan, A. (2022). A Novel Strategy for Waste Prediction Using Machine Learning Algorithm with IoT Based Intelligent Waste Management System. *Wireless Communications and Mobile Computing*, *2022*, 1–15. doi:10.1155/2022/2063372

Ukpabi, D. C., Aslam, B., & Karjaluoto, H. (2019). Chatbot adoption in tourism services: A conceptual exploration. In *Robots, artificial intelligence, and service automation in travel, tourism and hospitality*. Emerald Publishing Limited. doi:10.1108/978-1-78756-687-320191006

Ul Hasan, M. N., & Negulescu, I. I. (2020). Wearable technology for baby monitoring: A review. *Journal of Textile Engineering & Fashion Technology*, *6*(4). doi:10.15406/jteft.2020.06.00239

Ullah, R., Abbas, A. W., Ullah, M., Khan, R. U., Khan, I. U., Aslam, N., & Aljameel, S. S. (2021). EEWMP: An IoT-Based Energy-Efficient Water Management Platform for Smart Irrigation. *Scientific Programming*, *2021*, 1–9. doi:10.1155/2021/5536884

Umair, M., Cheema, M. A., Cheema, O., Li, H., & Lu, H. (2021). Impact of COVID-19 on IoT adoption in healthcare, smart homes, smart buildings, smart cities, transportation and industrial IoT. *Sensors (Basel)*, *21*(11), 3838. doi:10.339021113838 PMID:34206120

Umer, M. A., Junejo, K. N., Jilani, M. T., & Mathur, A. P. (2022). Machine learning for intrusion detection in industrial control systems: Applications, challenges, and recommendations. *International Journal of Critical Infrastructure Protection*, *38*, 100516. doi:10.1016/j.ijcip.2022.100516

UN. (2018). Food & Agriculture Organization of the United Nations; E-Agriculture in Action: DRONES for Agriculture, Bangkok, 2018. UN.

Urrea, C., & Agramonte, R. (2022a). Evaluation of Parameter Identification of a Real Manipulator Robot. *Symmetry*, *14*(7), 1446. https://www.mdpi.com/2073-8994/14/7/1446. doi:10.3390ym14071446

Urrea, C., Kern, J., & Álvarez, E. (2022b). Design of a generalized dynamic model and a trajectory control and position strategy for n-link underactuated revolute planar robots. *Control Engineering Practice*, *128*, 1–13. doi:10.1016/j.conengprac.2022.105316

Urrea, C., & Matteoda, R. (2020). Development of a virtual reality simulator for a strategy for coordinating cooperative manipulator robots using cloud computing. *Robotics and Autonomous Systems*, *126*, 103447. doi:10.1016/j.robot.2020.103447

Urrea, C., & Páez, F. (2021a). Design and Comparison of Strategies for Level Control in a Nonlinear Tank. *Processes (Basel, Switzerland)*, *9*(5), 735. doi:10.3390/pr9050735

Urrea, C., & Pascal, J. (2021b). Design and validation of a dynamic parameter identification model for industrial manipulator robots. *Archive of Applied Mechanics*, *91*(5), 1981–2007. doi:10.100700419-020-01865-2

Urrea, C., & Pascal, J. (2021c). Dynamic Parameter Identification based on Lagrangian Formulation and Servomotor-type Actuators for Industrial Robots. *International Journal of Control, Automation, and Systems*, *19*(8), 2902–2909. doi:10.100712555-020-0476-8

Us, A., & Gas, T. (2020). *5 Types of Toxic Gas & Their Health Effects*. GDS Corp. https://www.gdscorp.com/blog/gas-emission/5-types-of-toxic-gas-their-health-effects/

Utku Kose, V. B., Prasath, S., Mondal M., R., Podder P., & Bharati, S. (2023). Artificial Intelligence and Smart Agriculture Applications. Auerbach Publications.

Vadlamudi, S. (2020). Internet of Things (IoT) in Agriculture: The Idea of Making the Fields Talk. *Engineering International*, *8*(2), 87–100. doi:10.18034/ei.v8i2.522

Vaigandla, K. K., Azmi, N., Podila, R., & Karne, R. K. (2021). A Survey on Wireless Communications: 6g And 7g. *International Journal Of Science, Technology & Management*, *2*(6), 2018–2025. doi:10.46729/ijstm.v2i6.379

Valade, S., Ley, A., Massimetti, F., D'Hondt, O., Laiolo, M., Coppola, D., Loibl, D., Hellwich, O., & Walter, T. R. (2019). Towards global volcano monitoring using multisensor sentinel missions and artificial intelligence: The MOUNTS monitoring system. *Remote Sensing (Basel)*, *11*(13), 1–31. doi:10.3390/rs11131528

van der Veen, J. S., van derWaaij, B., & Meijer, R. J. (2012). Sensor data storage performance: SQL or NoSQL, physical or virtual. In *Proceedings of the IEEE 5th International Conference on Cloud Computing (CLOUD '12)*, (pp. 431–438). IEEE. 10.1109/CLOUD.2012.18

Van Houdt, G., Mosquera, C., & Nápoles, G. (2020). A review on the long short-term memory model. *Artificial Intelligence Review*, *53*(8), 5929–5955. doi:10.100710462-020-09838-1

Varlamov, O. (2021). "Brains" for Robots: Application of the Mivar Expert Systems for Implementation of Autonomous Intelligent Robots. *Big Data Research*, *25*, 100241. doi:10.1016/j.bdr.2021.100241

Varman, S. A. M., Baskaran, A. R., Aravindh, S., & Prabhu, E. (2017). Deep learning and IoT for smart agriculture using WSN. In *2017 IEEE International Conference on Computational Intelligence and Computing Research (ICCIC)* (pp. 1-6). IEEE.

Vectors, S. (n.d.). Support Vector Machine — Introduction to Machine Learning Algorithms SVM model from scratch Introduction What is Support Vector Machine ? *Hyperplanes and Support Vectors Cost Function and Gradient Updates*.

Veena, S., Mahesh, K., Rajesh, M., & Salmon, S. (2018). The survey on smart agriculture using IOT. *Int J Innov Res EngManag*, *5*(2), 63–66.

Veeraiah, V., Gangavathi, P., Ahamad, S., Talukdar, S. B., Gupta, A., & Talukdar, V. (2022, April). Enhancement of meta verse capabilities by IoT integration. In *2022 2nd International Conference on Advance Computing and Innovative Technologies in Engineering* (ICACITE) (pp. 1493-1498). IEEE. 10.1109/ICACITE53722.2022.9823766

Vélez, S., Vacas, R., Martín, H., Ruano-Rosa, D., & Álvarez, S. (2022). High-Resolution UAV RGB Imagery Dataset for Precision Agriculture and 3D Photogrammetric Reconstruction Captured over a Pistachio Orchard (Pistacia vera L.) in Spain. *Data*, *7*(11), 157. doi:10.3390/data7110157

Verma, P. & Umesh, K. (2021). "Smart Farming: using IoT and Machine Learning Techniques." Internet of Things and Machine Learning in Agriculture: Technological Impacts and Challenges, Jyotir Moy Chatterjee, Abhishek Kumar, Pramod Singh Rathore, & Vishal Jain, Berlin. De Gruyter. doi:10.1515/9783110691276-001

Verma, U., Garg, C., Bhushan, M., Samant, P., Kumar, A., & Negi, A. (2022). Prediction of students' academic performance using Machine Learning Techniques. *2022 International Mobile and Embedded Technology Conference (MECON)*, (pp. 151–156). IEEE. 10.1109/MECON53876.2022.9751956

Vidhate, D. A., & Kulkarni, P. (2017). Cooperative multi-agent reinforcement learning models (CMRLM) for intelligent traffic control. *2017 1st International Conference on Intelligent Systems and Information Management (ICISIM)*. IEEE. 10.1109/ICISIM.2017.8122193

Vijayakumar, P., Chang, V., Deborah, L. J., Balusamy, B., Shyn, P. G., Vijayakumar, P., Chang, V., Deborah, L. J., Balusamy, B., & Shyn, P. G. (2016). *Computationally Efficient Privacy Preserving Anonymous Mutual and Batch Authentication Schemes for Vehicular Ad Hoc Networks*. Future Generation Computer Systems. https://eprints.soton.ac.uk/403454/

Vijayarani, S, & Jothi, P. (2013). An efficient clustering algorithm for outlier detection in data streams. *International Journal of Advanced Research in Computer and Communication Engineering*, *2*(9), 3657–3665.

Villazon, L. (2022). Vertical farming: Why stacking crops high could be the future of agriculture. *Science Focus*. https://www.sciencefocus.com/science/what-is-vertical-farming/

Villegas-Ch, W., Román-Cañizares, M., & Palacios-Pacheco, X. (2020). Improvement of an online education model with the integration of machine learning and data analysis in an LMS. *Applied Sciences (Basel, Switzerland)*, *10*(15), 5371. doi:10.3390/app10155371

Vinayakumar, R., Alazab, M., Soman, K. P., Poornachandran, P., Al-Nemrat, A., & Venkatraman, S. (2019). Deep learning approach for intelligent intrusion detection system. *IEEE Access : Practical Innovations, Open Solutions*, *7*, 41525–41550. doi:10.1109/ACCESS.2019.2895334

Vu, M. T., Vuong, Q. H., Nguyen, V. T., Quach, C. H., Truong, N. T., & Pham, M. T. (2019). *Design and Implement Low-cost UAV for Agriculture Monitoring*.

Vujović, Ž. (2021). Classification model evaluation metrics. *International Journal of Advanced Computer Science and Applications*, *12*(6), 599–606. doi:10.14569/IJACSA.2021.0120670

Vunnava, S. L., Yendluri, S. C., & Dhuli, S. (2021). IoT based Novel Hydration System for Smart Agriculture Applications *10th IEEE Inter., Conf., on Communication Systems and Network Technologies*. IEEE. 10.1109/CSNT51715.2021.9509597

Wadhwani, P. (2022). Agriculture Technology Trends: Collaborating Tech with Agriculture. *Bacancy Technology*. https://www.bacancytechnology.com/blog/agriculture-technology-trends

Walczak, R., Koszewski, K., Olszewski, R., Ejsmont, K., & Kálmán, A. (2023). Acceptance of IoT Edge-Computing-Based Sensors in Smart Cities for Universal Design Purposes. *Energies*, *16*(3), 1024. doi:10.3390/en16031024

Wang, T., Zou, Y., Zhang, X., Liu, J., & Wu, J. (2023). *Automotive Mixed Criticality DAG Function Scheduling Optimization Based on Edge Computing*. 2023 6th World Conference on Computing and Communication Technologies (WCCCT), Chengdu, China. 10.1109/WCCCT56755.2023.10052398

Wang, F., Laili, Y., & Zhang, L. (2022). Multi-granularity service composition in industrial cloud robotics. *Robotics and Computer-integrated Manufacturing*, *78*, 102414. doi:10.1016/j.rcim.2022.102414

Wang, S., Wang, Y., Deying, L., & Zhao, Q. (2023). Distributed Relative Localization Algorithms for Multi-Robot Networks: A Survey. *Sensors (Basel)*, *23*(5), 2399. doi:10.339023052399 PMID:36904602

Wang, T., Wang, X., & Hong, M. (2018). Gas leak location detection based on data fusion with time difference of arrival and energy decay using an ultrasonic sensor array. *Sensors (Basel)*, *18*(9), 2985. doi:10.339018092985 PMID:30205433

Wang, Y., Yu, Y., Gao, S., Pan, H., & Yang, G. (2019). A hierarchical gravitational search algorithm with an effective gravitational constant. *Swarm and Evolutionary Computation*, *46*, 118–139. doi:10.1016/j.swevo.2019.02.004

Warden, P., & Situnayake, D. (2019). *Tinyml: Machine learning with tensorflow lite on arduino and ultra-low-power microcontrollers*. O'Reilly Media.

Waring, J. (2019, May 21). HKT selected for smart parking system. *Mobile World Live*. https://www.mobileworldlive. com/apps/news-apps/hkt-selected-for-smart-parking-system/

Wazid, M., Bera, B., Das, A., Mohanty, S. P., & Jo, M. (2022). Fortifying Smart Transportation Security Through Public Blockchain. *IEEE Internet of Things Journal*, *9*(17), 16532–16545. doi:10.1109/JIOT.2022.3150842

Wei, C., Ma, C., & Duan, H. (2021). Multi-robot olfactory system for odor mapping and plume tracking. *Measurement*, *178*, 109249. doi:10.1016/j.measurement.2021.109249

Wei, W., Wu, H., & Ma, H. (2019). An AutoEncoder and LSTM-Based Traffic Flow Prediction Method. *Sensors (Basel)*, *19*(13), 2946. doi:10.339019132946 PMID:31277390

WhatsNext. (2022, September 30). *Artificial Intelligence and Oil & Gas Industry – Partnership towards Sustainable Future*. Exponential Transformation Ecosystems. https://whatnextglobal.com/insights/artificial-intelligence-and-oil-gas-industry-partnership-towards-sustainable-future/

Who are America's independent producers? (n.d.). Independent Petroleum Association of America. https://www.ipaa.org/ independent-producers/#:~:text=There%20are%20about%209%2C000%20independent%20oil%20and%20natural,oil%20 and%2090%20percent%20of%20America%E2%80%99s%20natural%20gas

Wild, C. J., & Pfannkuch, M. (1999). Statistical thinking in empirical enquiry. *International Statistical Review*, *67*(3), 223–248. doi:10.1111/j.1751-5823.1999.tb00442.x

Wilkinson, F. (2022, June 2). *Industrial Revolution and Technology*. National Geographic. https://education.national-geographic.org/resource/industrial-revolution-and-technology/

Won, H., Lee, H., Song, G., Kim, Y., & Kwak, N. (2022). Reliable Data Collection Methodology for Face Recognition in Preschool Children. *Sensors (Basel)*, *22*(15), 5842. doi:10.339022155842 PMID:35957397

Wright, C., & Nyberg, D. (2017). An inconvenient truth: How organizations translate climate change into business as usual. *Academy of Management Journal*, *60*(5), 1633–1661. doi:10.5465/amj.2015.0718

Wu, C., Luo, S., Zuo, Y., & Chen, C. (2021). Decentralized multi-robot system for cooperative assembly. *Robotics and Autonomous Systems*, *133*, 103739. doi:10.1016/j.robot.2020.103739

Wu, K, Zhang, K, Fan, W, & ... (2014). RS-Forest: A rapid density estimator for streaming anomaly detection. *IEEE Int Conf Data Min.*, *2014*, 600–609. PMID:25685112

Wu, Y., Liu, T., Ling, S. H., Szymanski, J., Zhang, W., & Su, S. W. (2019). Air quality monitoring for vulnerable groups in residential environments using a multiple hazard gas detector. *Sensors (Basel)*, *19*(2), 362. Advance online publication. doi:10.339019020362 PMID:30658412

Wu, Z., Yang, Z., Yang, C., Lin, J., Liu, Y., & Chen, X. (2022, February). Joint deployment and trajectory optimization in UAV-assisted vehicular edge computing networks. *Journal of Communications and Networks (Seoul)*, *24*(1), 47–58. doi:10.23919/JCN.2021.000026

X. Wei, A. B. M. M. Rahman, D. Cheng D., & Wang, Y. (2023). Joint Optimization Across Timescales: Resource Placement and Task Dispatching in Edge Clouds. *IEEE Transactions on Cloud Computing, 11*(1), 730-744. . doi:10.1109/TCC.2021.3113605

Xhafa, F., & Naranjo, V. (2015). Processing and analytics of big data streams with Yahoo! *S4. IEEE 29th Int Conf Adv Inform Network Appl.* (pp. 263–270). IEEE. 10.1109/AINA.2015.194

Xiao, K., Gao, Z., Wang, Q., & Yang, Y. (2018). A Heuristic Algorithm Based on Resource Requirements Forecasting for Server Placement in Edge Computing. *2018 IEEE/ACM Symposium on Edge Computing (SEC)*, Seattle, WA, USA. 10.1109/SEC.2018.00043

Xiaoyi, Z., Dongling, W., Yuming, Z., Manokaran, K. B., & Benny Antony, A. (2021). IoT driven framework based on efficient green energy management in smart cities using multi-objective distributed dispatching algorithms. *Environmental Impact Assessment Review, 88*, 106567. doi:10.1016/j.eiar.2021.106567

Xing, L. (2020). Reliability in Internet of Things: Current status and future perspectives. *IEEE Internet of Things Journal, 7*(8), 6704–6721. doi:10.1109/JIOT.2020.2993216

Xiong, J., Guo, H., & Liu, J. (2019, March). Task Offloading in UAV-Aided Edge Computing: Bit Allocation and Trajectory Optimization. *IEEE Communications Letters, 23*(3), 538–541. doi:10.1109/LCOMM.2019.2891662

Xu, M., Peng, B., Zhu, X., & Guo, Y. (2022). Multi-Gas Detection System Based on Non-Dispersive Infrared (NDIR) Spectral Technology. *Sensors (Basel), 22*(3), 1–13. doi:10.339022030836 PMID:35161584

Xu, W., Lan, Y., Li, Y., Luo, Y., & He, Z. (2019). Classification method of cultivated land based on UAV visible light remote sensing. *International Journal of Agricultural and Biological Engineering, 12*(3), 103–109. doi:10.25165/j.ijabe.20191203.4754

Yadav, P. (2021). Evolution of Wireless Communications with 3G, 4G, 5G, and Next Generation Technologies in India. In Advances in Electronics, Communication and Computing. Springer. doi:10.1007/978-981-15-8752-8_35

Yan, F., Feng, S., Liu, X., & Feng, T. (2023). Parametric Dynamic Distributed Containment Control of Continuous-Time Linear Multi-Agent Systems with Specified Convergence Speed. *Sensors (Basel), 23*(5), 2696. doi:10.339023052696 PMID:36904911

Yang, Y., Hu, Y., & Gursoy, M. C. (2021). Deep Reinforcement Learning and Optimization Based Green Mobile Edge Computing, 2021 IEEE 18th Annual Consumer Communications & Networking Conference. CCNC. doi:10.1109/CCNC49032.2021.9369566

Yang, B., Luo, Y., & Wu, L. (2019). Challenges and Opportunities in IoT-Based Manufacturing Systems: A Review. *IEEE Access : Practical Innovations, Open Solutions, 7*, 35709–35723. doi:10.1109/ACCESS.2019.2903235

Yang, J. W., Sharma, A., & Kumar, R. (2021). IoT-Based Framework for Smart Agriculture. *International Journal of Agricultural and Environmental Information Systems, 12*(2), 1–14. doi:10.4018/IJAEIS.20210401.oa1

Yang, J., Wang, C., Jiang, B., Song, H., & Meng, Q. (2020). Visual perception enabled industry intelligence: State of the art, challenges and prospects. *IEEE Transactions on Industrial Informatics, 17*(3), 2204–2219. doi:10.1109/TII.2020.2998818

Yang, J., Yu, W., Yang, H., Lv, H., Zhang, X., & Xiao, L. (2019). Challenges and opportunities in IoT-based retail: A review. *IEEE Internet of Things Journal, 6*(5), 8165–8180. doi:10.1109/JIOT.2019.2910298

Yang, L., & Shami, A. (2022). IoT data analytics in dynamic environments: From an automated machine learning perspective. *Engineering Applications of Artificial Intelligence, 116*, 105366. doi:10.1016/j.engappai.2022.105366

Yang, Z., Xie, J., Gao, J., Chen, Z., & Jia, Y. (2020). Joint Optimization of Wireless Resource Allocation and Task Partition for Mobile Edge Computing. *2020 IEEE/CIC International Conference on Communications in China (ICCC),* Chongqing, China. 10.1109/ICCC49849.2020.9238820

Yan, J., Liu, J., & Tseng, F. M. (2020). An evaluation system based on the self-organizing system framework of smart cities: A case study of smart transportation systems in China. *Technological Forecasting and Social Change, 153,* 119371. doi:10.1016/j.techfore.2018.07.009

Yannuzzi, M., Milito, R., Serral-Gracià, R., Montero, D., & Nemirovsky, M. (2014, December). Key ingredients in an IoT recipe: Fog Computing, Cloud computing, and more Fog Computing. In *2014 IEEE 19th International Workshop on Computer Aided Modeling and Design of Communication Links and Networks (CAMAD)* (pp. 325-329). IEEE.

Ye, D., Wang, X., & Hou, J. (2022). An Edge Computing Offloading Algorithm Based on Second-Order Oscillatory Particle Swarm Optimization. *2022 3rd Information Communication Technologies Conference (ICTC),* Nanjing, China. 10.1109/ICTC55111.2022.9778592

Yeh, M-Y, Dai, B-R, & Chen, M-S. (2007). Clustering over multiple evolving streams by events and correlations. *IEEE Transactions on Knowledge and Data Engineering*, 1349–1362.

Yeole, A. S., & Kalbande, D. R. (2016, March). Use of Internet of Things (IoT) in healthcare: A survey. In *Proceedings of the ACM Symposium on Women in Research 2016* (pp. 71-76). 10.1145/2909067.2909079

Yi, S., Hao, Z., Qin, Z., & Li, Q. (2020). Fog computing: Platform andapplications. In 2020 Third IEEE Workshop on Hot Topics in WebSystems and Technologies (HotWeb), (pp. 73-78). IEEE.

Yi-Fei, T., Soon-Chang, P., Ooi, C. P., & Tan, W. H. (2023). Human activity recognition with self-attention. *Iranian Journal of Electrical and Computer Engineering, 13*(2).

Yilmaz, G. E., & Akar, A. R. (2020). Decentralized auction-based task allocation in swarm robotic systems. *Robotics and Autonomous Systems, 130,* 103538. doi:10.1016/j.robot.2020.103538

Yin, C., Li, T., Qu, X., & Yuan, S. (2020). *An optimization method for resource allocation in fog computing.* 2020 International Conferences on Internet of Things (iThings) and IEEE Green Computing and Communications (GreenCom) and IEEE Cyber, Physical and Social Computing (CPSCom) and IEEE Smart Data (SmartData) and IEEE Congress on Cybermatics (Cybermatics), Rhodes, Greece. 10.1109/iThings-GreenCom-CPSCom-SmartData-Cybermatics50389.2020.00139

Yin, C., Zhu, Y., Fei, J., & He, X. (2017). A deep learning approach for intrusion detection using recurrent neural networks. *IEEE Access : Practical Innovations, Open Solutions, 5,* 21954–21961. doi:10.1109/ACCESS.2017.2762418

Yogitaa, D.T. (2012). A framework for outlier detection in evolving data streams by weighting attributes in clustering. *2nd Int Conf Commun Comput Secur.* (pp. 214–222).

Yousefizadeh, S., Flores Mendez, J. D., & Bak, T. (2019). Trajectory adaptation for an impedance controlled cooperative robot according to an operator's force. *Automation in Construction, 103,* 213–220. doi:10.1016/j.autcon.2019.01.006

Yuan, W., Li, S., Xiang, L., & Ng, D. W. K. (2020). Distributed Estimation Framework for Beyond 5G Intelligent Vehicular Networks. *IEEE Open Journal of Vehicular Technology, 1,* 190–214. doi:10.1109/OJVT.2020.2989534

Yuan, Y, Cao, H, Zhang, Y, & ... (2016). Outlier mining based on neighbor-density-deviation with minimum hypersphere. *Information Technology and Control, 45*(3), 267–277.

Yuloskov, A., Bahrami, M. R., Mazzara, M., & Kotorov, I. (2021). Smart Cities in Russia: Current Situation and Insights for Future Development. *Future Internet, 13*(10), 252. doi:10.3390/fi13100252

Zaidan, A. A., & Zaidan, B. B. (2020). A review on intelligent process for smart home applications based on IoT: Coherent taxonomy, motivation, open challenges, and recommendations. *Artificial Intelligence Review*, *53*(1), 141–165. doi:10.100710462-018-9648-9

Zambon, I., Cecchini, M., Egidi, G., Saporito, M. G., & Colantoni, A. (2019). Revolution 4.0: Industry vs. Agriculture in a Future Development for SMEs. *Processes (Basel, Switzerland)*, *7*(1), 36. doi:10.3390/pr7010036

Zantalis, F., Koulouras, G., Karabetsos, S., & Kandris, D. (2019). A Review of Machine Learning and IoT in Smart Transportation. *Future Internet*, *11*(4), 94. doi:10.3390/fi11040094

Zhang, Y., Yao, X., Fei, Q., & Chen, Z. (2023). Smartphone sensors-based human activity recognition using feature selection and deep decision fusion. *IET Cyber-Physical Systems: Theory & Applications*.

Zhang, S., Wang, J., Luo, L., & Qiu, C. (2021). Aerial-robot-assisted multi-robot assembly system with real-time task allocation. *International Journal of Advanced Robotic Systems*, *18*, 1–12. doi:10.1177/1729881421993856

Zhang, X., Wang, L., & Su, Y. (2021). Visual place recognition: A survey from deep learning perspective. *Pattern Recognition*, *113*, 107760. doi:10.1016/j.patcog.2020.107760

Zhang, Y., Hamm, N. A. S., Meratnia, N., Stein, A., van de Voort, M., & Havinga, P. J. M. (2012). Statistics-based outlier detection for wireless sensor networks. *International Journal of Geographical Information Science*, *26*(8), 1373–1392. doi:10.1080/13658816.2012.654493

Zhang, Y., Niyato, D., Wang, P., & Kim, D. I. (2020). Optimal energymanagement policy of mobile energy gateway. IEEE T Veh. *Technol*, *65*(5), 3685–3699.

Zhang, Y., Zhang, Y., & Chen, Y. (2023). Interoperability Challenges in IoT-based Smart Homes: A Survey. *Journal of Ambient Intelligence and Humanized Computing*, *14*(2), 931–944. doi:10.100712652-022-03571-x

Zhang, Z., Wang, W., An, A., Qin, Y., & Yang, F. (2023). A human activity recognition method using wearable sensors based on convtransformer model. *Evolving Systems*, 1–17. doi:10.100712530-022-09480-y

Zhao, T., Liu, Y., Shou, G., & Yao, X. (2022, April). Joint optimization of latency and energy consumption for mobile edge computing based proximity detection in road networks. *China Communications*, *19*(4), 274–290. doi:10.23919/JCC.2022.04.020

Zhao, W., Lin, S., Han, J., Xu, R., & Hou, L. (2017). *Design and implementation of smart irrigation system based on LoRa. In 2017 IEEE Globecom Workshops (GC Wkshps)*. IEEE. doi:10.1109/GLOCOMW.2017.8269115

Zhou, C., Luo, H., Fang, W., Wei, R., & Ding, L. (2019). Cyber-physical-system-based safety monitoring for blind hoisting with the internet of things: A case study. *Automation in Construction*, *97*, 138–150. doi:10.1016/j.autcon.2018.10.017

Zhou, H., Zhang, Z., Li, D., & Su, Z. (2023, April 1). Joint Optimization of Computing Offloading and Service Caching in Edge Computing-based Smart Grid. *IEEE Transactions on Cloud Computing*, *11*(2), 1122–1132. doi:10.1109/TCC.2022.3163750

Zhou, X.-Y. (2007). A fast outlier detection algorithm for high dimensional categorical data streams. *Journal of Software*, *18*(4), 933–942. doi:10.1360/jos180933

Zhou, Y., Zhao, X., Zhao, J., & Chen, D. (2016). Research on fire and explosion accidents of oil depots. *Chemical Engineering Transactions*, *51*, 163–168. doi:10.3303/CET1651028

Zhu, B., Joseph, A., & Sastry, S. (2021). A Taxonomy of Cyber Attackson SCADA Systems. 2021 *International Conference on Internet of Things and 4th International Conference on Cyber, Physical and Social Computing, Dalian*. ACM.

Zhu, Q. (2015). Enhancing Reliability of Community Internet of Things Deployments with Mobility. In *34th International Symposium on Reliable Distributed System*, Montreal - Quebec, Canada

Zhu, W., Chebeir, J., & Romagnoli, J. A. (2020). Operation optimization of a cryogenic NGL recovery unit using deep learning based surrogate modeling. *Computers & Chemical Engineering*, *137*, 106815. doi:10.1016/j.compchemeng.2020.106815

Ziegler, S. (2013). Iot6–moving to an ipv6-based future iot. In The Future Internet Assembly. Springer. doi:10.1007/978-3-642-38082-2_14

Zolanvari, M., Teixeira, M. A., & Jain, R. (2019). Effect of ImbalancedDatasets on Security of Industrial IoT Using. *Machine Learning*.

Zulkernine, F., Martin, P., Zou, Y., Bauer, M., Gwadry-Sridhar, F., & Aboulnaga, A. (2013). Towards Cloud -Based Analytics-as-a-Service (CLAaaS) for Big Data Analytics in the Cloud. *2013 IEEE International Congress on Big Data*, Santa Clara, CA. 10.1109/BigData.Congress.2013.18

Зверева, К. (2022). Agribots are coming! How can Robots, Cobots, and Drones Contribute to Farming? (Part II). *East Fruit*. https://east-fruit.com/en/horticultural-business/blogs/agribots-are-coming-how-can-robots-cobots-and-drones-contribute-to-farming-part-ii/

About the Contributors

Neha Goel is working as Associate Professor in the department of Electronics & Communication Engineering at RKGIT, Ghaziabad. She has completed her Phd from SRM university, Chennai in 2019. She has 16 years of rich experience in teaching. Her area of interest on VLSI design, CMOS design, Internet of things, Machine learning. She has guided several B.tech Projects and she has published 36 papers in various National/International journals and conferences. She has granted & published 3 patents. She has also attended various workshops, seminars in various fields.

Ravindra Kumar Yadav is a Professor & Head in the Department of Electronics & Communication Engineering at RKGIT, Ghaziabad. He is B.E., M.E., and PhD in the field of Electronics & Communication Engg. He has 26 years of rich experience in Teaching, Research & Development activities, Administration and Managing & Upbringing of higher Educational Institutions. He has guided several B. Tech and M. Tech project and also guiding PhD students form IIT Dhanbad as co guide. He published 90 papers to his credit in international/national journals, conferences and symposiums. Prof. Yadav is Reviewer for several National/International Journals of high repute. He has chaired/participated technical sessions at multiple international and national conferences / seminars held throughout the country.

* * *

Alvis Abreo is a student of Christ University.

Manikandan Arunachalam completed his B.E in Electronics and Communication Engineering from Madurai Kamaraj University in 2004, Master of Engineering in Communication Systems from Anna University in 2006. He received his PhD from Anna University in 2018. He has 16 years of teaching experience and guided many UG & PG Projects. He is a Life member of IETE & ISTE. He has published papers in 11 International Journals, 6 International conferences and 6 National Conferences. Currently he is working in M2M communication. He is currently associated with Amrita School of Engineering, Amrita Vishwa Vidhyapeetham, Amritapuri, Kollam, Kerala, India.

Manoj B. E. is an MCA Student at Christ University, Bangalore.

Varun Barthwal is Professor in IT Department at H.N.B. Garhwal University, (A Central University) Uttarakhand, India. His research interest in cloud computing.

Sandeep Bhatia is a dedicated Assistant Professor with more than 13 years of experience teaching undergraduate and post-graduate courses in the branch of ECE and CSE. He has one-year industry experience as a hardware engineer at C-DAC Mohali, Punjab. Presently, He is working as an Assistant Professor in the Department of CSE, Galgotias University, Greater Noida. He is Ph.D. Pursuing from Jamia Millia Islamia, Central University, New Delhi in the field of IoT and Embedded System Design. He has completed his M.E from PEC University of Technology, Chandigarh in the field of Embedded System in the year 2010. He has completed his B.Tech from PTU Jalandhar in the year 2008. He has successfully Developed MOOC course for the subject "Microcontroller and Embedded Systems Design (KEC-061)" sponsored by AKTU Lucknow, UP at AKTU Digital Education Platform for the students. , (For Reference only). He has more than 7 published papers in Scopus Indexed Journals and 4 other reputed journals along with fifteen International/ National conferences attended. Also Published 7 patents, attended various FDPs and completed various online courses including NPTEL. He has guided 32 B.Tech projects along with 4 M.Tech thesis projects. He is comfortable in the field of IoT and its Applications, Embedded System, Microcontrollers and Sensors. He has taught value added course in IoT and Embedded System for graduate students.

Megha Bhushan is Associate Professor in the School of Computing, and Assistant Dean, Research & Consultancy at DIT University, Dehradun, India. She has received her ME and Ph.D. degrees from Thapar University, Punjab, India. She was awarded with a fellowship by UGC, Government of India, in 2014. In 2017, she was a recipient of Grace Hopper Celebration India (GHCI) fellowship. She has published 5 national patents and 1 international patent has been granted. She has published many research articles in international journals and conferences of repute. Further, she is the editor of many edited books with different publishers such as CRC Press, Taylor & Francis Group, Wiley-Scrivener and Bentham Science. Her research interests include Artificial Intelligence, Knowledge representation, Expert systems, and Software quality. She is also the reviewer and editorial board member of many international journals.

Annette Bolding is currently a DBA student at the University of Dallas in Irving, Texas, USA. She received both her BBA and MBA from Angelo State University in San Angelo, Texas. Annette has worked in the oil and gas industry for over 15 years and has extensive management experience in field operations. She has presented her work at various conferences and published research articles in a variety of journals. Primary research interests include strategy techniques within the oil and gas industry.

Mayur C. is an MCA Student of Christ University.

Ebin Deni Raj received the B.Tech. degree in computer science from the Cochin University of Science and Technology, Kochi, India, and the M.Tech. degree in information technology and the Ph.D. degree from the VIT University, Vellore, India. He is currently the head of the Department of Computer Science Engineering, Indian Institute of Information Technology at Kottayam, Kottayam, India. He is a Distinguished speaker of ACM in the domain of Data Science. He is a recipient of Microsoft AI for Earth grant in the year 2019. He also received the best thesis award by the Computer Society of India. He is right now working with various interdisciplinary international projects such as AIFORA. He has published more than 25 refereed research articles so far in various national, international journals and conferences. He has delivered technical talks at several national & international conferences. He is in the reviewer board of prestigious journals such as IEEE Transactions on Industrial Informatics, Cluster

Computing, Computer Communications, Knowledge Based Systems, Neurocomputing, IEEE Transactions on Computational Social Systems, International Journal of Remote Sensing, etc. His primary research interests include social big data analytics, machine learning and social computing.

Kunal Dhibar is an energetic and passionate college student working towards a B.Tech in CSE at the Bengal College of Engineering and Technology, Durgapur, West Bengal, India. Aiming to use my knowledge of Coding, Machine Learning, Software development, and Data Science strategies to satisfy the book chapter.

Ankur Gupta has received the B.Tech and M.Tech in Computer Science and Engineering from Ganga Institute of Technology and Management, Kablana affiliated with Maharshi Dayanand University, Rohtak in 2015 and 2017. He is an Assistant Professor in the Department of Computer Science and Engineering at Vaish College of Engineering, Rohtak, and has been working there since January 2019. He has many publications in various reputed national/ international conferences, journals, and online book chapter contributions (Indexed by SCIE, Scopus, ESCI, ACM, DBLP, etc). He is doing research in the field of cloud computing, data security & machine learning. His research work in M.Tech was based on biometric security in cloud computing.

Giri G. Hallur is working as an Associate Professor and Deputy Director in Symbiosis Institute of Digital and Telecom Management, Symbiosis International (Deemed University), Pune. His research areas include Digital Telecom, 5G, Regulatory Management.

Zainul Abdin Jaffery is at present Professor in the Department of Electrical Engineering and Honorary Director of University Training and Placement Jamia Millia Islamia (a Central university, Govt. of India). Dr Jaffery obtained his U.G degree in Electrical Engineering and P.G degree in Electronics and Communication Engineering from Aligarh Muslim University, Aligarh, India in 1987 and 1989 respectively. He obtained his PhD degree from Jamia Millia Islamia in 2004. Prof. Jaffery has published about 100 research papers in the area of Electronics and Electrical engineering in the peer reviewed international journals and conferences. His research area includes Digital signal processing, Digital image processing, biomedical image processing, soft computing techniques and their applications in power engineering and electronics engineering. Prof. Jaffery has delivered many invited lectures in various institutes and conferences. He is the reviewer of many international journals. Prof. Jaffery is a member of various academic societies of national and international repute. He is senior member of IEEE (USA). Prof Jaffery has wide administrative experience. Prof. Jaffery has served as Head of the Department of Electrical Engineering and as the Director of Centre for Innovation and Entrepreneurship, Jamia Millia Islamia. Prof. Jaffery has been General Chair of IEEE International conference ICPECA-19, Track Chair of IEEE International conference INDICON-2015 and technical committee member of many IEEE conferences. Prof Jaffery has been invited as subject expert in many public service commission and Industries.

K. Suresh Joesph received his Bachelor's Degree in Computer Science & Engineering from Bharathiar University and a Master's in Computer Science & Engineering from the University of Madras in 1999 and 2003, respectively. He received his Doctoral Computer Science & Engineering degree from Anna University in 2013. Presently he is working as an Associate Professor in the Department of Computer

Science at Pondicherry University. His areas of research interest include Services Computing and Image Processing. He has more than 50 research publications in reputed and peer-reviewed journals.

Sheetal Kalra did her Ph.D. in Computer Science & Engineering from Guru Nanak Dev University, Amritsar. She completed her MCA from the same University. She is currently working as an Assistant Professor in department of Computer Science and Engineering, G.N.D.U. Regional Campus, Jalandhar. She has 14 years of teaching experience and has more than 60 research publications in highly reputed journals to her credit. Her research areas are network & information security, cloud computing, big data and Internet of Things.

Saifullah Khalid obtained two Ph.D. degrees: one in Electronics and Communication Engineering and another in Electrical Engineering from Institute of Engineering and Technology, Lucknow. He worked at IIT BHU, Varanasi and different Universities before joining the dynamic job of Air Traffic Control officer at Airports Authority of India in 2008. Currently, he is working as a R& D Member at Civil Aviation Research Organization, India. His research interests are in the area of Optimization Algorithms applications for power quality improvement. He has developed three novel optimization algorithms Adaptive Spider Net search Algorithm, Adaptive Mosquito Blood Search Algorithm, and Adaptive blanket body cover search algorithm. Dr. Khalid has published around 100 research papers in various International Journals and Conferences including IEEE, Elsevier, Springer and many more in India and abroad. He is on the editorial board of many International Journals and also in the reviewer's panel of many International Journals (Including IEEE Transactions, Elsevier, Springer, Wiley, etc.). Dr. Khalid has authored four books published by University Science Press, USA.

A. V. Senthil Kumar is working as a Director & Professor in the Department of Research and PG in Computer Applications, Hindusthan College of Arts and Science, Coimbatore since 05/03/2010. He has to his credit 11 Book Chapters, 265 papers in International and National Journals, 25 papers in International Conferences, 5 papers in National Conferences, and edited Nine books (IGI Global, USA). He is an Editor-in-Chief for various journals. Key Member for India, Machine Intelligence Research Lab (MIR Labs). He is an Editorial Board Member and Reviewer for various International Journals. He is also a Committee member for various International Conferences. He is a Life member of International Association of Engineers (IAENG), Systems Society of India (SSI), member of The Indian Science Congress Association, member of Internet Society (ISOC), International Association of Computer Science and Information Technology (IACSIT), Indian Association for Research in Computing Science (IARCS), and committee member for various International Conferences.

Palanivel Kuppusamy holds a Ph.D. in Computer Science & Engineering (2023) from Pondicherry University (A Central University), Puducherry, India. He is a graduate of Engineering (1994) from Bharathiar University, Tamil Nadu, India, and a Post Graduate of Engineering (1998) from Pondicherry University, Puducherry, India. He has over 20 years of diverse professional and teaching experience at Pondicherry University. His research interests include smart education, software architecture, network management & analytics, and technological applications in the higher education sector.

Rohaya Latip is an Associate Professor at Faculty of Computer Science and Information Technology, University Putra Malaysia. She holds a Ph.D in Distributed Database in year 2009 and Msc. in

Distributed System in 2001 both from University Putra Malaysia, Malaysia. She graduated her Bachelor of Computer Science from University Technology Malaysia, Malaysia in 1999. She was the Head of Department of Communication Technology and Network in 2017-2022. She served as an Associate Professor at Najran University, Kingdom of Arab Saudi (2012-2013). She is the Head of HPC section in Institute for Mathematic Research (INSPEM)., University Putra Malaysia (2011-2012) and consulted the Campus Grid project and also the Wireless for hostel in Campus UPM project. She is also a Co-researcher at INSPEM. She is the editorial board of International Journal of Computer Networks and Communications Security (IJCNCS), editorial board of International Journal of Digital Contents and Applications (IJDCA) and editorial board for International Journal of Computer Networks and Applications (IJCNA). Her research interests include Big Data, Cloud and Grid Computing, Network management, and Distributed database. For her research work, she won Gold medal at ICANS 2019, Canada, and two medals at The World Inventor Award Festival (WIAF) 2014 organized by Korea Invention News. She was awarded Gold medal at IMIT-SIC Innovation Expo in 2018, Riau Indonesia, I-RIA 2018 (Best of the best award), Malaysia Technology Expo (MTE2014) and Malaysian Innovation Expo (MiExpo2013). She also won Silver medal at National Design, Research and Innovation Expo (PRPI) 2010 and Bronze medal at National Design, Research and Innovation Expo (PRPI) 2007 and 2006 respectively. She has published more than 80 papers in international and national journals, proceedings and posters.

Shabana Mehfuz (Senior Member, IEEE) received the B.Tech. degree in Electrical engineering from Jamia Millia Islamia, India, in 1996, the M.Tech. degree in Computer Technology from IIT Delhi, in 2003, and the Ph.D. degree in Computer Engineering from Jamia Millia Islamia, in 2008. She secured First position in the order of merit in B.Tech.(Electrical Engineering). She has been working at the Department of Electrical Engineering, Jamia Millia Islamia, for the past 24 years. She has published more than 100 articles in International Journals and Conferences. She has guided eight Ph.D.s and is currently supervising other ten candidates. She is a Life Member of ISTE, a member of the Institution of Engineers, and a Life Member of the Computer Society of India. She has received grants for research projects from agencies, such as AICTE and UGC. She has acted as the Track Chair for two flagship International IEEE conferences held in 2015 and 2019. She has been awarded the International Inspirational Women Award 2020 for Best Performer in Government Award by GISR Foundation and. She has been awarded national award for excellence in education 2021 in the category of university/college teachers by AMP organization. Her research interests include Cloud Computing, Internet of Things (IoT), Wireless Sensor Networks, Mobile ad-hoc Network, Artificial Intelligence and Machine Learning.

Sriya Mitra is pursuing her MBA in Digital and Telecom Management in Symbiosis Institute of Digital and Telecom Management at Symbiosis International (Deemed University). Her Research interests include IoT in Health care, Telecom, Management studies.

Jubaraj Mukherjee is a Master in Computer Applications student in Christ (Deemed to be University) and an avid IoT enthusiast.

N.A. Natraj is an Assistant Professor at SIDTM in Symbiosis International University, India. His academic chronicles include Ph, D. in Electronics and Communication Engineering. He has 11 years of Academic experience in various Countries. His research interests include Wireless Sensor Networks, the Internet of Things, 5G and Communication Networks. He holds publications in reputed Journals and has published patents in the mentioned research fields. He has completed certifications in domains like 5G, IoT, Digital Forensics and Blockchain from various organizations like Stanford, Europe Open University, Qualcomm, University of London, etc.

Arun Negi is currently a Manager at Deloitte USI, Hyderabad, India. He has completed a course in Business Management from IIM, Ahmedabad and has obtained B. Tech degree from Jawaharlal Nehru University, New Delhi, India. He has 13+ years of diverse experience in cyber risk services. He has worked on various network security technologies and platforms for Fortune 500 clients. His experience includes cyber security audits, gap assessments, network security audits, cloud migrations and project management. He is currently oriented towards developing multi cloud skills and has achieved certifications in Oracle Cloud and AWS. He has published one national patent and one international patent has been granted. He has published many research articles in international journals and conferences of repute. His research areas include Artificial Intelligence, Software Product Line, Cloud Computing, and Cyber Security.

Shahad P. received B.Tech. degree in computer science and engineering from Anna University, Chennai, Tamil Nadu, India and the M.Tech. degree from University of Calicut, Kerala, India. Currently he is a research scholar at Indian Institute of Information Technology, Kottayam, Kerala, India. He is currently a member of the Computer Society of India and his area of interest includes Data Mining, Data Science and Streaming Data Analysis, Machine Learning and Big Data Analysis. He has 13 years of teaching experience in various Engineering colleges. As a faculty in an engineering college he won three academic excellence awards for best teaching results from his college with reference to the result published by university at the state level. His goal is to share his knowledge with the students and apply it for the betterment of human life.

Man Mohan Singh Rauthan is Professor in Computer Science and Engineering Department at H.N.B. Garhwal University, (A Central University) Uttarakhand, India. He completed M.Sc. (Physics) from I.I.T., Delhi and M.Tech (Computer Science) from Roorkee University. He received Ph.D. degree in (Information Technology) from Kumaun University, Nainital (India). His research interest includes information retrieval and cloud computing.

Bidura Sarkar is a student studying in MCA Department of Christ University.

Suganthi Shanmuganananthan is working as an Assistant Professor in the Department of Genetics and Plant Breeding, Faculty of Agriculture, Annamalai University. She has more than ten+ years of teaching experience and more than fifteen years of research experience. She completed her Under Graduate in 1998, her Postgraduate in 2000, and a Doctoral degree in 2009 at Annamalai University. Her field of research specialization is heterosis breeding and resistance breeding. She has published 35 research articles in International and National Journals and three books and authored three book chapters. She has presented research papers at national and international seminars, conferences, and symposiums.

Meenakshi Sharma is a Ph.D. awarded in 2015, Assistant Professor-SS at University of Petroleum & Energy Studies, Dehradun having 13 years of experience in Academics and 6 years in the Industry. Her area of interest being Consumer Behavior & Market Research, Marketing Management and Customer

Sustainability. She is Six Sigma Green Belt. Expertise in SPSS, AMOS, Tableau. Published more than 16 publications including Research papers in Scopus, International and National Journals. Many papers presented in International and National Conferences. Conducted full day sessions in organizations like HPCL,Tata Power, UGVNL etc. Completed various certifications from Coursera.

Shyam R. Sihare completed his Ph.D. at Raksha Shakti University in Ahmedabad, India. He holds a Master's degree in Computer Science from Nagpur University, Nagpur, India, which he obtained in 2003. Additionally, he attained an M. Phil. in Computer Science from Madurai Kamraj University, Madurai, India. In 2011 and 2018, he successfully cleared the Professor Eligibility Test GSLET (Gujarat) and MS-SET (Maharashtra) in India, respectively. Furthermore, he completed his MCA from IGNOU in New Delhi, India, in 2011. Currently, Dr. Sihare serves as an Assistant Professor in Computer Science and Application at Dr. APJ Abdul Kalam Govt. College in Silvassa, Dadra & Nagar Haveli (UT), India. His research interests encompass a wide range of areas including Quantum Computing, Quantum Algorithms, Quantum Cryptography, and Classical Computer Algorithms.

Rajeshwari Sissodia is a Research Scholar in Computer Science and Engineering department at H.N.B. Garhwal University Uttrakhand, India. She completed B.Tech and M.Tech degree (Computer Science & Engineering) from J. B. Institute of Technology, Dehradun, India. Her research interest includes IoT & Cloud Computing.

Sanjay T. is currently an undergraduate student from the CSE department at Amrita Vishwa Vidyapeetham, Chennai. He participated in many hackathons under various themes like Healthcare, Education, Sustainability and Social Good. He has won prizes in around 9 international hackathons organised by Microsoft, MLH, Rapyd, UC Berkeley, IEEE, MongoDB and Iguazio. Recently he won the "Rapyd Demo Day Ticket" prize in Build The Best Super App challenge which is a part of Hack the Galaxy hackathon series.

Claudio Urrea was born in Santiago, Chile. He received the M.Sc.Eng. and the Dr. degrees from University of Santiago of Chile, in 1999 and 2003, respectively; and the Ph.D. degree from Institut National Polytechnique of Grenoble, France in 2003. Ph.D. Urrea has been Professor at the Electrical Engineering Department, University of Santiago of Chile, since 1998. He has developed and implemented a Robotics Laboratory, where intelligent robotic systems are developed and investigated. Currently, he is Full Professor and holds the position of Director of the Instrumentation Center and Electronic Development at the University of Santiago of Chile.

Soniya Verma, Assistant Professor in the Dept. of HSS, KIET, has a wide experience of more than 13 years in Academics. Her areas of specialization are Indian Literature in English, English Language Teaching, Fiction, Communication Skills and Value Education. Her publications have appeared in many National and International Journals of high repute and Presented many papers at National and International conferences. She has a passion for connecting students to learning modalities that incite their interest in the Humanities.

Index

W

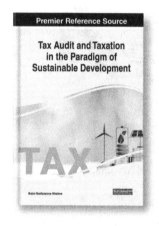

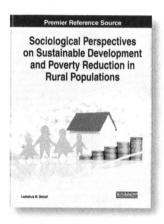

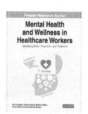

Printed in the United States
by Baker & Taylor Publisher Services